11th International LISA Symposium 2016

Journal of Physics: Conference Series Volume 840

Zurich, Switzerland
5 – 9 September 2016

ISBN: 978-1-5108-4185-7
ISSN: 1742-6588

Printed from e-media with permission by:

Curran Associates, Inc.
57 Morehouse Lane
Red Hook, NY 12571

CURRAN ASSOCIATES INC.
proceedings
.com

Some format issues inherent in the e-media version may also appear in this print version.

Printed by Curran Associates, Inc. (2017)

For permission requests, please contact the Institute of Physics
at the address below.

Institute of Physics
Dirac House, Temple Back
Bristol BS1 6BE UK

Phone: 44 1 17 929 7481
Fax: 44 1 17 920 0979

techtracking@iop.org

Additional copies of this publication are available from:

Curran Associates, Inc.
57 Morehouse Lane
Red Hook, NY 12571 USA
Phone: 845-758-0400
Fax: 845-758-2633
Email: curran@proceedings.com
Web: www.proceedings.com

Table of contents

Volume 840

11th International LISA Symposium
5–9 September 2016, Zurich, Switzerland

Accepted papers received: 3 May 2017
Published online: 1 June 2017

Preface

LISA Pathfinder - Overview

LISA - Overview

Astrophysics with Gravitational Waves

Cosmology with Gravitational Waves

Posters: LISA Pathfinder

Posters: LISA

Posters: Science with Gravitational Waves

After Dinner Speech

IOP Conf. Series: Journal of Physics: Conf. Series **840** (2017) 011001 doi:10.1088/1742-6596/840/1/011001

Editorial: XI International LISA Symposium

The XI International LISA Symposium took place at the University of Zürich (Irchel Campus) from 5 to 9 September 2016. Two years before, at the X LISA Symposium in Gainesville in Florida, it was decided to hold the following one in Zürich. At that time we could not imagine that in the following years the landscape in the field of gravitational waves would change in a dramatic way, and for the best. It started with the perfect flight of LISA Pathfinder on 3 December 2015, followed by the announcement of the first results on 7 June, showing that the performance of LISA Pathfinder was much better than expected and already at the level of the LISA requirements; a wonderful result, achieved after more than a decade of big efforts in preparing the mission. Then, on 11 February 2016 the LIGO Scientific Collaboration announced the first ground based detection of gravitational waves. These nice developments made 2016 a special year in the history of gravitational waves, 100 years after the first work of Einstein predicting the existence of gravitational waves within his theory of General Relativity.

Given these spectacular developments it came not as a surprise that the interest in the community for the LISA Symposium increased and indeed about 250 participants attended the meeting, which was in all aspects very successful. The format consisted of plenary talks, in total some 68 talks, and almost 100 posters. The main topics of the talks covered the first results from LISA Pathfinder, the further development of LISA and an overview of the results obtained so far by the ground based gravitational wave detectors (LIGO and VIRGO) as well as Pulsar Timing Arrays. Several talks covered different aspects of astrophysics and cosmology related to gravitational waves. One talk was devoted on the preliminary results of the satellite Microscope, whose aim is to test the equivalence principle in general relativity. A joint eLISA and L3ST (a NASA committee) consortium meeting was also organized. There were also programmatic talks given by Prof. A. Gimenez (Science Director of ESA) and by Dr. Paul Hertz (Astrophysics Division Director of NASA). The full program can be found in this volume as well as the lists of the Science Advisory Committee (SAC), the Local Organizing Committee (LOC) and the participants.

This volume of proceedings contains about 60 contributions, including talks and posters. Although the focus of this meeting was put on the first results of LISA Pathfinder, only few of the contributions in this volume are devoted to them, due to the fact that most of the material presented in the meeting was preliminary and will be published in peer reviewed journals after the end of the LISA Pathfinder mission. We thank all the contributors for their effort in sending a paper for the proceedings. We have a nice volume that will be of use for all participants and we are also very grateful that we could organize the meeting at such a special and probably unique period in the history of gravitational wave physics in Zürich.

Zürich is closely related to Einstein, who had relations with both the ETH, where he made his studies and later was a professor, and with the University, where he obtained his PhD and got his first academic position as a professor after some years at the Patent Office in Bern. Moreover,

IOP Conf. Series: Journal of Physics: Conf. Series **840** (2017) 011001 doi:10.1088/1742-6596/840/1/011001

substantial parts of his theory of General Relativity were developed during his stay in Zürich, and he received important mathematical help from Marcel Grossmann, a colleague at ETH.

We thank the members of the SAC and the conveners of the sessions who helped shaping the scientific program. A particular thank goes to Monica Colpi: she helped in a substantial way to define the program and in inviting many of the speakers. We thank all our collaborators in the LOC, who in a way or another helped in organizing the meeting and contributed in a crucial way to its success. Particularly our secretary Carmelina Genovese and Rafael Küng deserve a big credit for their effort. We acknowledge also the financial support of the University of Zürich, whose president Prof. Michael Hengartner held the welcome speech, and of the Pauli Center of ETH Zürich.

The future of our field is bright. ESA has already started the preparation for the L3 mission, the LISA mission proposal was submitted by the LISA Consortium in mid of January 2017 and is presently under review. LIGO and VIRGO are again collecting data and new events will surely be detected. After many years of planning and preparation efforts, gravitational wave astronomy has really started.

Domenico Giardini and Philippe Jetzer

Dedication
Pierre Binétruy (1955 - 2017): A scientific biography

Pierre Binétruy (1955 - 2017)

Pierre Binétruy, born in 1955, obtained his "state doctorate" in 1980 under the direction of Mary K. Gaillard, with a thesis on "Theoretical aspects and phenomenology of gauge theories". From 1979 to 1986, he held several positions as postdoctoral fellow at CERN and in the United States (University of California at Berkeley, University of Florida, University of Chicago). In 1986, he was recruited as a CNRS researcher at the LAPP, Annecy-le-Vieux and 4 years later as a professor at the University of Paris XI in the Laboratory of Theoretical Physics. In 2003, he was named professor at the University Paris Diderot.

His main interests evolved from high-energy particle physics (notably supersymmetry) to cosmology and gravitation, and he worked in particular on the interface between the study of the primordial Universe and the theories of fundamental interactions. Recent interests included inflation models, dark energy, and the cosmic gravitational wave background. During a prolific career, he has published seminal papers that have received more than 1000 citations each. For his research he has received several awards, such as the Thibaud Prize, the Paul Langevin Award

IOP Conf. Series: Journal of Physics: Conf. Series **840** (2017) 011002 doi:10.1088/1742-6596/840/1/011002

from the SFP, and was named Berkeley Miller Professor in 1999. He was arguably one of the most brilliant theorists of his time.

His memory will also remain because, to paraphrase André Malraux, he combined "spirit and courage". He knew that it is not sufficient to seek scientific truth but that one must also have the courage to organize the community for the scientific goals that this imposes and to also fight within the institutions to defend them.

Our more senior colleagues will remember the extraordinary intellectual atmosphere that animated the Supersymmetry Research Group (a CNRS "groupement de recherche"), which he initiated and directed from 1997 to 2004, transforming it into an unprecedented crossroads of experimenters and theorists, of new ideas both from theory and from experimental analysis.

He also had a fundamental intuition, at a time when the detection of gravitational waves was for many a distant dream (end of 2005), to involve France through CNES in the space program for the detection of gravitational waves, LISA-Pathfinder and LISA. To this scientific choice he devoted a large part of his dynamism and this until and during his hospitalization.

APC researchers and engineers also remember the energy and dynamism he put into the foundation of the Astroparticle and Cosmology Laboratory (APC) in 1999, following an incentive by Luc Valentin. He was director of the APC until 2013. He accompanied this original endeavor of the University Paris Diderot, the IN2P3 / CNRS, the CEA, and the Observatoire de Paris, with a never ending inventiveness.

We owe him to have embedded the laboratory in interdisciplinary and Earth Sciences, the realization of the importance of data sciences with the François Arago Centre, the foundation of the Centre of Cosmological Physics of Paris (PCCP) together with the Nobel Prize winner in Cosmology, George Smoot, and the immersion of the APC in a global network of similar centres (for example, the Associated International Laboratory with KIPAC at SLAC, relations with the University of Chicago and the Karlsruhe Institute for Technology).

He was also a professor who inspired hundreds of students, and through the MOOC "Gravity", in collaboration with George Smoot, his courses reached thousands (75,000 enrollees). The expression of gratitude from the students but also from all those who followed it touched Pierre enormously. This MOOC was for Pierre an avenue towards the future. Not only did he see it as a way to improve the visibility of the University but also as a revolution in the way knowledge is diffused, similar to the Gutenberg revolution. This revolution might mean a new type of universities, and through it an evolution of society and of mankind.

In parallel to these activities, four main responsibilities can be highlighted from among a multitude: Pierre found time to be president of the Fundamental Physics Advisory Group (2008-2010), of the Fundamental Physics Roadmap Committee (2009-2010) at ESA, the French consortium of the LISA space mission, of the Theory Division of the French Physical Society (1995-2003), the interdisciplinary section for Astroparticle (2003-2004), the theory section (2005-2008) of the CNRS, and director of the Endowment Fund "For Research and Training in the Physics of the Universe".

He was also a member of the IN2P3 Scientific Committee (1996-2000), the APPEC Scientific Advisory Committee, which brought together Astroparticle Agencies in Europe, the ApPIC working group of IUPAP (the International Union for Pure Physics

And Applied), which he helped to create in 2013, of which he was a key member until today, of the European Space Science Committee, the Scientific Program Committee (SPC) of the SLAC National Laboratory (Stanford, United States), the Evaluation Committee of the DOE Roadmap and the International Evaluation Committee of INFN (Italy) and NSERC (Canada). Finally, in recent years, he was a member of the CNRS Scientific Council.

This activity, pursued with enthusiasm and unfailing rigor, was accompanied by a great culture and sophistication, a profound knowledge of the arts, where he initiated and promoted several actions between art and science, always with a great human quality. This quality has

IOP Conf. Series: Journal of Physics: Conf. Series **840** (2017) 011002 doi:10.1088/1742-6596/840/1/011002

made that the news of his disappearance has been lived with great sadness throughout the world. As one of his eminent colleagues said of him: "Pierre was one of those very exceptional people who was at the top of the game and, at the same time, a remarkably simple and enjoyable colleague."

French science but also European and world science have lost one of its exemplary practitioners.

His colleagues at APC, and more precisely those of the LISA team which he put together, know how much we owe him. He was the one who brought us together and who was always pushing us to be active and to participate in all the facets of gravitational wave physics. He taught us everything about this field and was always one we could turn to in order to grasp the important and subtle aspects of this physics. Those working more on instrumental techniques also knew that he understood many aspects of this, particularly in terms of organization and interactions with the science and space agencies. We are all terribly indebted to him and are determined to use his legacy as a source of inspiration in order to push ahead with LISA.

Gerard Auger and Eric Plagnol, APC, Paris

Committees

Science Advisory Committee (SAC)

- Stanislav Babak
- Peter Bender
- Emanuele Berti
- Pierre Binetruy
- Chiara Caprini
- Monica Colpi
- John Conklin
- Neil Cornish
- Curt J. Cutler
- Karsten Danzmann
- Domenico Giardini
- Gabriela Gonzalez
- Gerhard Heinzel
- Martin Hewitson
- Allan Hornstrup
- Oliver Jennrich
- Philippe Jetzer
- Paul McNamara
- Guido Mueller
- Germano Nardini
- Gijs Nelemans
- Eric Plagnol
- Bernard Schutz
- Alberto Sesana
- David Shoemaker
- Carlos F. Sopuerta
- Robin Stebbins
- Tim Sumner
- James Ira Thorpe
- Stefano Vitale
- Harry Ward
- William Joseph Weber

Local Organizing Committee (LOC)

- **Domenico Giardini**
- **Philippe Jetzer**
- Ruxandra Bondarescu
- Yannick Bötzel
- Luigi Ferraioli
- Carmelina Genovese
- Elisabeth Läderach
- Rafael Küng
- Davor Mance
- Neda Meshksar
- Lionel Philippoz
- Jan Ten Pierick
- Andreas Schärer
- Peter Zweifel

IOP Conf. Series: Journal of Physics: Conf. Series **840** (2017) 011004 doi:10.1088/1742-6596/840/1/011004

Participants

Tal Alexander	Weizmann Institute of Science, IL
Pau Amaro Seoane	Max Planck Institute for Gravitational Physics AEI, DE
Stephen Apple	University of Florida, US
Michele Armano	ESAC, ESA
Heather E. Audley	Max Planck Institute for Gravitational Physics AEI, Hannover, DE
Gérard Auger	CNRS, FR
Stanislav Babak	Max Planck Institute for Gravitational Physics AEI, DE
Yanzheng Bai	Huazhong University of Science and Technology, CN
Jonathon Baird	Imperial College London, UK
John Baker	NASA Goddard Space Flight Center, US
Massimo Bassan	INFN, Università Tor Vergata, IT
Jean-Baptiste Bayle	APC, Université Paris Diderot, FR
Chris Belczynski	[none]
Peter L. Bender	JILA and Physics Dept., University of Colorado Boulder, US
Joel Bergé	ONERA, FR
Christopher Berry	University of Birmingham, UK
Pierre Binetruy	APC, Université Paris Diderot, FR
Roger Blandford	Stanford University, US
Yannick Boetzel	Physik Institut, Universität Zürich, CH
Ruxandra Bondarescu	Physik Institut, Universität Zürich, CH
Matteo Bonetti	Università dell'Insubria, IT
Michael Born	Max Planck Institute for Gravitational Physics AEI, Hannover, DE
Katelyn Breivik	Northwestern University, US
Christina Brugger	Airbus Defence and Space GmbH, DE
Ernst-Jan Buis	TNO, NL
Jordan Camp	NASA Goddard Space Flight Center, US
Chiara Caprini	IPhT, CEA Saclay, FR
Antonella Cavalleri	IFN Trento, CNR-Fondazione Bruno Kessler, IT
Cecile Cavet	APC, Université Paris Diderot, FR
Joan Centrella	NASA Goddard Space Flight Center, US
Massimo Cerdonio	INFN Padova, IT
Xiangsong Chen	Huazhong University of Science and Technology, CN
Andrew Chilton	Physics, University of Florida, US
Alvin Chua	Institute of Astronomy, Cambridge
Giacomo Ciani	University of Florida, US
Monica Colpi	INFN Milano Bicocca, IT
John Conklin	University of Florida, US
Ciarán Conneely	Imperial College London, UK
Neil Cornish	Montana State University, US
Tito Dal Canton	NASA Goddard Space Flight Center, US

Karsten Danzmann	Max Planck Institute for Gravitational Physics AEI, Hannover, DE
Francesco De Paolis	Università del Salento, IT
Lorenzo De Vittori	Physik Institut, Universität Zürich, CH
Rita Dolesi	TIFPA, Università di Trento / INFN Trento, IT
Katherine Dooley	University of Mississippi, US
Massimo Dotti	Università degli Studi di Milano-Bicocca, IT
Irina Dvorkin	Institut d'Astrophysique de Paris, FR
Carlos F. Sopuerta	Institut de Ciències de l'Espai, CSIC IEEC, ES
Fabio Favata	ESA
Germán Fernández Barranco	Max Planck Institute for Gravitational Physics AEI, Hannover, DE
Luigi Ferraioli	Institute of Geophysics, ETH Zürich, CH
Carlo Ferrigno	ISDC, Université de Genève, CH
Valerio Ferroni	University of Trento, IT
Ewan Fitzsimons	UK Astronomy Technology Centre, UK
Jonathan Gair	University of Edinburgh, UK
Antonio Garcia	OHB System AG
Juan Garcia-Bellido	Universidad Autonoma Madrid, ES
Martin Gehler	ESA
Carmelina Genovese	Physik Institut, Universität Zürich, CH
Oliver Gerberding	Max Planck Institute for Gravitational Physics AEI, Hannover, DE
Lluis Gesa Boté	Institut de Ciències de l'Espai, CSIC IEEC, ES
Alexandros Gianninas	University of Oklahoma, US
Domenico Giardini	Institute of Geophysics, ETH Zürich, CH
Ferran Gibert	Università di Trento / INFN Trento, IT
Alvaro Giménez Cañete	ESA
Roberta Giusteri	Università di Trento / INFN Trento, IT
Gabriela Gonzalez	Louisiana State University, US
A Gopakumar	TIFR, Mumbai, IN
Aniello Grado	INFN/INAF, IT
Philippe Grandclement	Observatoire de Paris, FR
Catia Grimani	Università di Urbino "Carlo Bo" / INFN Florence, IT
Felipe Guzman	ZARM & DLR, Universtiät Bremen, DE
Maria Chiara Guzzetti	Università di Padova / INFN Padova, IT
Hubert Halloin	APC, CNRS / Université Paris Diderot, FR
Maria Haney	Tata Institute of Fundamental Research, Mumbai, IN
Anna Heffernan	University of Florida, US
Gerhard Heinzel	Max Planck Institute for Gravitational Physics AEI, Hannover, DE
Thomas Hertog	KU Leuven, BE
Paul Hertz	NASA, US
Martin Hewitson	Max Planck Institute for Gravitational Physics AEI, Hannover, DE
Franz Georg Hey	Airbus Defence and Space GmbH, DE
Daniel Hillsberry	University of Florida, US
Kelly Holley-Bockelmann	Vanderbilt University, Nashville TN, US
Daniel Hollington	Imperial College London, UK
Ann Hornschemeier	NASA Goddard Space Flight Center, US
Allan Hornstrup	Danmarks Tekniske Universitet, DK
Steven Horowitz	NASA, US
Ming Hu	Institute of Geodesy and Geophysics, Chinese Academy of Sciences, CN
Martin C.E. Huber	Space Science Department, ESA (former)
Scott Hughes	Massachusetts Institute of Technology, US

Henri Inchauspe APC, CNRS / Université Paris Diderot, FR
Wako Ishibashi Institute of Geophysics, ETH Zürich, CH
Katharina-Sophie Isleif Max Planck Institute for Gravitational Physics AEI, Hannover, DE
Oliver Jennrich ESA
Philippe Jetzer Physik Institut, Universität Zürich, CH
Gang Jin Institute of Mechanics, Chinese Academy of Sciences, CN
Ulrich Johann Airbus Defence and Space GmbH, DE
Brittany Kamai Fermilab / Caltech, US
Nikolaos Karnesis Max Planck Institute for Gravitational Physics AEI, Hannover, DE
Fumiko Kawazoe Max Planck Institute for Gravitational Physics AEI, Hannover, DE
Christian Killow University of Glasgow, UK
Antoine Klein Institut d'Astrophysique de Paris, FR
William Klipstein JPL, California Institute of Technology, US
Natalia Korsakova Max Planck Institute for Gravitational Physics AEI, Hannover, DE
Rafael Kueng Physik Institut, Universität Zürich, CH
Martin Kunz Université de Genève, CH
Matthieu Laporte APC, IN2P3, CNRS, Université Paris Diderot, FR
Shane Larson Northwestern University / Adler Planetarium, US
Philipp Laufer Technische Universität Dresden, DE
Steve Lecomte CSEM, CH
Lindley Lentati Cambridge University, UK
Manbing Li Huazhong University of Science and Technology, CN
Maike Lieser Max Planck Institute for Gravitational Physics AEI, Hannover, DE
Li Liu Università di Trento, IT
Jeffrey Livas NASA Goddard Space Flight Center, US
Ivan Lloro Institut de Ciències de l'Espai, CSIC IEEC, ES
Lucas Lombriser University of Edinburgh, UK
Hudson Loughlin Princeton University, US
Andrew Lundgren Max Planck Institute for Gravitational Physics AEI, Hannover, DE
Juan Pedro López Zaragoza Institut de Ciències de l'Espai, CSIC IEEC, ES
Davor Mance Institute of Geophysics, ETH Zürich, CH
Oliver Mandel Airbus Defence and Space GmbH, DE
Etienne Marin-Matholaz APC, CNRS / Université Paris Diderot, FR
Joseph Martino APC, Université Paris Diderot, FR
Grégoire Martinon Observatoire de Paris, FR
Víctor Martín Institut de Ciències de l'Espai, CSIC IEEC, ES
Ignacio Mateos Institut de Ciències de l'Espai, CSIC IEEC, ES
Cristian Maureira-Fredes Max Planck Institute for Gravitational Physics AEI, DE
Lucio Mayer Institute for Computational Science, University of Zurich, CH
Paul McNamara ESA
Sean McWilliams West Virginia University
Neda Meshksar Institute of Geophysics, ETH Zürich, CH
Susanne Milde Milde Science Communication, DE
Li Ming DFH Satellite Co. Ltd, CN
Chiara M. F. Mingarelli Max Planck Institute for Radio Astronomy, DE
Christopher Moore University of Cambridge, UK
Didier Morancais Airbus Defence and Space GmbH, DE
Guido Mueller University of Florida, US
Mitsuru Musha Institute for Laser Science, University of Electro-communications Tokio, JP
Sourabh Nampalliwar Fudan University, Shanghai, CN

IOP Conf. Series: Journal of Physics: Conf. Series **840** (2017) 011004 doi:10.1088/1742-6596/840/1/011004

Germano Nardini	Universität Bärn, CH
Gijs Nelemans	Radboud University, NL
Marius Oltean	University of Orleans, FR / Autonomous University of Barcelona, ES
Sarah Paczkowski	Max Planck Institute for Gravitational Physics AEI, Hannover, DE
Yiyu Pang	Huazhong University of Science and Technology, CN
Daniel Penkert	Max Planck Institute for Gravitational Physics AEI, Hannover, DE
Michael Perreur-Lloyd	University of Glasgow, UK
Isabelle Petitbon	[none]
Antoine Petiteau	APC, Université Paris Diderot, FR
Lionel Philippoz	Physik Institut, Universität Zürich, CH
Mauro Pieroni	APC, Université Paris Diderot, FR
Paolo Pivato	TIFPA, Università di Trento / INFN Trento, IT
Eric Plagnol	CNRS, FR
Xuan Qin	Huazhong University of Science and Technology, CN
Detlef Reichardt	[none]
Jens Reiche	Max Planck Institute for Gravitational Physics AEI, DE
Fulvio Ricci	Università la Sapienza Roma, IT
Angelo Ricciardone	University of Stavanger, NO
Sascha Rieger	Milde Science Communication, DE
Francisco Rivas	Institut de Ciències de l'Espai, CSIC IEEC, ES
David Robertson	University of Glasgow, UK
Norna Robertson	Caltech, US
Elena Maria Rossi	Leiden Observatory, Universiteit Leiden, NL
Giuliana Russano	TIFPA, Università di Trento, IT
Rosmarie Rössel	Physik Institut, Universität Zürich, CH
Frédéric Safa	ESA
Rita Sambruna	NASA, US
Sotiris Sanidas	API, Universiteit van Amsterdam, NL
Josep Sanjuan	Deutsches Zentrum für Luft- und Raumfahrt, Bremen, DE
Shannon Sankar	NASA GSFC / CRESST / USRA, US
Bangalore Sathyaprakash	Penn State University / Cardiff University, US
Shuichi Sato	Hosei University, JP
Govert Schilling	[none]
Alexander Schleicher	Airbus Defence and Space GmbH, DE
Rainer Schoedel	Instituto de Astrofísica de Andalucía, CSIC, ES
Thilo Schuldt	Deutsches Zentrum für Luft- und Raumfahrt, Bremen, DE
Bernard Schutz	Max Planck Institute for Gravitational Physics AEI, DE
Thomas Schwarze	Max Planck Institute for Gravitational Physics AEI, Hannover, DE
Andreas Schärer	Physik Institut, Universität Zürich, CH
Alexander Sell	Airbus Defence and Space GmbH, DE
Geraldine Servant	DESY / University Hamburg, DE
Alberto Sesana	University of Birmingham, UK
Naoki Seto	Kyoto University, JP
David Shoemaker	Massachusetts Institute of Technology, US
Jacob Slutsky	University of Maryland, Baltimore County, US
Stephen Smartt	Queen's University Belfast, UK
Li Songming	DFH Satellite Co. Ltd, CN
Robin Stebbins	[none]
Aru Suemasa	Institute for Laser Science, University of Electro-communications Tokio, JP
Martin Tajmar	Technische Universität Dresden, DE

Nicola Tamanini	IPhT, CEA Saclay, FR
Jan ten Pierick	[none]
Damien Texier	ESA
Ira Thorpe	NASA Goddard Space Flight Center, US
Massimo Tinto	JPL, California Institute of Technology, US
Christian Trenkel	Airbus Defence and Space, UK
Hai-Bo Tu	Institute of Geodesy and Geophysics, Chinese Academy of Sciences, CN
Liangcheng Tu	Huazhong University of Science and Technology, CN
Akhilbabu Laxman Turai	SRM University, IN
Azita Valinia	NASA, US
Niels van Bakel	Nikhef, NL
Maarten van de Meent	University of Southampton, UK
Oana van der Togt	TNO, NL
Daniele Vetrugno	Università di Trento, IT
Stefano Vitale	Università di Trento, IT
Bin Wang	Huazhong University of Science and Technology, CN
Yan Wang	Huazhong University of Science and Technology, CN
Zhi Wang	Changchun Institute of Optics, fine Mechanics and Physics, Chinese Academy of Sciences, CN
Gudrun Wanner	Max Planck Institute for Gravitational Physics AEI, Hannover, DE
Henry Ward	University of Glasgow, UK
Peter Wass	Imperial College London, UK
William Joseph Weber	Università di Trento / INFN Trento, IT
David Weir	University of Helsinki, FI
Dennis Weise	Airbus Defence and Space GmbH, DE
Lennart Wissel	Max Planck Institute for Gravitational Physics AEI, Hannover, DE
Andreas Wittchen	Max Planck Institute for Gravitational Physics AEI, Hannover, DE
Peng Xu	Academy of Mathematics and Systems Science, Chinese Academy of Sciences, CN
Hao Yan	Huazhong University of Science and Technology, CN
Jun Yan	Huazhong University of Science and Technology, CN
Fangchao Yang	Huazhong University of Science and Technology, CN
Hang Yin	Huazhong University of Science and Technology, CN
Jianbo Yu	Huazhong University of Science and Technology, CN
Jingyi Zhang	Huazhong University of Science and Technology, CN
Zebing Zhou	Huazhong University of Science and Technology, CN
Peter Zweifel	Institute of Geophysics, ETH Zürich, CH

IOP Conf. Series: Journal of Physics: Conf. Series **840** (2017) 011005 doi:10.1088/1742-6596/840/1/011005

Programme

Monday, 5th of September 2016

Chair: M. Colpi

S. Vitale	The LTP experiment on LISA Pathfinder and its first results
K. Dooley	Advanced LIGO: news from the dark universe
B. Schutz	LIGOs Black Holes and Science in the LISA Era
P. McNamara	LISA Pathfinder: First steps to observing gravitational waves from space
M. Hewitson	In-flight Performance of the Optical Metrology System on LISA Pathfinder
W. Weber	Physical model of the LISA Pathfinder differential acceleration measurement and its application to LISA
K. Danzmann	The Status of LISA

Chair: O. Jennrich

J. Martino	On-flight characterization of Coldgas micro-Newton thrusters for Lisa Pathfinder
M. Nofrarias	Magnetic force noise in LISA Pathfinder
L. Ferraioli	The LISA Pathfinder GRS Front-End Electronics
D. Vetrugno	The Δg workflow: from measured displacements to the differential external acceleration.
A. Wittchen	Coupling of relative intensity noise and pathlength noise to the length measurement in the optical metrology system of LISA Pathfinder
S. Paczkowski	Laser frequency stabilisation and interferometer path length differences on LISA Pathfinder
R. Giusteri	The free-fall mode experiment on LISA Pathfinder: first results
P. Wass	Electrostatic test-mass force-noise disturbances as measured by LISA Pathfinder
R. Dolesi	The contribution of Brownian noise from viscous gas damping to the differential acceleration noise measured in LISA Pathfinder between two nominally free falling test masses
C. Trenkel	The engineering of LISA Pathfinder – the quietest laboratory ever flown in space

IOP Conf. Series: Journal of Physics: Conf. Series **840** (2017) 011005 doi:10.1088/1742-6596/840/1/011005

W. Klipstein	LISA-derived Flight Phasemeter on the GRACE Follow-On Mission
P. Laufer	MEMS Ion Thruster Chips to Significantly Enhance Cold Gas Thruster Lifetime for LISA
M. Tinto	Enhanced Gravitational Wave Science with LISA and gLISA

Tuesday, 6th of September 2016

Chair: P. McNamara

F. Ricci	Advanced VIRGO: State of the Art
C. Belczynski	Evolutionary predictions in the advanced LIGO/Virgo era
A. Sesana	The promise of multi-band gravitational wave astronomy
B. Sathyaprakash	Exploring the Universe with third generation interferometers
S. Sato	The status of DECIGO
G. Heinzel	Satellite interferometry from LTP and GRACE Follow-On to LISA
P. Binetruy	Cosmology with LISA

Chair: S. Vitale

A. Giménez Cañete	Beyond LISA Pathfinder
P. Hertz	NASA Astrophysics Program: Present and Future
F. Safa	ESA activities towards the Gravitation Waves Space Observatory
H. Ward	LPF Interferometry and route to LISA
J. Camp	Development of a US Laser System for the Gravitational Wave Mission LISA
T. Schwarze	Performance of an optical three-signal test for the LISA metrology
M. Laporte	Status of the eLISA on table (LOT) electro-optical simulator for space based gravitational waves observatories
S. Apple	UF Torsion Pendulum and GRS Technology Development
J. Livas	eLISA Telescope In-field Pointing and Scattered Light Study
K. Isleif	The LISA backlink – comparing optical phase reference systems

Wednesday, 7th of September 2016

Chair: C. Sopuerta

T. Alexander	EMRIs and the relativistic loss cone
R. Schoedel	The stellar cusp around the Milky Way's central black hole
J. Gair	Impact of eLISA design on EMRI Science
M. van de Meent	Modelling spinning EMRIs with Self-Force
A. Gianninas	A Progress Report on the Search for the Most Extreme Binary White Dwarfs

I. Thorpe Initial Results from ST7-Disturbance Reduction System on LISA Pathfinder

G. Jin Ongoing development of detection of gravitational waves in space in China

Thursday, 8th of September 2016

Chair: P. Binetruy

R. Blandford The Future of Black Hole Astrophysics in the LISA-VIRGO-LPF Era

C. Moore Observing black hole merger kicks directly with gravitational waves

N. Tamanini Late time cosmology with eLISA

I. Dvorkin A synthetic model of the gravitational wave background from evolving binary compact objects

A. Petiteau LISA science performance in the context of LISAPathfinder first results and simulation for LISA

N. Cornish Fast, fully precessing frequency domain waveforms for analyzing black hole mergers with LISA

A. Klein Supermassive Black Hole Observations with eLISA

S. Hughes Measuring EMRIs: A reality check

Chair: Ch. Caprini

L. Mayer Massive black hole mergers and their timescales: connecting galaxy formation with gravitational wave science

S. Babak Enchilada: analysis of simulated eLISA data set

A. Ricciardone Primordial Gravitational Waves and eLISA

S. Sanidas Can eLISA save the day? Hunting for cosmic strings in the SKA era.

G. Servant Probing cosmological phase transitions with eLISA

D. Weir Acoustic waves and the detectability of first-order phase transitions by eLISA

M. Bonetti Post-Newtonian Dynamics of Massive Black Holes Triplets

J. Garcia-Bellido Gravitational waves from primordial black holes as dark matter

T. Regimbau Searching for the stochastic gravitational-wave background with Advanced LIGO and Advanced Virgo

M. Pieroni Primordial GW from universality classes of pseudo-scalar inflation

IOP Conf. Series: Journal of Physics: Conf. Series **840** (2017) 011005 doi:10.1088/1742-6596/840/1/011005

Friday, 9th of September 2016

Chair: G. Mueller

K. Holley-Bockelmann	The Known Unknowns: predicting the landscape of LISA black hole sources
E. Rossi	Supermassive black hole seeds: updates on the quasi star model
C. Mingarelli	The Gravitational-Wave Universe seen by Pulsar Timing Arrays
S. Smartt	Searching for the electromagnetic counterparts to gravitational wave sources
J. Bergé	MICROSCOPE: the first few months in orbit
O. Jennrich	The preparation for L3 – building on LISA Pathfinder

IOP Conf. Series: Journal of Physics: Conf. Series **840** (2017) 011006 doi:10.1088/1742-6596/840/1/011006

IOP Conf. Series: Journal of Physics: Conf. Series **840** (2017) 011006 doi:10.1088/1742-6596/840/1/011006

11th LISA Symposium Group Photo
7th Sept. 2016; Irchel campus, University of Zurich, Switzerland
photograph by Rosmarie Rössel

IOP Conf. Series: Journal of Physics: Conf. Series **840** (2017) 011006 doi:10.1088/1742-6596/840/1/011006

Stefano Vitale with birthday cake for his 65 years
6th Sept. 2016; Irchel campus, University of Zurich, Switzerland
photograph by Rosmarie Rössel

IOP Conf. Series: Journal of Physics: Conf. Series **840** (2017) 011006 doi:10.1088/1742-6596/840/1/011006

Martin Gehler, Martin Huber, Paul McNamara, Philippe Jetzer, Rosmarie Rössel
7th Sept. 2016; Uto Kulm, Üetliberg Zurich, Switzerland
photograph by Fiorenza Bassetti

IOP Conf. Series: Journal of Physics: Conf. Series **840** (2017) 011006 doi:10.1088/1742-6596/840/1/011006

Paul Hertz, Philippe Jetzer, Martin Huber
7th Sept. 2016; Uto Kulm, Üetliberg Zurich, Switzerland
photograph by Fiorenza Bassetti

IOP Conf. Series: Journal of Physics: Conf. Series **840** (2017) 011006 doi:10.1088/1742-6596/840/1/011006

Martin Huber giving the after dinner speech
7th Sept. 2016; Uto Kulm, Üetliberg Zurich, Switzerland
photograph by Rosmarie Rössel

IOP Conf. Series: Journal of Physics: Conf. Series **840** (2017) 011007 doi:10.1088/1742-6596/840/1/011007

LISA Pathfinder author list

M Armano,[1] H Audley,[2] G Auger,[3] J Baird,[4] M Bassan,[5] P Binetruy,[3] M Born,[2] D Bortoluzzi,[6] N Brandt,[7] M Caleno,[8] A Cavalleri,[9] A Cesarini,[9] M Cruise,[10] K Danzmann,[2] M de Deus Silva,[1] R De Rosa,[11] L Di Fiore,[12] I Diepholz,[2] R Dolesi,[9] N Dunbar,[13] L Ferraioli,[14] V Ferroni,[9] E Fitzsimons,[15] R Flatscher,[7] M Freschi,[1] C García Marrirodriga,[8] R Gerndt,[7] L Gesa,[16] F Gibert,[9] D Giardini,[14] R Giusteri,[9] A Grado,[11] C Grimani,[17] J Grzymisch,[8] I Harrison,[18] G Heinzel,[2] M Hewitson,[2] D Hollington,[4] D Hoyland,[10] M Hueller,[9] H Inchauspé,[3] O Jennrich,[8] P Jetzer,[19] B Johlander,[8] N Karnesis,[2] B Kaune,[2] N Korsakova,[2] C Killow,[20] A Lobo,[16,*] I Lloro,[16] L Liu,[9] J.P. López-Zaragoza,[16] R Maarschalkerweerd,[18] D Mance,[14] V Martín,[16] L Martin-Polo,[1] J Martino,[3] F Martin-Porqueras,[1] S Madden,[8] I Mateos,[16] P W McNamara,[8] J Mendes,[18] L Mendes,[1] M Nofrarias,[16] S Paczkowski,[2] M Perreur-Lloyd,[20] A Petiteau,[3] P Pivato,[9] E Plagnol,[3] P Prat,[3] U Ragnit,[8] J Ramos-Castro,[21] J Reiche,[2] D I Robertson,[20] H Rozemeijer,[8] F Rivas,[16] G Russano,[9] P Sarra,[22] A Schleicher,[7] D Shaul,[4] J Slutsky,[23] C F Sopuerta,[16] R Stanga,[24] T Sumner,[4] D Texier,[1] J I Thorpe,[23] C Trenkel,[13] M Troebs,[2] D Vetrugno,[9] S Vitale,[9] G Wanner,[2] H Ward,[20] P Wass,[4] D Wealthy,[13] W J Weber,[9] L Wissel,[2] A Wittchen,[2] A Zambotti,[6] C Zanoni,[6] T Ziegler,[7] and P Zweifel[14]

[1] *European Space Astronomy Centre, European Space Agency, Villanueva de la Cañada, 28692 Madrid, Spain*
[2] *Albert-Einstein-Institut, Max-Planck-Institut für Gravitationsphysik und Universität Hannover, Callinstrasse 38, 30167 Hannover, Germany*
[3] *APC, Univ Paris Diderot, CNRS/IN2P3, CEA/lrfu, Obs de Paris, Sorbonne Paris Cité, France*
[4] *High Energy Physics Group, Physics Department, Imperial College London, Blackett Laboratory, Prince Consort Road, London, SW7 2BW, UK*
[5] *Dipartimento di Fisica, Università di Roma "Tor Vergata", and INFN, sezione Roma Tor Vergata, I-00133 Roma, Italy*
[6] *Department of Industrial Engineering, University of Trento, via Sommarive 9, 38123 Trento, and Trento Institute for Fundamental Physics and Application / INFN*
[7] *Airbus Defence and Space, Claude-Dornier-Strasse, 88090 Immenstaad, Germany*
[8] *European Space Technology Centre, European Space Agency, Keplerlaan 1, 2200 AG Noordwijk, The Netherlands*
[9] *Dipartimento di Fisica, Università di Trento and Trento Institute for Fundamental Physics and Application / INFN, 38123 Povo, Trento, Italy*
[10] *Department of Physics and Astronomy, University of Birmingham, Birmingham, UK*
[11] *INAF Osservatorio Astronomico di Capodimonte, I-80131 Napoli, Italy and INFN sezione di Napoli, I-80126 Napoli, Italy*
[12] *INFN - Sezione di Napoli, I-80126, Napoli, Italy*
[13] *Airbus Defence and Space, Gunnels Wood Road, Stevenage, Hertfordshire, SG1 2AS, UK*
[14] *Institut für Geophysik, ETH Zürich, Sonneggstrasse 5, CH-8092, Zürich, Switzerland*
[15] *The UK Astronomy Technology Centre, Royal Observatory, Edinburgh, Blackford Hill, Edinburgh, EH9 3HJ, UK*
[16] *Institut de Ciències de l'Espai (CSIC-IEEC), Campus UAB, Carrer de Can Magrans s/n, 08193 Cerdanyola del Vallès, Spain*
[17] *DISPEA, Università di Urbino "Carlo Bo", Via S. Chiara, 27 61029 Urbino/INFN, Italy*
[18] *European Space Operations Centre, European Space Agency, 64293 Darmstadt, Germany*
[19] *Physik Institut, Universität Zürich, Winterthurerstrasse 190, CH-8057 Zürich, Switzerland*
[20] *SUPA, Institute for Gravitational Research, School of Physics and Astronomy, University of Glasgow, Glasgow, G12 8QQ, UK*
[21] *Department d'Enginyeria Electrònica, Universitat Politècnica de Catalunya, 08034 Barcelona, Spain*
[22] *CGS S.p.A, Compagnia Generale per lo Spazio, Via Gallarate, 150 - 20151 Milano, Italy*
[23] *Gravitational Astrophysics Lab, NASA Goddard Space Flight Center, 8800 Greenbelt Road, Greenbelt, MD 20771 USA*
[24] *Dipartimento di Fisica ed Astronomia, Università degli Studi di Firenze and INFN - Sezione di Firenze, I-50019 Firenze, Italy*

Peer review statement

All papers published in this volume of *Journal of Physics: Conference Series* have been peer reviewed through processes administered by the proceedings Editors. Reviews were conducted by expert referees to the professional and scientific standards expected of a proceedings journal published by IOP Publishing.

LISA Pathfinder: First steps to observing gravitational waves from space

LISA Pathfinder collaboration

Reference to author list in volume

E-mail: paul.mcnamara@esa.int

Abstract. LISA Pathfinder, the European Space Agency's technology demonstrator mission for future spaceborne gravitational wave observatories, was launched on 3 December 2015, from the European space port of Kourou, French Guiana. After a short duration transfer to the final science orbit, the mission has been gathering science data since. This data has allowed the science community to validate the critical technologies and measurement principle for low frequency gravitational wave detection and thereby confirming the readiness to start the next generation gravitational wave observatories, such as LISA.

This paper will briefly describe the mission, followed by a description of the science operations highlighting the performance achieved.

Details of the various experiments performed during the nominal science operations phase can be found in accompanying papers in this volume.

1. Introduction

LISA Pathfinder (LPF), the European Space Agency's second SMART [1] mission, was instigated to demonstrate technologies for future spaceborne gravitational wave observatories, for example, the Laser Interferometer Space Antenna (LISA) mission, recently proposed to ESA [1].

LISA has continually been ranked as one of the most scientifically important missions under study [2]. However the very concept of low frequency gravitational wave detection, *i.e.* that a particle falling under the influence of gravity alone follows a geodesic in spacetime, has never been demonstrated to the required precision. LISA Pathfinder has been designed to demonstrate that low frequency gravitational wave detection from space is indeed possible.

LPF mimics one arm of the LISA constellation by shrinking the 2.5 million-kilometre armlength down to a few tens of centimetres, and hosting the hardware inside a single spacecraft. This results in giving up the sensitivity to gravitational waves, however, it maintains (and in some cases worsens) the instrumental noise which could disturb the eventual LISA measurement. The concept is to measure, using laser interferometry, the relative separation of two free-falling test masses nominally following their own geodesics (Figure 1), and thereby determining the relative residual acceleration between them over a measurement bandwidth from 1-30 mHz, about a decade above the lowest frequency required by LISA.

[1] SMART = Small Missions for Advanced Research in Technology

Figure 1: *Left*:Artist's impression of the LISA Technology Package Core Assembly showing the gold:platinum test masses housed in their respective electrode housings. Credit: ESA/MediaLabs, *Right*: LTP flight hardware during integration. Credit: Airbus Defence and Space

LISA Pathfinder was launched on 3 December 2015 on-board a dedicated VEGA launch vehicle, with the spacecraft and expendable propulsion module being injected into a low earth orbit (200 km x 1540 km). After a series of 6 apogee raising manoeuvres, LPF entered a short duration transfer trajectory towards the final science orbit, an 800,000 km x 500,000 km Lissajous orbit around the first Sun-Earth Lagrange point (L1) (see Figure 2). Following the initial on-orbit check-out and instrument calibration, the in-flight validation of the LISA technology began on 1 March 2016.

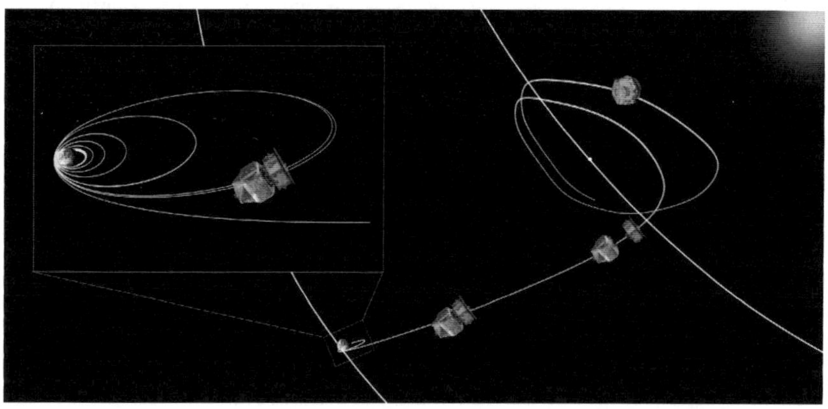

Figure 2: Artist's impression of the Launch and Early Orbit Phase (LEOP) of the mission, resulting in the final science orbit around the first Sun-Earth Lagrange point (L1). Credit: ESA/MediaLabs

2. The LISA Technology Package

Unlike traditional observatory or planetary missions, the payload in LISA Pathfinder cannot be considered as a discrete piece of hardware carried by the spacecraft. Instead, during science operations, the payload and the spacecraft act as a single unit: the attitude control of the spacecraft is driven by the payload. LISA Pathfinder carries two payloads: the LISA Technology Package (LTP), provided by a consortium of European national space agencies

(France, Germany, Italy, Spain, Switzerland, the Netherlands, and the United Kingdom) and ESA; and the NASA provided Disturbance Reduction System (DRS), part of NASAs New Millennium Program. Only the LTP will be described in this paper.

The LTP consists of two major subsystems; the Inertial Sensor Subsystem, and the Optical Metrology Subsystem. Both are described in further detail in the following sections.

2.1. Inertial Sensor Subsystem

The Inertial Sensor Subsystem (ISS) is at the heart of the LISA Pathfinder mission; the development and on-orbit testing of this subsystem are the main reasons for ESA implementing the mission. *The ISS of LISA Pathfinder is the ISS of LISA* - the relaxation in the requirements of LPF comes from the relaxation in the environmental conditions of the LPF spacecraft compared to LISA.

The inertial sensor subsystem comprises the test masses and all systems interacting directly with the test masses, *i.e.* the electrode housing, front-end electronics, vacuum system, charge management, and caging mechanism (see Figure 3). This section will describe each of these subsystems in turn.

The test masses consist of a 1.93 kg cube of Gold:Platinum mono-phasic alloy of dimension 46 mm on a side. The alloy is formed from 73% gold and 27% platinum, chosen as this alloy has an extremely low magnetic susceptibility ($\chi_m \approx 10^{-5}$) and high density $\approx 2 \times 10^4\,\mathrm{kgm^{-3}}$. The combination of both greatly reduces the effect of external forces on the test mass.

The test masses' position is readout by two means: high resolution laser interferometry, and electrostatic (capacitive) sensing. The former only senses the test mass position along the sensitive axis (the line joining the two test masses) and the angles of rotation around the axes perpendicular to the sensitive axis, whereas the capacitive sensor measures the position of the test mass in all six degrees of freedom. The capacitive sensor comprises a hollow cubic molybdenum housing with gold coated sapphire electrodes mounted in the faces (see Figure 3). The housing is sized to allow for a $\approx$4 mm gap between the electrode faces and the test mass. The size of the gap is a trade off between reducing the effects of noise sources, *e.g.* from uncontrolled potentials on the electrodes, and being able to meet the capacitive sensing requirement of $1.8\,\mathrm{nm}/\sqrt{\mathrm{Hz}}$ over the measurement bandwidth.

The capacitive readout system, known as the *Inertial Sensor Subsystem Front End Electronics* (ISS FEE), is arranged such that electrodes facing opposing faces of the test mass are combined via a capacitive bridge. A change in the position of the test mass gives a differential, bi-polar, signal at the output of the bridge, which is used as an input to the drag-free control system. As well as sensing the position of the test masses, the ISS FEE can also be used to actuate (force) the test mass.

The test mass and electrode housing are mounted inside a dedicated vacuum enclosure. To meet the mission requirements, the vacuum around the test mass must be maintained, throughout the mission lifetime, to less than 10^{-5} Pa. In order to limit the pressure increase due to outgassing or virtual leaks, the enclosure was vented to space once the spacecraft reached its operational science orbit.

As there is no physical contact between the test mass and the surrounding environment, one issue that must be dealt with is charging of the test mass due to cosmic ray and solar energetic particle impacts. A build up of charge on the test mass, coupled with the potentials on the electrodes, creates a force, resulting in additional noise in the test mass position. The charge is controlled using a non-contact discharge system based on the photo-electric effect. UV light from Mercury vapour lamps is channelled to the electrode housing via fibre optic cables. Depending on the sign of the charge on the test mass, the light is either shone onto the test mass or the

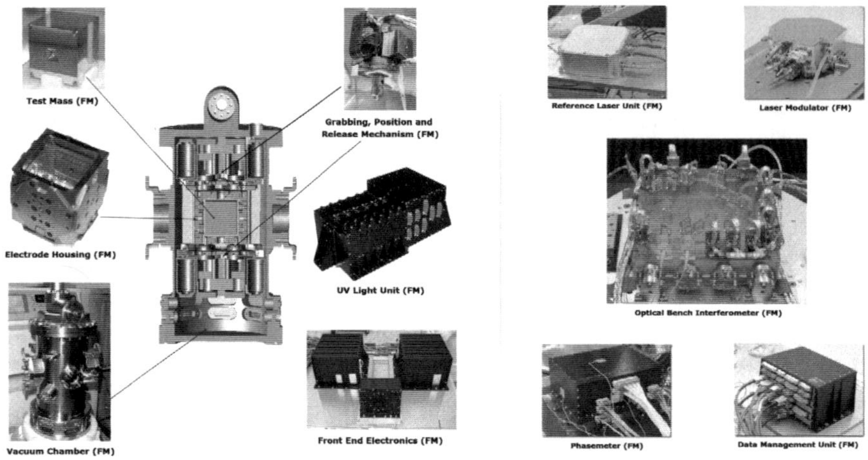

Figure 3: *Left*: Photographs of the Inertial Sensor Subsystem. From bottom left: Vacuum Chamber (CGS), Electrode Housing (CGS), (uncoated) Test Mass (Thales Alenia Space), Caging Mechanism (RUAG), UV Lamp Unit (Imperial College London), Front-End Electronics (ETH, Zurich). *Right*: Photographs of the Optical Metrology System hardware. From top left: Reference Laser Unit (Tesat), Laser Modulator (APC/Contraves), Data Management Unit (ICE), Phasemeter (Uni Birmingham). Centre: Optical Bench Interferometer (Uni Glasgow)

electrode housing, thereby extracting electrons from either surface, providing bi-polar charge management.

Several other challenges must also be solved in order to meet the requirements of the LTP. These include: balancing of the differential gravitational force and gradient at the test mass positions - achieved by mounting compensation masses inside, and external to, the vacuum enclosure; creating a thermally quiet environment around the test mass - a temperature stability of $10^{-5}\,\mathrm{K}/\sqrt{\mathrm{Hz}}$ over the measurement bandwidth - and associated with the thermal stability requirement is the need to have thermometers with a resolution better than $10^{-5}\,\mathrm{K}/\sqrt{\mathrm{Hz}}$.

2.2. Optical Metrology Subsystem

The Optical Metrology Subsystem (OMS) is the high resolution laser interferometric readout of the test masses' positions. The OMS comprises several subsystems, namely; the reference laser unit, the laser modulator, the optical bench interferometer, the phasemeter, and the data management unit (Figure 3).

The *Reference Laser Unit* (RLU) comprises a 40 mW Nd:YAG non-planar ring oscillator [3] of the same design commonly used in metrology labs around the world. This laser design is ideal for space applications due to its small size, high electrical to optical efficiency and inherent low noise operation. The RLU is baselined as the master oscillator in the LISA laser system.

The RLU output is fibre coupled using single-mode, polarisation-maintaining (sm/pm) fibre. The fibre couples the light to the subsequent component in the optical chain, the *Laser Modulator* (LM). The LM consists of a beam splitter, two acousto-optic modulators, and optical pathlength actuators. The light from the laser is split into two paths, each path is passed through an acousto-optic modulator. One modulator is driven at 80 MHz, while the other is driven at 80 MHz + 1 kHz, thereby creating two beams with a frequency difference of 1 kHz. The beams are then passed through the optical pathlength difference (OPD) actuator. The OPD is used to stabilise the optical pathlength of the fibre optic cables leading to the optical bench. After the

OPD, the beams are transmitted, again via sm/pm fibre, to the *Optical Bench Interferometer* (OBI).

The main function of the OBI is to direct the beams to the relevant positions in 3-dimensional space, without adding any significant noise to the measurement path. The optical bench is constructed from a block of Zerodur ceramic glass with fused silica mirrors and beamsplitters bonded to the bench using hydroxy catalysis bonding [4] . The mirrors and beamsplitters are used to direct the two beams to form four interferometers: the x_{12} interferometer which measures the differential motion of the two test masses - this is the primary science measurement of the mission; x_1 interferometer which measures the position and angles of test mass 1 with respect to the optical bench (and therefore, the spacecraft) - equivalent to the LISA local test mass interferometer; the *Frequency* interferometer which is an unequal arm Mach-Zehnder interferometer, the output of which is sensitive to laser frequency fluctuations, and therefore can be used to stabilise the laser frequency; and the *Reference* interferometer which is a rigid equal arm interferometer which provides the system noise floor, and is used to stabilise the optical pathlengths via the OPD. The light from each fibre is also sent directly to a photodiode which is used to monitor the laser intensity noise. The signal from these photodiodes is used to stabilise the intensity of both beams by feeding back to the acousto-optic modulator drive signal.

The signals from the (quadrant) photodiodes of each interferometer are sent to the *Phasemeter Assembly*. The phasemeter samples the data at 50 kHz and performs a Single Bin Discrete Fourier Transform [5] to measure the phase of the signal at the heterodyne frequency. This technique is used due to the efficiency of the algorithm. The phasemeter not only outputs the longitudinal phase from the respective interferometers, but also outputs the angles between the wavefronts interfering on the photodetectors - commonly known as *differential wavefront sensing* (DWS) - at 100 Hz. The DWS signals from the x_1 and x_{12} interferometers are used to align the test mass to the interferometer. The longitudinal signals from the interferometers are used to stabilise the laser frequency, the optical pathlength, and (with the DWS signals) as inputs for the Drag-Free and Attitude Control System (DFACS) [6].

As mentioned above, the phasemeter outputs the data at 100 Hz. However, the 100 Hz samples are not required for routine operation, and so the data is downsampled to 10 Hz prior to transmission to the on-board computer (and hence the DFACS). The downsampling is performed inside the *Data Management Unit* (DMU) - a 12 MHz ERC32 processor. The DMU is also responsible for the interface to the LTP subsystems, routing telecommands and timing information to the units, and collecting and transmitting telemetry to the on-board computer.

3. Operations Timeline

Following the launch and early operations, the spacecraft was put on a transfer towards the first Sun-Earth Lagrange point (L1). After $\approx$50 days cruise, the propulsion module was jettisoned, and LPF entered its final science orbit. This triggered the formal start of the spacecraft commissioning phase.

After initial check-out of the hardware subsystems, the launch locks holding the test masses during launch were released, however the test masses were still mechanically held in place via piezo actuated fingers. In addition to performing the launch lock function the actuators were also used to open a valve, allowing the interior of the vacuum enclosure to be vented to space. This activity took place on 2 Feb 2016. The test masses were released from their mechanical actuators on 15 & 16 February, after which they were free floating with no mechanical contact to the satellite.

LISA Pathfinder science operations began on 1 March 2016, with the nominal mission ending on 26 June 2016.

The goal of LISA Pathfinder is to not only demonstrate the required free-fall performance

Table 1: List of the high level experiments to be performed during the LISA Pathfinder nominal operations. The colour corresponds to the colours in the timeline in Figure 4.

Label	Experiment type
	Differential acceleration noise run
	Measurement of dynamical coefficients by system identification
	Measurement of cross-talk
	Measurement of stray DC potentials
	Dedicated optical metrology experiments
	Measurement of the thermal environment of the system
	Measurement of the magnetic environment of the system

for future gravitational wave observatories, but also to understand the physical model of the forces which perturb the test masses from pure free-fall. For this reason, the operations of LISA Pathfinder were designed to probe different aspects of the system, and thereby validate the system noise model. Table 1 lists the high level experiments run during the mission. Each of the experiments are then sub-divided in to a number of individual investigations which are run on the spacecraft.

Figure 4 graphically displays the dates on which each experiment was run. As can be seen, long periods of the operations are dedicated to noise runs. This is essential in order to understand the system at the lowest frequencies of the measurement bandwidth. As will be seen in Section 5, this becomes even more apparent when the system is used to investigate the lowest measurement frequencies of interest to LISA ($\approx 20\,\mu$Hz) where each noise run lasts a minimum of two weeks of operational time.

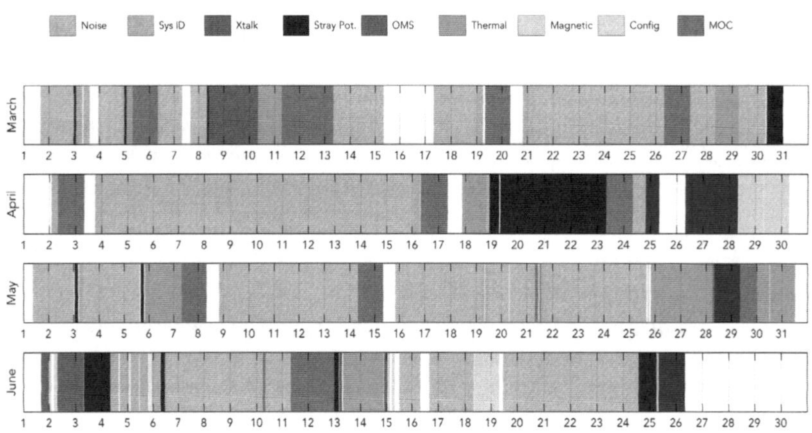

Figure 4: Timeline of the nominal science operational phase of LISA Pathfinder showing the experiments being run on a given day. The colours relate to the experiments listed in Table 1. Note: the blocks labelled *Config* relate to changes in the instrument configuration to prepare the system for the upcoming experiment, while the blocks labelled *MOC* are windows reserved for the Mission Operations Centre (MOC) team to perform routine activities such as station keeping.

Details of each of the experiments and associated investigations can be found in the accompanying papers in this volume.

4. First Results

As mentioned previously, the primary science measurement of LISA Pathfinder is the measure of differential acceleration between the test masses. Being a precursor mission, the corresponding differential acceleration requirement was relaxed with respect to LISA, however, the relaxation comes from the spacecraft environment and the nature of the LISA Pathfinder mission (two test masses sharing the same sensitive axis).

Figure 5 shows the performance of the system on the first day of operations, before any tuning of the system parameters had taken place. It is clearly seen that the primary performance requirement was already met. The performance was limited by three main noise sources: 1) at frequencies >20 mHz, the limiting noise source is the interferometer sensing noise, 2) mid-band frequencies are limited by residual gas damping (Brownian noise), 3) the low frequencies (<1 mHz) are limited by actuation noise of the second test suspension control system.

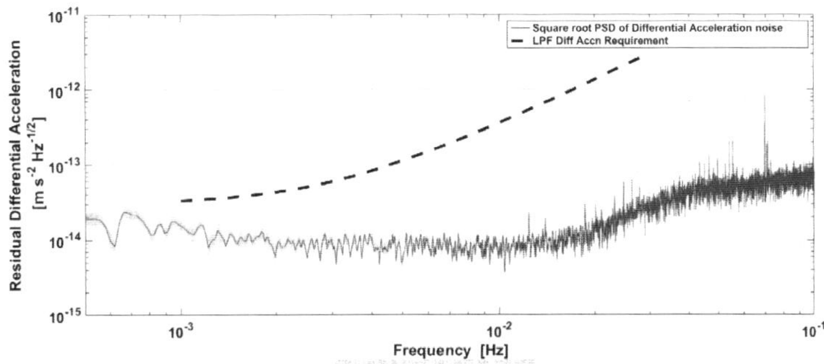

Figure 5: Differential acceleration, Δg, performance between the two free floating test masses, as measures on the first day of science operations.

The evolution of each noise source will be described in more details in the following paragraphs

The interferometer sensing noise has been stable since the beginning of science operations, with a noise floor of $\approx 35\,\mathrm{fm}/\sqrt{\mathrm{Hz}}$ - this is approximately 250 times lower than the sensing noise requirements, demonstrating the ability to monitor the position of free-falling test bodies to sub-picometre precision.

Concerning the mid-band frequencies; as the interior of the electrode housing was vented to space, the pressure, and hence Brownian noise, reduced over time, with the noise following a exponential decrease with the t_0 being centred around the opening of the venting valve on 2 February. This has resulted in the mid-band noise continuously reducing as science operations have progressed.

As mentioned above, the low frequency noise was dominated by the voltage noise of the electrostatic actuation system of the second test mass suspension control system. The required force (and hence voltage) to be applied is driven by the differential gravity experienced by test mass two with respect to test mass one. The LPF requirement was such that the differential gravity, coming from the self-gravity of the spacecraft, shall be less than 2000 pN. On the first day of operations, the differential gravity was measured, by looking at the force required to keep TM2 following TM1, to be ≈40 pN, a factor of 50 lower than requirements. This allowed the science operations team to lower the required maximum applied force, and thereby reduce the effect of the voltage noise in the actuation system [7]. In so doing, the low frequency noise was significantly reduced.

Figure 6 shows the differential acceleration performance published in the LPF first results paper [7]. As can be seen, the performance has significantly improved, approaching the levels required for a full gravitational wave observatory such as LISA. The data shown in the first results paper is from April 2016. Details of the performance can be found in [7]. The instrument performance continued to improve over the lifetime of the mission, and more so during the extended mission [1].

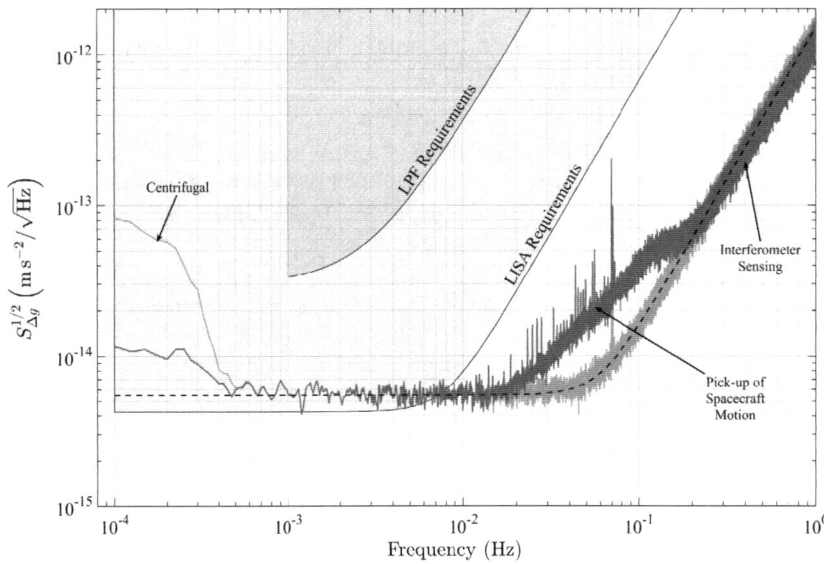

Figure 6: LISA Pathfinder differential acceleration performance result as published in the first results paper ([7])

5. Extended Mission

Following the LPF nominal science mission, the control of the satellite was handed over to the NASA Disturbance Reduction System (DRS) payload.

The DRS nominal operations concluded on 7 December 2016, signalling the start of the mission extension. This phase of the mission will run until 30 June 2017. The mission extension operations are focused on the investigation of the low frequency noise, especially at lowest frequencies of interest to LISA (20 μHz to 1 mHz) - this necessitates long noise runs (typically 2-3 weeks) in order to build statistics of the noise content. The performance of the LTP now exceeds the LISA performance requirement, however, quantitative analysis of the data is ongoing.

6. Conclusions

The LISA Pathfinder satellite has proven the concept of low frequency gravitational wave detection from space is possible. The performance of the LISA Technology Package has demonstrated residual differential acceleration noise between free floating test masses to over an order of magnitude lower than the mission level requirements, fulfilling the LISA performance goals over the entire LISA measurement bandwidth.

The success of LISA Pathfinder has paved the way for LISA. The European Space Agency have selected the LISA concept for further study, with a launch date slated for 2034.

References

[1] Laser Interferometer Space Antenna (LISA), Proposal to ESA for Cosmic Vision L3, "The Gravitational Universe" mission slot (`https://www.elisascience.org/files/publications/LISA_L3_20170120.pdf`)

[2] National Academy of Sciences, *New Worlds, New Horizons in Astronomy and Astrophysics*, Report of the National Research Council's Decadal survey of Astronomy and Astrophysics (2010)

[3] Kane, T.J., and Byer, R.L., *Monolithic, unidirectional single-mode Nd:YAG ring laser*, Optics Letters, **10**, 2, 65-67 (1985)

[4] Middleton, K., *et al.*,*Prototype optical bench instrument in the interferometer for the LISA-Pathfinder space mission*, Optical Engineering, **45** (2006)

[5] Heinzel, G, *et al.*, *Interferometry for the LISA Technology Package, LTP: an update*, Journal of Physics Conference Series, **32**, 132 (2006)

[6] Fichter, W., *et al.*, *Drag-Free Control Design with Cubic Test Masses*, In "Lasers, Clocks, and Drag-Free Control", Astrophysics and Space Science Library, Vol. 349, Dittus, H.J.; Lammerzahl, C.;Turyshev, S. (Eds.), December 2007, Springer, ISBN: 978-3-540-34376-9

[7] Armano, M, *et al.*,*Sub-Femto-g Free Fall for Space-Based Gravitational Wave Observatories: LISA Pathfinder Results*, PRL, **116**, 231101 (2016)

Calibrating LISA Pathfinder raw data into femto-g differential accelerometry

Daniele Vetrugno[1] and Nikolaos Karnesis[2] on behalf of the LPF collaboration

[1] Dipartimento di Fisica, University of Trento and Trento Institute for Fundamental Physics and Application / INFN, 38123 Povo, Trento, Italy
[2] Albert-Einstein-Institut, Max-Planck-Institut für Gravitationsphysik und Universität Hannover, 30167 Hannover, Germany

E-mail: daniele.vetrugno@unitn.it, karnesis@aei.mpg.de

Abstract. LISA Pathfinder is an in-flight test of the local sources of acceleration noise in LISA. The acceleration noise level in LISA Pathfinder is measured by the residual differential acceleration Δg between the two test masses once the coupling to the spacecraft motion has been removed. The full process from raw data to Δg passes through a series of calibration experiments and different data elaboration procedure which are thoroughly used during the mission and represent the baseline for any other further investigation.

1. Introduction

LISA Pathfinder[1] (LPF) is an European Space Agency (ESA) satellite launched on December 3, 2015 aiming to test the technological readiness of the local key hardware and software elements for the Laser Interferometer Space Antenna [2] (LISA) detector, the future space-based gravitational wave observatory. LISA is formed by three spacecraft in a triangular quasi-equilateral constellation around the Sun in a heliocentric orbit. Each satellite is separated by about 5×10^6 km from the other two and encloses two Test-Masses (TMs) which must be in principle freely falling. A passing gravitational wave would make then the TMs oscillate perpendicularly to the direction of gravitational wave propagation. However, this effect is a very tiny one, of the order of $h \sim 10^{-21}$ and can be similarly caused by any other non-gravitational spurious force acting on the TMs. To be able to measure gravitational waves it is necessary not only to measure the relative displacement between TMs with the required precision but also to be able to put the TMs in the most quiet possible free-fall. This last requirements for LISA is encoded in the request for the residual acceleration of each TM, g, to have an amplitude spectral density $S_g^{1/2} < 3$ fm s$^{-2}/\sqrt{\text{Hz}}$ at 0.1 mHz, the lowest frequency of its measurement band. The quality of the free fall depends both on the the drag-free control and on the spurious forces that act on the TMs, which are local sources of noise in LISA and hardly testable or measurable on ground. For this reason ESA decided to launch LPF, a technological demostrator of the local sources of noise on LISA. In particular, LPF is one arm of LISA enclosed in one SC, reducing the arm-length from 5×10^6 km to 0.376 m and placing the two TMs in the same spacecraft. The LISA acceleration noise requirement was relaxed for LPF to $S_{\Delta g}^{1/2} < 30$ fm s$^{-2}/\sqrt{\text{Hz}}$ at 1 mHz, where Δg represents the differential residual acceleration between the two TMs. To properly measure Δg, the full dynamics of the three body system represented by LPF needs to be calibrated and the contribution of the commanded forces of the controllers and the coupling of TMs motion to the spacecraft need to be subtracted.

In Section 2, the equations of motion of LPF are defined and a model for the calibration and the calculation of Δg is introduced. In Section 3, the Δg workflow from raw data to the ultimate estimation of the residual differential acceleration is presented. Finally, in Section 4, some conlusion about the results of Δg calculation are reported.

2. LPF dynamics

LPF is a dynamical system composed by two 46 mm quasi-cubic TMs, and a SC which encloses and protects them from external disturbanses [3]. Each TM is surrounded by an Electrode Housing [4] (EH) whose electrodes both sense the TM position and actuate forces on all the translational and angular degrees of freedom. Each TM plus EH forms the Gravity Reference Sensor (GRS). A three body dynamical system as LPF cannot be stable. Indeed, any static difference of force between the TMs will eventually push them to hit on their housing walls. Three control loops have been implemented in the Drag Free and Attitude Control System (DFACS) on-board software to make LPF dynamics stable. The Drag-free control [5] leaves TM1 freely falling along x and forces the SC to follow it by means of μ-Newton thrusters driven by the interferometer measurement of the distance between SC and TM1, o_1. The electrostatic suspension control forces TM2 to follow TM1 by means of electrostatic actuation[6] which are driven by a second laser interferometer measurement, o_{12}, which measures the relative displacement between the two TMs. Finally, the spacecraft angular degrees of freedom $\{\Phi_{SC}, H_{SC}, \Theta_{SC}\}$ are controlled by the attitude control loop which is fed by the Autonomous Star Tracker (AST) signal in order to make the solar array panel always pointing towards the Sun and the antenna towards the Earth. All the control loops act to maintain the system stable and to reduce as much as possible the noise contribution, in particular on the main measurement axis, the x-axis in Figure 1.

The quality of the free fall is measured by Δg, the differential acceleration noise between the two TMs, given by

$$\Delta g(t) = \ddot{x}_{12}(t) + \omega_2^2 x_{12}(t) + \Delta\omega^2 x_1(t) - A_{sus} F_{cmd,x2}(t) \tag{1}$$

where x_1 is the coordinate of TM1 along the x-axis and $x_{12} = x_2 - x_1$ is the displacement of TM2 with respect to TM1 aong the same axis. $F_{cmd,x2}$ is the commanded force on TM2. ω_2^2 is the stiffness of TM2 and $\Delta\omega^2 = \omega_2^2 - \omega_1^2$ is the differential stiffness which couples the SC motion to the differential acceleration. Finally, A_{sus} is the gain of the electrostatic commanded force on TM2.

The actual application of eq. (1) to the data of the LPF mission requires some manipulation. First of all, what is measured is not the coordinate x but the interferometric read-out signal o. Using the readouts entails that a whole set of possible sensing cross-talk has to be considered, whose the most important is the common mode cross-coupling, where the o_1 signal is leaking into the o_{12} measurement, through a coupling factor δ_{ifo}. Moreover, eq. (1) assumes that, at any given instant, the coordinates, the readouts and the actuation forces are synchronous, which is simply not true. Instead, they acquire a relative delay τ due to the phasemeter measurement process and to DFACS calculation. It means that any applied force $F_{cmd,x2}(t)$ has been in reality calculated at a previous time $(t-\tau)$, and the corrected force to be considered for the equation of Δg is $F_{cmd,x2}(t-\tau)$, where τ is a free parameter. The delayed force can be linearised for small delay values, as $F_{cmd,x2}(t-\tau) \simeq F_{cmd,x2}(t) - \tau\frac{d(F_{cmd,x2}(t))}{dt}$. Then, the model for $\Delta g(t)$ can be written as

$$\Delta g(t) = \ddot{o}_{12}(t) + \omega_2^2 o_{12}(t) + \Delta\omega^2 o_1(t) - A_{sus} F_{cmd,x2}(t) + C_1 \dot{F}_{cmd,x2}(t) + \delta_{\text{ifo}} \ddot{o}_1(t) \tag{2}$$

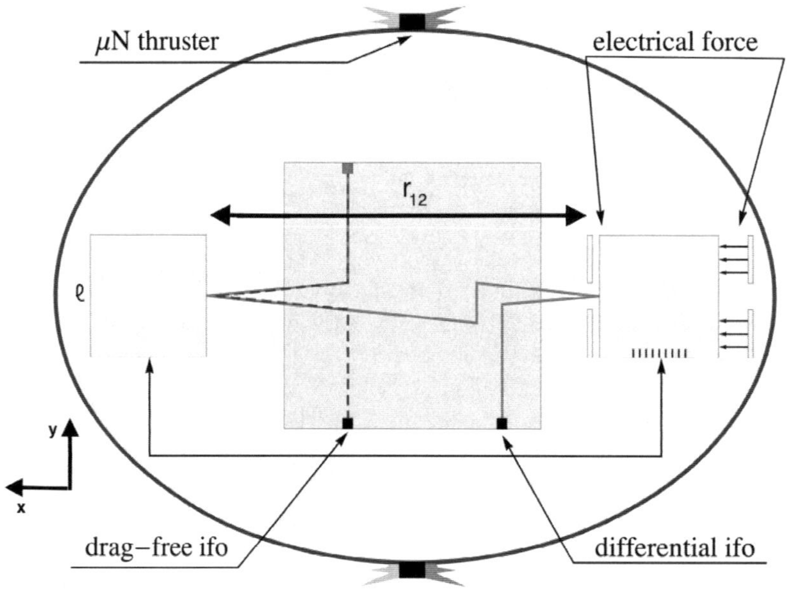

Figure 1: A schematic illustration of LPF. The two TMs in yellow are enclosed by the SC, which is equipped by μNewton thrusters. The TM on the left is called TM1 and is freely falling along the x axis. The drag-free interferometer reads the distance between TM1 and the SC and feeds the μ-Newton thruster to keep the SC centered around TM1. The differential interferometer reads the relative displacements between the TMs and feeds electrodes surrounding the other TM, the TM2, which apply a force on it such to keep the two TMs at a fixed distance.

where $C_1 = \tau A_{sus}$. The values of the free parameters in eq. (2) can be estimated by means of a calibration procedure which consists in exciting the system along the x-axis by injecting fake signals in both the drag-free and electrostatic suspension control loops.

Then, two more additional steps in the calculation of Δg must be considered. The first one is the subtraction of the centrifugal force contribution, Δg_Ω, to Δg at frequencies below 1 mHz. The second one is the subtraction of the high frequencies ($>$ 20 mHz) optical crosstalk. The calibration procedure together with the two additional steps in Δg calculation and the adopted fitting techinque are described in the next section.

3. The Δg workflow

The whole procedure from raw data to the final $\Delta g(t)$ time-series passes through three main steps of data handling, a process that henceforth we can call Δg workflow. The first step requires system calibration via large signal injection, in order to get an estimation of the free parameters of eq. (2) with high precision. With those parameters, a "calibrated" version of the $\Delta g(t)$ time-series can be produced. The second step is then needed to reduce the noise at frequencies below 1 mHz by direct subtraction of the centrifugal forces. Finally, the SC pick-up motion contribution at high frequencies is subtracted.

3.1. System Identification and fitting technique

In order to calibrate the dynamics of LPF described in eq. (2), the system identification experiments must be regularly performed during the duration of the mission, because different working configurations of the system yields different calibration parameters. The system

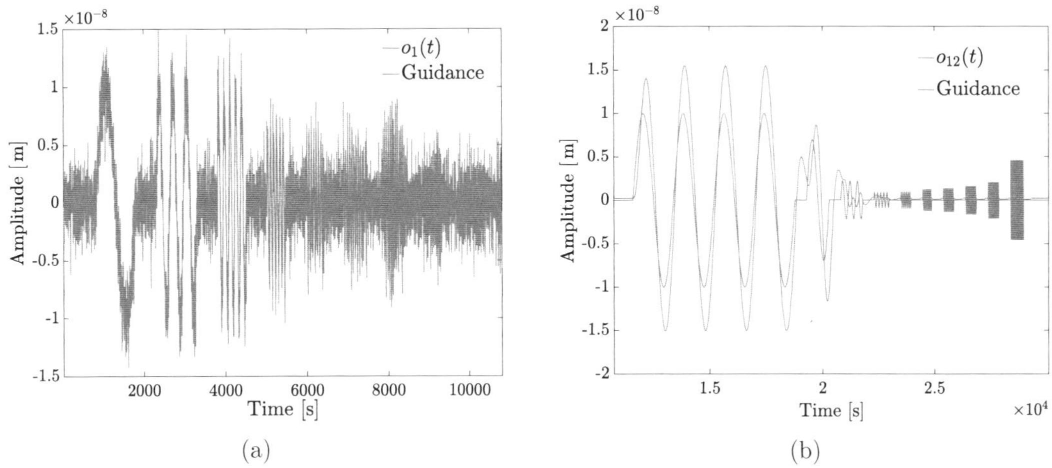

Figure 2: Batteries of injections during hte system identification experiment. *Left*: Injection sinusoidal signals (red) into the drag-free loop. The response of the system as recorded by the o_1 interferometer is shown in grey. *Right*: The same philosophy applies to the suspension loop as well. The measurement of the o_{12} interferometer is again shown in grey.

identification experiment consists of two trains of guidance injections into the drag-free and the electrostatic suspension loops. The experiment is designed to balance the desire for high signal to noise ration (SNR) for each parameter, and to avoid stimulation of non-linearities of the system. The effect of the guidance injections into o_{12} and o_1 are shown in Figure 2.

The linearised model of eq. (2) allows us to use a Iterative Reweighed Least Square algorithm to fit the data in the frequency domain. The application of this technique to LPF data and its theoretical and empirical justification were firstly introduced in [7]. For each experiment, and in the absence of spurious effects, such as glitches, we were able to subtract all of the induced signal down to comparable levels with the noise runs. The results of the full campaign of system identification experiments will be presented in a future work in prepation.

3.2. Centrifugal force contribution

The Δg time-series calibrated in the previous step and calculated during a pure noise period, that is when no other experiments were running, showed a amplitude spectral density of the residuals noise acceleration exceeding the expected level between $0.1 - 1$ mHz. At those frequencies, the inertial forces associated to SC rotation are expected to contribute to $\Delta g(t)$. Indeed, the attitude control loop driven by the noisy AST, makes the SC rotate. The SC rotation introduces an apparent forces on TMs, that the suspension control tries to compensate in order to keep the TMs at a fixed distance of 0.376 m. Therefore, the commanded force used to calibrate Δg at the previous step contains this contribution which is necessary to subtract from the acceleration noise in eq. (2). A simple calculation shows how this contribution affects Δg and how the relative signal can be obtained.

An observer in a reference system attached to the rotating SC measures an apparent acceleration of one TM due to centrifugal forces along the x-axis given by

$$\ddot{x}(t) = (\vec{\omega} \times (\vec{\omega} \times \vec{r})) \cdot \hat{x}. \tag{3}$$

Assuming the x-axis orientation in Figure 1, the centrifugal acceleration on TM1 and TM2 are respectively

$$\ddot{x}_1 = (-\omega_\phi^2 - \omega_\eta^2)x_1 \tag{4}$$

$$\ddot{x_2} = (\omega_\phi^2 + \omega_\eta^2)x_2 \tag{5}$$

where ω_ϕ and ω_η are the SC angular velocities around z and y, respectively. The centrifugal contribution to the differential acceleration between the two TMs is then given by

$$\Delta g_\Omega = \ddot{x_2} - \ddot{x_1} = (\omega_\phi^2 + \omega_\eta^2)(x_2 + x_1) \tag{6}$$

In case of a perfect centering between TMs and spacecraft, $(x_2 + x_1) = 2x_1 = 2x_2 = 0.376$ m, the distance between the two TMs. This would reduce the acceleration due to centrifugal forces to

$$\Delta g_\Omega = (\omega_\phi^2 + \omega_\eta^2)2x_1. \tag{7}$$

If ω_ϕ and ω_η were measurable parameters, the term in eq. (7) can be subtracted directly from Δg in eq. (2).

However, the SC angular velocity is not directly available in LPF telemetry. But we are able to synthesise it by combining the Autonoumous Star Tracker signal in the form of quaternions with the mean applied torques on the TMs. But even if the quaternions are directly linked to the SC angular velocity, they are very noisy between $0.1 - 1$ mHz and cannot be used directly. A polynomial fit to the equivalent angular velocity extracted from the quaternions allows to calculate the DC angular velocity of the spacecraft, ω_{DC}. Then, by integrating the commanded torques on the TMs, which are the driving signal to the attitude control loop which in turn makes the SC rotate, the in-band angular velocity, ω_{noise}, can be calculated. The final SC angular velocity for each angular degree of freedom can be calculated as $\omega = \omega_{DC} + \omega_{noise}$ and the relative contribution of the centrifugal forces, Δg_Ω subtracted from $\Delta g(t)$.

3.3. SC pick-up motion subtraction

In [8] it was shown that in the frequency band between 200 mHz - 20 mHz the pick-up of the SC motion leaking into the main differential measurement can be detected. In order to measure this contribution a fit to the following linear model has been performed

$$\Delta g_{x-talk} = C_1\ddot{\bar{\phi}}[t] + C_2\ddot{\bar{\eta}}[t] + C_3\ddot{\bar{y}}[t] + C_4\ddot{\bar{z}}[t] + C_5\bar{y}[t] + C_6\bar{z}[t] + \delta_{\text{ifo}}\ddot{o}_1[t], \tag{8}$$

where each $k \in \{\phi, \eta, y, z\}$ is a different coordinate, and the $\bar{k}$ denotes the mean displacement or rotation of both test-masses along the coordinate k, or equivalently SC motion along $-k$. The δ_{ifo} term refers to signal leakage from o_1 to the o_{12} readout. The model of eq. (8) performs really well, subtracting the excess bulge-signal down to the white noise levels of the interferometer readout noise. As expected the parameter estimates depend on the geometrical configuration of the three-body system. However, different combinations of signals can yield very similar and satisfactory results, and that means that the model provides no direct physical interpretation of the coefficients C_i. More information on the given cross-talk model, as well as for an alternative approach to suppress the cross coupling by realigning the test masses, is given in Wanner *et al* in the same proceedings.

4. Conclusions

In this paper, the main steps of the Δg workflow were presented and the full post-processing procedure to get the final product of the noise acceleration Δg is described. This procedure has been routinely performed during the whole mission and has been adopted as the baseline for any other investigation and results, such as those presented for the first time in [8]. The robustness and reliability of the whole Δg workflow has been scrupulously tested and the results of these tests will be presented in following papers which are right now in preparation.

IOP Conf. Series: Journal of Physics: Conf. Series **840** (2017) 012002 doi:10.1088/1742-6596/840/1/012002

References

[1] P. McNamara, S. Vitale, and K. Danzmann, *Classical Quantum Gravity* **25** (2008) 114034.
[2] K. Danzmann et al., *LISA: Unveiling a Hidden Universe Report* No. ESA/SRE2011(2011)3.
[3] S. Anza et al., *Classical Quantum Gravity* **22** (2005) S125.
[4] R. Dolesi et al., *Classical Quantum Gravity* **20** (2003) S99.
[5] W. Fichter, P. Gath, S. Vitale, and D. Bortoluzzi, *Classical Quantum Gravity* **22** (2005) S139.
[6] F Antonucci et al., *Classical Quantum Gravity* **28** (2011) 094002.
[7] S. Vitale et al.,*Phys. Rev. D* **90** (2014) 042003 .
[8] M. Armano et al. *Phys. Rev. Lett.* **116** (2016) 231101.

IOP Conf. Series: Journal of Physics: Conf. Series **840** (2017) 012003 doi:10.1088/1742-6596/840/1/012003

Coupling of relative intensity noise and pathlength noise to the lcngth measurement in the optical metrology system of LISA Pathfinder

Andreas Wittchen, for the LPF Collaboration

Max Planck Institute for Gravitational Physics (Albert Einstein Institute), and Leibniz Universitt Hannover

E-mail: andreas.wittchen@aei.mpg.de

Abstract. LISA Pathfinder is a technology demonstration mission for the space-based gravitational wave observatory, LISA. It demonstrated that the performance requirements for the interferometric measurement of two test masses in free fall can be met. An important part of the data analysis is to identify the limiting noise sources.[1] This measurement is performed with heterodyne interferometry. The performance of this optical metrology system (OMS) at high frequencies is limited by sensing noise. One such noise source is Relative Intensity Noise (RIN). RIN is a property of the laser, and the photodiode current generated by the interferometer signal contains frequency dependant RIN. From this electric signal the phasemeter calculates the phase change and laser power, and the coupling of RIN into the measurement signal depends on the noise frequency. RIN at DC, at the heterodyne frequency and at two times the heterodyne frequency couples into the phase. Another important noise at high frequencies is path length noise. To reduce the impact this noise is suppressed with a control loop. Path length noise not suppressed will couple directly into the length measurement. The subtraction techniques of both noise sources depend on the phase difference between the reference signal and the measurement signal, and thus on the test mass position. During normal operations we position the test mass at the interferometric zero, which is optimal for noise subtraction purposes. This paper will show results from an in-flight experiment where the test mass position was changed to make the position dependant noise visible.

1. Introduction

In the first mission studies and noise projections Relative Intensity Noise did not play a major role, and many steps were taken to suppress the other noise sources researched. This was very succesfull, which is why Relative Intensity Noise was found to be one of the limiting noise sources at higher frequencies. RIN is a property of the laser, and these intensity fluctuations are frequency dependant. The impact of this noise source is reduced with different methods, RIN couples to the measurement signal in three ways.

RIN at low frequencies couples directly into the phase measurement, but this contribution is so low that it can be disregarded.

RIN at the heterodyne frequency, ω_{het}, is suppressed with the fast amplitude control loop, and subtracted with balanced detection.

RIN at two times the heterodyne frequency also couples, and the impact of this noise is also suppressed with the fast amplitude control loop.

But balanced detection does not suppress this noise. When the Drag Free Attitude Control system (DFACS) transitions from Inertial Sensor control to OMS control we have two choices for the setpoint of the test mass: Stay at the position the test mass currently has, and use this as setpoint. For each transition the offset is different, but this reduces the transition time between DFACS settings. Alternatively we can move the test mass so that the phase difference between the measurement signal and reference signal is always the same. This increases the transition time depending on the TM position at the beginning, but it has the advantage that the measurement is more reproducible. Zero phase difference between the two signals was chosen since this subtracts RIN from the measurement signal.

In the following we will give a short overview of the coupling mechanism, and of the experiment that was perfomed to explore the noise at higher frequencies for different test mass positions in preparation for the LISA mission. [2]

2. Relative Intensity Noise

The phase measurement is performed with heterodyne interferometry, the interferometer is bonded on a Zerodur baseplate and sits at the heart of the system in between the test masses. The photodiode current generated by the interferometer signal also contains RIN. From the electric signal the phasemeter calculates the phase change, and thus the motion of the test masses, as well as the DC laser power.

We add a noise term to the laser signal on the photodiode, and calculate the normalised photodiode signal $y_{\mathrm{A/B}}(t)$:

$$y_{\mathrm{A/B}(t)} = (1 + r(t))\,(1 \pm c\,\cos(\omega_{\mathrm{het}}\,t - \varphi)),$$

where $r(t)$ is a frequency dependant noise term, and the noise is small compared to the total power, $r(t) \ll 1$. The coupling of RIN into the phase measurement signal depends on frequency. The contribution can be derived analytically, the relevant noise frequencies are the following:

RIN at DC is pure amplitude noise, and to first order this noise does not couple into the phase.

RIN at the heterodyne frequency is near the same frequency as our signal, and ends up in the same phasemeter bin. Since this bin covers only a small frequency band around the heterodyne frequency, RIN couples as a white noise into the phase measurement. Per interferometer we use two photodiodes. These two signals are subtracted from each other to create the interferometer signal. This subtraction is called balanced detection since we use the symmetric and antisymmetric port of the beam splitter to remove noise sources common to both signals.

For RIN at two times the heterodyne frequency the situation is different. It also couples into the phase measurement, but the coupling to the phase has a sign difference between the two ports. When the two photodiode signals are now subtracted this noise is not removed from the interferometer signal, and therefore is still present in the phase measurement. The same noise is also present in the reference interferometer signal. This interferometer has the same path length difference as the measurement interferometers, and also uses balanced detection. The signal is not only used as an error point for the optical path length difference control loop (OPD loop), but also to subtract remaining noise from the other signals. RIN at $2\omega_{\mathrm{het}}$ is one of the noise sources that is subtracted with the reference signal. For normal operations this subtraction works very well. If the test mass is removed from its nominal position this subtraction does not work properly, and some amount of RIN remains and leads to an increase in noise at higher frequencies. One example where the test mass is moved many nanometers is the free fall experiments, a description of these experiments can be found in "The free-fall mode experiment on LISA Pathfinder: first results" in this issue.

3. In-flight experiment

To understand the coupling, and the level at which the noise is increased for different test mass positions, we moved Test Mass 2 to various offsets and measured the noise. This measurement showed the position dependency of this high frequency noise, and confirms the model derived mathematically from a calculation of the phase with RIN present at different frequencies.

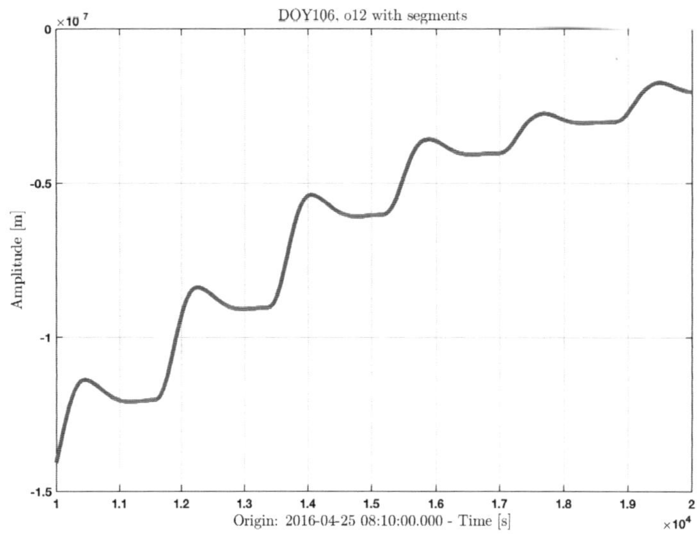

Figure 1. The test mass is moved in steps, at every position a noise measurement is performed. This plot shows a short segment of a longer experiment, marked are the timespans where the test mass is stable enough to calculate a noise level.

For the analysis of these steps the timespans of each step where the test mass was stable was cut into smaller parts, for each part the spectrum was calculated. From these many parts the spectra with glitches are removed, and then the average of all of these are used as a mean noise level from this step. The distribution of the noise levels are used for the error calculation.

From this noise over position measurement a fit is made with different models to combine the RIN to other measurement noise sources, and to calculate the amount of RIN present in the system for specific test mass positions. Other noise sources of interest in this frequency range are cross coupling from the angular degrees of freedom, and the noise floor consisting of ADC quantisation noise, electronic noise and shot noise.

For the mission extension an additional experiment is planned to explore not only the contribution of RIN from $2\omega_{\text{het}}$, but also the noise behaviour with RIN at $1\omega_{\text{het}}$ and $2\omega_{\text{het}}$ present at the same time. This is accomplished by turning one phasemeter side off and therefore removing balanced detection.

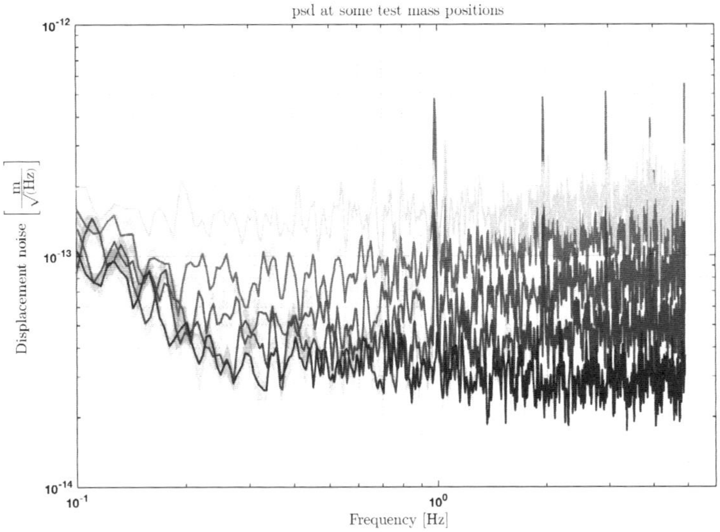

Figure 2. This is the high frequency spectrum for a few steps, the difference in the noise level is visible. For each step an average noise level is calculated, and plotted over the mean test mass position. We use a BH window with 50% overlap,

4. Acknowledgments
The Albert-Einstein-Institut acknowledges the support of the German Space Agency, DLR. The work is supported by the Federal Ministry for Economic Affairs and Energy based on a resolution of the German Bundestag (FKZ 50OQ0501 and FKZ 50OQ1601).

5. References
[1] Armano M *et al.* 2016 *Physical Review L* **116** 231101
[2] Danzmann K *et al.* 2017 *A proposal in response to the ESA call for L3 mission concepts* URL `https://www.lisamission.org/proposal/LISA.pdf`

Laser Frequency Noise Stabilisation and Interferometer Path Length Differences on LISA Pathfinder

Sarah Paczkowski on behalf of the LPF collaboration

[1] Albert-Einstein-Institut, Max-Planck-Institut für Gravitationsphysik und Universität Hannover, Callinstrasse 38, 30167 Hannover, Germany

E-mail: `sarah.paczkowski@aei.mpg.de`

Abstract. The LISA Pathfinder mission is a technology demonstrator for a LISA-like gravitational wave observatory in space. Its first results already exceed the expectations. This is also true for the optical metrology system which measures the distance in between the two free-floating test masses with unpreceded precision. One noise source that can possibly affect the measurement is the laser frequency noise. It is measured with a dedicated interferometer and suppressed with a control loop. We measured the laser frequency noise and characterised the control loop in flight. The coupling of laser frequency noise into the measured phase is directly proportional to the path length difference in the respective interferometer. Dedicated experiments have been performed to estimate the path length difference in flight. In addition, this frequency stabilisation scheme is also a possible solution for the LISA mission.

1. Introduction

On LISA Pathfinder, key technologies for the LISA mission are being tested. One of them is the optical metrology system (OMS).

From this point of view, LISA Pathfinder is a unique optics laboratory in space. It experiences only extremely low temperature gradients and mechanical vibrations. In addition, the free-floating test masses produce an outstanding alignment. For more information on the OMS, we refer to the corresponding proceedings article [1].

Our aim is to understand as good as possible the factors that result in the current measured OMS noise level. This means we account for all known effects to find the fundamental limits. This knowledge is essential for the future development of interferometers for LISA and other space missions.

One possible noise contribution which has been known for a long time is the laser frequency noise. In general, laser frequency noise, δf, produces a phase noise, $\delta\phi$, in an interferometer if the light in the measurement beam does not travel exactly the same distance as in the reference beam. The difference of these two lengths is called a path length difference or an arm length mismatch, denoted Δs. This relation is summarised in Equation 1[2]:

$$\delta\phi = 2\pi \frac{\Delta s}{c} \delta f \,. \tag{1}$$

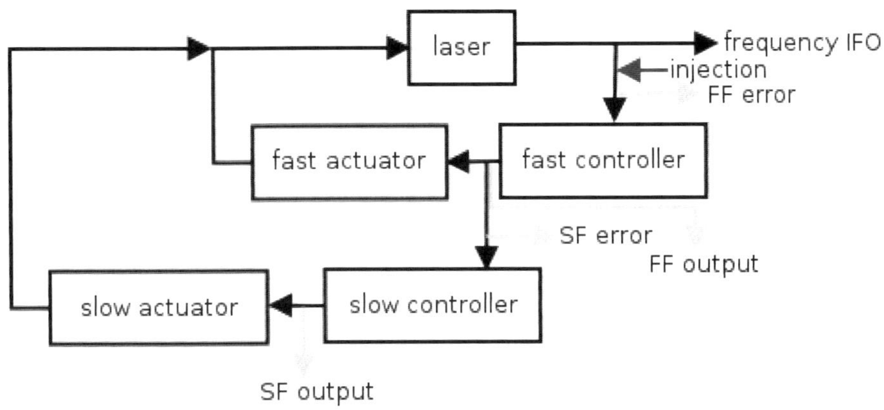

Figure 1. A simplified diagram of the laser frequency control loop. The quantities with the green arrows are housekeeping data channels and the frequency interferometer is a science data channel. The injected signal (blue arrow) cannot be downloaded.

In a real interferometer, the path length difference is never exactly zero. This is also the case on LISA Pathfinder. Equation 1 is valid for all four interferometers and this phase noise contribution cannot be distinguished from any other phase signal and directly affects the measurement. To minimise this undesired noise, there are two approaches, which are both used on LISA Pathfinder. We try to stabilise the laser frequency with a dedicated control loop to minimise the laser frequency noise, δf, and the utmost care was taken during the construction and integration of the satellite to have minimal path length differences, Δs.

Accordingly, the measured value for Δs provides a way to check the integration quality. In addition this number is required to estimate the contribution of laser frequency noise to the total noise in the differential measurement. Details will be given in Section 3.

The frequency noise measurements on LISA Pathfinder, and the performance of the stabilisation scheme, are especially interesting because a similar approach is being discussed for LISA. Other possibilities are the arm-locking technique [3] and the use of a cavity.

2. Laser Frequency Noise Stabilisation

In LISA Pathfinder, the laser frequency noise is measured by a dedicated interferometer on the highly stable optical bench. Its most important characteristic is the intentional path length difference of 0.382 m [4]. It occurs in the fibres between the laser modulator unit and the optical bench. On the optical bench itself, the path of the two beams in the frequency interferometer is equally long. Following Equation 1, the intentional path length mismatch amplifies the frequency noise.

The measurement of the frequency interferometer is the input to the frequency stabilisation control loop, as shown in Figure 1. This measurement has already the measurement from the reference interferometer subtracted, which we will assume implicitly in the remainder of the article. The frequency control loop is implemented digitally in the data management unit (DMU) and operates at 100 Hz. As shown in Figure 1, it is a nested loop containing both a fast frequency controller and actuator and a slow frequency controller and actuator. Both controllers are infinite impulse response filters. The nominal filter equations for the fast frequency controller are $y_i = -1.58 \cdot (\Delta x + \delta x_i - x_i) + 1 \cdot y_{i-1}$ and for the slow frequency controller $y_i = -5 \cdot 10^{-5} \cdot (\Delta x + \delta x_i - x_i) + 1 \cdot y_{i-1}$. In the filter equations, y_i is a selected sample of the controller output. This value depends on any commanded constant offsets Δx as well as on the value of a possible time-varying probing function, denoted as δx, and on the current input value, x_i, and previous output value, y_{i-1}. The fast frequency actuator is a piezo which acts on the laser crystal and the slow frequency actuator is a heater [5]. In Figure 1, all loop channel quantities that can be downloaded are marked with the channel names. We also marked the point of the injection with a blue arrow even though this channel cannot be downloaded. All internal measurements are only available at 1 Hz for most of the time, but the measurement of the frequency noise is provided at 10 Hz during most of our dedicated OMS experiments. These data rates are a function of the design of the DMU. In the DMU, data is treated differently depending on whether it is part of the so-called 'science data', as is, for example, the frequency noise measurement, or whether it is part of the so-called 'housekeeping data', as are all the loop parameters [6]. The DMU offers a special feature to deviate from this procedure which is called the interferometer data log (IDL). This allows us to collect data at 100 Hz from nominal science or housekeeping channels. This is only possible for very limited durations, of the order of minutes.

The aforementioned control loop has been characterised in flight during dedicated experiments. The aim was to ensure that the loop performs as expected. In these experiments, a sinusoidal probing function was injected at the point marked by the blue arrow in Figure 1 into the fast frequency controller error signal. The selected amplitude was 0.05 rad and eight frequencies from 0.011 Hz to 1.123 Hz were chosen. Each sinusoid lasted for 20 cycles at the respective frequency. For frequencies above 0.5 Hz, the IDL was used to record the data. Moreover, open-loop and closed-loop measurements of the laser frequency noise have been undertaken. During the nominal LPF mission, the frequency noise is downloaded at 1 Hz for most of the time to minimise the telemetry load. Thus, the long term monitoring can only be done with this data.

3. Measurements of Interferometer Pathlength Differences

As mentioned in the introduction, interferometer path length differences or arm length mismatches occur in all realistic interferometers. On LISA Pathfinder, this means they are present in all of the four interferometers. In the frequency interferometer, the path length difference is known to some accuracy. In the differential interferometer, X12, and in the spacecraft interferometer, X1, the path length difference depends on the absolute position of one or both test masses.

We can measure the arm length mismatch in the interferometers X1, X12 and Ψ_{R} from the coupling of laser frequency noise into the respective interferometer. Indeed, as we measure, δf, we can solve Equation 1 for Δs and calculate it. The path length difference is measured best when we amplify the frequency noise at known frequencies. To this end, we use sinusoidal injections similar to the loop characterisation experiment but at frequencies of 1.123 Hz and 2.879 Hz. The signals are injected for approximately half an hour each to improve the signal-to-noise-ratio. In the first version of the arm length mismatch experiment, the aforementioned signals have been injected two times. During the first set of injections, the reference test mass was at its nominal position and during the second set of injections, the same test mass was offset by approximately 9.5 µm. This allows us to check the two measurements for consistency and to get the sign of the mismatch. In the arm length mismatch experiment, these two sets of two injections each are followed by an open-loop measurement of the laser frequency noise at the original position of the reference test mass. From the increased noise, we can estimate again the arm length mismatch but with larger errors. The strategy of this experiment is summarised in Figure 2. The experiment was successfully executed on June 14th 2016. The arm length mismatch can be estimated best with this dedicated experiment.

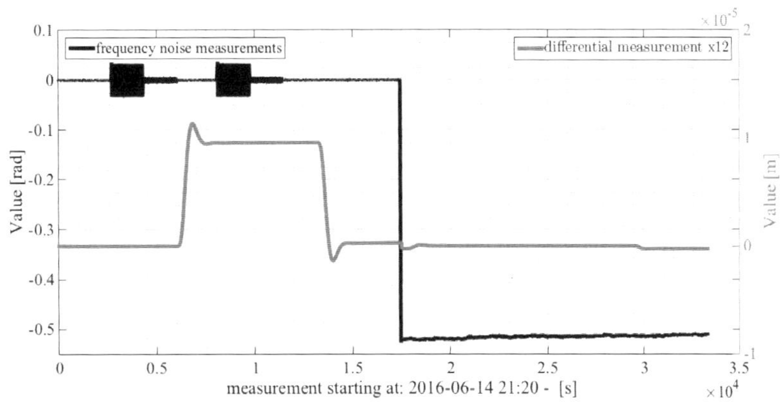

Figure 2. The concept of the arm length mismatch experiment. The black trace (corresponding to the left y-axis) shows the laser frequency noise measurement. The two sets of two injections are clearly discernible. The DC value of the signal is at a lower level during the open-loop measurement. The distance between the two test masses is shown in red (refer to right y-axis). The second set of modulations is taking place at the offset position. Some transients are discernible.

4. Conclusion

On LISA Pathfinder, there is a dedicated interferometer to measure the laser frequency noise. This signal is used by a digital nested control loop which stabilises the laser frequency. A similar scheme is also being discussed for LISA. The interferometer path length differences on LISA Pathfinder are measured with a dedicated experiment which was successfully executed on June 14th 2016. Analysis of this experiment and the control loop characterisation is ongoing and the results will be reported in a future publication.

5. Acknowledgements

The Albert-Einstein-Institut acknowledges the support of the German Space Agency, DLR. The work is supported by the Federal Ministry for Economic Affairs and Energy based on a resolution of the German Bundestag (FKZ 50OQ0501 and FKZ 50OQ1601).

References

[1] Audley H for the LPF Collaboration *Journal of Physics: Conference Series to be published in this issue*
[2] Heinzel G 2002 SMART-2 interferometer - S2-AEI-TN-3010 Tech. rep. AEI Hannover
[3] Sheard B S, Gray M B, McClelland D E and Shaddock D A 2003 *Physics Letters A* **320** 9 – 21
[4] Robertson D 2013 3OB As Built OptoCAD Model - S2-UGL-TN-3045 Tech. rep. University of Glasgow
[5] Kersten M 2011 OMS- control loop stability analysis and filter design - S2-ASD-TN-3107 Tech. rep. Astrium
[6] Denskat U 2009 Phasemeter Processing and Laser Control- S2-ASD-RS-3018 Tech. rep. Astrium

IOP Conf. Series: Journal of Physics: Conf. Series **840** (2017) 012005 doi:10.1088/1742-6596/840/1/012005

The free-fall mode experiment on LISA Pathfinder: first results

Roberta Giusteri[1] on behalf of the LPF collaboration

[1] Dipartimento di Fisica, University of Trento and Trento Institute for Fundamental Physics and Application / INFN, 38123 Povo, Trento, Italy

E-mail: roberta.giusteri@unitn.it

Abstract. The LISA Pathfinder space mission is testing the critical experimental challenge for LISA by measuring the differential acceleration between two free-falling test masses inside a single co-orbiting spacecraft at a level of sub-femto-g for frequencies down to 0.1mHz. In LPF it is necessary that one test mass (TM) is electrostatically forced to follow the orbit of the other TM. This force represents a noise source in differential acceleration at frequencies below 1mHz. The free-fall mode experiment has been performed in order to reduce this source of noise: the actuation is limited to short impulses on one TM, so that it is in free fall between two successive kicks, while the other TM is drag-free. The free-fall mode thus provides a different technique for measuring the differential TM acceleration without the added force noise and calibration issues introduced by the actuator. Data analysis challenge is related to the presence of the kicks: they represent a high-noise contribution and need to be removed, thus leaving short gaps in data. This article presents preliminary data of the LPF free-fall measurement campaign and describes the three data analysis techniques developed to mitigate the presence of gaps.

1. Introduction

LISA Pathfinder (LPF) is an European Space Agency mission launched on December 3rd 2015 with the objective of measuring the nearly-pure free-fall motion close to the level needed for future space-based gravitational wave observatories, like LISA [1]. The main observable of LPF is the differential acceleration between two TMs measured along the x sensitive degree of freedom joining the TMs and arising from spurious random forces, Δg (see Figure 1). While the TMs in LISA will be hosted separately in two spacecrafts on average 2.5 million kilometers apart [2], in LPF the acceleration measurement is evaluated between two test masses that are enclosed in the same spacecraft. To fullfill the specifications, the differential acceleration noise on LPF must be maintained within 30 $\mathrm{fm\,s^{-2}\,Hz^{-1/2}}$ at 1 mHz, in terms of Amplitude Spectral Density (ASD).

Because it is not possible, for a single spacecraft, to follow both the trajectories of the TMs along the same sensitive degree of freedom x, one TM (TM2) must be electrostatically controlled to follow the other. The electrostatic control on TM2 is based on the differential interferometric readout o_{12} and is principally needed to compensate the gravity imbalance experienced between the TMs along the sensitive x axis. The continuous control along x for TM2 is typical of the standard science measurement in LPF and it will not present in LISA. Actually, in LPF all degrees of freedom for both TMs, except x for the free-falling TM (TM1), are controlled.

The electrostatic suspension loop acts at frequencies $\leq$ 1 mHz and is based on a capacitive sensor composed of 18 electrodes surrounding the TMs. Each set of opposite electrodes controls

IOP Conf. Series: Journal of Physics: Conf. Series **840** (2017) 012005 doi:10.1088/1742-6596/840/1/012005

the TM motion along two degrees of freedom (i.e. x_2 and ϕ_2, as depicted in Figure 1). Unavoidable fluctuations in the applied voltages induce fluctuations in the force exerted along x_2 which add acceleration noise to the TM2 motion and thus to Δg. The resulting noise in force for a given relative fluctuation of voltage on a certain electrode is proportional to the force applied by that electrode:

$$F \propto V_{act}^2 \Rightarrow \frac{\delta F}{F} \propto 2 \cdot \frac{\delta V_{act}}{V_{act}} \tag{1}$$

where V_{act} is the voltage applied to the electrode [3]. If the actuation were limited to a single force to compensate the DC differential acceleration Δg_{DC}, applied with a single electrode, we would have a noise contribution due to actuation of $S^{1/2}_{\Delta g_{act}} \approx 2\Delta g_{DC} S^{1/2}_{\delta V_{act}/V_{act}}$ in ASD terms. The true actuation noise is complicated by the simultaneous application of actuation torques in ϕ and the "constant stiffness" actuation scheme [4] which pulls on both sides of the TM. The ASD of the relative voltage amplitude measured on ground and confirmed in flight, takes a value in a range between 3 and 8 ppm/$\sqrt{\text{Hz}}$ at 1 mHz [1, 4]. Considering the requirement levels for $\Delta g_{DC} < 0.65\,\text{nm}\,\text{s}^{-2}$ and rotational ϕ angular acceleration below 2 nrad s^{-2}, the pre-flight estimates of actuation noise were roughly 7.5 fm s$^{-2}/\sqrt{\text{Hz}}$ at 1 mHz [4]. This source of noise is thus expected to be the dominant contribution at low frequency in LPF.

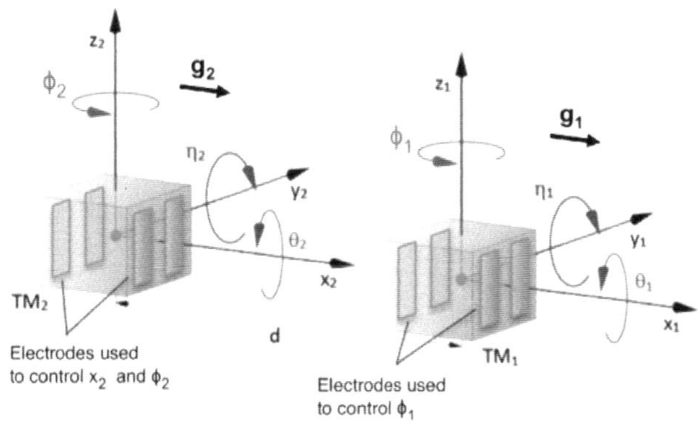

Figure 1: Schematic representation of the coordinate systems and the capacitive actuation for the x-ϕ configuration in LPF. g_1 and g_2 represent the stray acceleration experimented by the TM1 and TM2 respectively.

It is important to underline that the actuation noise depends also on the force/torque authorities. As a consequence, the noise changes with the actuation configuration. On LPF the authority was reduced with respect to the nominal values. Indeed, the time-series of Δg measured in flight within the first months of science operations, was characterized by a quasi-static part, of a few tens of pm s^{-2}, with fluctuations of (5.57 ± 0.04) fm s$^{-2}/\sqrt{\text{Hz}}$ for frequencies between 0.7 and 20 mHz [1]. Because of this low static gravitational acceleration level compared with the prediction of 650 pm s^{-2}, the maximum force along x_2 has been reduced from 2200 pN, typical of the *Nominal* authority, to 50 pN. The corresponding lower actuation configuration is called *URLA* (Ultra Reduced Low Authority).

As anticipated, the sensitive axis (x_2) actuation control, that potentially could limit sensitivity in LPF, is not needed in LISA, where each TM is drag-free along its sensitive interferometry axis. However, an experimental investigation of the actuation force noise contributes to the characterization of the capacitive sensor with effect on the LISA instrument design, especially for the x-ϕ scheme as in LISA the ϕ degree of freedom will be controlled. In this context, the LPF free-fall mode is a control scheme that aims at eliminating the noise contribution due to x-axis actuation: the electrostatic control on the coordinate x_2 is applied by means of short electric impulses. In this way the actuation is limited to brief kicks, so that TM2 is in free fall between two successive kicks. The actuation-free motion is then analyzed for the remaining sources of acceleration noise. In fact this mode solves, at least partially, the problem of actuation noise, as the ϕ control remains. However, this experiment can provide an upper limit on the LISA sensitivity, by assuming the same ϕ-gravitational balance level of LPF. In any case, the free-fall experiment constitutes an alternative technique for measuring the differential TM acceleration without the added force noise and calibration issues introduced by the actuator.

It is possible to predict the actuation noise contributions, in standard science and free-fall mode, on basis of a full actuation noise model and a dedicated actuation noise test campaign that will be discussed in detail in a separate article [5]. The projections, for both URLA and Nominal authorities, are depicted in Figure 2, together with the first LPF results in URLA and the LISA and LPF requirements [1]. The figure evidences that, in Nominal case, the implementation of an intermittent control on LPF is expected to subtract a considerable contribution to the total noise measured in a standard run. Differently, in the lowest authority configuration, a small fraction of actuation noise is subtracted (roughly 15% in ASD terms, as shown in Figure 2). This result derives from the fact that actuation is already dominated by ϕ-control in URLA standard measurement and removing x-actuation does not reduce much noise at low frequency. However, if the predictions in URLA are confirmed, the free-fall data could validate the accuracy of the actuator calibration achieved in the standard measurement and reported in [1]. Indeed, in this case, an agreement between the intermittent and the continuous control is expected, in terms of low frequency performance [5]. It is necessary to remark, finally, that this result is related to the low control levels effectively applied in flight, thanks to the unexpected low gravity imbalance measured on the spacecraft.

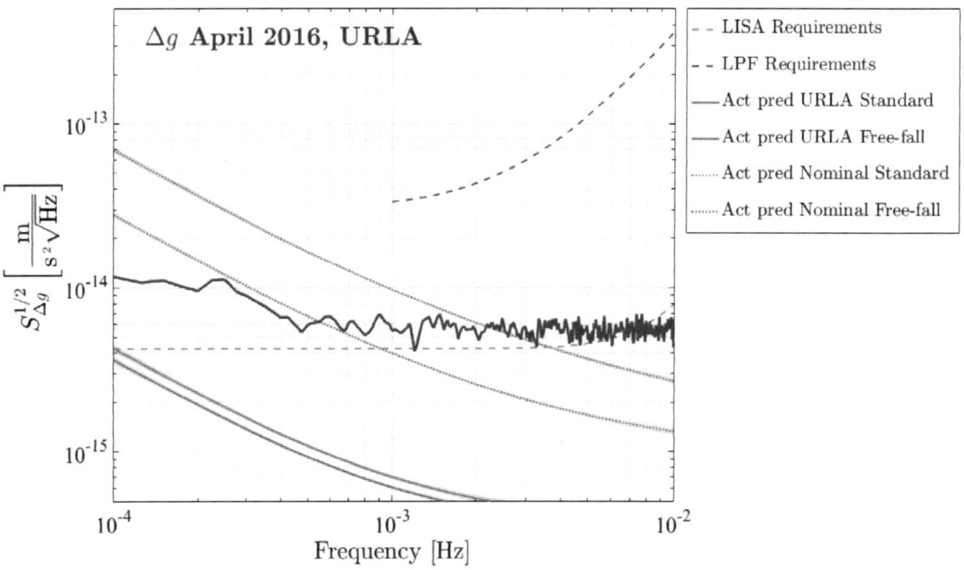

Figure 2: Blue curve: ASD of Δg measured for 6.5 days after 127 days from launch in URLA authority (F_{max} = 50 pN) and compared with LPF and LISA requirements (dashed curves) [1]. The ASD is the result of averaging 26 periodograms of 40000-s each with 50% overlap, according to the standard Welch's averaged periodogram method [6]. The figure includes the actuation noise contribution in URLA and Nominal (F_{max} = 2200 pN) authority for both standard (red curves) and free-fall mode measurements (green curves).

2. Eliminating the x-axis actuator: the free-fall mode experiment

The free-fall mode experiment has been implemented in LPF according to the configuration indicated in Figure 3a: the flight and kick durations are fixed to $\sim$ 350 s and 1 s respectively [7]. These values have been set to limit the displacement to reduce non-linearity issues: in presence of the maximum gravitational imbalance predicted ($\sim 0.65\,\mathrm{nm\,s^{-2}}$), the TM exceeds the sensing range ($\pm 5\,\mu$m [8]) in $\sim$ 400 s. A kick controller is used to produce periodic kicks: it keeps track of the motion of TM2 during the free-phase and estimates the impulse needed to "kick it back" on the other side. The amplitude of the following kick is then set to apply this impulse and the kick-and-drift scheme is repeated. The amplitude of the kick force is thus strictly related to the DC gravity imbalance to be compensated but it is also set by the duty cycle of the experiment, $\chi \equiv T_{kick}/T_{exp}$: $F_{kick} = F_{DC}/\chi$, where T_{kick} is the kick duration and $T_{exp} = T_{flight} + T_{kick}$ the experimental time.

To estimate the spectrum at frequencies below the experimental frequency, $f_{exp} = 1/T_{exp} \sim$ 2.8 mHz, data from successive flights are combined. Figure 3b shows the differential displacement time-series measured by the interferometer $o_{12} = x_2 - x_1$, during the free-fall experiment performed in URLA in June 2016 and compared with the kick force.

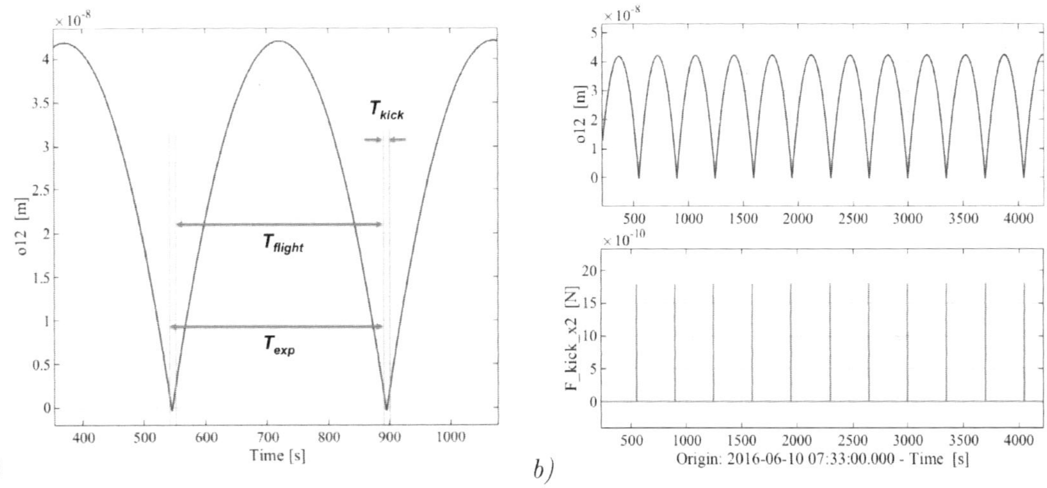

Figure 3: *a)* Free-fall mode configuration in terms of relative displacement between the TMs, with 349.2 s flight and 1 s kick durations. The data comes from the first free-fall experiment performed in URLA authority, when the differential DC acceleration was roughly $-2.7\,\mathrm{pm\,s^{-2}}$ and thus, the kick force amplitude was $\sim$ 1.8 nN. The corresponding flight amplitude is $\sim$ 41 nm. *b)* Time-series of the differential readout of the interferometer, $o_{12} = x_2 - x_1$ and the kick force on TM2.

In the period to which this article refer, this experiment has been executed four times on LPF (from June to August 2016). Here we present the June measurements only, performed in Nominal and URLA authority. In all cases the kick control has been achieved successfully and maintained stable over the mission.

3. Data analysis challenge and proposed strategies

The differential acceleration time-series measured during the free-fall mode experiment is characterized by periodic kicks which represent an high noise configuration, as shown in Figure 4a. The kicks reveal as spikes in the spectrum at frequencies multiple of the experimental frequency, $f_{exp} = 1/T_{exp} = 2.85$ mHz, right in the LPF measurement bandwidth (Figure 4b).

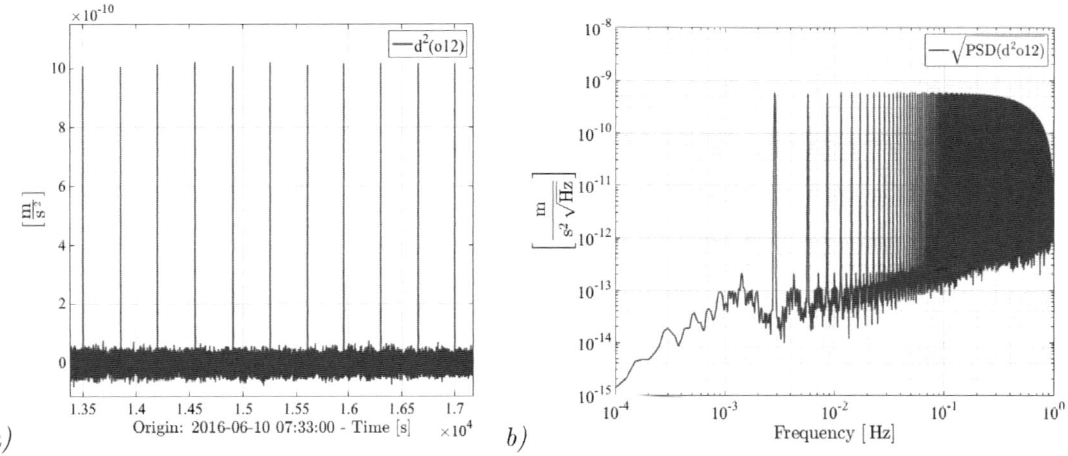

a) *b)*

Figure 4: Effect of the presence of kicks on the differential acceleration time-series and on the relative spectrum. The data refer to the experiment executed in June in URLA authority.

Because the purpose of the experiment is to consider only data relative to the free-phases, the kicks must be removed. As a result, Δg data contain gaps. Gaps may corrupt the spectral estimation in form of *spectral leakage* both from low frequencies and high frequencies, thus introducing systematic *bias* in the underlying spectrum. The nature of this bias is related to the kick characteristics (duration, repetition) and to the original spectrum of the signal. The understanding of this effect is fundamental, as it might corrupt the spectrum especially at frequencies below 1 mHz, where actuation noise could be the main contribution to the total noise and needed to be investigated.

Three data analysis approaches, which will be described in the following, have been developed within the LPF collaboration to mitigate the presence of gaps. These methods are applied to Δg, which, in turn, is estimated after having "calibrated" the free-fall data, that is once the dynamical parameters of the experiment are extracted, such as the elastic couplings.

3.1. Δg estimation with free-fall data

The method followed to estimate Δg in this specific experiment consists in fitting the numerical second derivative of the relative displacement $\ddot{o}_{12}$, to the following model:

$$\ddot{o}_{12} = \Delta g - \omega_2^2 o_{12} - \Delta\omega^2 o_1 = \Delta g_{DC} + \dot{g}_0\, t - \omega_2^2 o_{12} - \Delta\omega^2 o_1 \tag{2}$$

where $o_{12} = x_2 - x_1$ and $o_1 = x_1 - x_{SC}$ are the interferometric readout, ω_1^2 and $\omega_2^2 = \Delta\omega^2 + \omega_1^2$ the elastic couplings between each TM and the spacecraft and Δg is disentangled into its static part Δg_{DC}, and its time-drift observed over the mission, $\dot{g}_0$ [1]. Δg_{DC}, $\dot{g}_0$, ω_2^2 and $\Delta\omega^2$ are thus the fit parameters. This fit is performed in acceleration flight by flight, by means of a dedicated routine implemented in LTPDA[1]. To reduce the high frequency noise, the telemetry channels included in the model are low-pass filtered ($\ddot{o}_{12}$, o_{12}, o_1). This is achieved by means of an anti-aliasing Finite Impulse Response filter with coefficients of a Blackmann-Harris window and with duration equal to $\mathcal{T}_{win} = T_{exp}/4$ s. Next, to reduce the number of samples per flight time, the data are downsampled at ~ 57 mHz, which correspond to have 20 samples per flight. The parameter configuration of filtering and downsampling has been chosen to limit aliasing

[1] LTPDA (LISA Technology Package Data Analysis) is the MATLAB© toolbox used for the data analysis within the LPF collaboration [9].

effects. At the end, as many sets of parameters as flights are obtained. Each parameter is then averaged over its N-values to get a single estimate. (N is the total number of flights in the experiment)[2]. Using the averaged dynamical parameters, the time-series of Δg can be evaluated as follows from equation 2, with the original telemetry channels sampled at 10 Hz:

$$\Delta g(t) = \ddot{o}_{12}(t) + \omega_2^2 o_{12}(t) + \Delta\omega^2 o_1(t) - \dot{g}_0\, t \tag{3}$$

The Δg obtained in this experiment shows noisy data during the kicks, which must be removed. Similarly to what performed to the standard science data, the resulting time-series is corrected for centrifugal effects, as discussed in [1] and then analyzed by means of one of the following techniques proposed to fill the kick data by limiting the unavoidable bias that this operation would produce on spectral estimation.

3.2. Blackmann-Harris low-pass technique with kicks set to zero

The purpose of this method is to reduce the spectral leakage at frequencies within the LPF frequency band that derives from high frequency noise [10]. It essentially consists of a low-pass filtering and downsampling applied to Δg data, after which the kick samples are set numerically to zero. A schematic of the method is depicted in Figure 5. The name of the technique refers to the shape of the low-pass filter chosen, that is a normalized Blackmann-Harris window.

The first step of this technique consists in setting up the correct number of samples per experimental time ($T_{exp} = 350.2\,\mathrm{s}$) once Δg data, initially sampled at 10 Hz, are decimated. This number must be an integer factor of the total number of samples per experimental time in the original data such that each experimental segment still contains, after decimation, a fixed number of data points. It is straightforward that this number, n_{tot}, will fix the new sampling time T_{samp} after the downsampling:

$$T_{samp}\, n_{tot} = T_{exp} = T_{flight} + T_{kick} \tag{4}$$

The flight duration considered in the analysis is reduced to $T_{flight} - 2T_{cut}$ ($T_{cut} = 2\,\mathrm{s}$) in order to avoid transients which may be close to the kicks. The low-pass filter length T_{win}, is set up in such a way as to have an integer number of finite windows per flight time:

$$T_{win} = T_{flight} - 2T_{cut} - (n_{keep} - 1)T_{samp} \tag{5}$$

where n_{keep} is the number of samples maintained per flight time.

The free-fall data have been downsampled considering 34 samples per experimental time, as it is an integer factor of the total number of samples per experimental time ($N_{exp} = T_{exp} \cdot 10\,\mathrm{Hz} = 3502$). The decimation factor N_{dec} is thus equal to 103. According to equations 4 and 5, it follows that the new sampling time is T_{samp} = 10.3 s and the filter length is T_{win} = 98 s, as n_{keep} is set to 25. Data points corresponding to windows that overlap with the kick period are set to zero ($n_{gap} = n_{tot} - n_{keep} = 9$ samples).

The low-pass filtering is achieved by means of an anti-aliasing Finite Impulse Response filter, in order to not mix in the data in the gap, with coefficients of a Blackmann-Harris window (to be precise we use the "minimum 4-term Blackman-Harris window", also known as "BH92"). The choice of such a window is justified by the requirements in terms of spectral leakage

[2] An alternative fit performed to estimate Δg, that can be applied to original 10 Hz data, is based on a time domain logarithmic likelihood function with the noise level being a parameter of the fit. After a first search using the Nelder-Mead simplex algorithm, an adaptive Markov Chain Monte Carlo algorithm (MCMC) which samples the parameter space with the Metropolis-Hastings algorithm, is applied to estimate proper errors.

performances[3] [6]. The filter is applied by sliding it over the Δg time-series with a rate equal to the new sampling time: filtering and decimation are thus applied at the same time, as shown in Figure 5. After that, the spectrum can be estimated and normalized for the BH window transfer function. Finally it must be corrected for the bias related to the "gap ratio" n_{tot}/n_{keep} because of presence of zeros in data[4].

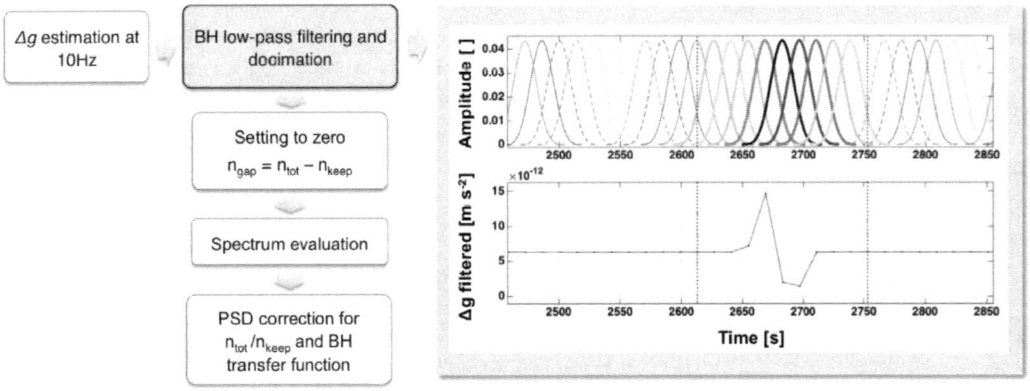

Figure 5: Block diagram of the BH low-pass approach. The figure on the right shows the effect of the technique: the BH filter is slid over the Δg data stretch with a rate equal to the new sampling time, 10.3 s. The bold coloured windows are those contaminated by the kick interval, as it is shown in the filtered and decimated Δg time-series below. The corresponding samples in will be set to zero. The vertical dashed lines delimit the resulting gap.

3.3. Windowing technique

The basic idea of this approach is to zero out the kicks in Δg by means of a spectral window [11]. The concept is similar to spectral estimation in general, where the time-series is multiplied by a normalized windowing function before computing the spectrum. In this case the windowing ensures that the data stretch smoothly approaches zero at its ends, avoiding artifacts in the spectrum caused by the unavoidable truncation of the data series to a finite length.

In the approach proposed here, the choice of the window is crucial because it must guarantee two conditions: it has to take the value zero at both the extremities of the time-series and also during each gap and furthermore it has to minimize the spectral leakage in the LPF bandwidth. The window used here is an "high frequency" window, in the form of an Hahn spectral window, that smoothly reaches zero at each kick. The effect of the gap-windowing is visible in Figure 6, where the kick force time-series is compared with the window profile. The resulting Δg data is thus multiplied by this window and then the spectrum is extracted. Note that this analysis is independent of the sampling frequency of the data. If desired, the data can be downsampled prior to the analysis itself with standard LTPDA methods. The only quantity the user has to specify is the position or the duration of the kicks which can either be detected automatically or be derived from the time-series of the pulsed force. The estimation of the spectral bias, introduced by the data reduction procedure, is under investigation also for this approach.

[3] 4-term Blackman-Harris window has a small lobe adjacent to the main peak in the transfer function (92 dB below the main peak) with the first zero located at frequency $f = \pm 4.00$ bins. Thus the power within the side lobes, which contributes to the spectral estimation, is reduced of 92 dB, making the BH92 one of the best-performing available window in terms of spectral leakage suppression.

[4] In fact, this factor corrects the bias on the *white* contribution to the spectrum only, as can be calculated analytically. In presence of a "coloured" noise spectrum, as that in LPF, the spectral bias estimation is more complicated. We refer this discussion to upcoming publications as it is strictly related to the technique calibration which is still under consolidation.

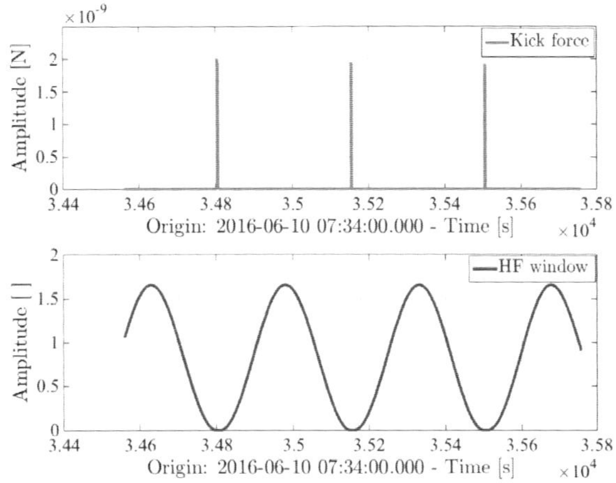

Figure 6: Time-series of the window used to zero out the kicks in Δg and comparison with the kick force.

3.4. Constraint-Gaussian gap patching

Another strategy proposed to analyze the free-fall data consists of filling the gaps with proper random noise that has the same statistics as the rest of the free-fall data. The purpose of the approach is to generate a reconstructed data series with a spectrum as close as possible to that one would obtain if the data had been taken continuously but without the application of force kicks. To minimize the bias resulting from the patches, the data in the patches should have the same spectral content as the existing data. The technique, that is called "Constrained-Gaussian Gap Patching", is described in detail in [11]. Once the gaps are filled with the synthetic noise, standard approaches can be followed to estimate the spectrum of the whole data set.

A comparison among the three approaches described above, in terms of Δg with gaps filled, is shown in Figure 7. Data refer to the free-fall measurement executed in June in URLA authority.

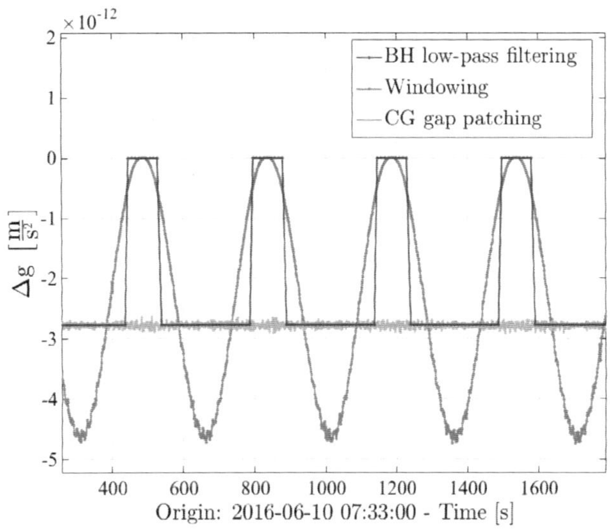

Figure 7: Resulting Δg data after having applied the three approaches.

4. First results of free-fall experiment and discussion

The free-fall experiment described in this article was performed in June 2016 and it lasted two days: it was implemented in Nominal authority on the first day, then the maximum forces and torques were reduced to URLA values and maintained until the end of the measurement.

As the initial mean value of the static differential acceleration was about $-2.5\,\mathrm{pm\,s^{-2}}$, according to what estimated in the preceding noise-only measurement (see Figure 8), the flights have a downward concavity and amplitude of the order of roughly 40 nm, as already shown in the previous figures. Moreover, because of the negative sign of Δg_{DC}, TM2 moves away from TM1. As a consequence, the kick force applied on TM2 is expected to be positive with respect to the TM2 reference frame (see Figure 1) as it forces TM2 toward TM1. We see a confirmation of this in Figure 3*b*.

The overall free-fall time-series estimated in June and analyzed for instance with the BH low-pass filtering approach, is compared with that measured in the preceding science measurement in Figure 8. The configuration of the free-fall measurements performed in June, in terms of duration, static and time-dependent differential acceleration, dynamic range in displacement and control force, is summarized in Table 1. A comparison with the previous standard noise measurement is included in the table.

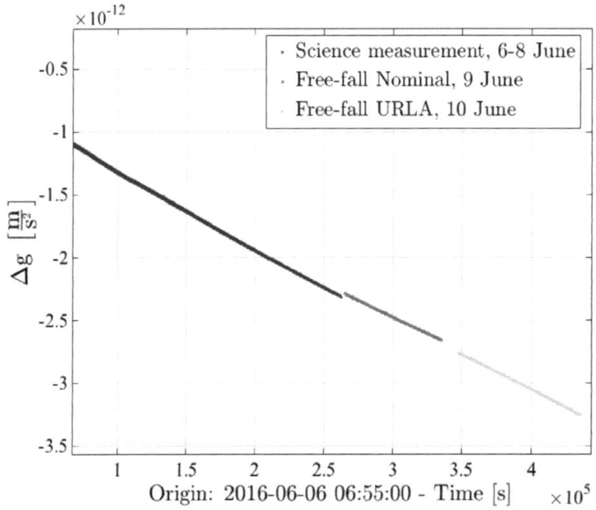

Figure 8: Δg time-series measured in June 6-10 during a standard science measurement (blue) and the free-fall experiment performed in Nominal (red) and URLA (green) authority. The data in blue are sampled at 10 Hz and low-passed with a cut-off frequency of 0.6 mHz. The free-fall time-series are, for instance, the result of the BH low-pass approach, sampled at $\sim$ 97 mHz before subtracting the time-drift: they correspond to the flight samples only (the kick samples are set numerically to zero).

To conclude, the free-fall mode control has been achieved and maintained successfully over the mission. The experiment dynamics is in agreement with what expected in terms of flight amplitude and kick force, considering the differential acceleration effectively measured between the TMs. In addition, the estimate of Δg is consistent with that evaluated in the preceding science measurement, as shown in Figure 8.

As regards the data analysis, all the approaches described here allow to extract the power spectral density of Δg measured in this specific experiment. However, the spectral bias estimation is still an ongoing analysis. The final results will be reported in a future publication.

Run	DOY	duration (h)	Δg_{DC} ($\mathrm{pm\,s^{-2}}$)	$\dot{g}_0$ ($\mathrm{pm\,s^{-2}/day}$)	Δx (nm)	F_{x_2} (nN)
Standard noise URLA	158-160	54	-1.7	-0.6	$\sim$ 0.04	$\sim$ 0.010
Free-fall Nominal	161	20	-2.5	-0.4	38	1.6
Free-fall URLA	162	24	-3.0	-0.5	46	1.9

Table 1: Characteristic parameters of the standard science measurement and free-fall experiments executed in June 2016. The table reports the averaged value of Δg_{DC} and $\dot{g}_0$, as well as the mean value of the dynamic range (flight amplitude in case of free-fall experiment) and control force.

References

[1] M Armano et al. In: *Physical Review L* **116**, 231101 (2016).
[2] Danzmann K et al. In: *A proposal in response to the ESA call for L3 mission concepts* (2017). URL: `https://www.lisamission.org/proposal/LISA.pdf`.
[3] Anza S et al. In: *Physical Review L* **22**, S125-S138 (2005).
[4] Antonucci F et al. In: *Physical Review L* **28**, 094002 (2011).
[5] LPF collaboration. In: (to be published).
[6] Vitale S et al. In: *Physical Review D* **90**, 042003 (2014).
[7] Grynagier A et al. In: *Class. Quantum Grav.* **26**, 094007 (2009).
[8] Schleicher A et al. In: *LISA Pathfinder Project Documentation, S2-ASD-TN-2001* (2008).
[9] Antonucci F and et al. In: *Class. Quantum Grav.* **28**, 094006 (2011).
[10] Vitale S. In: *LISA Pathfinder Project Documentation, S2-UTN-TN-3091* (2012).
[11] Armano M et al. In: *Journal of Physics: Conference Series* **610**, 012006 (2015).

IOP Conf. Series: Journal of Physics: Conf. Series **840** (2017) 012006 doi:10.1088/1742-6596/840/1/012006

The Engineering of LISA Pathfinder – the quietest Laboratory ever flown in Space

Christian Trenkel[1], Dave Wealthy[1], Neil Dunbar[1], Carl Warren[1], Alexander Schleicher[2], Tobias Ziegler[2], Nico Brandt[2], Rüdiger Gerndt[2]

[1]Airbus D&S, Stevenage SG1 2AS, UK
[1]Airbus D&S, Friedrichshafen, Germany

christian.trenkel@airbus.com

Abstract. We review the engineering approach adopted to ensure the required gravitational, magnetic, thermal and residual acceleration stability on-board LISA Pathfinder, and present the in-flight results that have been achieved. Arguably, this stability makes LISA Pathfinder the quietest laboratory ever flown in space. The implications for LISA are also discussed.

1. Introduction

The mission objective for LISA Pathfinder is to demonstrate the technology required for LISA, as far as the following two performance aspects are concerned:

(1) The suppression of residual forces acting on the freely-floating test masses on-board the spacecraft

(2) The local test mass metrology, ie the relative spacecraft-to-test mass metrology entering the overall LISA metrology chain

In this paper, we focus and review the engineering approach adopted to ensure the success of (1), ie how LISA Pathfinder as a laboratory is made stable enough to guarantee sufficiently low residual force noise acting on the test masses.

In particular, the following sources of disturbance will be discussed:

(1) Gravitational forces
(2) Residual accelerations
(3) Test Mass residual charge
(4) Thermal effects
(5) Magnetic forces

In each case, the engineering approach will be presented, the predicted performance will be compared to the observed in-flight performance, and implications for LISA will be discussed.

In order to avoid duplication with many other papers even in this volume, only a brief description of the main couplings will be provided. Detailed description of the physical effect, or a derivation of the requirements placed on individual parameters, can be found elsewhere.

The overall LPF performance observed during the first few weeks of nominal science operations has been described in [1].

2. Control of Force Noise Disturbances acting on Test Masses

2.1. Gravitational environment

Many parameters characterising the gravitational environment of the Test Masses needed to be controlled, such as

- DC absolute linear accelerations
- DC differential linear accelerations
- DC angular accelerations
- Gravitational stiffness (linear acceleration gradient)
- Differential linear acceleration fluctuations (at low frequency)

2.1.1. Engineering Approach Unfortunately, a direct verification by test of the gravitational environment was not feasible. Instead, the engineering approach was based on the following elements:

(1) Spacecraft design – eg no moving components, material choice in order to minimise thermoelastic distortions, etc

(2) Verification by analysis – modelling of the entire spacecraft hardware in order to predict the gravitational field, field gradient etc at the test mass location. This relied, heavily, on measured inputs such as location and mass of all items on-board the spacecraft. Just as one example of many, the mass model of one of the Low Gain Antennas (LGA) is shown below:

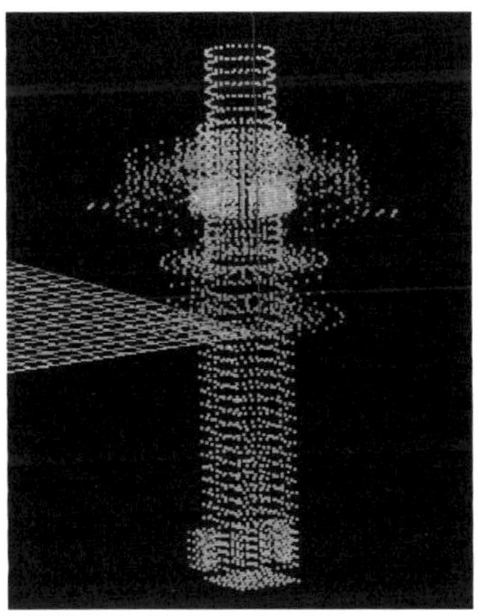

Figure 1 LGA gravitational model

(3) Strict gravitational control during manufacture, assembly and integration, throughout the project – both at payload and spacecraft level. For example at Airbus D&S UK, of order 10^4 individual mass measurements were carried out over an AIT phase that lasted around seven years.

IOP Conf. Series: Journal of Physics: Conf. Series **840** (2017) 012006 doi:10.1088/1742-6596/840/1/012006

ASTRIUM

LISA Pathfinder
SC AIV Mass Tracking Log

Figure 2 Representative Mass Log sheet used at Airbus D&S for gravitational model during AIT

(4) Final mass balancing of residual imbalance upon AIT completion

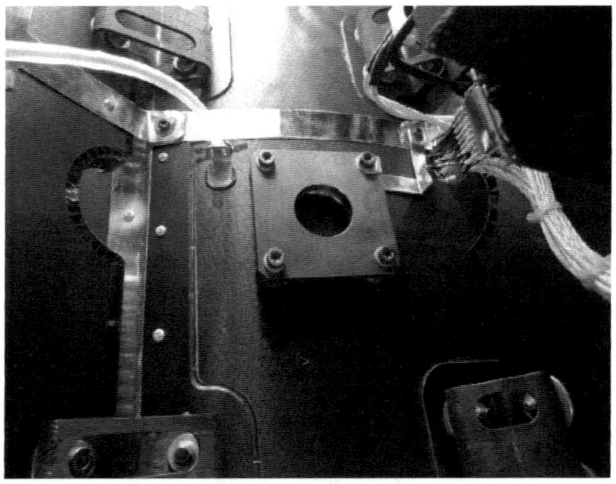

Figure 3 Mass balances installed to compensate residual imbalance following AIT

2.1.2. Comparison of predicted vs in-flight performance Where possible, a comparison of the predicted and measured parameters has been carried out. Not all of the gravitational parameters lend themselves to this comparison. The following table summarises the comparison that was done in time for the in-orbit commissioning review, ie approximately one week into science operations:

Parameter	Pre-flight Estimate	In-flight Measurement	Requirement	Requirement Met
Ax	-1.7e-10 -1.8e-10	(1.4±5.0)e-10 (TM1) (-3.9±2.2)e-10 (TM2)	<1.0e-8	Yes
Ay	-1.5e-9 -1.5e-9	(-1.0±0.2)e-9 (TM1) (-1.9±0.2)e-9 (TM2)	<1.0e-8	Yes
dAx	<5.5e-10	<1.0e-10	<6.5e-10	Yes
dAy	<3.9e-10	5.0e-10	<1.1e-9	Yes
dAz	<2.8e-10	0.1e-10	<1.85e-9	Yes
θ	-0.4e-9 +0.4e-9	-0.6e-9 (TM1) -0.1e-9 (TM2)	<13.5e-9	Yes
η	+3.1e-9 -1.2e-9	+2.9e-9 (TM1) -1.3e-9 (TM2)	<11.5e-9	Yes
ϕ	+0.8e-9 -0.2e-9	+1.0e-9 (TM1) -0.1e-9 (TM2)	<8.0e-9	Yes

Table 1 Comparison of predicted and in-flight observed gravitational parameters

As can be seen, all requirements have been met with margin. Along the y direction, the differential acceleration requirement is met with a margin of more than 100%, but this is still highlighted here as it is, in fact, the "worst" performance, relatively speaking. Unfortunately it has not been possible to verify the absolute linear acceleration along z with any useful accuracy, due to the degeneracy with the Solar Radiation Pressure (SRP).

The effect of the continuous cold gas depletion on the differential acceleration between the test masses had been predicted before the mission, and was found to be acceptable. It is interesting to note how the predicted change in differential gravitational acceleration as a result of changing the feed branch from which cold gas is used, agrees to within better than 10% with the one that was observed:

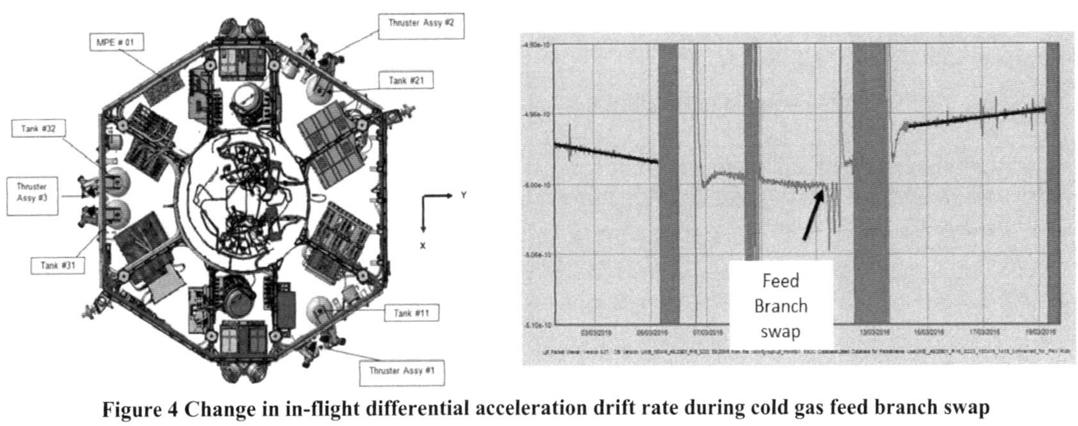

Figure 4 Change in in-flight differential acceleration drift rate during cold gas feed branch swap

The predicted change in slope was approx. $6.6\text{x}10^{-13}\text{ms}^{-2}$/day, and the observed one was approx. $6.2\text{x}10^{-13}\text{ms}^{-2}$/day.

Finally, it is worth noting that the worst-case pre-flight noise estimate related to electrostatic actuation fluctuations coupling to DC differential linear test mass acceleration, along X, was not realised in flight:

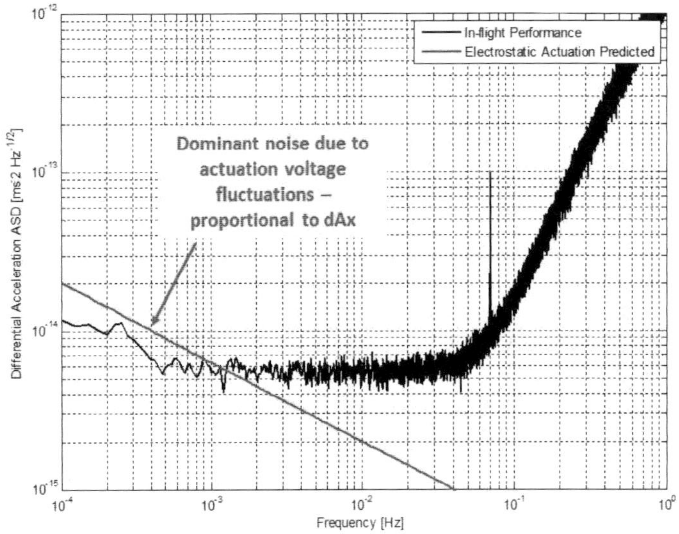

Figure 5 Pre-flight worst-case estimate of actuation voltage fluctuations coupling to differential gravitational acceleration, together with in-flight noise performance

2.1.3. Implications for LISA As far as LISA is concerned, the following can be concluded regarding gravitational control:

(1) No improvements to the approach adopted for LPF is necessary (in fact LISA is not as susceptible as LPF to differential gravitational accelerations

(2) The effect of the depletion of large amounts of cold gas propellant is well understood and manageable – a micropropulsion system based on cold gas is certainly an option for LISA

Even so, the following considerations apply:

(1) Moving parts (eg periodic repointing of HGA) need careful assessment for LISA
(2) Partial verification of gravitational requirements by test could result in time & cost savings, and reduce risks.

2.2. *Residual Accelerations*
The LPF shields the test masses on-board from external disturbances, primarily SRP. This shielding does not come completely for free, and now many residual spacecraft Test Mass couplings need to be considered.

No matter what the detailed coupling mechanism is (electrostatic, magnetic or gravitational), the relative spacecraft – test mass motion (acceleration) needs to be kept to a minimum.

2.2.1. Engineering Approach A relatively complex and sophisticated Drag-Free Attitude Control System (DFACS), comprising a set of algorithms that control one spacecraft and two free-flying test masses simultaneously along 15DOFs, is used to minimise the relative motion between LPF spacecraft and test masses.

DFACS relies on a number of low noise sensors (optical and electrostatic test mass sensing, star tracker sensing of spacecraft attitude) and actuators (electrostatic test mass actuation, micropropulsion system acting on spacecraft) to achieve its goal, namely the robust control from the initial test mass release through to the nominal science operation modes.

The following schematic shows most of the main hardware elements used by the DFACS control loops:

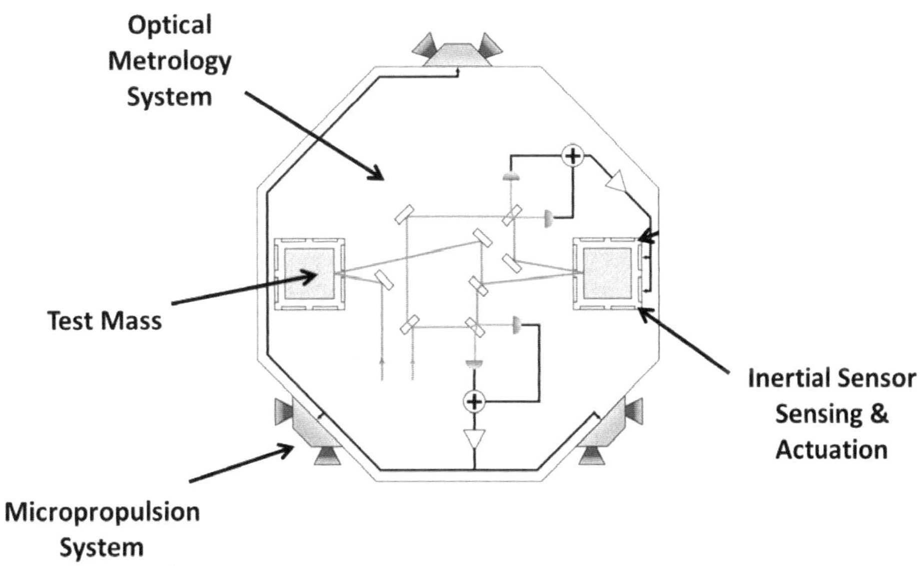

Figure 6 Sensors and actuators used by DFACS to control spacecraft and two test masses

2.2.2. Comparison of predicted vs in-flight performance As far as the DFACS in-flight performance is concerned, compared to the pre-flight predictions, three main conclusions can be drawn:

(1) Test Mass initial release conditions in orbit were much worse than expected

   a. Initial offsets and velocities exceeded by factors of up to 60 and 8, respectively
   b. Nevertheless DFACS managed to capture TMs eventually

(2) DFACS handled all transitions to science mode very reliably and with excellent repeatability – the figure below shows several transitions from ACC3 mode to SCI1_2 mode:

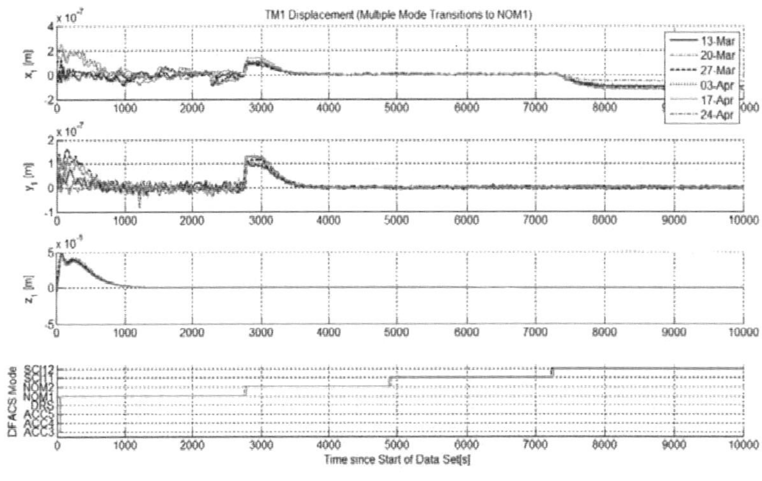

Figure 7 Multiple DFCAS transitions from Acc3 to SCI1_2 mode

(3) The actual science mode performance was better than predicted, as shown by data taken during commissioning

a. Sensor and actuator noise models were conservative

b. Offsets and misalignments were conservative

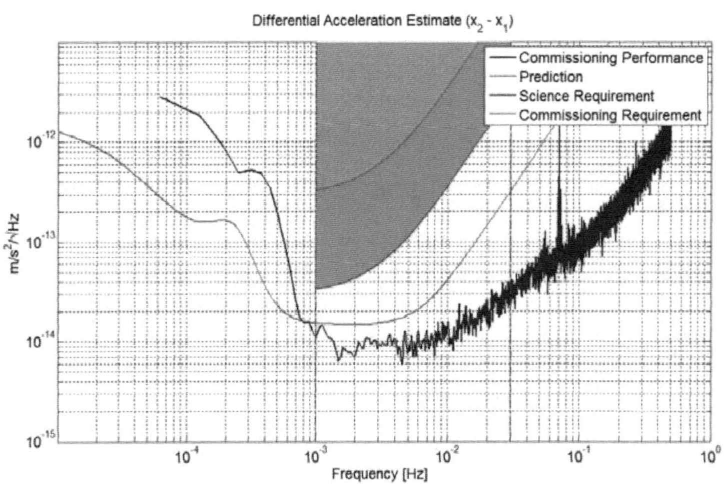

Figure 8 SCI1_2 mode Performance during Commissioing

Finally, the relative spacecraft – test mass motion can be taken as proxy for residual spacecraft accelerations, assuming that the test masses provide a perfect free-fall reference frame. The plot below shown this relative motion as a function of frequency, together with typical seismic vibrations experienced on ground:

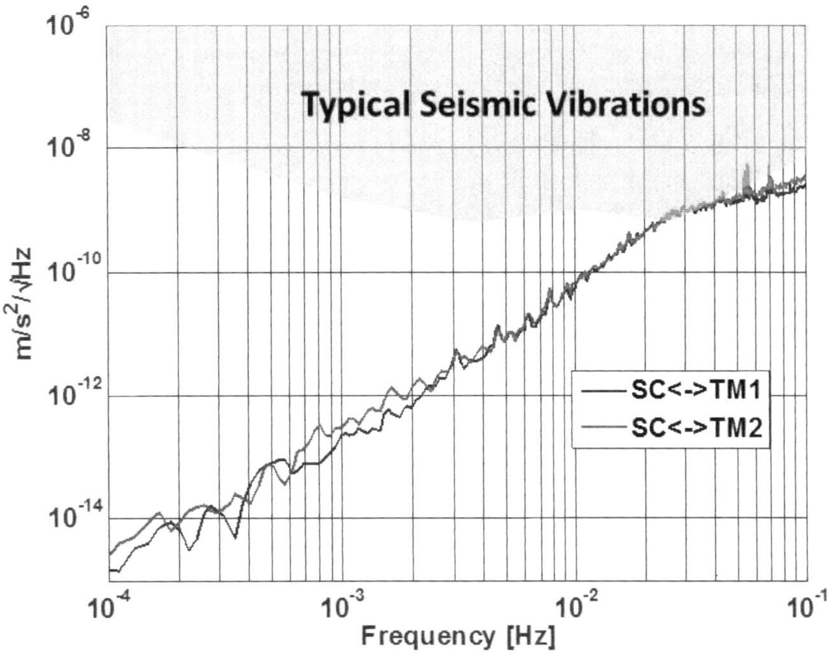

Figure 9 Relative spacecraft – TM acceleration spectrum, compared to typical seismic noise on ground

The above plot is effectively a perfect "vindication" of the approach to use space as the only environment where low frequency gravitational waves become observable. Seismic vibrations at low frequencies simply cannot be attenuated sufficiently. Arguably, this plot also justifies calling LPF "the quietest laboratory ever flown in space".

2.2.3. Implications for LISA As far as LISA is concerned, the following can be concluded regarding the required attitude control system:

(1) LPF DFACS science mode performance goes a long way towards LISA requirements

(2) The DFACS performance model has been verified in-flight, and can be extended for LISA

(3) The robustness of mode transitions could be improved further:

  a. Uni-directional thruster configuration is efficient, but limits control authority durig critical phases. Additional thrusters would enhance margins.
  b. Margins for electrostatic suspension actuation should be increased – this would account for disturbance uncertainties

2.3. Test Mass Charge control to minimise electrostatic couplings
The freely floating test masses on-board LPF can become electrostatically charged, primarily from radiation and cosmic rays continuously bombarding the spacecraft. Any such net charge will couple electrostatically to the electrode housing surrounding the test masses, resulting in unwanted interactions. The test mass charge level therefore needs to be actively controlled.

2.3.1. Engineering Approach A dedicated, active charge management system (CMS) has been developed for LPF. Based on the photoelectric effect, is uses UV light to release electrons either from the test mass or from the surrounding electrodes, depending on the polarity of the test mass charge that needs to be reduced.

The schematic discharge principle is illustrated below:

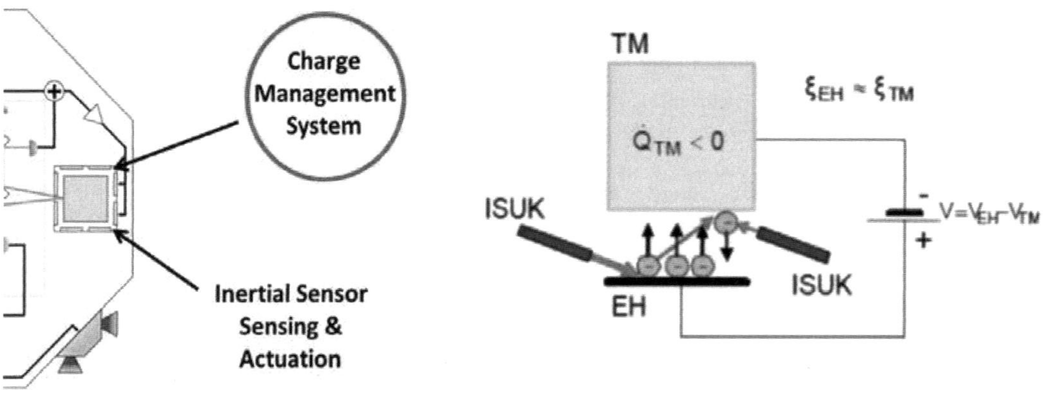

Figure 10 Schematic showing Charge Management System elements. EH: Electrode Housing, ISUK: Inertial Sensor UV Kit (UV feedthroughs)

The CMS provides a robust way to reduce unwanted charges on the test masses, and comprises automatic on-board algorithms to achieve regular semi-autonomous test mass discharge.

2.3.2. Comparison of predicted vs in-flight performance The in-flight performance of the CMS on-board LPF can be summarised as follows:

(1) The on-board charge estimation performance is in line with pre-launch predictions

(2) The closed loop discharge performance is also in line with predictions

(3) On-board closed loop fast discharge is used regularly for LTP and DRS operations

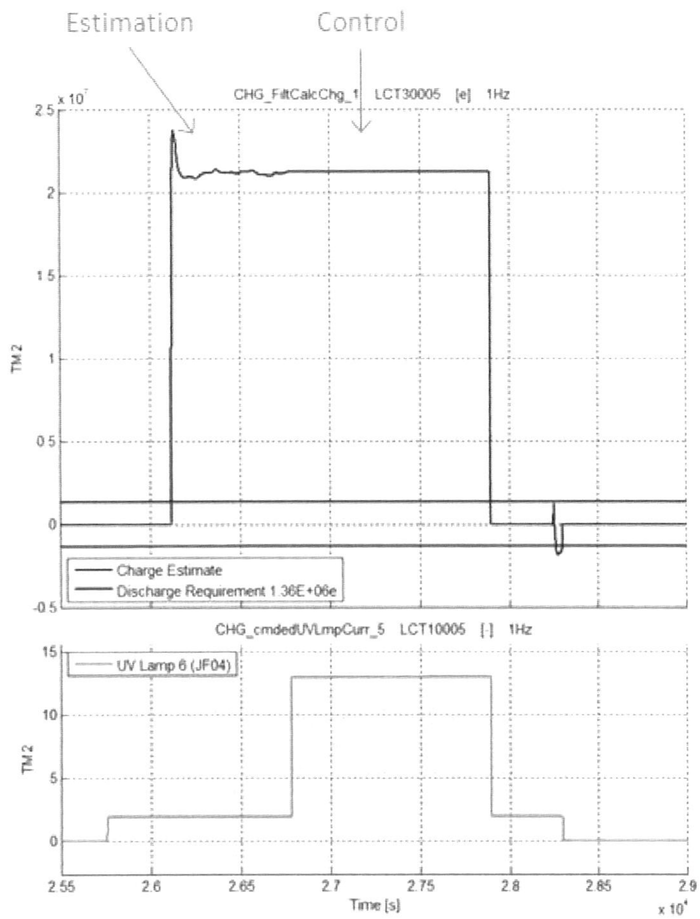

Figure 11 TM2 Closed Loop Discharge performance

The above plot shows just one example of successful in-flight charge control on test mass 2.

2.3.3. Implications for LISA As far as LISA is concerned, the following can be concluded regarding the charge management system:

(1) On-board charge estimation has been verified
  a. Highly flexible and can be adjusted for the use of different DOFs
  b. DOFs with optical readout are preferred

(2) Closed loop discharge control has been verified

(3) The robustness of charge control could potentially be improved:
  a. Optimization of light injection would avoid the need for DC biasing

2.4. *Thermal Environment*

An ultra-stable thermal environment, at low frequencies, is essential to achieve the required LPF performance – numerous physical effects couple to temperature, and fluctuations in temperature can then result in fluctuating forces acting on the test masses.

2.4.1. Engineering Approach The nominal LPF orbit around L1, together with the constant sun-pointing attitude of the spacecraft, already provides a highly stable external environment.

Nevertheless, a careful thermal spacecraft design is still essential to reach the required levels of stability. The schematic diagram below illustrates the main thermal features of the nested design that has been implemented for LPF:

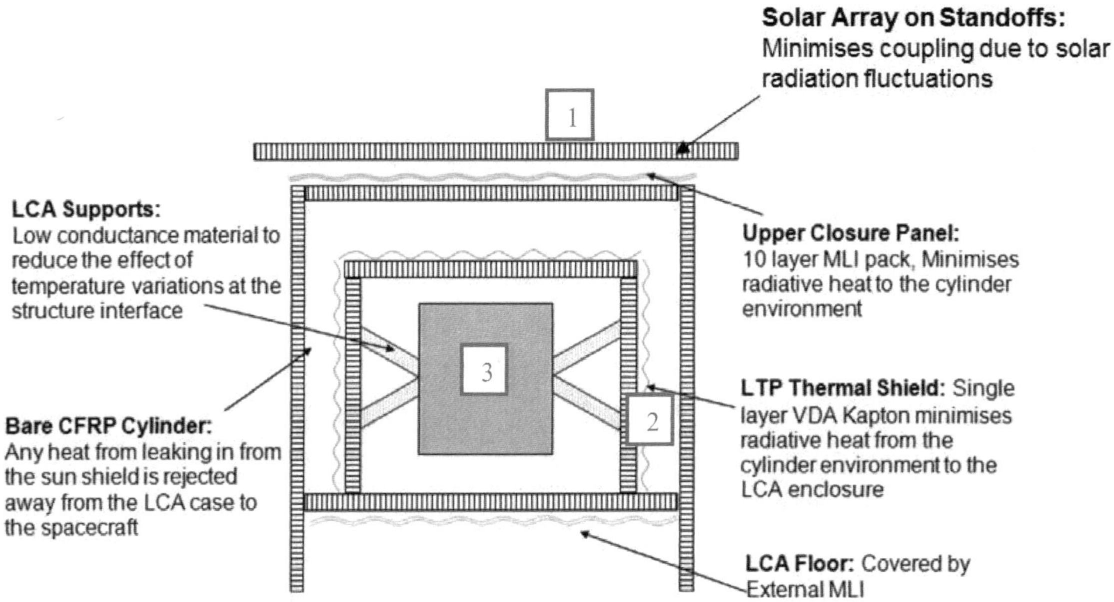

Figure 12 Nested LPF spacecraft design with thermal features necessary to reach required thermal stability. Numbers correspond to locations where thermal stability has been measured below.

In addition, the thermal design was based on the principle that no unit or heater switching was allowed during nominal operations, so as not to disturb the stable environment. One of the consequences of this was a purely passive thermal control – all the heaters are either permanently ON or OFF.

Finally, extensive thermal test campaigns were carried out to verify as many requirements by test as was possible, thus strongly reducing the risk of in-flight surprises.

2.4.2. Comparison of predicted vs in-flight performance The thermal stability achieved in flight is shown below for the following locations indicated in the figure above:

(1) Solar array temperatures – a very slow drift of approx. -0.4K over 6 days is apparent due to the increasing Sun distance at the time of the measurement.

IOP Conf. Series: Journal of Physics: Conf. Series **840** (2017) 012006 doi:10.1088/1742-6596/840/1/012006

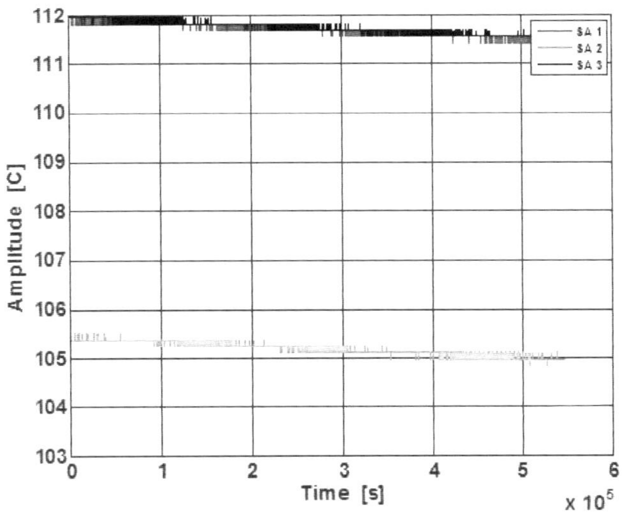

Figure 13 Solar Array temperature measurement

(2) LTP Composite Assembly Cylinder – a stability of ≈10^{-3}K/√Hz is achieved down to 1mHz:

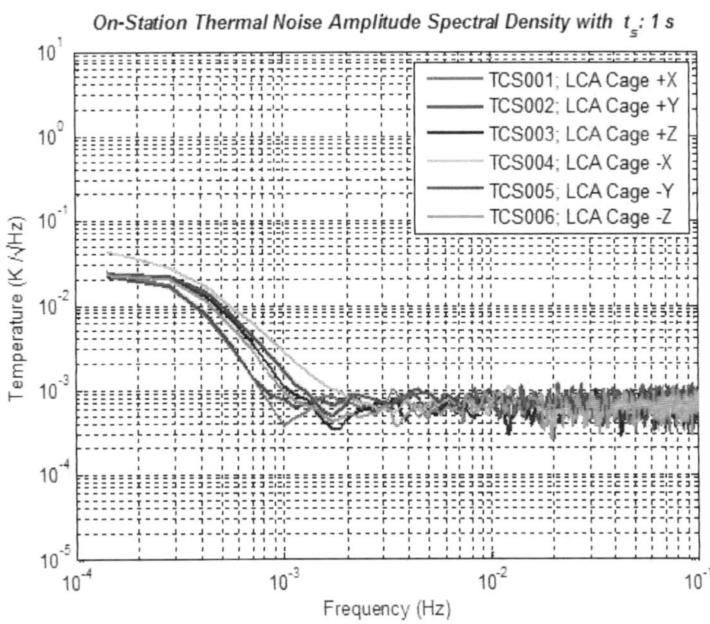

Figure 14 LCA temperature stability

(3) Optical Bench – a stability of <3x10^{-5}K/√Hz is achieved down to 0.1mHz:

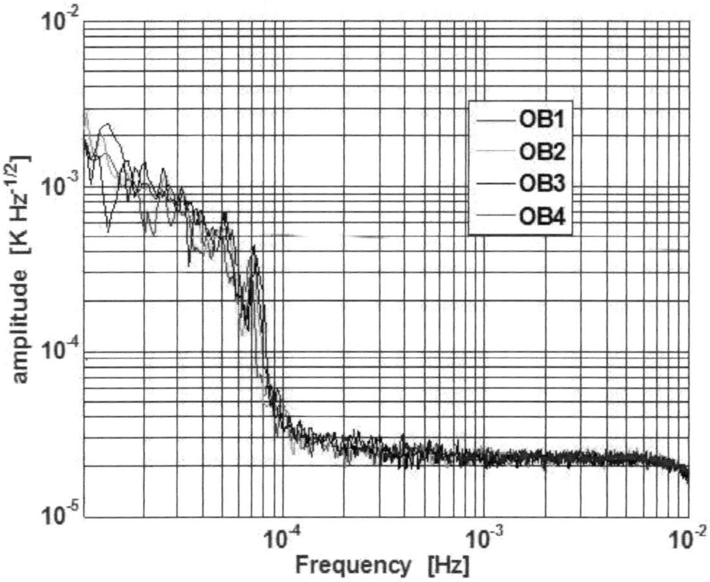

Figure 15 Optical Bench temperature stability

2.4.3. Implications for LISA LPF has shown that the stringent temperature stability requirements at low frequencies can be met through careful spacecraft design. Specifically, the following conclusions can be drawn:

(1) A stable thermal external environment certainly helps – LISA will also benefit from this

(2) The Solar Array shadowing of the spacecraft body is essential for thermal stability

(3) The following considerations, lessons learnt and potential improvements apply:
   a. The nested LPF spacecraft design helps, but will not be directly transferable to LISA – LISA will have large telescope apertures
   b. Thermal control using only fixed (trim) heaters is not as flexible as would be desirable – quiet PID control should be possible

Equipment like a PCDU should not be placed near thermally sensitive equipment

2.5. Magnetic Environment

Non-zero magnetic test mass properties will couple to the local magnetic environment, generating force noise as a result. A significant number of magnetic environmental parameters need to be controlled, such as:

- DC magnetic field and magnetic field gradients
- Fluctuating magnetic fields and field gradients

Unlike most magnetically sensitive missions, these parameters do not have to be controlled on a long instrument boom, at a large distance from the spacecraft, but right in the centre of the LPF spacecraft.

IOP Conf. Series: Journal of Physics: Conf. Series **840** (2017) 012006 doi:10.1088/1742-6596/840/1/012006

2.5.1. Engineering Approach The approach adopted to ensure that magnetic requirements were met was based on a combination of:

(1) Mitigation by design – avoid magnetic parts and establish clear EMC design guideline
(2) Little heritage information due to unusually low frequency range – extensive test campaigns were conducted at unit and spacecraft level.

The pictures below show a selection of magnetic tests carried out at various stages of the LPF project:

Figure 16 Selection of LPF magnetic tests. Top left: Spacecraft test at IABG, top right: IS FEE test at Airbus DS, bottom centre: OBC test at Airbus DS

2.5.2. Comparison of predicted vs in-flight performance Local absolute DC magnetic fields have been found to be of order 1μT, as predicted and shown below.

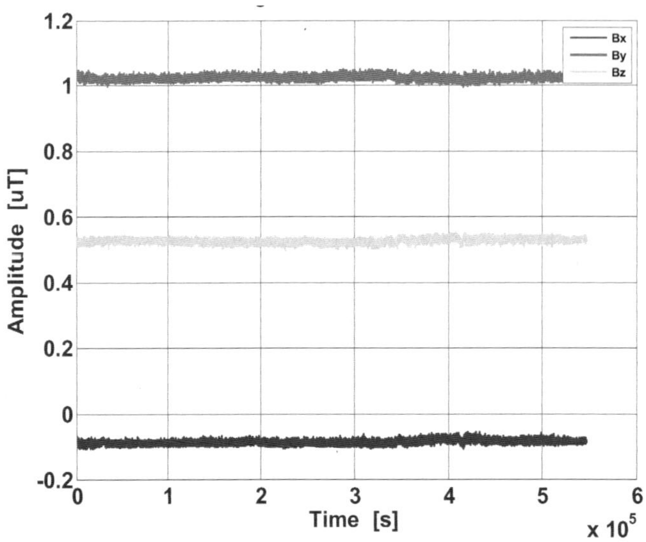

Figure 17 Magnetic field recorded by magnetometer on-board LPF

Low frequency magnetic field fluctuations of order 30nT/√Hz were measured at 1mHz:

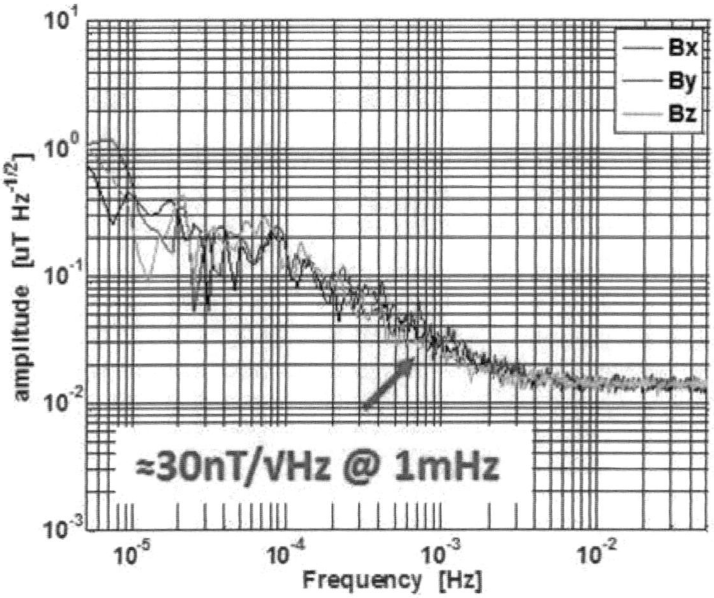

Figure 18 Low frequency magnetic field fluctuations on-board LPF

Above about 4mHz, the noise becomes white at a level of about 13nT/√Hz , and is not thought to correspond to real magnetic field fluctuations, but instead is due to intrinsic DAQ noise.

This is corroborated by the spectrum of magnetic field gradient fluctuations, shown below:

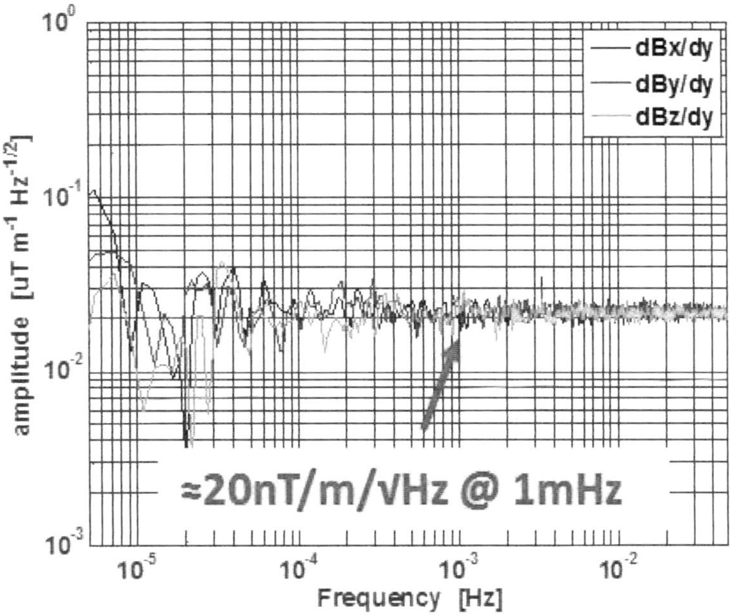

Figure 19 Low frequency magnetic field gradient fluctuations on-board LPF

The white noise level of just over 20nT/m/√Hz not only looks "unphysical", indicating that it is not due to real magnetic field gradients, but the level also agrees very well with that expected from combining two uncorrelated magnetic field noise streams of 13nT/√Hz, separated by a baseline of around 0.8m: (13nT/√Hz x √2 / 0.8m) ≈ 23nT/m/√Hz.

The conclusion is that the magnetic field fluctuations observed above correspond, in fact, to fluctuations in a field that is uniform across the spacecraft – and this can only be the external, interplanetary magnetic field.

Finally, it is interesting to note that the worst-case pre-flight estimate of the magnetic noise has not been realised in flight:

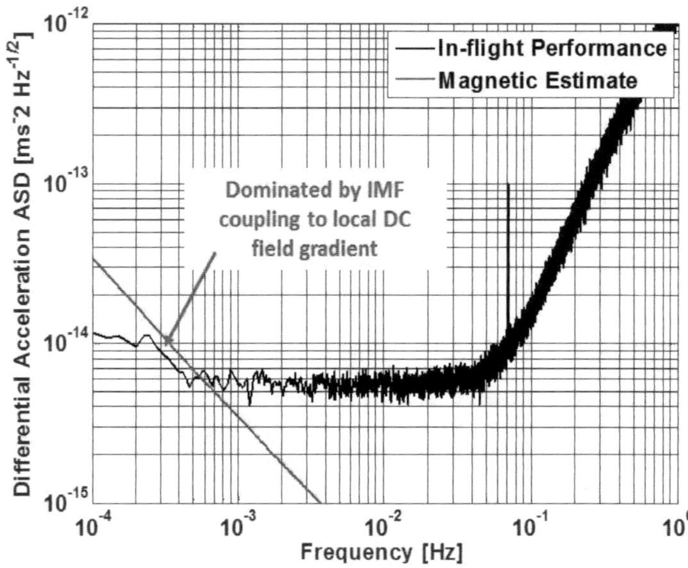

Figure 20 Worst-case pre-flight magnetic noise estimate, compared to overall LPF performance

The dominant contribution to the pre-flight estimate was the coupling of interplanetary field fluctuations of up to 55nT/√Hz at 1mHz, to local DC gradients of up to 18μT/m, limited by the measurement accuracy of the magnetic testing.

2.5.3. Implications for LISA No showstoppers, or real problems, have been identified for LISA. Nevertheless, the following should be improved:

(1) DC magnetic gradient testing (in particular for payload elements in close proximity)

(2) The low frequency behaviour (eg amplitude modulation) of high frequency AC lines should be characterised by test

(3) Needless to say, for LISA equipment without LPF heritage (eg TWTA), magnetic testing is still essential.

3. Summary and conclusions

In this article, we have presented the engineering approach, as well as compared in-flight performance with pre-flight estimates, of several aspects absolutely crucial to the success of LPF. The overall noise performance has been met with margin, and inevitably worst-case assumptions about offsets, noise levels etc have proved to be conservative.

LPF has shown itself to be an extremely stable platform to allow fantastically precise experiments to be carried out. In particular as far as residual low frequency accelerations are concerned, it can justifiably be called the quietest laboratory ever flown in space. The techniques and experience that have been developed in the process is sure to find applications in addition to the main target, LISA.

No real showstopper for LISA has been identified in any of the areas, and on the contrary, some concerns in certain areas have been laid to rest.

4. Acknowledgements

We would like to thank the LPF collaboration as a whole, from ESA through academia to industrial partners, as well as to all the national funding agencies, for the opportunity to work as a team on a mission as exciting and challenging as LPF. We should collectively be very proud of what we have achieved, and we look forward to working on the future observation of gravitational waves from space!

References

[1] Armano, M. et al. (2016) Sub-femto-g free fall for space-based gravitational wave observatories: LISA Pathfinder results. *Physical Review Letters*, 116(23), 231101

IOP Conf. Series: Journal of Physics: Conf. Series **840** (2017) 012007 doi:10.1088/1742-6596/840/1/012007

LISA Pathfinder as a Micrometeoroid Instrument

J.I. Thorpe[a], T.B. Littenberg[b], J. Baker[a], and J. Slutsky[a], for the The LISA Pathfinder Team

[a] NASA Goddard Space Flight Center, 8800 Greenbelt Road, Greenbelt, MD 20771, USA

[b] NASA Marshall Space Flight Center, Redstone Arsenal, Huntsville, AL 35812, USA

E-mail: james.i.thorpe@nasa.gov

Abstract. LISA Pathfinder is perhaps the most precise accelerometry instrument ever flown in space. The drag-free control system can sense and react to external disturbances of an extremely small magnitude. One class of such disturbances are the impacts of micrometeoroids or dust. A simple model of the LPF system suggests that individual impacts with transferred momentum exceeding a few tens of nanoNewton-meters are detectable. Furthermore, the ability of LPF to resolve both the linear and angular momentum transfer as vector quantities allows information such as the sky location and the impact location of the impactor to be reconstructed. This novel approach to micrometeoroid detection and characterization, as well as the location of LPF at L1, provide an opportunity to improve our understanding of the dust environment in the inner solar system. Here we present some preliminary findings from LPF, including four candidate impact events.

1. Introduction

LISA Pathfinder (LPF) [1]is a technology demonstrator mission for future space-based gravitational wave observatories. Led by the European Space Agency and with contributions from a consortium of European member states and NASA, the mission was launched on December 3rd, 2015 and began science operations on March 1st, 2016. The operational orbit is a 0.8 Mkm × 0.5 Mkm Lissajous orbit around the first Earth-Sun Lagrange point (L1), a gravitational balance point that lies approximately 1.5 Mkm in the sunward direction from Earth. LPF has been operating in a variety of modes through baseline operations of its European payload (March-June 2016), baseline operations of its NASA payload (July-November 2016), and extended mission operations (December 2016 - present).

The primary purpose of these operations is to demonstrate drag-free control as a technique for producing a low-disturbance inertial reference that could be used as the basis for a long-baseline gravitational wave observatory. In LPF, the reference is a 4 cm cube of Au-Pt alloy with a mass of approximately 1.92 kg. This 'test mass' is contained within an electrode housing that is fixed to the spacecraft that can be used to both measure the position and attitude of the test mass relative to the spacecraft as well as to apply forces and torques between the test mass and spacecraft. A control system maintains gaps between the spacecraft and test mass by monitoring the separation and applying force or torque commands to the test mass via the electrode housing or to the spacecraft via a micropropulsion system. In typical drag-free operations the spacecraft is commanded to follow the test mass in one axis (denoted as x) and the test mass is forced to follow the spacecraft in other degrees-of-freedom (DoF). Additionally, an interferometric metrology system is employed to measure the relative separation between the test mass and spacecraft along the $x-$axis with sub-picometer precision and rotations along the $y-$axis and $z-$axis with nanoradian precision.

LPF carries a second test mass, located approximately 38 cm from the primary test mass in the x direction. This test mass is used as a witness to characterize the residual disturbances of the drag-free test mass. This differential acceleration measurement has the advantage that it common-mode rejects external disturbances on the spacecraft, such as fluctuations in the solar radiation pressure, solar wind, magnetic environment, and particle impacts. This strategy is effective for characterizing drag-free test masses as references for gravitational-wave instruments and was successfully employed to demonstrate residual acceleration noise significantly below the LPF mission requirements and consistent with levels required for LISA-like gravitational wave observatories [2].

In addition to this primary science goal, the external disturbances present in the single-test mass signal represent an opportunity to characterize the space environment of LPF in a unique and novel way - by measuring acceleration of the spacecraft. In a previous work [3] it was shown that impacts of dust and micrometeoroids should be detectable in LPF if they transferred momentum to the satellite of a few tens of $\mu\mathrm{N}\cdot\mathrm{s}$. It was estimated that such events should occur relatively frequently and that LPF should be able to measure properties including the total transferred momentum, momentum direction, and location of impact on the spacecraft. In this paper we present four impact candidates from the operations period of March 1, 2016 through June 28, 2016. This does not represent a complete catalog during this time period but rather an example of typical events.

In section 2, we present one event and describe the process of reconstructing the external force from the measured positions and attitudes of and the commanded forces and torques. In section 3 we describe the 6-DoF model used to fit the impact candidates. In section 4 we present the four impact candidates and the model fits. Section 5 summarizes our results and describes ongoing and future work with this data set.

2. Event Reconstruction

The goal of the reconstruction process is to remove the effect of the LPF control loops and estimate the dynamics of the spacecraft and test mass as if they were kinematicly free bodies. The signals available for this are the measured position and attitude of the test mass (6 DoFs from electrostatic measurements, 3 DoFs from interferometry), the commanded force and torque to the spacecraft, and the commanded force and torque to the test mass. The reconstruction is done in the acceleration domain to avoid integration constants that do not contribute to the micrometeoroid signal. For example, in the x-axis, in which the spacecraft test mass separation is controlled by forcing the spacecraft, the external spacecraft acceleration can be estimated using the following equation:

$$a_{ext}(t) = \frac{d^2x(t)}{dt^2} - \frac{1}{M}C\,F(t-\tau) - \omega^2 x(t), \tag{1}$$

where $a_{ext}(t)$ is the external acceleration, $x(t)$ is the measured position, $F(t)$ is the commanded force, M is the spacecraft mass, C is the gain in the micropropulsion system, τ is a delay in the response of the micropropulsion system, and ω^2 is a term describing the "stiffness", a spring-like coupling between the test mass and spacecraft that arises from electrostatic, magnetic, and gravitational effects. Quantities such as C, τ, and ω^2 are measured using signal injections as part of the normal LPF measurement campaign.

Figure 1 shows an example of an impact candidate and the reconstructed external acceleration in the x-axis. In the top panel of Figure 1a, the position of the test mass relative to the spacecraft exhibits a sharp rise that is indicative of an impact. The bottom panel of Figure 1a shows the corresponding force command, which includes the control system's response to this disturbance. The control system does its job of rejecting the disturbance and settles into steady-state after one or two oscillations. Figure 1b shows the reconstructed external acceleration, which is an approximate delta function, as would be expected for an impact. A slight overshoot is believed to be a reconstruction error associated with inaccuracies in the parameters used in (1) or interpolation errors when applying sub-sample delays.

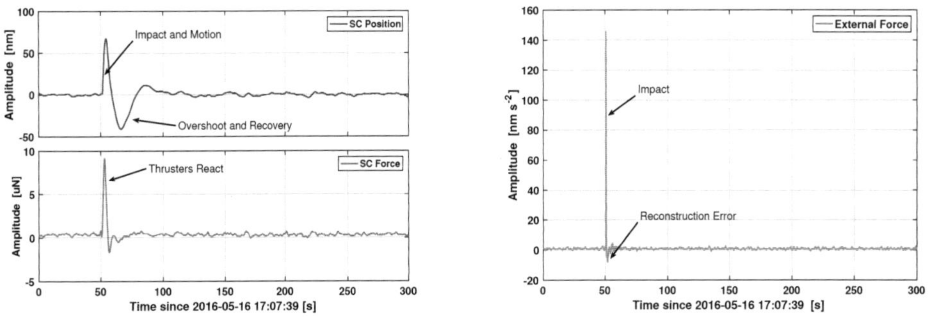

(a) SC position and Force (b) Reconstructed External Force

Figure 1: x-axis dynamics for a candidate impact event. Left panel top shows relative motion of test mass and spacecraft along x as measured by the interferometer. Left panel bottom shows force commanded by drag-free system to the spacecraft along x. Right panel shows external force along x as reconstructed from the measured position, commanded force, and knowledge of the spacecraft mass and stiffness.

When extending this reconstruction process to all six DoFs, care must be taken to account for couplings between the linear and angular DoFs as measured by the test masses, which are offset from the spacecraft center of mass in both the $+z$ direction and along the x-axis. To address this coupling, the applied forces and torques on the spacecraft are first used to estimate the linear and angular acceleration in the spacecraft body frame (a cartesian frame with the origin at the spacecraft the center of mass). These accelerations are then used to compute the contribution to the acceleration at the test masses using the known location of the housings relative to the spacecraft body frame.

3. Models

To identify impact candidates and estimate the parameters of interest, a model of the system's response is needed. Since the event reconstruction described in Section 2 produces the equivalent acceleration of the spacecraft as a free body, the impact model only needs to account for the 3D kinematics of the spacecraft. The impact model assumes that the impact timescale is short relative to the sample cadence of the LPF data (typically 0.1 s) such that the impact can be represented as a delta-function in acceleration. The geometric coupling of the impact momentum to different DoFs is identical to that for the micropropulsion system and is computed in two steps. First, the acceleration in the body frame is computed for both linear and angular DoFs:

$$\vec{a}_{x,B}(t) = P\, M^{-1}\delta(t-\tau)\hat{e}, \tag{2}$$

$$\vec{a}_{\theta}(t) = P\,\mathbf{I}^{-1}\delta(t-\tau)\,(\vec{r}\times\hat{e})\,, \tag{3}$$

where $\vec{a}_{x,B}$ is the acceleration of the spacecraft body frame in the linear DoFs, $\vec{a}_{\theta,B}$ is the acceleration of the spacecraft body frame in the angular DoFs, P is the total transferred momentum, τ is the impact time, $\hat{e}$ is the unit-vector in the direction of the transferred momentum, M is the mass of the spacecraft, $\mathbf{I}$ is the spacecraft moment of inertia, and $\vec{r}$ is the location of the impact relative to the center of mass. The angular accelerations at the housing frame are the same as described in (3) but the linear accelerations pick up an additional term due to the offset of the test mass from the center of mass:

$$\vec{a}_{x,TM}(t) = \vec{a}_{x,B} + (\vec{r}_{TM} \times \vec{a}_{\theta}) \,, \tag{4}$$

where $\vec{a}_{x,TM}$ is the acceleration in the linear DoFs as measured in the test mass frame and $\vec{r}_{TM}$ is the location of the test mass in the body frame.

4. Results

The impact model in Section 3 was used in a Markov-Chain Monte-Carlo tool to both search for and characterize potential impacts in stretches of LPF data [4, 5]. Sections of data with no known signal injections were identified and passed through the event reconstruction pipeline described in Section 2. The output of this pipeline was then divided into segments 1638.4 s in length and passed to the MCMC tool, which used frequency-domain versions of the models in (2) - (4) to fit for the parameters P, τ, $\hat{e}$ (represented by two sky angles), and $\vec{r}$. The impact location was constrained to lie of the surface of the spacecraft and also to be consistent with the measured sky angle (e.g. the impact location must have an unobstructed view of the origin of the momentum vector). For segments with an identified candidate, the MCMC was allowed to run for 10^5 iterations, including a burn-in period of 10^4 iterations.

While the results of a systematic search will be the subject of a future publication, here we present four impact event candidates that occurred in the timeframe March 1, 2016 through June 28, 2016. Table 1 summarizes the recovered parameters for these events. Figure 2 contains posterior distribution functions for the transferred momenta, Figures 3a - 3d show sky maps of the transferred momentum vector's origin (direction opposite to the transferred momentum), and Figure 4a - 4d show scatter plots of the MCMC samples of the recovered impact location on the spacecraft colored by the natural log of the likelihood.

Table 1: Sample impact event candidates from the first portion of LTP operations. Values represent the median of the posterior distribution and intervals represent the upper and lower limits of the 90% confidence intervals. Colatitude represents the angle of the impact vector from the $+z$ direction. Azimuth represents the angle in the $x - y$ plane from the $+x$ axis.

Day	GPS Time [s]	Amplitude [μN · s]	colatitiude [rad]	azimuth [rad]
2016-04-09	$1144229907.61^{+0.001}_{-0.0003}$	495^{+10}_{-9}	$1.576^{+0.003}_{-0.002}$	$0.002^{+0.15}_{-0.12}$
2016-05-04	$1146429821.79^{+0.04}_{-0.03}$	55^{+44}_{-23}	$1.60^{+.04}_{-0.04}$	$0.2^{+0.6}_{-0.8}$
2016-05-16	$1147453725.596^{+0.005}_{-0.005}$	419^{+23}_{-15}	$1.568^{+0.005}_{-0.006}$	$2.99^{+0.10}_{-0.24}$
2016-06-21	$1150511110.27^{+0.01}_{-0.02}$	101^{+41}_{-31}	$1.59^{+.02}_{-0.02}$	$-0.77^{+0.31}_{-0.17}$

Not surprisingly, the recovered parameter accuracy is highly dependent on the amplitude of the event, which is proportional to its signal-to-noise. For the two larger events, transferred momentum is measured to within a few percent, sky angles to a few tenths of a radian (both 90% confidence intervals), and impact location restricted to a patch of a few tens of centimeters in diameter. For the quieter events, the variation in recovered momenta rises to 50%, sky angles are uncertain by as much as 0.5 rad, and the impact location is only weakly constrained and in one case bimodal. In all cases, the impact is localized in time to better than 0.1 s, the sample cadence of the measurement.

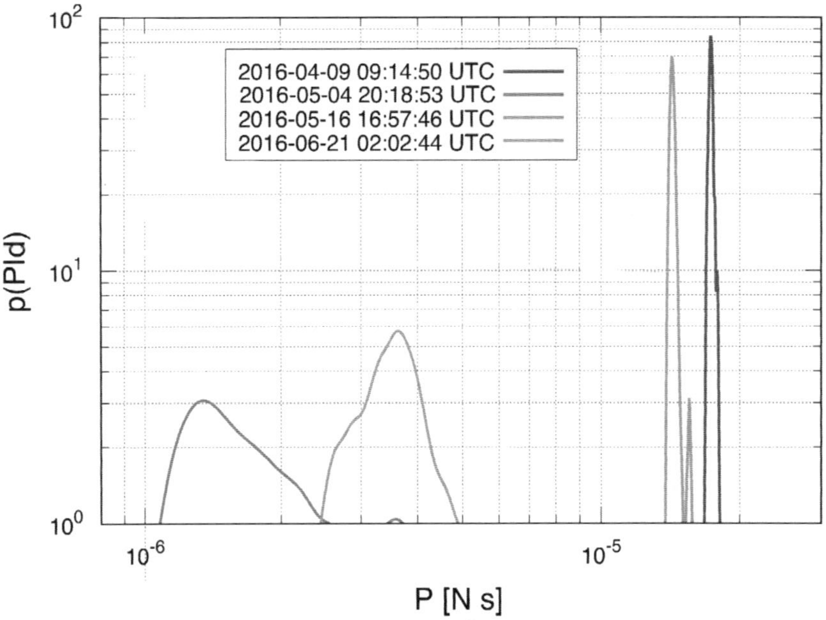

Figure 2: Posterior distribution functions of reconstructed amplitudes for the four example impact candidates.

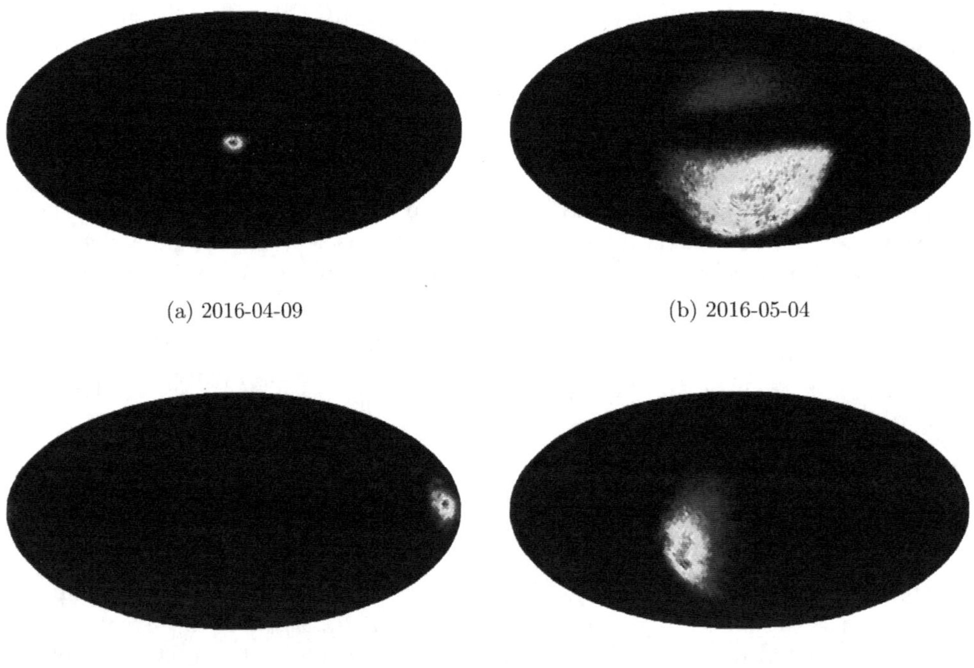

(a) 2016-04-09

(b) 2016-05-04

(c) 2016-05-16

(d) 2016-06-21

Figure 3: Reconstructed sky position histograms for the four example impact candidates. Sky maps are Mollweide projections with zero longitude along the $+x$ axis of the SC and $+90°$ latitude along the $+z$ direction.

IOP Conf. Series: Journal of Physics: Conf. Series **840** (2017) 012007 doi:10.1088/1742-6596/840/1/012007

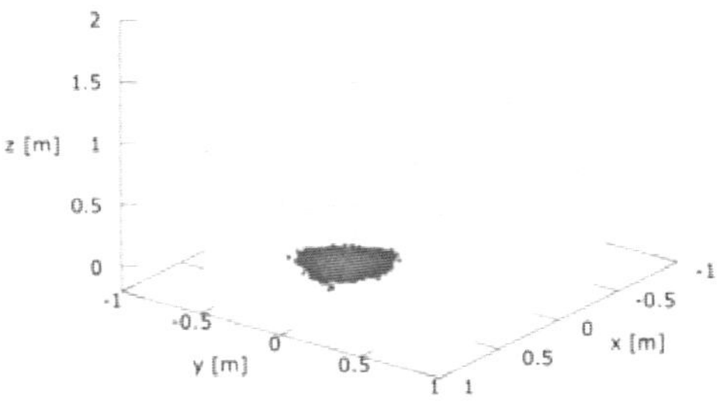

(a) 2016-04-09

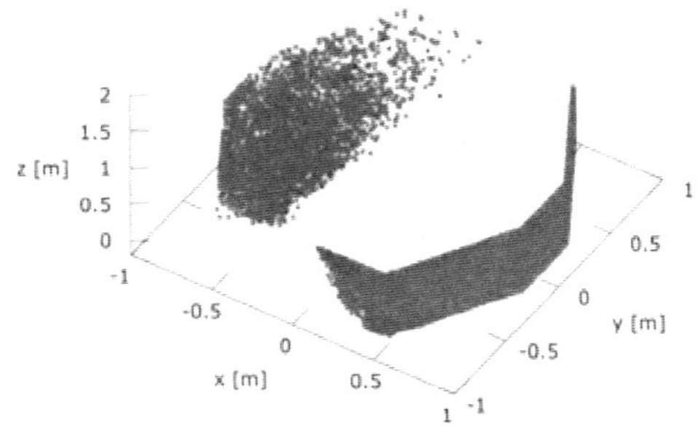

(b) 2016-05-04

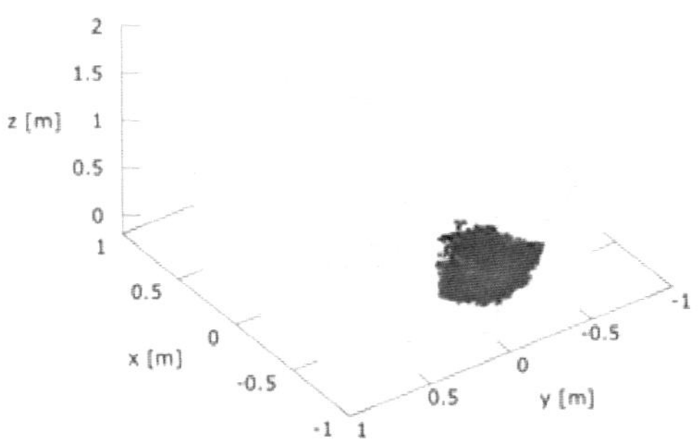

(c) 2016-05-16

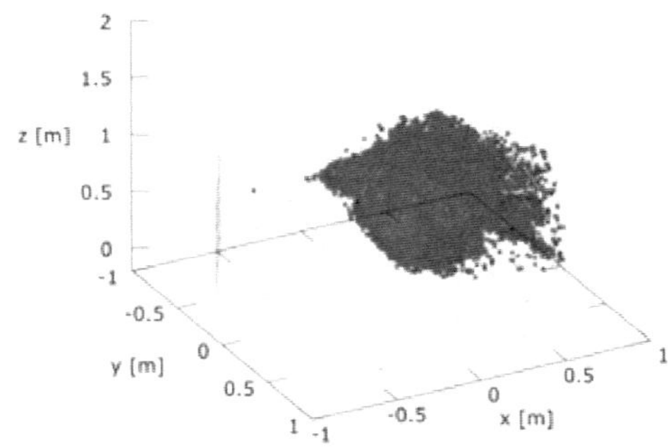

(d) 2016-06-21

Figure 4: Reconstructed impact locations for the four example impact candidates. Points are MCMC chain locations after burn-in, color coded by log likelihood with warmer colors representing higher likelihood.

5. Discussion

The impact candidates presented in Section 4 demonstrate that detecting and characterizing impact events with LPF is achievable in practice. However, additional steps are needed before this work is complete.

First, the impact candidates need to be vetted and vetoed against any possible non-impact transients that fool the search process. There are two general classes of these transients. The first is a transient in the test mass system, either a force applied on the test mass such as an outgassing event or a sensor glitch. Our approach to guarding against these types of transients will be to utilize both the reference and non-reference test masses aboard LPF to simultaneously search for impact events. Spurious transients of this type will either show up in only one test mass or will be fit with inconsistent parameters between the two test masses.

The second kind of transient is an external force on the spacecraft that is not arising from a micrometeoroid impact. The most likely culprit is one of the two micro-propulsion systems, which may occasionally produce a thrust transient which would by definition be common mode in the two test masses. These transients will be vetoed by comparing the recovered location and sky angles of the impacts with the thruster location and directions as well as by examining thruster diagnostic signals in the time period around an impact. Other non-impact disturbances are not expected to have the short period impulse in acceleration that is characteristic of an impact.

Another avenue of research is to compare the number and type of impacts detected with models of the dust environment in the inner solar system. This will require a more complete catalog of events as well as a model that can predict the flux of impacts near L1 as a function of particle mass and velocity. If a sufficient number of events are detected, a detailed comparison of such a model with data can be made, thus using a gravitational wave technology demonstrator to inform our knowledge of the dust environment in the inner solar system.

References

[1] Armano M, et al., 2015 *Journal of Physics: Conference Series* **610** 012005
[2] Armano M, et al., 2016 *Phys. Rev. Lett.* **116**(23) 231101
[3] Thorpe J I, Parvini C and Trigo-Rodriguez J 2016 *Astron. Astrophys.* **586** A107
[4] Cornish N and Littenberg T 2015 *Classical and Quantum Gravity* **32**
[5] Littenberg T and Cornish N 2015 *Physical Review D* **91**

NASA Astrophysics Program: Present and Future

Paul Hertz and Rita Sambruna

Astrophysics Division, Science Mission Directorate
NASA Headquarters, 300 E Street SW, Washington DC 20546 USA

paul.hertz@nasa.gov, rita.m.sambruna@nasa.gov

Abstract. NASA's astrophysics program includes support of space-based gravitational wave research in the U.S. The future includes plans for NASA's contributions as a minority partner to the ESA-led L3 Gravitational Wave Observatory.

1. Astrophysics at NASA

Astrophysics at NASA encompasses a broad range of topics, from the birth of the universe and its evolution and composition, to the processes leading to the development of planets and stars and galaxies, to the physical conditions of matter in extreme gravitational fields, and to the search for life on planets orbiting other stars. In seeking to understand these phenomena, NASA's astrophysics science embodies some of the most enduring quests of humankind. [1]

The U.S. science community sets the science priorities for astrophysics at NASA through advisory committees and the Decadal Surveys, which occur every ten years and are conducted by the National Academy of Sciences. The 2010 Decadal Survey report, *New Worlds, New Horizons in Astronomy and Astrophysics* [2], recommends as its highest priority the Wide-Field Infrared Survey Telescope (WFIRST) [3], a mission aimed at addressing the origin of dark energy, a search for exoplanets, and an infrared survey of the universe. The second and third priorities of the 2010 Decadal Survey among large missions are the Laser Interferometer Space Antenna (LISA) space-based gravitational wave observatory and the International X-ray Observatory (IXO).

Astrophysics at NASA is managed by the Astrophysics Division in the Science Mission Directorate. [4] Figure 1 shows NASA's current portfolio of astrophysics missions, including operating missions and mission in formulation and development. Among the latter is the James Webb Space Telescope [5], the top priority of the 2000 Decadal Survey [6] and a NASA Agency priority; the Webb Telescope remains on track for a launch in 2018. WFIRST entered the formulation phase in February 2016 and the WFIRST project is working toward a launch in the mid-2020s. As shown in Figure 1, NASA's suite of operating astrophysics missions includes small (Swift, NuSTAR), medium-size (Spitzer, Fermi, Kepler), and large (Hubble, Chandra, SOFIA) missions. An update on current and future NASA astrophysics missions can be found in the *Astrophysics Implementation Plan* and its updates. [7]

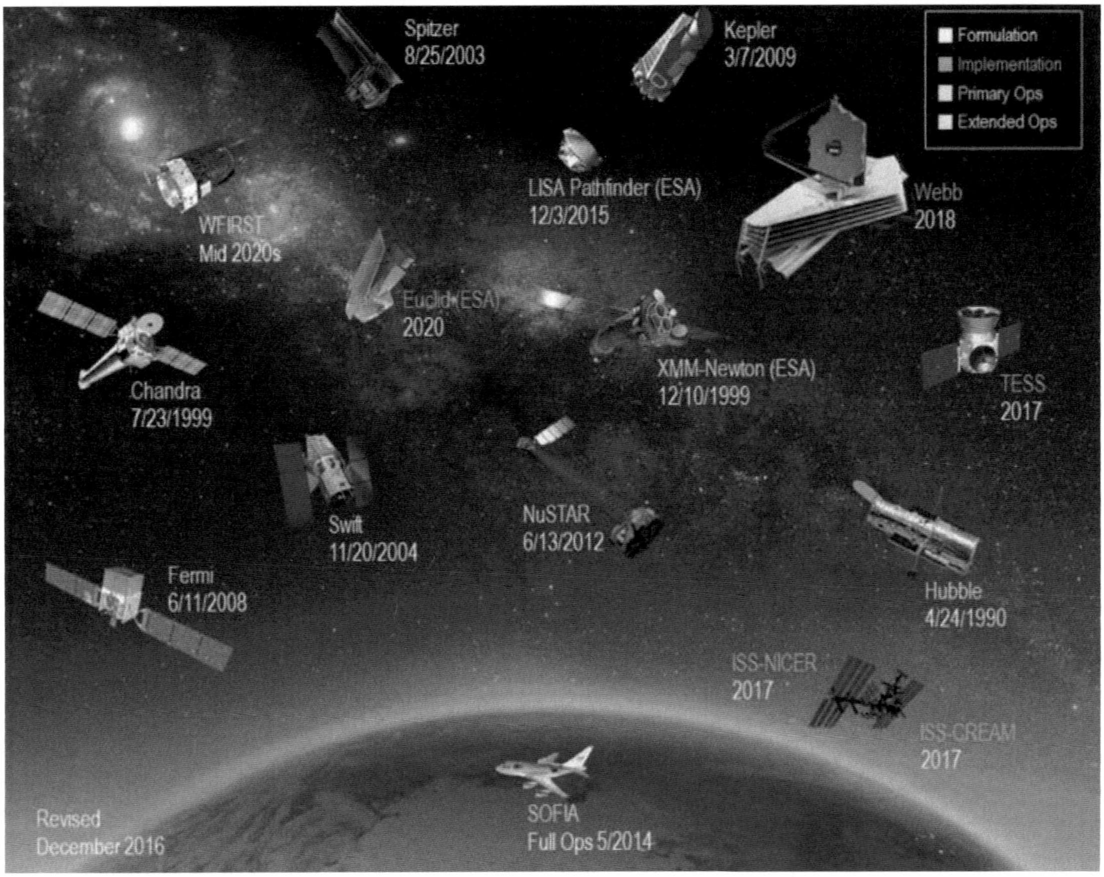

Figure 1: NASA Astrophysics Missions as of December 2016

2. Gravitational Wave Astrophysics at NASA

As part of NASA's objective to understand the scientific principles governing the universe, the new discipline of gravitational wave astrophysics is an important component of astrophysics at NASA. NASA's activities in this field go back to the early 2000s with the support of Gravity Probe B [8], a satellite-based gyroscope aimed at directly measuring space-time curvature near Earth, and even earlier to the 1970s with Gravity Probe A [9], which aimed at testing Einstein's equivalence principle.

NASA has a history of working with ESA on space-based gravitational wave missions. During the 1990s and 2000s, NASA and ESA jointly formulated the LISA mission. Another, most recent example is the successful partnership between NASA and ESA for the LISA Pathfinder mission [10]. NASA provided the Disturbance Reduction System experiment [11], which includes a set of micro-thrusters that control the spacecraft's position to within a millionth of a millimeter. The NASA-provided Disturbance Reduction System successfully met its science requirements, and NASA extended its operations through the extended LISA Pathfinder mission.

Additionally, NASA continues to invest in additional activities in support of gravitational wave (GW) research in the U.S. These include:

- **Searching for the electromagnetic counterparts of LIGO sources**. NASA's *Fermi* and Swift observatories, with their large sky coverage, are particularly suited to observe possible electromagnetic counterparts of LIGO sources in gamma-rays and X-ray/optical/UV radiation, respectively, allowing a better understanding of their nature. A recent example is the tentative

Fermi detection of a possible gamma-ray counterpart to GW150914 [12]. Less than half a second after the LIGO detection, the Gamma-ray Burst Monitor on *Fermi* picked up a brief, weak burst of high-energy light consistent with the same part of the sky. Analysis of this burst suggests just a 0.2% chance of simply being random coincidence. Gamma-rays arising from a black hole merger would be a landmark finding because black holes are expected to merge "cleanly," without producing light.

- **Developing GW technology for a future space-based GW observatory**. Through its Strategic Astrophysics Technology program, NASA is supporting development of four main technologies for GW astrophysics: 1) lasers, developed at NASA's Goddard Space Flight Center (GSFC); 2) telescope, also developed at GFSC; 3) phasemeters, developed at NASA's Jet Propulsion Laboratory (JPL); and 4) microthrusters, similar to those flown on LISA Pathfinder, also developed at JPL. In addition, NASA-supported activities for the optical bench are ongoing at the University of Florida. Detailed reports on the state-of-the-art of these technologies can be found in this Volume by J. Livas, J. Camp, W. Klipstein, and I. Thorpe.

- **Supporting theory studies and data analysis,** through grants from NASA's Astrophysics Theory Program and NASA's Astrophysics Research and Analysis program.

Looking to the future, and building on the successful LISA Pathfinder collaboration, NASA has started discussions with ESA for a possible role as a minority partner in ESA's Large 3 (L3) Gravitational Wave Observatory. To this end, NASA has drawn from the U.S. astrophysics community to establish the L3 Study Team (L3ST) [13]. NASA has directed the L3ST to provide an analysis of possible NASA hardware contributions and to collaborate with the European community in responding to the ESA L3 Call for Mission Proposals. Also part of the charge to the L3ST is developing a science case for the 2020 Decadal Survey on behalf of the U.S. astrophysics community, to reinforce the case for NASA's participation in L3.

The L3ST has recently released an Interim Report [14] with an analysis of possible hardware contributions and a plan for NASA to develop these technologies to a sufficient level of maturation for ESA mission adoption.

The Midterm Assessment of progress in implementing the 2010 Decadal Survey was conducted by the National Academy of Sciences [15]. The Midterm Assessment recommended that NASA seek an increased participation level on the L3 mission, with the goal of enabling the U.S. to be "a strong technical and scientific partner in the ESA-led L3 [gravitational wave] mission." [16] As a result, NASA is studying an increase in its contribution to L3, within the ESA-set contribution cap of 20% of the mission cost.

3. Summary and Conclusions

Within the broad swath of NASA's objectives for astrophysics, advancing the new field of gravitational wave astrophysics plays a fundamental role in directly addressing the question of how our universe works. NASA is currently supporting a large variety of GW-related activities, from technology development to modelling and data analysis. Building on the successful LISA Pathfinder partnership, NASA is currently discussing with ESA its role in ESA's L3 Gravitational Wave Observatory. NASA has established the L3 Study Team to assist NASA in identifying potential hardware contributions and to collaborate with the European community in the development of the L3 mission concept. The future of space-based gravitational wave astrophysics is bright, and NASA looks forward to being a part of it.

References

[1] *NASA 2014 Science Plan* (NASA, 2014), https://science.nasa.gov/about-us/science-strategy.
[2] *New Worlds, New Horizons in Astronomy and Astrophysics* (National Academies, 2010), https://www.nap.edu/download/12951.
[3] Wide-Field Infrared Survey Telescope (WFIRST) mission homepage at https://wfirst.gsfc.nasa.gov/.
[4] NASA Astrophysics Division homepage at https://science.nasa.gov/astrophysics/.
[5] James Webb Space Telescope mission homepage at http://jwst.nasa.gov/.
[6] *Astronomy and Astrophysics in the New Millennium* (National Academies, 2000), https://www.nap.edu/download/9839.
[7] *Astrophysics Implementation Plan* and Updates (NASA, 2012/2014/2016), http://science.nasa.gov/astrophysics/documents.
[8] Gravity Probe B mission homepage at http://einstein.stanford.edu/.
[9] Gravity Probe A experiment summary page at http://funphysics.jpl.nasa.gov/technical/grp/grav-probea.html.
[10] LISA Pathfinder mission homepage at http://sci.esa.int/lisa-pathfinder/.
[11] Disturbance Reduction System experiment homepage at http://www.jpl.nasa.gov/missions/disturbance-reduction-system-drs/.
[12] Connaughton et al 2016, *Astrophysical Journal (Letters)*, vol 826, p L6.
[13] NASA L3 Study Team homepage at https://pcos.gsfc.nasa.gov/studies/L3/.
[14] *L3 Study Team Interim Report* at https://pcos.gsfc.nasa.gov/studies/L3/.
[15] *New Worlds, New Horizons: a Midterm Assessment* (National Academies, 2016), https://www.nap.edu/download/23560.
[16] *Midterm Assessment*, ibid, Recommendation 4-4.

IOP Conf. Series: Journal of Physics: Conf. Series **840** (2017) 012009 doi:10.1088/1742-6596/840/1/012009

Program in space detection of gravitational wave in Chinese Academy of Sciences

G Jin[1,2*]

[1] Institute of Mechanics, Chinese Academy of Sciences, Beijing 100190, China;
[2] School of Engineering Science, University of Chinese Academy of Science, Beijing 100049, China.
* Corresponding author: gajin@imech.ac.cn

Abstract. On behalf of Space Gravitational Wave Detection Working Group in Chinese Academy of Sciences (CAS), ongoing development of gravitational wave detection in space in China has been presented in this talk. The preliminary mission design, primary science drivers, program for technological developments and the road-map will be described.

1. Space GW Detection mission design in China

A feasibility study of gravitational wave detection in space in China has been launched since 2008, meanwhile an investigation group in CAS for Space GW Detection was organized. In 2009 Space GW Detection was listed in "2050 Development Project" of CAS. The Space GW Detection Working Group in CAS was founded in 2012. It includes about ten institutions of CAS, Huazhong Univ. of Science and Technology, Dong Fang Hong satellite Co., and etc. Taiji [1,2] was proposed in 2015 and supported by CAS in 2016, who uses three Spacecrafts in a triangle that orbits the Sun. Tianqin [3] orbits the Earth was initiated by Zhongshan Univ. and supported by local government.

The road map of Space GW Detection in China could be summarized into three stages.

2016-2020: Technique Prototype developments and ground testing.

2021-2025: Technological developments and a pathfinder mission.

Two options are under consideration in this stage. One is gravity satellite mission which is similar to Grace-follow-on but with a laser interferometer as the primary ranging instrument. The other is to send two satellites into deep space (likely to be L1or L2). A 105 km long interferometer will be established between these two satellites. Apart from inertial sensor, single arm laser interferometry (telescope, pointing, metrology, ...etc) will be tested.

2026-2035: Implementation and launching of Chinese mission for GW detection in Space.

There is also a double track for mission development. One is to join eLISA or LISA. Chinese part likely to contribute over 20% with technical assemblies, such as Micro-Thruster, Telescope, Laser, Launcher, and others ... Another option is to develop an independent Chinese Mission with international collaboration.

For the latter one, we prefer the mission design of Taiji. From the sensitivity curve of the simulation, we may see that the most sensitive band of the design is around 0.01 Hz. Together with a slight improvement of the position noise, it enables an enhanced detection to light seed black holes [2]. This sensitivity curve will be the blueprint for our prototype development in the next five years.

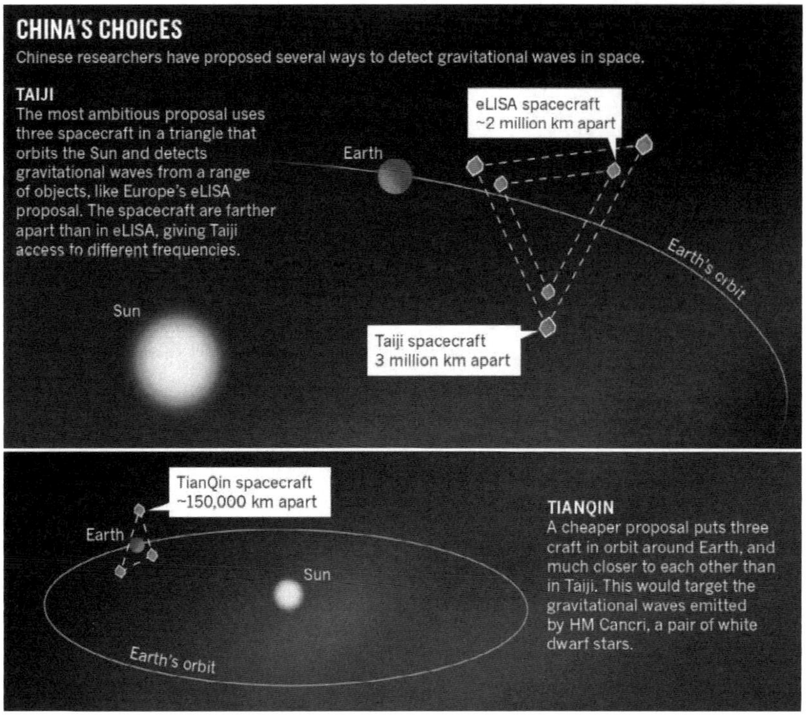

Fig.1 The programme design of Taiji and Tianqin missions

As we may see from the figures [2,4], the Taiji mission will be able to probe light seeded black holes at the earlier epoch and this will be important in our understanding of the structural formation of our Universe. Event rate estimates of MBH mergers at earlier cosmological epoch and Cosmological MBH merger simulation based on Monte Carlo realization of EPS formalism, Equal mass light seed black hole (PopIII remnants) seeding, Semi-analytical dynamics —prolonged and chaotic accretion models, coalescence spin and recoil determination: numerical relativity fitting formula, etc. Taiji mission is going to be an integral part of high redshift astronomy in China, complementary to the Infrared and radio astronomy.

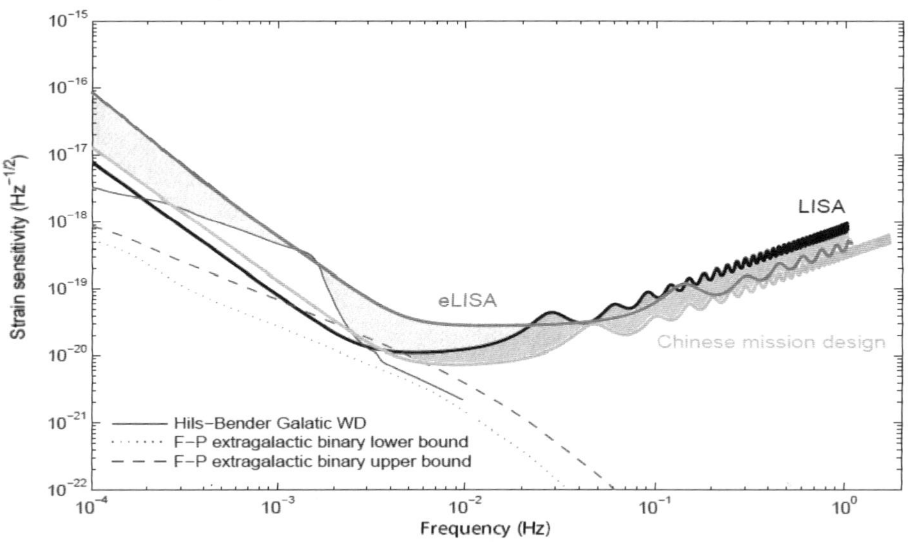

Fig.2 The sensitivity design of the Chinese space gravitational wave detection project

Table 1 Baseline design parameters

Armlength (m)	Telescope diameter (m)	Laser power (W)	1-way position noise $(\frac{\mathrm{pm}}{\sqrt{\mathrm{Hz}}})$	Acceleration noise $(\frac{\mathrm{m\ s^{-2}}}{\sqrt{\mathrm{Hz}}})$
3×10^9 (Chinese mission option)	0.46	2	8	3×10^{-15}(> 0.1mHz)
1×10^9 (eLISA)	0.2	2	11	5×10^{-15}(> 0.1mHz)
5×10^9 (LISA)	0.4	2	18	3×10^{-15}(> 0.1mHz)

2. Key technology development

Some key techniques have been developed for this mission in recent years. Experiments on frequency locking of Nd:YAG lasers have been performed in Wuhan, Institute of Physics and Mathematics, CAS. The primary solution to the laser source is to develop an on-satellite frequency and intensity stabilized Nd:YAG laser that covers a 5-year space mission. The laser consists pump laser diodes and a monolithic non-planer ring oscillator (NPRO), being capable of delivering more than 2 W optical power at 1064 nm without additional gain stage. The frequency noise of the pre-stabilized laser will be on the order of 10 Hz/√Hz at Fourier frequency of 10 mHz. We are also looking into the possibilities of a fiber laser system that is promising in terms of reliability and simplicity. Currently, the frequency pre-stabilization system has been developed based on a 10-cm ultra-stable space-borne optical reference cavity. In addition, vibration tests have been performed for several vital components and a radiation test is scheduled in order to evaluate the susceptibility of the pump laser and the NPRO to the proton beam. In the future we will further investigate the Nd:YAG approach, focusing on its output power, lifetime, and space qualification. Laser interferometers with 10 pm/√Hz have been developed in Institute of Mechanics, CAS [5] and Huazhong Univ. of Sci. and Tech. [6] with the aid of AEI on how to simulate and build the experimental interferometer to achieve the required sensitivity of the mission. Accelerometer development is mainly in Huazhong Univ. of Sci. and Tech. and Lanzhou Institute of Physics, CAST, and the flight mode reaches 10 ng/√Hz. The telescope is another key element of Interferometric Ranging System (IRS). Each optical assembly in each Spacecraft(SC) includes a telescope pointing towards a far SC. The aim of the telescope is: (1) it gathers the light coming from the far SC(~100 pW) and (2) it expands and collimates the small outgoing beam(~1 W) and send it to the far SC. Due to the very demanding noise requirements, crucial stability must be taken in the design and validation of the telescope not to degrade the IRS performance. For instance, any fluctuation in the distance and angle between the primary and secondary mirrors of the telescope will translate directly into phase noise in the IRS. This fact implies that the path-length noise in the telescope must be less than 1 pm/√Hz in the LISA band and M1-M2 divergence angle must be less than 1 nrad/√Hz. All SiC telescope (stop diameter: 200mm) was designed and analyzed and the simulation result of stability of all-SiC telescope can reach less than 10 pm/√Hz according to the environment of the mission orbit. Space gravitational wave detection also requires low noise (0.1 μN/√Hz), long life (about 5 years) micro propulsion system. To meet this target, we need to study the mechanism, the principle prototype and the integration scheme of micro propulsion system. The working mechanisms of field emission electric thruster and radio frequency ion micro thruster were studied by National Microgravity Laboratory, Institute of mechanics, CAS, and the principle prototypes were developed. Further optimization is under research. In addition, other parts of the micro propulsion system, such as the power processing unit, the propellant supply unit and neutralizer etc. will be entrusted to the companies who have relevant qualifications. We will work with Lanzhou Institute of Space Physics CAST, and Beijing Institute of Control Engineering on the micro propulsion system integration, engineering transformation and system testing. In order to reduce the disturbance applied on the inertial reference caused by non- conservative forces, the statically and stable satellite platform need to be studied, including technology of statically and stable satellite platform system design, inertial measurement sensors, drag free control algorithm, micro thruster , highly stable structure and precision temperature control technology. Major research institutions in China including DFH Satellite Co Ltd, Shanghai Engineering Center of Small Satellite, CAS, China Academy of space technology (CAST), CAS (Institute of mechanics etc.), Sun Yat-sen University (SYSU), Huazhong University of Science and Technology (HUST) etc. Launch Vehicle Technology is established in China as an escaped orbit launch service for the mission. ChangZheng-5(CZ-5) rocket will launch the main part of Chinese space station load of 25 tons in about 2020, and it is able to launch 3 satellites orbiting the Sun with a Load of 5.7 tons.

3. Some international activities

Chinese mission of Space gravitational wave detection in space will open for collaboration with international society, especially with ESA scientists. Chinese mission can develop and work with eLISA for collaboration and data comparison. With international collaboration we can employ more techniques in same models and types for high efficiency and low cost. Here we propose several international scientific activities to push on the collaboration.

- International Symposium on Gravitational Waves, (ISGW2017) Beijing, May 25-29, 2017
- Taiji Union for Gravitational Wave Detection in Space is set up and all researchers and scientists all over the world are welcome.
- Joint Albert Einstein Institute for Radio Astronomy and Gravitational Physics between Max Planck Society and Chinese Academy of Sciences to be discussed in MPG-CAS Meeting to be held in Bonn, Germany in March, 2017.
- Chinese pathfinder would be launched in 10 years to test key technology for GWD in Space, International collaboration is welcome.
- Joint research proposal to be submitted to the DFG and National Science Foundation, China after the Sino-German meeting in 2017.

References

[1] Gong X F *et al* 2011 *Class. Quantum Grav.* **28** 094012
[2] Gong X F *et al* 2015 Journal of Physics: Conference Series **610** 012011
[3] Luo J *et al* 2016 *Class. Quantum Grav.* **33** 035010
[4] Gong X F 2015 *et al Chinese Astronomy and Astrophysics* **39(4)** 411-446
[5] Li Y Q et al 2015 *Appl. Phys. B: Lasers and Optics* **118** 309-317
[6] Yan H et al 2015 *Rev. Sci. Instrum.* **86** 123102

The status of DECIGO

Shuichi Sato[1], Seiji Kawamura[18], Masaki Ando[3], Takashi Nakamura[4], Kimio Tsubono[3], Akito Araya[5], Ikkoh Funaki[6], Kunihito Ioka[7], Nobuyuki Kanda[8], Shigenori Moriwaki[9], Mitsuru Musha[10], Kazuhiro Nakazawa[3], Kenji Numata[11], Shin-ichiro Sakai[6], Naoki Seto[2], Takeshi Takashima[6], Takahiro Tanaka[12], Kazuhiro Agatsuma[3], Koh-suke Aoyanagi[27], Koji Arai[2], Hideki Asada[13], Yoichi Aso[14], Takeshi Chiba[15], Toshikazu Ebisuzaki[44], Yumiko Ejiri[16], Motohiro Enoki[45], Yoshiharu Eriguchi[35], Masa-Katsu Fujimoto[2], Ryuichi Fujita[46], Mitsuhiro Fukushima[2], Toshifumi Futamase[43], Katsuhiko Ganzu[4], Tomohiro Harada[47], Tatsuaki Hashimoto[6], Kazuhiro Hayama[48], Wataru Hikida[38], Yoshiaki Himemoto[49], Hisashi Hirabayashi[17], Takashi Hiramatsu[18], Feng-Lei Hong[19], Hideyuki Horisawa[20], Mizuhiko Hosokawa[21], Kiyotomo Ichiki[3], Takeshi Ikegami[19], Kaiki T. Inoue[22], Koji Ishidoshiro[3], Hideki Ishihara[8], Takehiko Ishikawa[6], Hideharu Ishizaki[2], Hiroyuki Ito[21], Yousuke Itoh[50], Nobuki Kawashima[22], Fumiko Kawazoe[23], Naoko Kishimoto[6], Kenta Kiuchi[27], Shiho Kobayashi[24], Kazunori Kohri[25], Hiroyuki Koizumi[6], Yasufumi Kojima[51], Keiko Kokeyama[16], Wataru Kokuyama[3], Kei Kotake[2], Yoshihide Kozai[26], Hideaki Kudoh[3], Hiroo Kunimori[21], Hitoshi Kuninaka[6], Kazuaki Kuroda[18], Kei-ichi Maeda[27], Hideo Matsuhara[6], Yasushi Mino[14], Osamu Miyakawa[14], Shinji Miyoki[18], Mutsuko Y. Morimoto[6], Tomoko Morioka [3], Toshiyuki Morisawa[4], Shinji Mukohyama[39], Shigeo Nagano[21], Isao Naito[57], Kouji Nakamura[2], Hiroyuki Nakano[52], Kenichi Nakao[8], Shinichi Nakasuka[31], Yoshinori Nakayama[29], Erina Nishida[16], Kazutaka Nishiyama[6], Atsushi Nishizawa[30], Yoshito Niwa[30], Taiga Noumi[31], Yoshiyuki Obuchi[2], Masatake Ohashi[18], Naoko Ohishi[2], Masashi Ohkawa[32], Norio Okada[2], Kouji Onozato[3], Kenichi Oohara[32], Norichika Sago[33], Motoyuki Saijo[34], Masaaki Sakagami[30], Shihori Sakata[2], Misao Sasaki[12], Takashi Sato[32], Masaru Shibata[35], Hisaaki Shinkai[53], Kentaro Somiya[14], Hajime Sotani[36], Naoshi Sugiyama[37], Yudai Suwa[3], Rieko Suzuki[16], Hideyuki Tagoshi[38], Fuminobu Takahashi[39], Kakeru Takahashi[3], Keitaro Takahashi[12], Ryutaro Takahashi[2], Ryuichi Takahashi[37], Tadayuki Takahashi[6], Hirotaka Takahashi[54], Takamori Akiteru[5], Tadashi Takano[40], Keisuke Taniguchi[50], Atsushi Taruya[3], Hiroyuki Tashiro[4], Yasuo Torii[2], Morio Toyoshima[21], Shinji Tsujikawa[41], Yoshiki Tsunesada[55], Akitoshi Ueda[2], Ken-ichi Ueda[10], Masayoshi Utashima[56], Yaka Wakabayashi[16], Hiroshi Yamakawa[42], Kazuhiro Yamamoto[23], Toshitaka Yamazaki[2], Jun'ichi Yokoyama[3], Chul-Moon Yoo[8], Shijun Yoshida[43], Taizoh Yoshino[28],

[1]Faculty of Science and Engineering, Hosei University, kajinocho, Tokyo 184-8584, Japan,
[2]National Astronomical Observatory Japan, Osawa 2-21-1, Tokyo 181-8588, Japan,
[3]Department of Physics, The University of Tokyo, Tokyo 113-0033, Japan,
[4]Department of Physics, Kyoto University, Kyoto 606-8502, Japan,
[5]Earthquake Research Institute, The University of Tokyo, Tokyo 113-0032, Japan,
[6]Institute of Space and Astronautical Science, Japan Aerospace Exploration Agency (JAXA), Kanagawa 229-8510, Japan,
[7]Institute of Particle and Nuclear Studies, High Energy Accelerator Research Organization (KEK), Ibaraki 305-0801, Japan,
[8]Department of Physics, Osaka City University, Osaka 558-8585, Japan,
[9]Department of Advanced Materials Science, The University of Tokyo, Chiba 277-8561, Japan,
[10]Institute for Laser Science, The University of Electro-Communications, Tokyo 182-8585, Japan,
[11]NASA Goddard Space Flight Center, Code 663, 8800 Greenbelt Rd., Greenbelt, MD20771, USA,
[12]Yukawa Institute for Theoretical Physics, Kyoto University, Kyoto 606-8502, Japan,
[13]Department of Earth and Environmental Sciences, Hirosaki University, Aomori 036-8560, Japan,
[14]California Institute of Technology, 1200 E. California Blvd. MC 18-34, Pasadena, CA 91125, USA,
[15]Nihon University, Setagaya, Tokyo 156-8550, Japan,
[16]Ochanomizu University, 2-1-1, Tokyo 112-0012, Japan,
[17]Space Educations Center, Japan Aerospace Exploration Agency (JAXA), Kanagawa 229-8510, Japan,
[18]Institute for Cosmic Ray Research, The University of Tokyo, Chiba 277-8582, Japan,
[19]National Institute of Advanced Industrial Science and Technology (AIST), Ibaraki 305-8563, Japan,
[20]Department of Aeronautics and Astronautics, Tokai University, Kanagawa 259-1292, Japan,
[21]National Institute of Information and Communications Technology (NICT), Tokyo 184-8795, Japan,
[22]School of Science and Engineering, Kinki University, Osaka 577-8502, Japan,
[23]Max-Planck-Institute for Gravitational Physics (Albert-Einstein-Institute), Callinstr. 38 D-30167 Hannover, Germany,
[24]Astrophysics Research Institute, Liverpool John Moores University, Twelve Quays House, Egerton Wharf, Birkenhead L41 1LD, UK,
[26]Gunma Astronomical Observatory, Agatsuma-gun, Gunma 377-0702, Japan,
[27]Department of Physics, Waseda University, Tokyo, 169-8555, Japan,
[28]Nakamura-minami Nerima, Tokyo 176-0025, Japan,
[29]Department of Aerospace Engineering, National Defense Academy, Yokosuka 239-8686, Japan,
[30]Faculty of Intergrated Human Studies, Kyoto University, Kyoto 606-8501, Japan,
[31]Department of Aeronautics and Astroautics, The University of Tokyo, Tokyo 113-8656, Japan,
[32]Niigata University, Niigata 950-2181, Japan,
[33]Highfield, Southampton SO17 1BJ, United Kingdom,
[34]Department of Physics, Rikkyo University, Tokyo 171-8501, Japan,
[35]Department of Earth Science and Astronomy, the University of Tokyo, Tokyo 153-8902, Japan,
[36]Theoretical Astrophysics,Institute for Astronomy and Astrophysics, Eberhard Karls University of Tuebingen, Auf der Morgenstelle 10, 72076 Tuebingen, Germany,
[37]Nagoya University, Graduate School of Science, Aichi 464-8601, Japan,
[38]Department of Earth and Space Science, Osaka University, Osaka 560-0043, Japan,
[39]Institute for Physics and Mathematics of the Universe (IPMU), The University of Tokyo, Chiba 277-8568, Japan,
[40]Department of Electronics and Computer Science, Nihon University, Funabashi 274-8501 Japan,
[41]Department of Physics, Tokyo University of Science, Tokyo, 162-8601, Japan,
[42]Research Institute for Sustainable Humanosphere, Kyoto University, Kyoto 611-0011, Japan
[43]Astronomical Institute, Tohoku University, Sendai 980-8578, Japan,

[44]RIKEN, 2-1 Hirosawa, Wako 351-0198, Japan,
[45]Faculty of Business Administration, Tokyo Keizai University, Tokyo 185-8502, Japan,
[46]Theoretical Physics,Raman Research Institute, Sir C.V.Raman Avenue, Sadashivanagar P.O., Bangalore 560 080, India,
[47]Department of Physics, Rikkyo University, Tokyo 171-8501, Japan,
[48]University of Texas, 80 Fort Brown, Brownsville 78520, Texas, U.S.A.,
[49]Center for Educational Assistance, Shibaura Institute of Technology, Saitama 337-8570, Japan,
[50]Department of Physics, University of Wisconsin-Milwaukee, 1900 East Kenwood Blvd. Milwaukee, WI 53211, USA,
[51]Graduate School of Science, Hiroshima University, Hiroshima 739-8526, Japan,
[52]Rochester Institute of Technology, 78 Lomb Memorial Drive, Rochester, NY 14623, USA,
[53]Dept of Information Systems, Osaka Institute of Technology, Hirakata 573-0196, Japan,
[54]Department of Management and Information Systems Science, Nagaoka University of Technology, Niigata 940-2188, Japan,
[55]Graduate School of Science and Engineering / Physics, Tokyo Institute of Technology, Tokyo 152-8550, Japan,
[56]Tsukuba Space Center, Japan Aerospace Exploration Agency (JAXA), Ibaraki 305-8505, Japan,
[57]Numakage, Saitama-shi, Saitama 336-0027 Japan,

E-mail: sato.shuichi@hosei.ac.jp

Abstract. DECIGO (DECi-hertz Interferometer Gravitational wave Observatory) is the planned Japanese space gravitational wave antenna, aiming to detect gravitational waves from astrophysically and cosmologically significant sources mainly between 0.1 Hz and 10 Hz and thus to open a new window for gravitational wave astronomy and for the universe. DECIGO will consists of three drag-free spacecraft arranged in an equilateral triangle with 1000 km arm lengths whose relative displacements are measured by a differential Fabry-Perot interferometer, and four units of triangular Fabry-Perot interferometers are arranged on heliocentric orbit around the sun. DECIGO is vary ambitious mission, we plan to launch DECIGO in era of 2030s after precursor satellite mission, B-DECIGO. B-DECIGO is essentially smaller version of DECIGO: B-DECIGO consists of three spacecraft arranged in an triangle with 100 km arm lengths orbiting 2000 km above the surface of the earth. It is hoped that the launch date will be late 2020s for the present..

1. Introduction

The first direct detection of gravitational wave (GW) has been done with aLIGO [1]. The first result of LISA pathfinder showed demonstration of surprisingly low-noise-level free fall exceeding expectations before launch [2]. The gravitational wave physics and astronomy took the new step to next stage. As astronomy using electromagnetic wave, gravitational wave is also expected in various frequency band to have wide frequency spectrum. Terrestrial detecters like aLIGO, aVIRGO, GEO, KAGRA and ET are most sensitive at audio frequency band around 10 to 1kHz, on the other hand, space born detector LISA is at low frequency region around mHz. At further low frequency band, PPTA (Pulsar Timing Array) and polarized CMB (Cosmic Microwave Background) are also interesting option to access to the unique information of physics and universe. Japanese DECIGO, planned space gravitational wave antenna, might be able to provide another new way to observe universe, because only DECIGO will be sensitive to deci-Hz gravitational wave signals.

2. DECIGO

DECIGO (DECi-hertz Interferometer Gravitational wave Observatory) is the planned Japanese space gravitational wave antenna [3, 4, 5], which was originally proposed by Seto, Kawamura and Nakamura [6] to measure the acceleration of the universe through GWs from binary NS–NS at $z \sim 1$. DECIGO is targeting to observe gravitational waves from astrophysically and cosmologically significant sources mainly between 0.1 and 10 Hz, thus, to open a new window of observation for gravitational wave astronomy, and also for the universe. The scope of DECIGO is to bridge (Fig.2) the frequency gap between LISA [7] band and terrestrial detectors band such as advanced LIGO, advanced VIRGO, GEO and KAGRA.The major advantage of DECIGO specializing in this frequency band is that the expected confusion limiting noise level caused by irresolvable gravitational wave signals from many compact binaries, such as white dwarf binaries in our Galaxy, is quite low above 0.1 Hz [8], therefore there is a potentially extremely deep window in this band. Thus, as DECIGO will have sensitivity in the frequency range between LISA and terrestrial detectors band, DECIGO can serve as a follow-up for LISA by observing inspiraling sources that have moved above the LISA band, or as a predictor for terrestrial detectors by observing inspiraling sources that have not yet moved into the terrestrial detectors band.

2.1. Pre-conceptual design

The pre-conceptual design of DECIGO consists of three drag-free spacecraft which keep triangular configuration with formation flying technique. The separation of each spacecraft is designed to be 1,000 km, whose relative displacements are measured by a differential Fabry-Perot (FP) interferometer (Fig.1). The laser source is supposed to be frequency-doubled Yb:YAG laser with $\lambda = 515$ nm yielding output power of 10 W. The mass of the mirror is 100 kg with 1 m diameter, with low-loss high-reflectivity coatings, which enables the finesse of FP cavity to reach 10 with green light. Three sets of such interferometers sharing the mirrors as arm cavities comprise one cluster of DECIGO. As shown in Fig.1, four clusters of DECIGO, located separately in the heliocentric orbit with two of them nearly at the same position, form the constellation DECIGO.

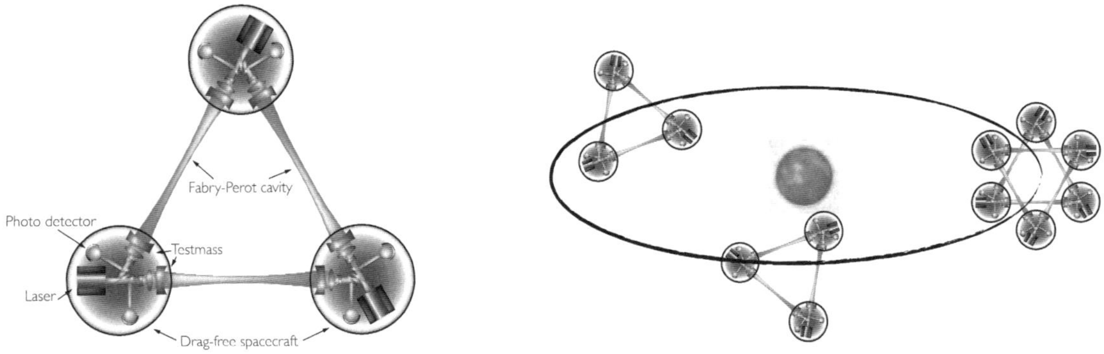

Figure 1. Image of DECIGO and constellation around the heliocentric orbit.

2.2. Sensitivity goal and science

The target sensitivity of DECIGO, as shown in Fig.2, is supposed to be limited by quantum noise in all frequency band: by the radiation pressure noise below 0.15 Hz, and by the shot noise above 0.15 Hz. In order to reach this sensitivity, all the practical noise should be suppressed well below this level. This imposes more stringent requirements than LISA for some subsystems of DECIGO, especially in the acceleration noise and frequency noise, therefore rigorous investigations are supposed to be indispensable for attainment of design sensitivity. Nonetheless, full success of DECIGO is expected to extract fruitful sciences.

As shown in Fig.2, the sensitivity goal of DECIGO is better than 10^{-23} in terms of strain between 0.1 and 10 Hz. To achieve this sensitivities, all the practical noises have to be suppressed below the stringent requirement, especially on the acceleration noise of the mirror and frequency noise of the light.

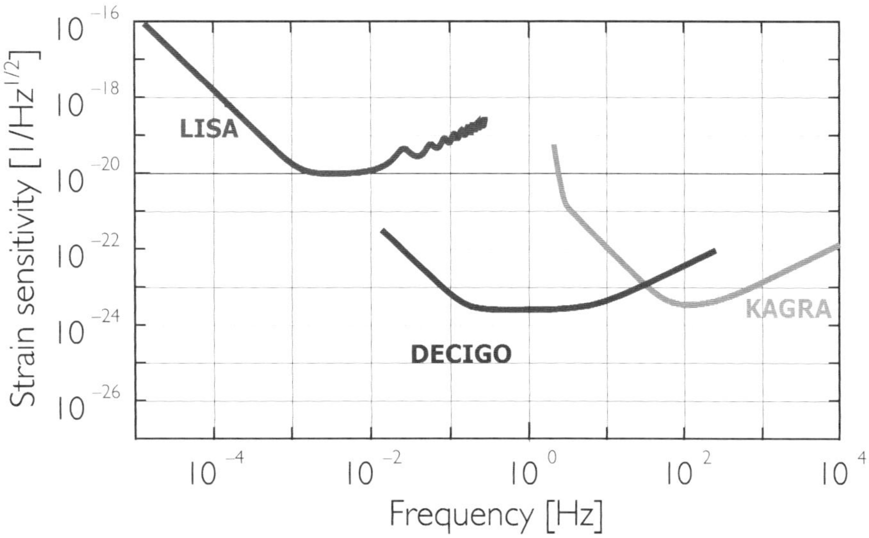

Figure 2. Expected sensitivity of DECIGO in terms of strain in comparison with LISA and terrestrial detectors, like KAGRA

2.3. Roadmap

DECIGO is expected to be launched in the era of 2030s, before that, we plan to launch a precursor satellites, B-DECIGO. Major objective of B-DECIGO is to detect astrophysical GW signals to extract scientific results, in addition to demonstration of key technologies required for DECIGO just as LISA pathfinder [2] did for LISA. The technical objectives of B-DECIGO are demonstration of accurate formation flying, precision laser metrology with long baseline FP cavity and drag-free control for multiple spacecraft, based on several fundamental precision measurement technologies like drag-free control of the spacecraft, stabilized laser system in space, precision laser metrology in space and test mass lock mechanism. B-DECIGO is basically a small version of DECIGO, but will have 100km-scale FP cavity, therefore, it is supposed to have reasonable sensitivity to detect gravitational waves with minimum specifications. We hope that it will be launched around 2020s.

3. B-DECIGO [1]

B-DECIGO is re-defined space GW antenna mission as first precursor satellite for DECIGO, succeeding former Pre-DECIGO [11]. The objectives of B-DECIGO are scientifically to detect

[1] B-DECIGO was formerly called Pre-DECIGO, whicih was second precursor satellite for DECIGO after first precursor satellite, DPF (DECIGO pathfinder) [9, 10]. The science objectives and design of Pre-DECIGO had not been defined so clearly, while DECIGO has a definite design and clear targets to access to the information of the inflation. Recently, expected astrophysical science targets and pre-conceptual design were defined for Pre-DECIGO, then which was renamed as B-DECIGO as first precursor satellite for DECIGO on this occasion.

gravitational waves from promising astrophysical sources with modest optical parameters, and also technologically to demonstrate the formation flight using three spacecraft, which is one of key technologies for DECIGO. B-DECIGO is designed to have a sensitivity that is conservative compared with DECIGO by about factor of 10 in all frequency band. Accordingly, the optical parameters and the noise requirements of B-DECIGO are less stringent than DECIGO, whereas the required acceleration noise level is still challenging compared with LISA pathfinder and LISA. B-DECIGO consists of three drag-free spacecraft containing freely-falling mirrors, whose relative displacement is measured by a differential FP Michelson interferometer.

Figure 3. Image of B-DECIGO which is smaller size DECIGO consisting of three spacecraft arranged in an equilateral triangle with 100 km arm lengths orbiting 2000 km above the surface of the earth.

3.1. Pre-conceptual design

Each spacecraft holds a couple of test-mass mirrors of 30 kg in weight and 30 cm in diameter, freely floating on the spacetime. One test-mass mirror in one spacecraft and the another test-mass mirror in the other spacecraft are connected by laser beam, forming 100 km Fabry–Pérot cavity, with finesse of 100 resulting in a cavity cut-off frequency around 20 Hz. Therefore, three spacecraft are connected with three 100 km Fabry–Pérot cavities to maintain 100 km triangular formation flight. Frequency-doubled, Iodine-stabilized Yb:fiber DFB laser, with wavelength of 515 nm will be used as a light source, The laser light from Yb:fiber DFB laser with wavelength of 1030 nm is amplified with YDFA (Yb-Doped Fiber Amplifier), then frequency-doubled with nonlinear crystal to have enough power to illuminate each Fabry–Pérot cavities with 1 W. The frequency-doubled green light, then, frequency-stabilized in reference to the saturated absorption of iodine molecules to have low enough frequency noise contribution in an observational band of B-DECIGO.

In order to make test-masses freely floating in inertial spacetime as a probe of GW, and also to avoid an external force fluctuation on the test-mass caused by the unwanted coupling from spacecraft motion, the spacecraft is drag-free controlled with a couple of test-mass mirrors inside spacecraft as inertial reference. The position and attitude of the spacecraft with respect to theses test-masses are drag-free controlled by feeding error signals back to the spacecraft. The formation flight of the three spacecraft to keep triangular shape is realized by continuous feedback control. The laser interferometers measure the deviation of the cavity-length, which are fed back to the position of test-mass mirrors to maintain the length of the cavities. Since the spacecraft follows the test-mass positions inside it using drag-free control scheme, as a result, exact 100-km length triangular formation is realized. One of candidate orbit for B-DECIGO is LISA-like cart-wheel orbit around the earth. (Fig.3) If the altitude of the spacecraft formation and inclination angle of orbital plane are selected properly, the reference orbit, the orbit of the center of the mass of the three spacecraft, could be a sun-synchronized dawn-dusk circular orbit. In addition, it is possible to design the dawn-dusk orbit so that there will be no eclipse in these spacecraft, by selecting the altitude between 2,000-3,000km ?, which is beneficial to avoid thermal shock and drift in the spacecraft, and also to keep continuous power supply from the sun. The orbital period of the formation-flight interferometer unit around the earth is about 124 min. for the altitude of 2,000km. Assuming this orbit, orbital motion of formation and the earth's annual orbital motion around the sun make the antenna pattern of B-DECIGO to observe GWs change in time scale of 100 min. Owing to this variation of antenna pattern, parameter estimation accuracy for the GW sources, such as sky localization, is expected to be improved.

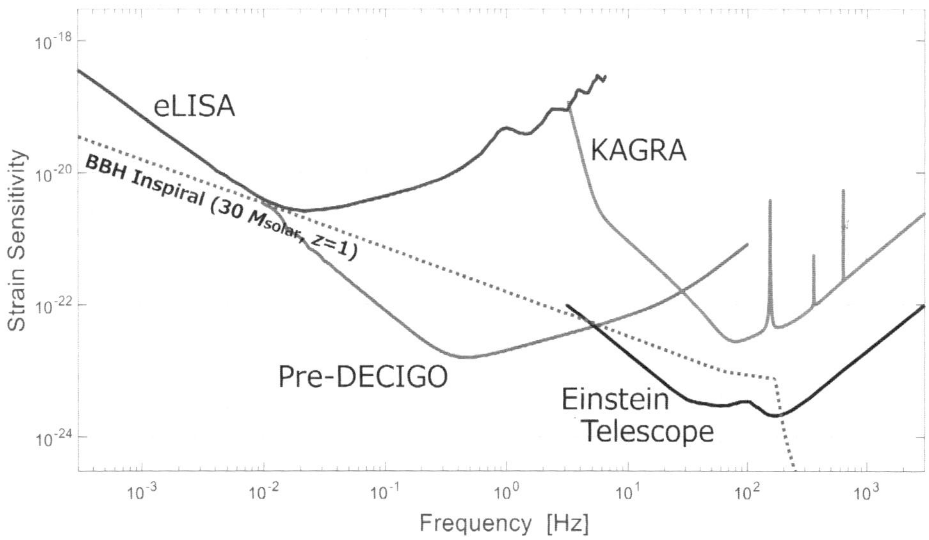

Figure 4. Strain sensitivity of B-DECIGO. Sensitivity curves for 2nd-generation terrestrial GW antenna (KAGRA [?]), 3rd-generation antenna (ET [?]), and space antenna (eLISA [?]) are shown together for references. The dashed curve shows the signal amplitude from BBH merger with masses of $30M_{\odot}$ at a distance of $z = 1$.

3.2. Sensitivity goal and science

Using above essential parameters, the target sensitivity of B-DECIGO is set to be $2 \times 10^{-23}\,\mathrm{Hz}^{-1/2}$ in strain in the current design (Fig. 4). The noise curve is basically limited by fundamental noise, optical quantum noises of the interferometer; laser shot noise and radiation

IOP Conf. Series: Journal of Physics: Conf. Series **840** (2017) 012010 doi:10.1088/1742-6596/840/1/012010

pressure noise in high and low frequency bands, respectively. The external force noises level on the test-mass mirrors are set not to exceed these optical quantum noises level, which place critical requirements; the requirement is $1 \times 10^{-16}\,\mathrm{N/Hz^{1/2}}$. With this sensitivity, mergers of BBHs at $z = 10$ will be within the observable range of B-DECIGO, assuming optimal direction and polarization of the source, and detection SNR of 8.

References

[1] B. P. Abbott et al., Observation of Gravitational Waves from a Binary Black Hole Merger, Phys. Rev. Lett. 116, 061102 (2016)

[2] Armano M. et al, Sub-Femto-g Free Fall for Space-Based Gravitational Wave Observatories: LISA Pathfinder Results, Phys. Rev. Lett. 116 231101 (2016)

[3] Kawamura S et al 2006 The Japanese Space Gravitational Wave Antenna - DECIGO *Class. Quantum Grav.* **23** S125

[4] Kawamura S et al 2008 The Japanese Space Gravitational Wave Antenna - DECIGO *Journ. of Phys.: Conf. Ser.* **120** 032004

[5] Kawamura S et al 2008 The Japanese Space Gravitational Wave Antenna; DECIGO *Journ. of Phys.: Conf. Ser.* **122** 012006

[6] Seto N, Kawamura S and Nakamura T 2001 Possibility of direct measurement of the acceleration of the universe using 0.1 Hz band laser interferometer gravitational wave antenna in space *Phys. Rev. Lett.* **87** 221103

[7] LISA: System and Technology Study Report, ESA document ESA-SCI (2000)

[8] Farmer A J and Phinney E S 2003 The gravitational wave background from cosmological compact binaries *Mon. Not. R. Astron. Soc.* **346** 1197

[9] Ando M et al 2008 DECIGO pathfinder *Journ. of Phys.: Conf. Ser.* **120** 032005

[10] Ando M et al 2008 DECIGO pathfinder *in this volume*

[11] Nakamura T. et al, Pre-DECIGO can get the smoking gun to decide the astrophysical or cosmological origin of GW150914-like binary black holes, Prog. Theor. Exp. Phys. 2016, 093E01

IOP Conf. Series: Journal of Physics: Conf. Series **840** (2017) 012011 doi:10.1088/1742-6596/840/1/012011

Flight phasemeter on the Laser Ranging Interferometer on the GRACE Follow-On mission

B Bachman, G de Vine, J Dickson, S Dubovitsky, J Liu, W Klipstein, K McKenzie, R Spero, A Sutton, B Ware, and C Woodruff

Jet Propulsion Laboratory (JPL), California Institute of Technology, 4800 Oak Grove Dr., Pasadena, CA 91109 USA

E-mail: Robert.Spero@jpl.nasa.gov

Abstract. As the first inter-spacecraft laser interferometer, the Laser Ranging Interferometer (LRI) on the GRACE Follow-On Mission will demonstrate interferometry technology relevant to the LISA mission. This paper focuses on the completed LRI Laser Ranging Processor (LRP), which includes heterodyne signal phase tracking at μcycle/$\sqrt{\text{Hz}}$ precision, differential wavefront sensing, offset frequency phase locking and Pound-Drever-Hall laser stabilization. The LRI design has characteristics that are similar to those for LISA: 1064 nm NPRO laser source, science bandwidth in the mHz range, MHz-range intermediate frequency and Doppler shift, detected optical power of tens of picoWatts. Laser frequency stabilization has been demonstrated at a level below 30 Hz/$\sqrt{\text{Hz}}$, better than the LISA requirement of 300 Hz/$\sqrt{\text{Hz}}$. The LRP has completed all performance testing and environmental qualification and has been delivered to the GRACE Follow-On spacecraft. The LRI is poised to test the LISA techniques of tone-assisted time delay interferometry and arm-locking. GRACE Follow-On launches in 2017.

1. Introduction, LISA and GRACE Follow-On synergy

The Laser Ranging Interferometer (LRI) for GRACE Follow-On will measure changes in the separation of earth-orbiting satellites with nanometer-scale precision. It is the product of a joint US-German collaboration, with management shared between the Jet Propulsion Laboratory (JPL) and the Albert Einstein Institute (AEI). The US[1] subsystems are the phasemeter, laser and optical cavity. Subsystems from Germany[2] are the optical bench, photodetectors, triple mirror assembly, and baffles.

The science from earth-observing missions GRACE[1] and its successor GRACE Follow-On have no connection to science from the gravitational-wave detector LISA[2]. Nonetheless, the LRI[3] overlaps in both function and design parameters with LISA's laser metrology. Concepts essential to LISA interferometry that were developed at JPL include Time Delay Interferometry (TDI)[4], post-processing interpolation for TDI[5], the arm-locking technique for laser stabilization[6][7], and velocity-correcting TDI[8]. LISA Technology developments from JPL include the invention of the first digital phasemeter meeting the LISA performance requirements [9][10] and a demonstration of clock noise suppression[11]. These efforts culminated in the first experimental implementation of LISA-like TDI[12]. Also demonstrated on the TDI

[1] NASA/JPL

[2] GFZ Research Center for Geosciences/AEI; Spacetech Gmbh, Immenstaad; DLR

IOP Conf. Series: Journal of Physics: Conf. Series **840** (2017) 012011 doi:10.1088/1742-6596/840/1/012011

testbed were absolute metrology with accuracy of 0.2 m rms and optical communications on the laser link at a rate of 20 kbps[13].

Much of our experience with LISA technology was transferred to the LRI. LISA and the LRI have similar phase tracking/signal readout performance, received optical power, lasers, photodetectors, and laser frequency control by either offset-frequency phase locking or reference-cavity based stabilization. The development of a frequency reference cavity suitable for flight[14] was an extension of LISA efforts. There are plans to use the LRI to test two LISA-specific techniques on orbit: an arm-locking experiment[15] and a TDI experiment[16].

2. GRACE Follow-On LRI measurement concept

The GRACE Follow-On mission will comprise a pair of satellites in identical low-earth orbit, separated by approximately $\rho = 200\,\mathrm{km}$, the same as the GRACE mission that has been operating successfully since 2002. Variations in Earth's gravitational potential measured by GRACE or GRACE Follow-On are proportional to changes in $\Delta\rho/\rho$ resulting from the satellite pair passing over regions of nonuniform gradient[17]. The measurement sensitivity of the GRACE K-band microwave ranging system [18], a copy of which will be the primary science instrument on GRACE Follow-On, is approximately $\tilde{\rho}(f) = 1\,\mu\mathrm{m}/\sqrt{\mathrm{Hz}}$, where $\tilde{\rho}(f)$ designates a root power spectral density. A similar microwave system was used to map the moon's gravity in the GRAIL mission between 2011 and 2012[19]. In addition, GRACE Follow-On includes the LRI as a technology demonstrator to perform the same measurement as the microwave system but with better sensitivity. The 1×10^4 times smaller laser wavelength enables an improvement of a factor of 10 to $\tilde{\rho}(f) = 80\,\mathrm{nm}/\sqrt{\mathrm{Hz}}$ in the fourier frequency band $2\,\mathrm{mHz} < f < 100\,\mathrm{mHz}$. Long-term variations in the absolute distance ρ are known at the level of approximately 1 cm.

Figure 1 shows how LRI1 and LRI2 on separate spacecraft are used to measure $\Delta\rho$ ($= \delta L$ in the figure). The two LRI's are identical, and provide for laser frequency to be controlled either by a local cavity or by incoming light from the opposite LRI. As shown, the LRI1 laser frequency is locked to its cavity and LRI2 phase is phase-locked to the LRI1 laser with typical offset frequency $f_o = 10\,\mathrm{MHz}$. A heterodyne signal is formed on the DWS photodetector, using leakage light of approximately 1 mW from the local laser interfering with pW-level light from the distant spacecraft. Approximately 20 mW of signal power is available at the main output. The heterodyne frequency is $f_h = f_o + f_D$ where f_D is the Doppler shift from the LRI1-LRI2 relative velocity. The phases from the 4 segments of the photodetector are separately measured by the phasemeter in the Laser Ranging Processor (LRP). Changes in the sum phase $\Delta\phi$ on LRI1 are proportional to $\Delta\rho$, and are transponded to ground as the primary science signal. The difference phases give misalignment in pitch and yaw of the local beam relative to the incoming light. A control system in the LRP uses the difference phases to control the Fast Steering Mirror (FSM), aligning the outgoing beam to the incoming beam to μrad level precision. The Triple Mirror Assembly (TMA) routes the beam within the spacecraft such that the optical vertex coincides with the spacecraft center of mass.

3. Hardware

The photographs in Figure 2 are of the LRI1 during laboratory testing and while integrated on the spacecraft at Airbus.

Below we discuss the LRI subsystems and their relevance to LISA.

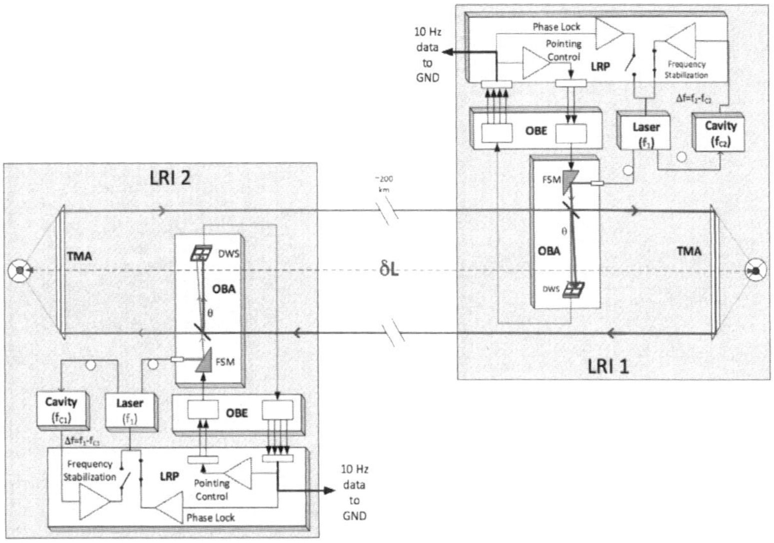

Figure 1. Block diagram of the LRI. OBA = Optical Bench Assembly, TMA = Triple Mirror Assembly, FSM = Fast Steering Mirror, DWS = Differential Wavefront Sensor, LRP = Laser Ranging Processor.

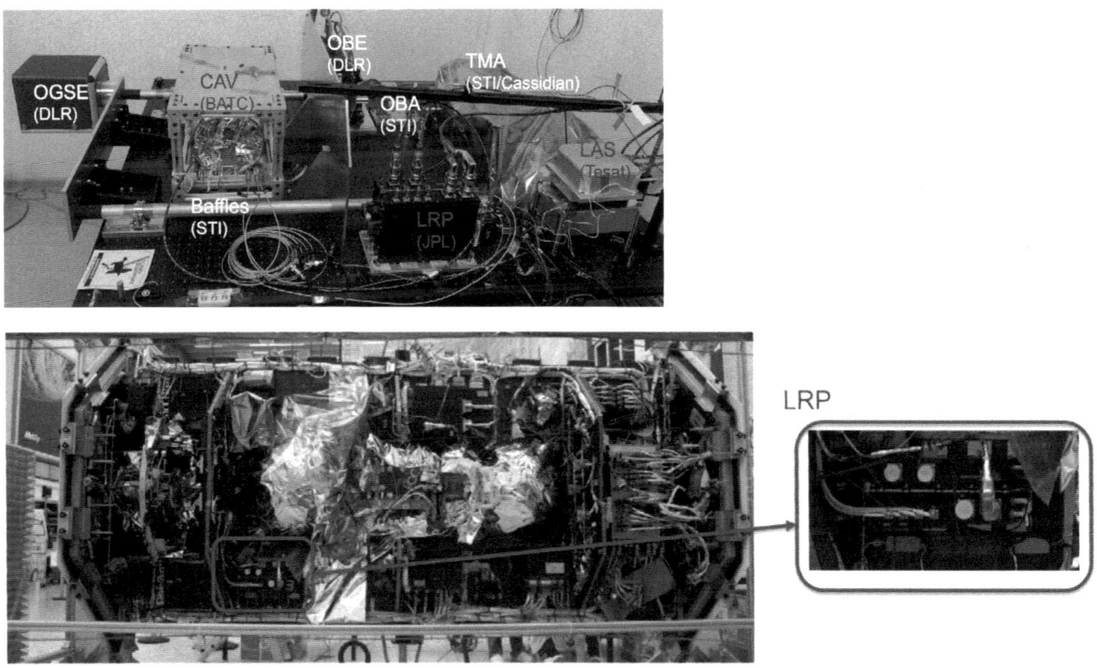

Figure 2. *Upper*: LRI flight hardware for Spacecraft 1 at Spacetech Gmbh (STI), Immenstaad, just prior to spacecraft integration. OGSE = Optical Ground Support Equipment, CAV = cavity manufactured by Ball Aerospace and Technology Corporation (BATC), OBE = Optical Bench Electronics, TMA = Triple Mirror Assembly manufactered by STI and Cassidian Oberkochen DE, LAS = Laser. *Middle*: Spacecraft 1 during integration at Airbus Defence and Space Gmbh, Immenstaad. *Inset*: Laser Ranging Processor (LRP). Photographs courtesy of Airbus and JPL.

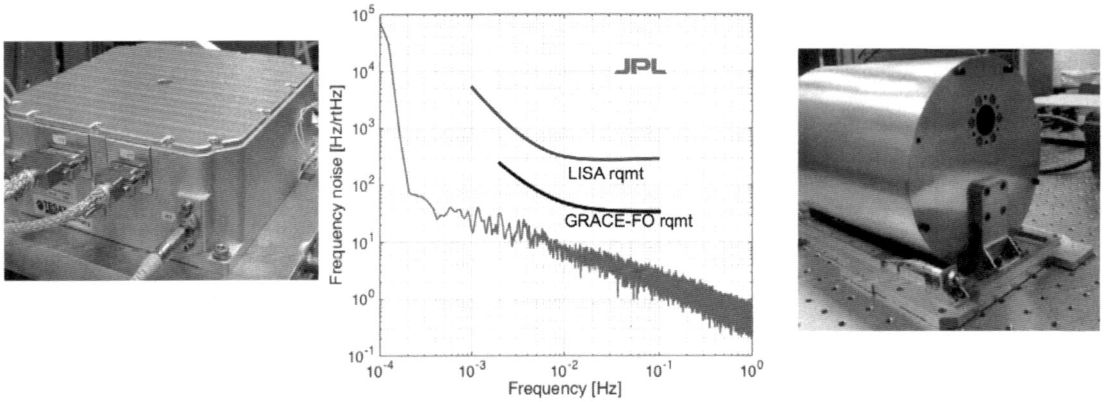

Figure 3. *Left*: NPRO laser manufactured by Tesat Spacecom, Gmbh, Backnang *Mid*: Measured frequency stability is better than requirements for both LISA and GRACE Follow-On. Right Reference cavity assembly from Ball Aerospace, Boulder Colorado.

3.1. *Laser and laser stabilization*

The laser, Figure 3, is a Nd:YAG Non-Planar Ring Oscillator (NPRO) of wavelength of 1064 nm, similar to the laser on LISA Pathfinder. Its output power is adjustable between 20 and 30 mW. The output frequency is stabilized to a resonance of a thermally isolated reference cavity with free spectral range = 1.9 GHz using the Pound-Drever-Hall technique implemented in the LRP. The residual frequency noise $\tilde{\nu}(f)$ limits performance according to $\tilde{\rho}(f) = \rho\tilde{\nu}(f)/\nu_0$ where $\nu_0 = 281$ THz. $\tilde{\nu}(f)$ was measured by stabilizing two flight lasers to flight cavities and measuring the interference phase between the lasers with a flight phasemeter, showing performance that meets the LRI requirement with margin.

3.2. *Phase tracking*

Figure 4 shows performance of the digital part of the phasemeter in response to synthesized noise that simulates a free-running laser. The noise in the difference between two such stimuli is below the phasemeter requirement, demonstrating that the dynamic range and the noise floor meets the LRI requirements. The residual is equal to the more stringent requirement on the LISA phasemeter: no accident, since the LRI phasemeter is based on the LISA design.

Table1 summarizes phase tracking performance of a heterodyne signal at the limit of low signal power. The carrier-to-noise ratio is defined as $\mathrm{CNR} = V_{\mathrm{carrier}}/\tilde{V}(f)$ where V_{carrier} and $\tilde{V}(f)$ are respectively the signal (carrier) amplitude and the noise floor at frequencies adjacent to the signal. CNR of 68 dB-Hz results from optical signal power of 3 pW with nominal laser intensity noise and photoreceiver electronic noise. The LRI phasemeter satisfies the LISA requirement of no cycle slipping for CNR as low as 75 dB-Hz, with margin of 7 dB, in the presence of laser frequency noise at the requirement level of $150\,\mathrm{kHz}/\sqrt{\mathrm{Hz}}\cdot(\mathrm{Hz}/f)$.

4. LRI and LISA phase measurement comparison

Table 2 compares parameters for the LRI phase measurement system to those for LISA. The top level phase noise requirement in LRI is relaxed compared to LISA, and the laser frequency

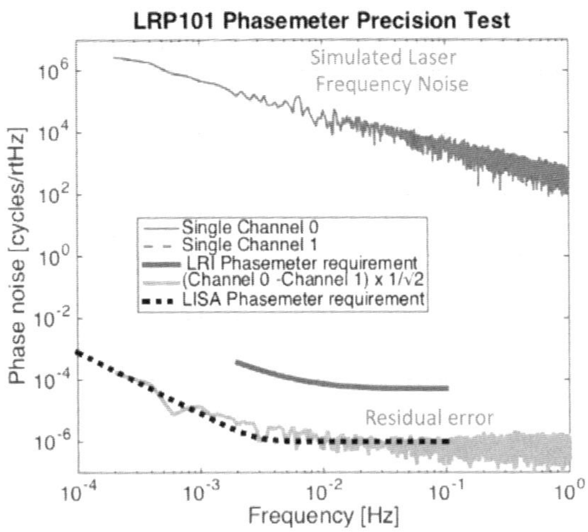

Figure 4. Electronic test of the digital part of the LRI phasemeter shows suppression of laser frequency noise to the requirement level for LISA.

Table 1. Phasemeter performance at CNR limits, tracking a flight laser.

CNR	Performance
61 dB-Hz	Acquire DWS with cycle slips ∼ 1 per second. DWS algorithm is resilient to cycle slipping
64 dB-Hz	Lock and track for hours with some cycle slips, ∼ 1 per few hundred seconds
66 dB-Hz	Lock and track for hours without cycle slips

stabilization is tightened. The Doppler shift, signal Intermediate Frequency (IF), received optical power, and science frequency band are all similar. Both LRI and LISA require phase tracking with low optical power and a DWS system to maintain alignment. The similarities demonstrate that the LRI provides a relevant technology demonstration for LISA and represents a valuable step towards LISA technology development.

5. Acknowledgment

The LRI instrument development at the Jet Propulsion Laboratory, California Institute of Technology, is performed under contract with NASA. The authors greatfully acknowledge the extensive contributions of the rest of the LRP team at JPL and the rest of the LRI team in the US. The LRI has been a satsifying and productive partnership with our German colleagues.

References

[1] Tapley B, Bettadpur S, Ries J, Thompson P and Watkins M 2004 *Science* **305** 503
[2] Bender P et al. 1998 *LISA Pre-Phase A Report, 2nd Ed,* MPQ 233.
[3] Sheard B, Heinzel G, Danzmann K, Shaddock D, Klipstein W and Folkner W 2012 *J Geod* **86** 1083
[4] Armstrong J Estabrook F, and Tinto M 1999 *Ap J* **527** 814
[5] Shaddock D, Ware B, Spero R and Vallisneri M 2004 *Phys Rev D* **70** 08110
[6] Sheard B, Gray M, McClelland D and Shaddock D 2003 *Phys Lett A* **320**
[7] McKenzie K, Spero R and Shaddock D 2009 *Phys Rev D* **80** 102003
[8] Shaddock D, Tinto M, Estabrook F and Armstrong W 2003 *Phys Rev D* **58** 061303
[9] Halverson P, Ware B, Shaddock D and Spero R 2007 US Patent 7411469
[10] Shaddock D, Ware B, Spero R and Klipstein W, LIMAS2007-002, LISA Project document (2007, updated 2009)

[11] Shaddock D, Ware B, Halverson P, Spero R and Klipstein W 2006 *AIP Conf Proc* **873** 654
[12] de Vine G, Ware B, McKenzie K, Spero R, Klipstein W and Shaddock D 2010 *Phys Rev Lett* **104** 211103
[13] Sutton A, McKenzie B, Ware B and Shaddock D 2010 *Optics Express* **18** 20759
[14] Folkner W et al. 2010 *Earth Science Technology Forum*, Arlington VA
[15] Thorpe J and McKenzie K 2016 *Phys Rev D* (accepted for publication)
[16] Francis S, Shaddock D and Sutton A 2015 *Phys Rev D* **92** 012005
[17] Wolf M 1969 *J Geophys Res* **74** 5295
[18] Dunn C et al. 2002 *Proc 15th Int. Tech Meeting of Satellite Division of Institute of Navigation,* Portland, OR, 724
[19] Klipstein W, Arnold W, Enzer D, Ruiz A, Tien J, Wang R and Dunn C 2013 *Space Science Reviews* **178** 57
[20] Sheard B, Heinzel G, Danzmann K, Shaddock D, Klipstein W and Folkner M 2012 *J. Geod* **86** 1083

Table 2. Comparison of LRI and LISA phase measurement parameters

Parameter	LRI	LISA
Measurement noise total	$80\,\mathrm{nm}/\sqrt{\mathrm{Hz}}$	$20\,\mathrm{pm}/\sqrt{\mathrm{Hz}}$
Shot noise	$0.01\,\mathrm{nm}/\sqrt{\mathrm{Hz}}$	$7\,\mathrm{pm}/\sqrt{\mathrm{Hz}}$
Phasemeter noise $\cdot\lambda$	$1\,\mathrm{nm}/\sqrt{\mathrm{Hz}}$	$1\,\mathrm{pm}/\sqrt{\mathrm{Hz}}$
Optical pathlength noise	$30\,\mathrm{nm}/\sqrt{\mathrm{Hz}}$	$3\,\mathrm{pm}/\sqrt{\mathrm{Hz}}$
Laser frequency noise	$35\,\mathrm{nm}/\sqrt{\mathrm{Hz}}$	$1\,\mathrm{pm}/\sqrt{\mathrm{Hz}}$
USO noise	$1\,\mathrm{nm}/\sqrt{\mathrm{Hz}}$	$1\,\mathrm{pm}/\sqrt{\mathrm{Hz}}$
Phasemeter noise	$1 \times 10^{-3}\,\mathrm{cycles}/\sqrt{\mathrm{Hz}}$	$1 \times 10^{-6}\,\mathrm{cycles}/\sqrt{\mathrm{Hz}}$
Satellite separation	170–270 km	5×10^{9} m
Satellite relative velocity	$< \pm 3$ m/s	$< \pm 15$ m/s
Laser Wavelength λ	1.065×10^{-6} m	1.065×10^{-6} m
Nominal carrier-to-noise ratio	> 75 dB-Hz (single channel)	> 75 dB-Hz (single channel)
IF signal frequency	4–16 MHz	2–18 MHz
IF signal dynamics (at 1 Hz)		
Before frequency stabilization	$5 \times 10^{3}\,\mathrm{Hz}/\sqrt{\mathrm{Hz}}$	$5 \times 10^{3}\,\mathrm{Hz}/\sqrt{\mathrm{Hz}}$
After frequency stabilization	$30\,\mathrm{Hz}/\sqrt{\mathrm{Hz}}$	$30\,\mathrm{Hz}/\sqrt{\mathrm{Hz}}$
Science bandwidth	2 mHz–100 mHz	0.1 mHz–1 Hz
Received optical poser	79–625 pW	80 pW
Number of phase channels	4	44+
ADC clocking rate	38.656 MHz	50 MHz
Time coordination	GPS	Laser ranging code
Frequency-offset laser phase locking	Required	Required
Beam pointing sensing	Wavefront sensing	Wavefront sensing
Pointing noise	$1\,\mu\mathrm{rad}/\sqrt{\mathrm{Hz}}$	$80\,\mathrm{nrad}/\sqrt{\mathrm{Hz}}$

IOP Conf. Series: Journal of Physics: Conf. Series **840** (2017) 012012 doi:10.1088/1742-6596/840/1/012012

Chip based MEMS Ion Thruster to significantly enhance Cold Gas Thruster Lifetime for LISA

M. Tajmar, P. Laufer, D. Bock

Institute of Aerospace Engineering, Technische Universität Dresden, Germany

martin.tajmar@tu-dresden.de

Abstract. Micropropulsion is a key component for ultraprecise attitude and orbit control required by the eLISA mission. LISA pathfinder uses cold gas micro thrusters that are accurate but require large tanks due to their very low specific impulse, which in turn limits the possible mission duration of the follow up eLISA mission. Recently, we developed a compact MEMS ion thruster on the chip with a size of only 1cm² that can be simply attached to a gas feeding line like the one used for cold gas thrusters. It provides a specific impulse greater than 1000 s and only requires a single DC voltage. Since the operating principle is based on field emission, very low thrust noises similar to FEEP thrusters are expected but with gas propellants. The MEMS ion thruster chip could be mounted in parallel to the existing gold gas system providing high Isp and therefore long mission durations while leaving the cold gas system in place. To enable a possible mission extension, the MEMS ion thruster could take over from the cold gas system as a backup while maintaining the existing micropropulsion thruster system with its heritage therefore minimum risk.

1. Introduction

Micropropulsion systems are of high interest for the LISA mission due to precise thrust control in a wide range as well as exact and reliable attitude and orbit control. LISA pathfinder uses 6 cold gas thrusters with an accurate thrust in a range of 10 to 500µN which contains a mass flow sensor with a piezo-controlled valve [1]. Concerning the very low specific impulse of only 50s already 40% of the high-pressure Nitrogen propellant have been used during the first few months of LPF mission [1]. The low efficiency of the described propulsion system and the very large tanks will in turn limit the lifetime of the follow-up eLISA mission. NASA provided the DRS package which features a Colloid thruster with a larger specific impulse of up to 1500s, however the maximum thrust range is only 35µN which is not sufficient for the full eLISA mission [2]. Various micropropulsion systems have been studied as possible candidates for eLISA.

Here, we are going to present a new and simple design called MEMS ion thruster [3] that can use the cold gas infrastructure and provide ion thruster performance. It uses the well-established principle of field emission combined with gas propellant thus eliminating all propellant-related wetting and contamination issues that are common for FEEP or Colloid thruster. Based on the recent development of our thruster chip with a size of less than 1cm², our system could significantly increase the lifetime of eLISA and create the possibility for mission extensions. As an add-on to the existing cold gas thruster units, the heritage and reliability of the system are retained while the possible specific impulse is

IOP Conf. Series: Journal of Physics: Conf. Series **840** (2017) 012012 doi:10.1088/1742-6596/840/1/012012

increased to the 2000s-range. The principle of function responses to electrostatic field emission using the same gas propellant like the cold gas system and providing an ionization efficiency up to 30%. A low thrust noise similar to the ones of FEEP of Colloid thrusters is expected. By switching the polarity of the required single DC power supply, the emitter chips can be simultaneously used as a neutralizer. The MEMS ion thrusters can employ the existing gas feeding system as well as the gas propellant and uses the cold gas thrusters as a backup. Therefore eLISA benefits from high specific impulse, high propulsion efficiency and a significantly enhanced lifetime with minimal risk.

2. MEMS Gas Field Ionization Source Thruster

This chapter will outline the principle of function of the MEMS field ionization thruster and introduce previous works. A review of the novel manufacturing method, as well as first results of MEMS ion thruster prototypes, are given.

2.1. Review of the Concept

The MEMS field ionization thruster is taking advantage of the basic gas field ion source (GIFS) which already has been invented in 1951. The gas is ionized in an electric field between an anode (tip) and a cathode (grid) and the ions are accelerated in the field (Figure 1). The substitution of the vintage anode tip geometry by an aligned CNT forest could lead to high ionization volume and consequently higher ionization efficiencies. The opportunity to emit both gas ions and electrons depending on the polarity of the DC power supply is another positive aspect of GFIS which eliminates the need for a separate neutralizer.

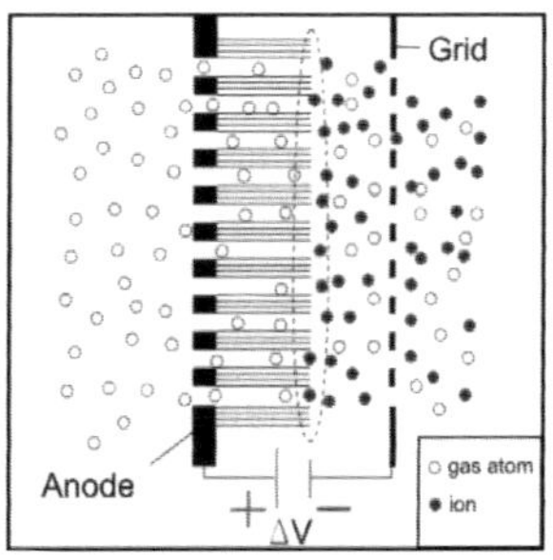

Figure 1. MEMS Gas Field Ionization Source (GFIS) Thruster with illustrated high ionization area in gas ion emission mode.

Previous attempts in the development of MEMS GIFS have been done at MIT and NASA Ames Research Center [4,5]. Both evolutions proved the functionality of MEMS GFIS but they struggled in various details of the manufacturing approach. As shown in Figure 2 the flow-through silicon μFoam with CNTs of MIT have serious restrictions due to poor CNT homogeneity which corresponds to poor ionization efficiency. In consequence of the strong field enhancement between the single CNTs, the MIT obtained ionization potentials of 300V only. NASA abandoned the development of a flow through GFIS and concentrated on flow-past patterned electrodes (Figure 3) as a result of unexpected challenges in the etching procedure of the silicon chip. The developed electrodes achieved continuous CNT growth leading to ion currents up to 100μA and electron currents up to 400μA.

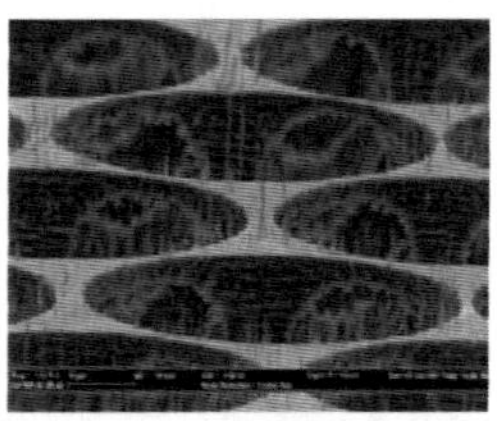

Figure 2. Flow Through Electrode by MIT [4]

Figure 3. Flow Past Electrode designed by NASA Ames [5]

2.2. Chip Design and Manufacturing Process

The field ionization source is an assembly of silicon chips which are easily scalable to fit in the needed application. We manufactured both round and square chips with slits or holes working as gas feed through. For the mechanical treatment, we partnered up with the Laserzentrum Mittweida for laser manufacturing of the chips. The laser drilling process enables a fast alteration of the shape parameters to investigate various arrangements and represents a less risky process to generate small hole diameters to attain high aspect ratios for high ionization efficiencies. In cooperation with the Fraunhofer Institut für Keramische Technologien und Systeme (IKTS), the CNT deposition is implemented with the atmospheric pressure chemical vapor deposition (APCVD) process. The manufacturing procedure is shown in Figure 4.

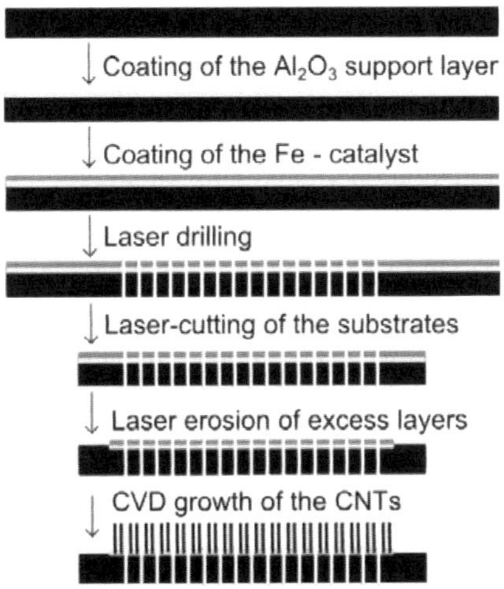

Figure 4. Manufacturing approach for flow through electrode

Figure 5. Results of Laser Drilling and CNT-Deposition

Highly n-doped low resistance 100mm silicon wafers are used as base material. In a first and second stage, the wafers are coated with an Al_2O_3 and a Fe-catalyst layer. Both are crucial for the following CNT deposition since the CNT growth site is defined by the nm-sized Fe-particles. To prevent a possible damage of the CNTs the laser manufacturing is performed as the third stage. Therefore, a pulsed UV laser CNC drilling machine is used processing the silicon chips from the back to minimize the pollution of the coated surface. In a finale stage, the APCVD-process is accomplished at IKTS. With a process duration of 80min and temperatures of 700°C CNTs with a maximum length of 130µm, high density, a very close growth near the holes and a high uniformity are achieved as shown in Figure 5.

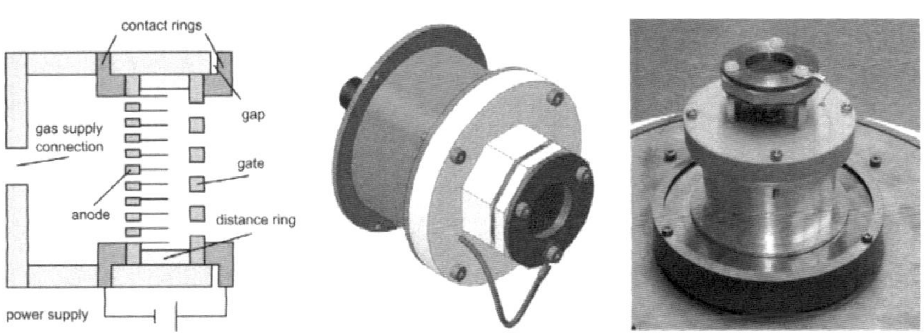

Figure 6. Test module design and laboratory model

2.3. Test module design and test results

The design of the developed test module for field emission and field ionization is shown in Figure 6. A modular setup with a PEEK case and copper contact electrodes has chosen for performance tests of various configurations.

Initial tests were performed with a chip design with 20mm outer diameter and Argon was chosen for the gas field ionization test due to its availability at the lab. First experiments are realized in the electron-emission mode at a pressure of 10^{-6}mbar. The initial voltage for electron emission is in the area of 220V only, with electron currents up to 1mA are easily reachable. As shown by the results in Figure 7, no higher electron currents are attempted to prevent the thruster from early damage.

Further experiments were performed in the ion-emission mode by switching the polarity of the DC power supply and enabling the gas feed through. I-V characteristics for mass flow rates between 0.6sccm and 2.2sccm are plotted in Figure 8. As higher mass flow rates will increase the possibility of arc discharges inside of the plasma generator the tests were operated at low voltages to prove and demonstrate its functionality and true bipolar capability.

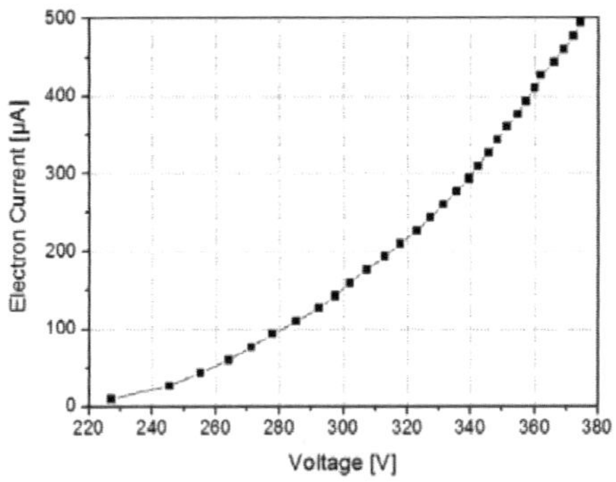

Figure 7. Results of I-V characteristics in electron emission mode.

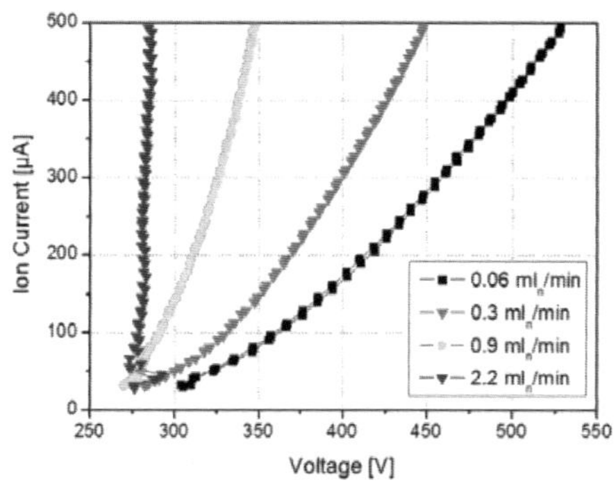

Figure 8. Results of ion emission with various gas flow rates.

Subsequently, the highest achievable ionization efficiency was determined by a gas flow rate of 0.06sccm Argon. The higher the applied voltage the more gas will be ionized inside of plasma generator for a good efficiency. The limit was set just below micro discharges occur. As plotted in Figure 9 the maximum was found at 740V with an ion current of about 1.2mA. The measurement data enables the calculation of the ionization efficiency to 28% with a specific impulse of 1700s and a total thrust of close to 30µN (Figure 10).

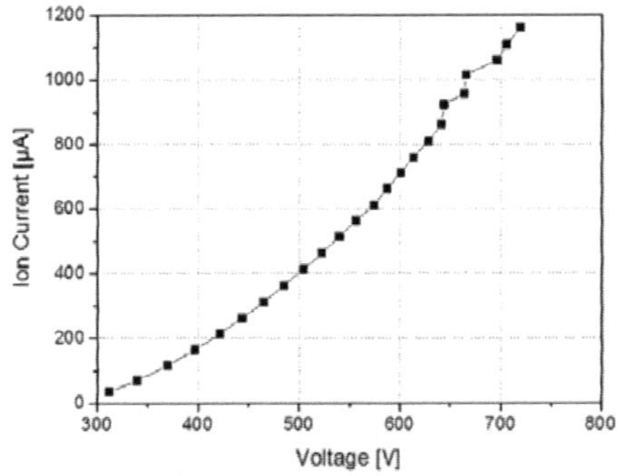

Figure 9. I-V characteristics with Argon at 0.06sccm for maximum achievable current

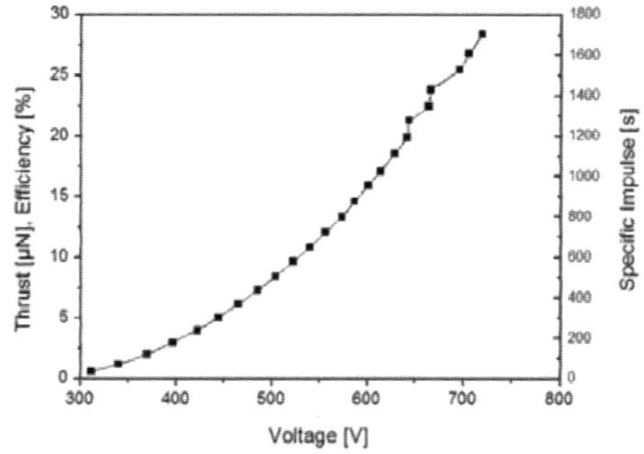

Figure 10. Thruster-related performance for a flow rate of 0.06sccm Argon propellant.

IOP Conf. Series: Journal of Physics: Conf. Series **840** (2017) 012012 doi:10.1088/1742-6596/840/1/012012

The MEMS gas field ionization thruster can be used with any kind of gas propellant. eLISA will use Nitrogen which is lighter and has a lower ionization potential than Argon. Accordingly, an even higher specific impulse is expected. With the flow rates of LPF, thrusts even in the mN-range are possible.

3. Integration of MEMS Ion Thruster to eLISA

The integration of MEMS ion thruster involves minimum risk to the eLISA mission while enabling the lifetime enhancement of the cold gas thruster system and the possibility for mission extensions. At an electric potential of 500V the estimated specific impulse for Nitrogen is 2150s, which increases the lifetime of the propellant by a factor of 48. By keeping the cold gas thrusters, the heritage and reliability of the overall thruster system for eLISA will be retained. Two considerable integration methods are illustrated in Figure 11.

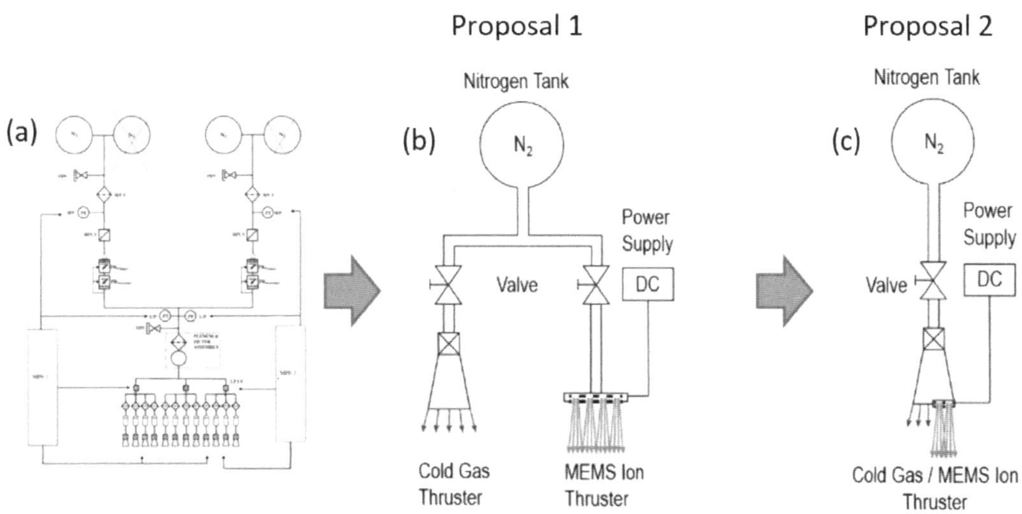

Figure 11. Considerable integration methods of MEMS ion thruster by keeping the heritage of the cold gas system. The existing cold gas system (a) could benefit from high specific impulse of electric propulsion with parallel integration (b) or hybrid integration (c).

Both configurations abandon separate neutralizers as well as magnetic fields for gas ionization. For the generation of plasma only one DC power supply is needed. The MEMS ion thruster could be mounted in parallel to the existing cold gas thruster system using the same gas feeding line and gas flow sensor. This option requires an additional piezo valve for enabling the gas flow to the MEMS ion thruster. The second option presents an integration of the MEMS chip into the existing cold gas thruster nozzle. Without applied voltage to the MEMS ion thruster, the cold gas thruster can operate as usual. By activating an electric potential, it will benefit from high efficiency and improved specific impulse.

4. Summary

A novel silicon chip based MEMS ion thruster has developed and tested. An ion beam is generated at the tip of a forest of carbon nanotubes with combines the simplicity of field emission electric propulsion with ion thruster technology thus only a single DC power supply is required. Moreover, by switching the polarity, the thruster can operate in an electron emission mode as well which eliminates the necessity of a separate neutralizer. Magnetic fields or heater units are not needed for the thruster operation. Expecting very low thrust noise an integration to eLISA mission should be considered to benefit from high specific impulse as well as high propellant efficiency. The MEMS ion thruster could increase the propellant lifetime of eLISA by a factor of 48, improve the maximum thrust capability and enable possible mission extensions.

Two proposals were presented for a possible integration to eLISA mission. Both retain the heritage and reliability of the existing cold gas system by keeping it in place. The MEMS ion thruster could be operated in parallel or in a hybrid mode. The case of complicacy is considered by switching back to the heritage cold gas system thus a minimum risk to eLISA mission.

References

[1] J. Martino and E. Plagnol, *LISA Pathfinder Coldgas Thrusters*, LISA Symposium 2016

[2] J. K. Ziemer et al. *Colloid micro-Newton thruster development for the ST7-DRS and LISA missions*. Pasadena, CA: Jet Propulsion Laboratory, National Aeronautics and Space Administration, 2005.

[3] M. Tajmar and S. Stämm: *Micro Ion Thruster Concept based on a MEMS Gas-Field-Ionization Source*, Proceedings of the 4th Spacecraft Propulsion Conference, Cologne, May 19-22 (2014)

[4] L. F. Velasquez-Garcia and A. I. Akinwande: A pecvd cnt-based open architecture field ionizer for portable mass spectrometry. in *Micro Electro Mechanical Systems, 2008. MEMS 2008.* IEEE, 2008, 742–745.

[5] T. C. Hicks: Carbon nanotube pillar array ionizer for miniature ion thruster applications. Master's thesis, Naval postgraduate school, Monterey, California (2008)

IOP Conf. Series: Journal of Physics: Conf. Series **840** (2017) 012013 doi:10.1088/1742-6596/840/1/012013

Progress and Plans for a US Laser System for LISA

J Camp, K Numata, and M Krainak
NASA Goddard Space Flight Center, Greenbelt, Maryland 20771, USA

E-mail: jordan.b.camp@nasa.gov

Abstract. A highly stable and robust laser system is a key component of the space-based LISA mission architecture. We describe our progress and plans to demonstrate a TRL 5 LISA laser system at Goddard Space Flight Center by 2020. The laser system includes a low-noise oscillator followed by a power fiber amplifier. The oscillator is a low-mass, compact external cavity laser, consisting of a semiconductor laser coupled to an optical cavity, built by the laser vendor Redfern Integrated Optics. The amplifier is a diode-pumped Yb fiber with 2.5 W output, built at Goddard. We show noise and reliability data for the full laser system, and describe our plans to reach TRL 5.

1. Introduction

The past year has seen the opening of the field of Gravitational Wave Astronomy, due to the LIGO discovery of gravitational radiation from a merging Black Hole Binary system [1]. In addition the results of the LISA Pathfinder mission have shown it is possible to achieve drag-free control at the required LISA levels [2]. These results have re-focused the LISA community on LISA technology development. We report here on the status and plans to provide a TRL 5 LISA laser by 2020.

We are pursuing an all-fiber/waveguide space laser solution based on the MOFA (master oscillator fiber amplifier) configuration, which is a waveguide-based oscillator followed by a preamplifier and a power amplifier (Fig. 1). These components naturally fit into the high precision LISA laser system because of their high mechanical robustness, excellent reliability, compact form factor, and high wall-plug efficiency. Our research has included development of 1064 nm Planar-Waveguide external cavity laser (PW-ECL), PW-ECL reliability tests, and amplifier development, including its noise measurements and reliability studies. For the preamplifier, we are looking into two options, a fiber-based amplifier (as in Fig. 1) and a semiconductor optical amplifier (SOA).

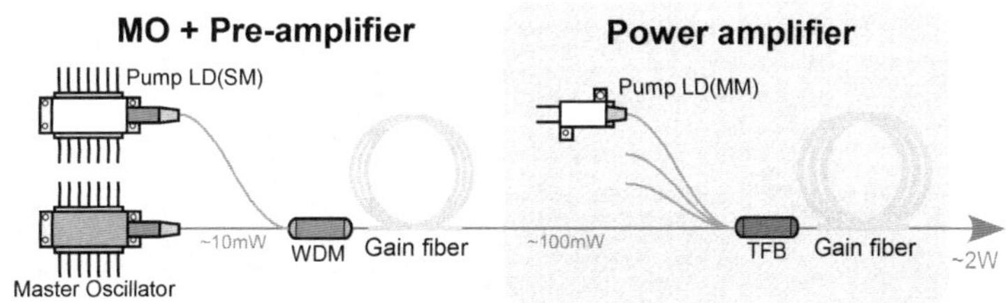

Fig. 1 Concept of the MOFA configuration. Optical isolators, modulators, and redundant LDs are not shown. SM: single-mode; MM: multi-mode

IOP Conf. Series: Journal of Physics: Conf. Series **840** (2017) 012013 doi:10.1088/1742-6596/840/1/012013

2. Master Oscillator

We have developed a fiber ring laser [3] and a fiber DBR laser for space interferometry. Although these lasers perform like the non-planar ring oscillator (NPRO) at low (<10 kHz) frequency, they have a larger relaxation oscillation peak, around 1 MHz, which could affect the heterodyne interferometry typically operated near this frequency. Therefore, we shifted our focus to the development of the planar-waveguide external-cavity diode laser (PW-ECL). The Telecordia-qualified PW-ECL, built by Redfern Integrated Optics (RIO), offers advantages over solid-state lasers, including simpler design, more compact size, lower mass, and less consumption of electrical power. The narrow reflection peak of the Bragg reflector in the planar lightwave circuit (PLC) enables stable, low-noise, single-mode lasing at a selected wavelength. The design may be modified to support a 1 GHz bandwidth, integrated intra-cavity phase modulator [4]. Originally, the PW-ECL was only available at the telecom C-band (1550 nm); based on experience with this product, RIO has successfully built 1064 nm version of PW-ECL by changing the gain chip and the PLC design. Maximum output power of PW-ECL is 15 mW.

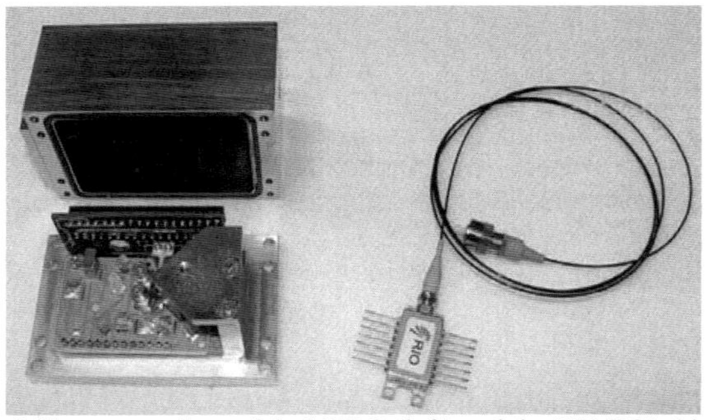

Fig. 2 Size comparison of the NPRO (left) and the PW-ECL (right)

Figure 2 shows a comparison of its size with the non-planar ring oscillator (NPRO). The PW-ECL's package is much more compact than that of the NPRO, in which a strong magnet limits the size. We based our choice of the PW-ECL on an investigation of number of lasers [5].

Figure 3 shows the frequency noise of the 1064 nm (blue) and 1550 nm (black and red) PW-ECL, and Figure 4 shows the PW-ECL intensity noise in comparison to the NPRO. The frequency noise of the 1064 nm ECL must be lowered by a factor of 5 at 100 kHz so that cycle slips do not occur within the LISA phasemeter; this work is on-going at RIO. We have decided to build a space-qualified NPRO (already developed by several groups), in case the frequency noise requirement of the PW-ECL cannot be met. Figure 4 shows the PW-ECL intensity noise is significantly lower than the NPRO at frequencies around 1 MHz due to the presence of the NPRO relaxation oscillation.

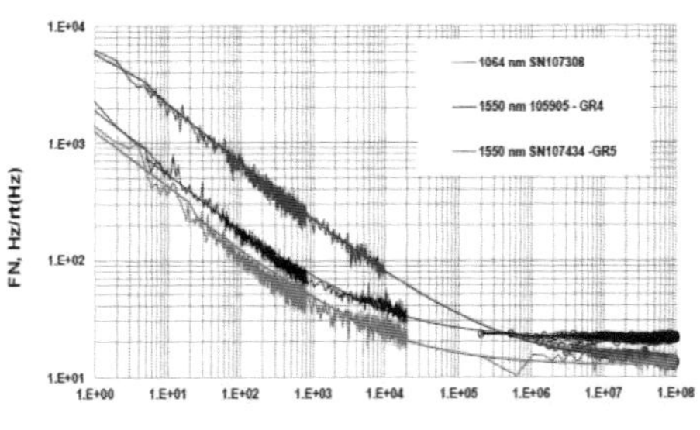

Fig. 3 Frequency noise of 1064, 1550 nm PW-ECLs

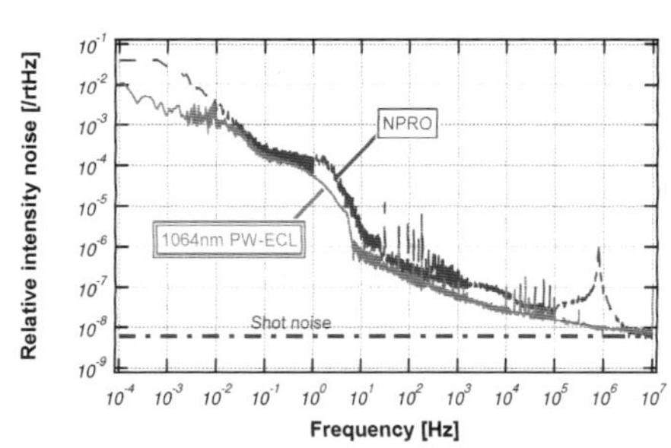

Fig. 4 Intensity noise of PW-ECL and NPRO

IOP Conf. Series: Journal of Physics: Conf. Series **840** (2017) 012013 doi:10.1088/1742-6596/840/1/012013

We made a detailed study of the mechanical, thermal, and radiation robustness of the PW-ECL, and found it to be qualified for use in space. Lucent Government Solutions (LGS) planned and oversaw the tests, which involved vacuum thermal cycling, hermiticity, radiation, and accelerated aging [6]. For these reasons, the 1550-nm PW-ECL was adopted as the metrology laser for the OpTIIX mission [7] on the International Space Station. We will revisit some of these reliability studies for the 1064 nm PW-ECL in FY17. We believe the risk of failing reliability testing is low, since it is using the same packaging as the 1550 nm PW-ECL and the 1064 nm gain-chip vendor (Eagleyard) data indicates high reliability.

3. Amplifier and Preamp Build

Our recently constructed 2.5 W fiber amplifier is shown in Fig. 5. It includes a pump diode to provide power, a tapered fiber bundle (TFB) to allow redundant power input, and a 2.3 m length, 10-μm core, double-clad, large-mode-area gain fiber that converts the pump power to amplification gain. The forward-pumped design and optical isolator minimize potential sources of feedback. The amplifier uses a robust mechanical design and temperature stabilization to suppress fiber-length variations. Its 2.5 W output power is shown in Fig. 6, along with a linearly-increasing backscattered power indicating negligible Stimulated Brillouin Scattering (SBS). The same package contains the PW-ECL as a maser oscillator, a SOA as a preamplifier, and a phase modulator, making it a complete opto-mechanical package of the LISA laser system. A clean output spectrum was observed with low Amplitude-Stimulated Emission (ASE) floor.

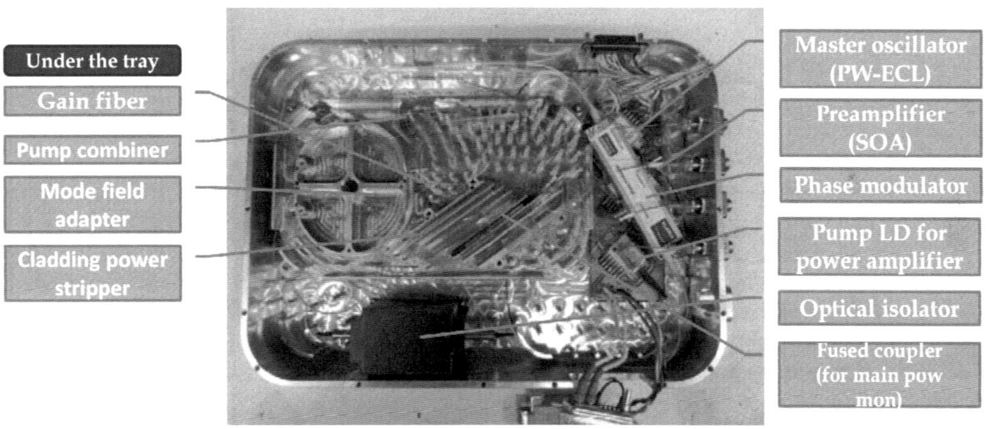

Fig. 5 Laser amplifier system including ECL, SOA preamp, phase modulator, gain fiber, and pump diode

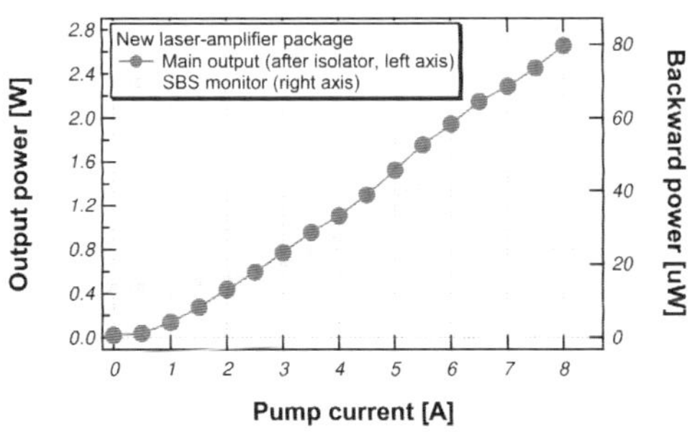

Fig. 6 Output of laser system, showing 2.5 W output power. No nonlinear increase in backward power indicates no sign of SBS buildup.

IOP Conf. Series: Journal of Physics: Conf. Series **840** (2017) 012013 doi:10.1088/1742-6596/840/1/012013

4. Laser System Noise Testing: We have performed laser system noise tests by amplitude- and frequency-stabilizing the PW-ECL and amplifier (Fig. 7). By stabilizing the amplifier pump-diode current, an amplitude noise attenuation of ~30 was achieved at a frequency of 0.1 mHz (a factor of ~10 from the LISA requirement). This noise will decrease once temperature stabilization of the amplifier and the detection system is applied. The laser system was also frequency-stabilized by locking a small fraction of the amplifier's output to a hyperfine absorption line of iodine molecule, using the PW-ECL's injection current as a frequency actuator. Fig. 8 shows that the differential phase noise of a 2 GHz sideband transmitted through the fiber amplifier meets the LISA requirements.

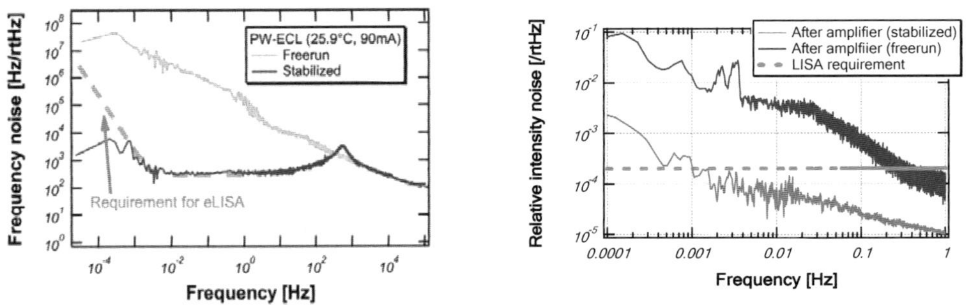

Fig. 7 Frequency and intensity noise of laser system: ECL, preamp and amplifier

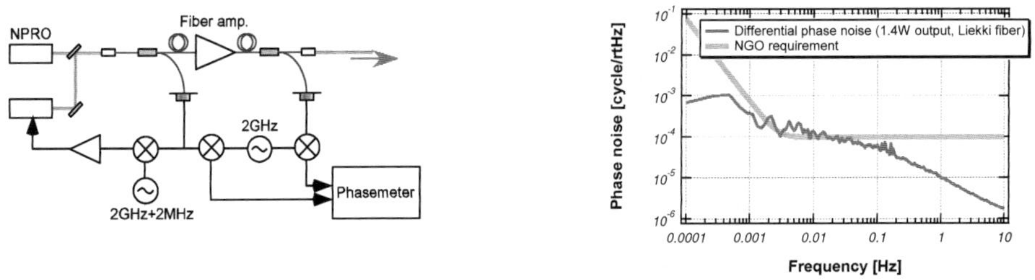

Fig. 8 Differential Phase Noise of the Laser System (right), determined with 2 GHz modulation frequency sidebands. The figure on the left shows the measurement apparatus

5. Laser Preamp Environmental Testing: a critical aspect of this work is environmental testing to prove flight readiness, which involves thermal vacuum cycling, vibration testing, and exposure to radiation levels that will be seen in the LISA space environment. These tests have been applied to the PW-ECL oscillator and a fiber preamp, as shown below. The fiber preamp (Fig. 9) contained a Yb doped gain fiber of 2 m length, and key components including optical isolator, polarization combiner, wavelength multiplexer, pump diodes, and fiber splices.

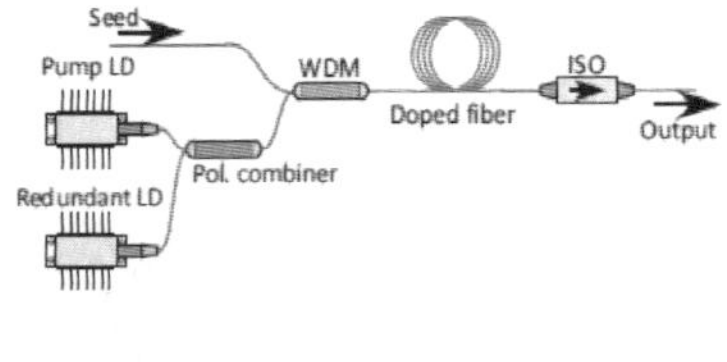

Fig. 9 Fiber preamplifier

Fig. 10 Thermal vacuum cycling apparatus

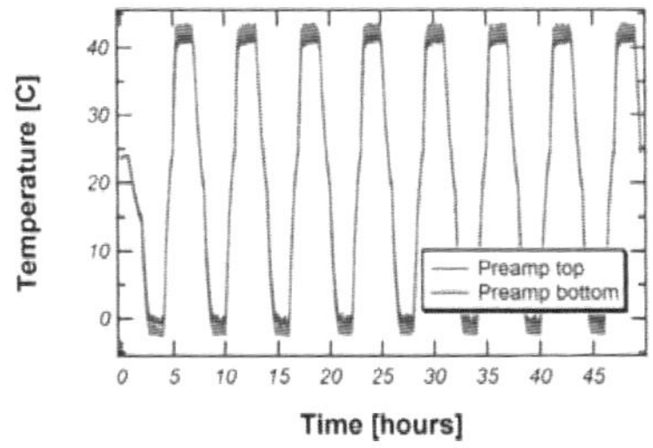

Fig. 11 Thermal levels in preamp/PW-ECL

Fig. 10 shows the vacuum cycling apparatus deployed in our laser lab, and Figure 11 shows the thermal levels and duration used in the thermal cycling of the preamp. Fig. 12 shows the vibration ("shaker") apparatus used to vibrate the preamplifier and other components under test, while figure 13 shows the acceleration time-series that the shaker provides.

Fig. 12 vibration instrument shaking preamp/ECL

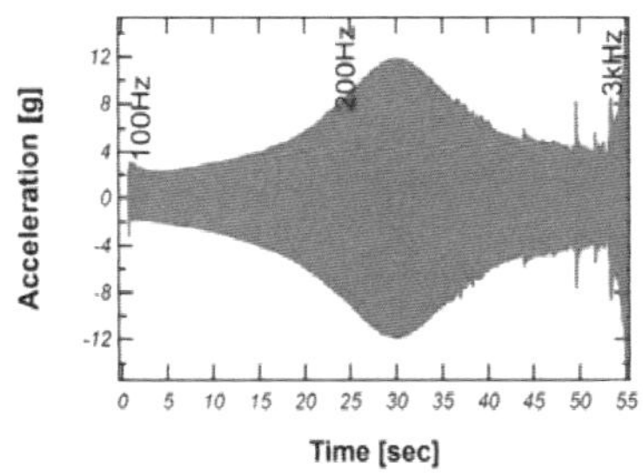

Fig. 13 acceleration from vibration apparatus

The results of the preamp environmental testing are shown in Fig. 14. There was no effect on the output power from thermal cycling or vibration. There was a ~30% drop in output power from gamma-ray radiation exposure, with exposure at the full LISA level of 45 krad in 8 hours, which represents a rate 10^4 times the LISA rate experienced in a 5 year orbit. Since radiation damage is related to the both the total exposure and the rate, another run with higher exposure is needed to separate the two effects and determine whether radiation shielding should be considered.

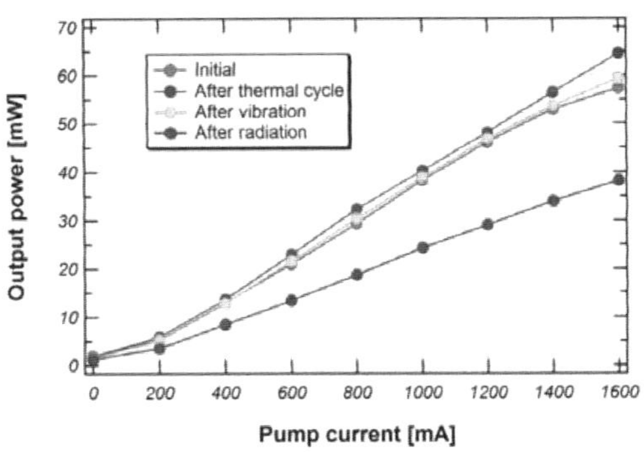

Fig. 14 Results of environmental testing on ECL + preamp

IOP Conf. Series: Journal of Physics: Conf. Series **840** (2017) 012013 doi:10.1088/1742-6596/840/1/012013

The environmental testing facility will be used in the future to test the full laser system, with the thermal cycling temperature range and the acceleration levels tuned to match the LISA mission launch parameters.

6. Future Plans

Remaining work on the US laser system for LISA includes: 1) reduction of the 1064 nm PW-ECL frequency noise, which will involve a redesign of the gain chip to address thermal issues that are causing ellipticity in the laser beam; 2) design and construction of a space-qualified NPRO as a backup oscillator option; 3) reliability testing of individual laser system components to show compliance with a 5 year lifetime requirement; 4) accelerated aging and reliability testing of the full laser system to demonstrate the 5 year lifetime requirement is met. Tasks 1) and 2) can be carried out in parallel. After the final oscillator is downselected Task 3) will be done, followed by Task 4) which will demonstrate TRL 5.

The above tasks are achievable by 2020 provided sufficient funding to the laser system development is allocated. Another possibility may be a funding level consistent with delivery of a TRL 5 laser system by 2024. Importantly, NASA has recently received input from the 2016 Mid Decadal Review [8] emphasizing the need for increased US technology funding to allow the Agency to play an important role in LISA.

7. Summary

NASA/GSFC has been involved in research on spaceborne lasers since the 1990s. Taking advantage of its space laser experience and the emerging telecom laser technology, we are developing a full laser system for the LISA mission. Our research has included both master laser and amplifier developments, and their environmental testing and reliability for space flight. We are awaiting decisions on funding that will allow us to deliver a TRL 5 laser system in the 2020-2024 timeframe.

References

[1] Abbott B et al (LIGO Scientific Collaboration and Virgo Scientific Collaboration), PRL **116** 061102 (2016)
[2] Armano M et al, PRL **116** 231101 (2016)
[3] Numata K and Camp J 2012 *Laser Phys. Lett.* **9** 575
[4] Koren U et al, *Applied Phys Lett* **53**, 2132-2134
[5] Numata K, Camp J, Krainak M A and Stolpner L 2010 *Opt. Express* **18** 22781
[6] Numata K and Camp J 2012 *J. Phys.: Conf. Ser.* **363** 012054
[7] Postman M, Sparks W B, Liu F, Ess K, Green J, Carpenter K G, Thronson H and Goullioud R, 2012 *Proc. SPIE* **8442** 84421T
[8] www.nap.edu/catalog/23560/new-worlds-new-horizons-a-midterm-assessment

Status of the LISA On Table experiment: a electro-optical simulator for LISA

M Laporte[1], H Halloin[1], E Bréelle,[1] C Buy[1], P Grüning[1,2], P Prat[1]

[1]APC, AstroParticule et Cosmologie, Université Paris Diderot, CNRS/IN2P3, CEA/Irfu, Observatoire de Paris, Sorbonne Paris Cité, F-75205 Paris Cedex 13, France

[2]Laboratoire de l'Accélérateur Linéaire, Univ. Paris-Sud et IN2P3/CNRS, Orsay, France

Matthieu.laporte@apc.univ-paris7.fr

Abstract. The LISA project is a space mission that aim at detecting gravitational waves in space. An electro-optical simulator called LISA On Table (LOT) is being developed at APC in order to test noise reduction techniques (such as Timed Delayed Interferometry) and instruments that will be used. This document presents its latest results: TimeDelayed Interferometry of 1st generation works in the case of a simulated white noise with static, unequal arms. Future and ongoing developments of the experiment are also addressed.

Acknowledgements. This work has been funded by the French Space Agency (CNES), under grants RS07/SU-0001-012 and R-S08/SU-0001-012. The authors wish to thank G Heinzel and I Bykov (Albert Einstein Institut, Hanover, Germany), who provided us the phasemeter and adapted it to our needs.

1. The Laser interferometer Space Antenna project (LISA)

The LISA project, like others, is born of the will to detect gravitational waves. Those local, propagating deformations of space-time are so weak **Erreur ! Source du renvoi introuvable.** that their detection represents one of the greatest scientific and technical challenges. Whereas several projects are based on ground (LIGO and VIRGO for instance), LISA chooses space as an alternative.

The mission consists of 3 satellites, separated by a few millions kilometers with a 1064nm laser beam propagating between them. The constellation would fly around the Sun on the same orbit as the Earth, 20° behind it. The whole system acts as a giant laser interferometer, allowing one to monitor the tiny variations (down to 1 pm) of the distance between the satellites due to a possible gravitational waves passing by. Figure 1 [1] presents the principle of LISA in the case of a mother satellite and two daughter satellites (eLISA design). The final design will have the same laser link between the two daughter satellites. The mother satellite sends a laser beam towards the daughter satellite. It is then injected into a phase locked loop (PLL) so that it can be sent back to the mother satellite. At this point it interferes with a local laser. This interference, heterodyne, generates a beat signal whose phase gives the information of the distance between the satellites.

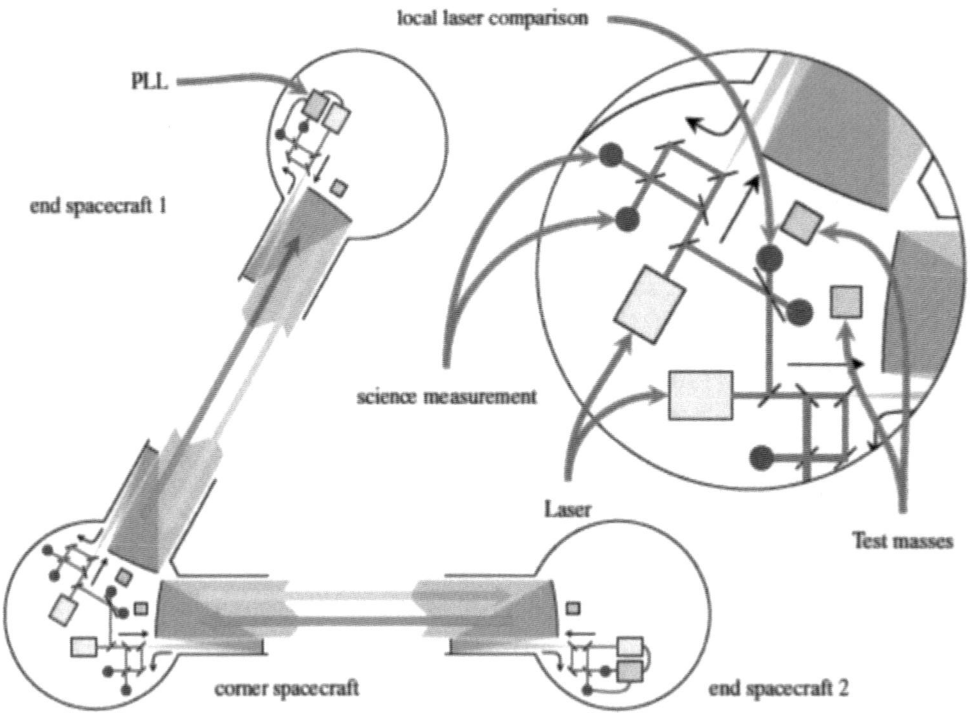

Figure 1: Principle of the measurement in LISA [1]

Such sensitivity goals require a deep knowledge of the noise sources in LISA and effective noise reduction techniques. Two of them, the most noticeable, will be discuss here: the drag-free principle, because of its originality and the timed delayed interferometry, because it is the main technique tested by the experiment that is to be presented here.

1.1. Drag-free principle

Such a sensitivity requirement requires a deep knowledge of the noise sources in LISA and effective noise reduction techniques. The drag-free principle allows to get rid of external perturbations such as radiation pressure from the Sun. It consists of test-masses insides cavities in the satellites. Those test-masses are subject only to gravitation; a local interferometer monitors the position of the satellites with respect to their test-masses, so that one can deduce, in the end, the distance variations between the test-masses in two satellites, hence getting information on the effect of the gravitational wave. This principle have been test and verified by the LISA Pathfinder in 2015. A $5.2 \pm 0.1\, fm.s^{-2}/\sqrt{Hz}$ residual acceleration noise of test-masses of has been achieved, which is far below requirements for LISA-Pathfinder and almost at the level of LISA requirements [3].

1.2. Time delayed interferometry (TDI)

TDI is a data-processing technique [4] that can reduce the noise level of LISA by a factor 10^8, thus meeting the requirements for LISA. The distances between the satellites change over time (up to 30 000 km) since each spacecraft is independent, which induce a phase shift. Thus, noises do not compensate each other when signals interfere. TDI consist in recombining the data at the right time, knowing the positions of the satellites, which virtually create the situation of equal arm length in the interferometer.

IOP Conf. Series: Journal of Physics: Conf. Series **840** (2017) 012014 doi:10.1088/1742-6596/840/1/012014

2. LISA on Table (LOT)

The LOT is an experiment that has been developed for several years at the APC laboratory. It is an electro-optical simulator of LISA, the goals of which are to test experimentally the noise reduction techniques (TDI mainly) and various instruments (photodiodes, phasemeter...) in a representative acquisition chain. In this way, it can be seen halfway between a purely numerical simulator and a hardware prototype.

2.1. Principle

The principle of the LOT is diagrammed in Figure 2:

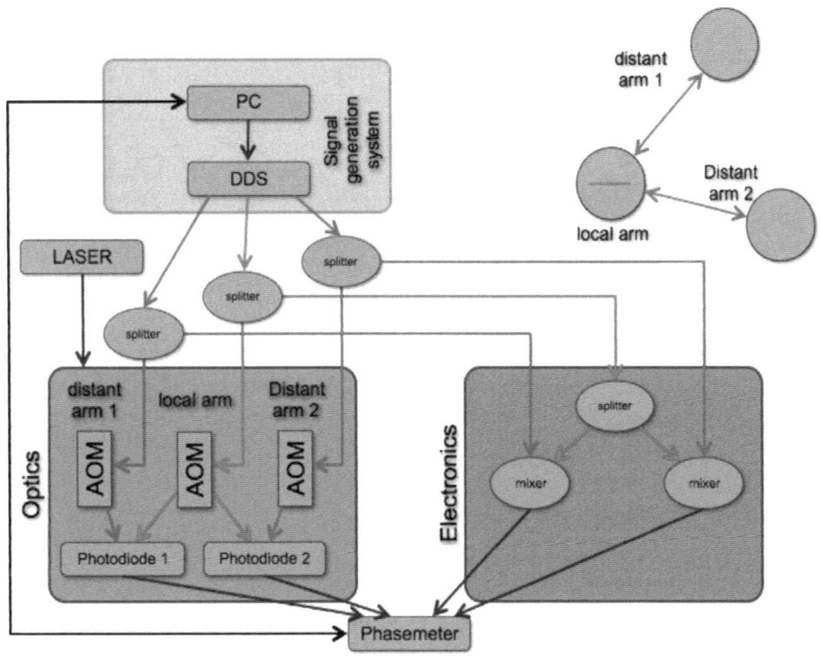

Figure 2: principle of the LISA On Table experiment

A signal generation system using direct digital synthesizers (DDS) produce three signals, that simulate the modulations imprinted on a local laser and two lasers coming from the distant satellites (their frequencies are around 110 MHz: 108, 112.5 and 112.7 MHz). Those modulations correspond to noises, gravitational wave signals, Doppler effect, or in general, anything one wants to simulate. The three signals are split between an optical simulator and an electronic simulator. In the optical one, the signals are imprinted on laser beams through acousto-optic modulators (AOM). The beam representing the local laser then interferes with each "distant arm" signals on photodiodes, hence reproducing the actual interference signals that are to occur in LISA. The electronic part makes the signals interfere in a same way, and every interference signals are monitored via a phasemeter.

2.2. Optical layout

The optical layout of the optical part of the LOT is presented in Figure 3:

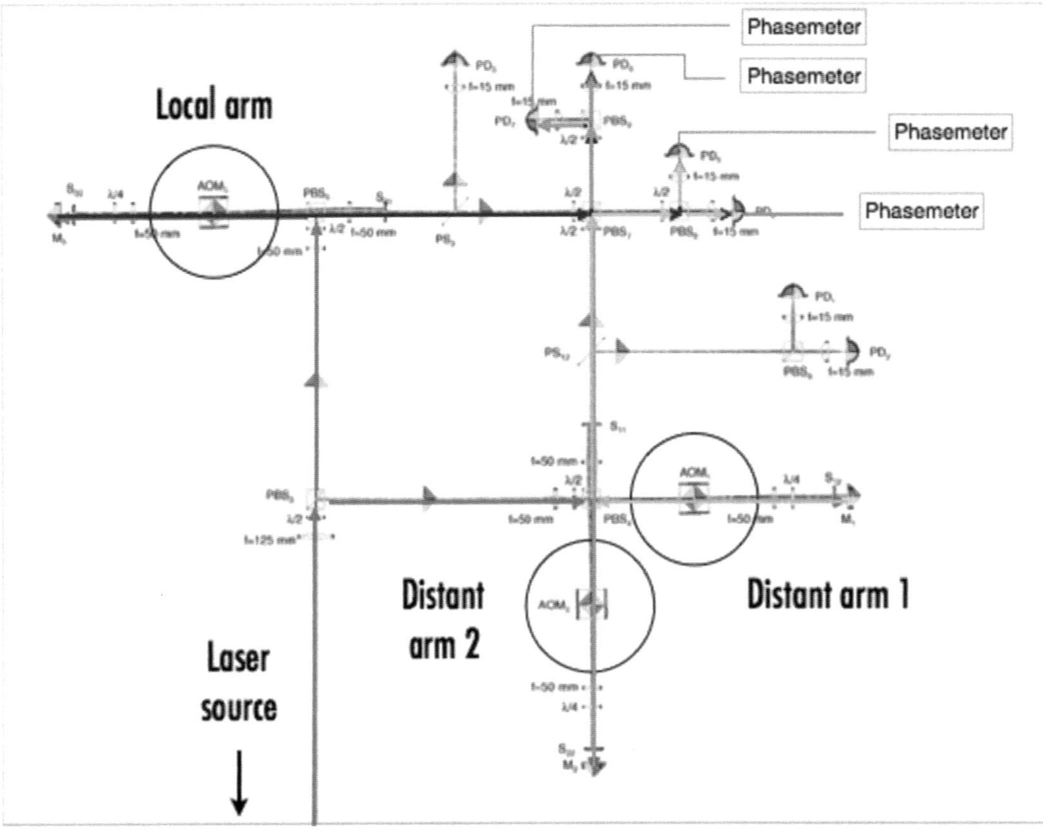

Figure 3: optical layout of the LOT

This layout corresponds to a modified Mach-Zehnder interferometer. A polarizing beam splitter splits a 1064nm laser beam. One part goes entirely through a first AOM corresponding to the local arm, whereas the other is split between two AOMs corresponding to the distant arms. Every AOMs are in a so-called cat's eye configuration: the beam goes twice through the AOM, and stay parallel in the end. Using a well-chosen configuration of λ/2 and λ/4 plates, one can have the beam to be completely transmitted or reflected in the different polarizing beam splitters. In the end, 2 interferences occur: the local arm with the first distant arm on one hand, the local arm with the second distant arm on the other hand.

3. Latest results and future developments

3.1. Latest results

Pierre Grüning has obtained the presented results during his PhD work [5]. They have been obtained using the following configuration: TDI first generation, static uneven arms. The simulated noise is a white noise. Blue lines are the noises from the optical part, red line the ones from the electronic part. Both upper lines are the signals before TDI is applied.

After TDI, the noise of the optical part goes down by a factor 5.10^7 and meet the reference noise level (obtained when there is no simulated noise). It means that in this case TDI works, all the remaining noise comes from the system itself.

The noise in the electronic part goes down by a factor 10^8 but is still 20 times higher than the reference. It is thought that there is a remaining jitter noise between the DDS channels; a solution to this problem is under development at the laboratory.

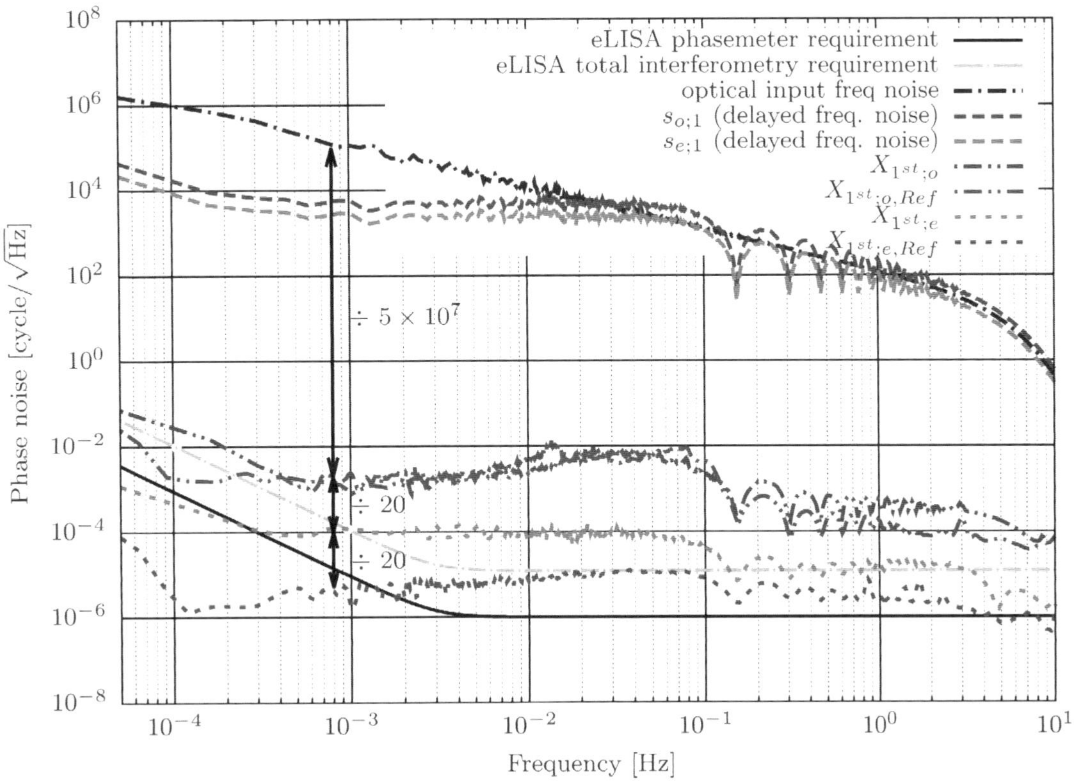

Figure 4: Latest results obtained by P Grüning with TDI 1st generation, static uneven arms, white noise

3.2. Future development

Important developments of the optical part of the LOT are ongoing. First, one wants to lower the system noise level by operating the instrument under primary vacuum. For this purpose, the LOT has been installed in a large vacuum chamber, as seen in Figure 5. Preliminary studies let expect a one order of magnitude noise reduction. The optics still has to be tuned to the optimal operating point.

Figure 5: The optical part of the LOT in its vacuum chamber

Secondly, one wants to perform new simulations, such as Doppler effect and clock noise transfer. The latter needs another modulation of the signal that is to be generated by electro-optical modulators (EOM) placed right after the first beam splitter (before the AOMs). Those EOM have been installed and are ready to be used.

References

[1] P Amaro-Seoane et al., eLISA: Astrophysics and cosmology in the millihertz regime. arXiv:1201.3621 [astro-ph, physics :gr-qc], January 2012. arXiv : 1201.3621.

[2] A Einstein, Die Grundlage der allgemeinen Relativitätstheorie, Annalen der Physik 354, 769

[3] M Armano et al, PRL 116, 231101

[4] Michele Vallisneri. Geometric time delay interferometry. Phys Rev D, 72(4) :042003, 2005.

Michele Vallisneri. Erratum : Geometric time delay interferometry [phys. rev. d 72, 042003(2005)]. Physical Review D, 76(10) :109903, 2007.

D A Shaddock, M Tinto, F B Estabrook, and J W Armstrong. Datacombinations accounting for lisa s pacecraft motion. Phys Rev D, 68 :061303, Sep 2003.

James Ira Thorpe. Lisa long-arm interferometry. Classical and Quantum Gravity, 27(8) :084008, 2010.

S V Dhurandhar. Time-delay interferometry and the relativistic treatment of lisa optical links. Journal of Physics : Conference Series, 154(1) :012047, 2009.

S V Dhurandhar, W-T Ni, and G Wang. Numerical simulation of time delay interferometry for a lisa-like mission with the simplification of having only one interferometer. Advances in Space Research, 51(1) :198 – 206, 2013.

D A Shaddock, B Ware, R E Spero, and M Vallisneri. Postprocessed time-delay interferometry for lisa. Phys Rev D, 70 :081101, Oct 2004.

M Tinto and S V Dhurandhar. Time-delay interferometry. Living Reviews in Relativity, 8(4), 2005.

M Tinto, D A Shaddock, J Sylvestre, and J W Armstrong. Implementation of time-delay interferometry for lisa. Phys Rev D, 67 :122003, Jun 2003.

[5] P Grüning, Développement et exploitation d'un simulateur électro-optique pour le futur détecteur spatial à ondes gravitationnelles eLISA, PhD thesis (2015)

eLISA Telescope In-field Pointing and Scattered Light Study

J Livas[1], S Sankar[1,2], G West[1], L Seals[1], J Howard[1], E Fitzsimons[3]

[1] NASA Goddard Space Flight Center, 8800 Greenbelt Road, Greenbelt, MD 20771, USA
[2] CRESST/USRA, Greenbelt, MD 20771 , USA
[3] UK Astronomy Technology Centre, Royal Observatory, Blackford Hill, Edinburgh EH9 3HJ

E-mail: Jeffrey.Livas-1@nasa.gov

Abstract. The orbital motion of the three spacecraft that make up the eLISA Observatory constellation causes long-arm line of sight variations of approximately ± one degree over the course of a year. The baseline solution is to package the telescope, the optical bench, and the gravitational reference sensor (GRS) into an optical assembly at each end of the measurement arm, and then to articulate the assembly. An optical phase reference is exchanged between the moving optical benches with a single mode optical fiber ("backlink" fiber). An alternative solution, referred to as in-field pointing, embeds a steering mirror into the optical design, fixing the optical benches and eliminating the backlink fiber, but requiring the additional complication of a two-stage optical design for the telescope. We examine the impact of an in-field pointing design on the scattered light performance.

1. Introduction

The LISA reference mission concept consists of a constellation of three identical spacecraft oriented in a planar equilateral triangle orbiting the sun in the same orbit as the Earth, but trailing in orbital phase by approximately 22 degrees (figure 1) [1]. The plane of the triangle is oriented at 60 degrees to the ecliptic. The triangular constellation rotates once per year about the center of the triangle, and the interior angles vary during the course of the mission. Figure 2 shows a representative variation of the angles with time for one choice of orbits [2].

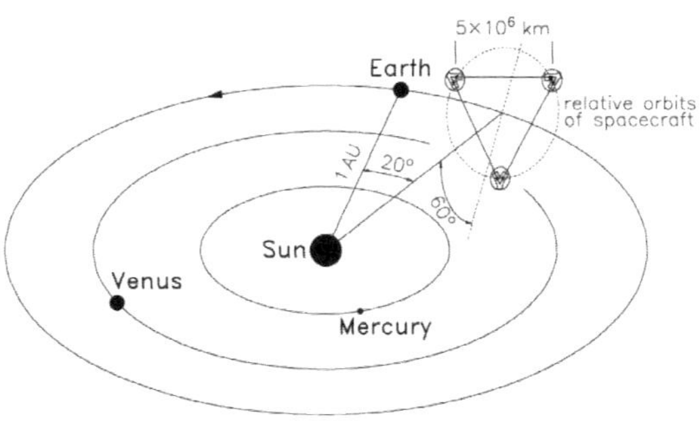

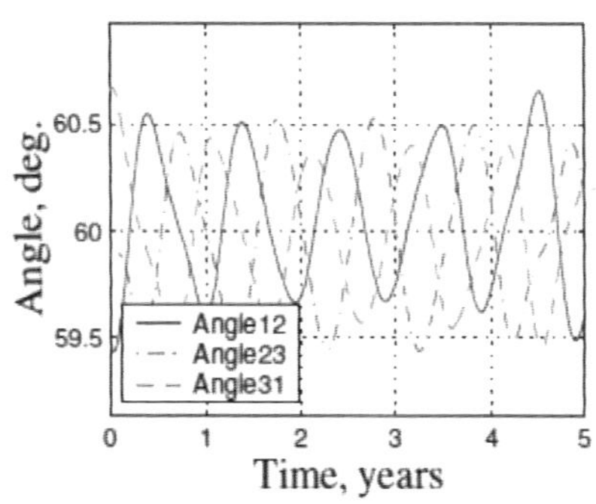

Figure 1: LISA reference mission constellation **Figure 2**: Representative interior angle variation

The variation in the internal angles of the constellation triangle means that the line of sight pointing of the telescopes will vary and must be compensated. There are two proposed solutions: telescope articulation, and in-field pointing [3,4,5]. These two solutions require different telescope designs. The articulation solution is the baseline for the LISA reference mission. Figure 3 shows the two proposed solutions. Each solution is discussed in more detail below.

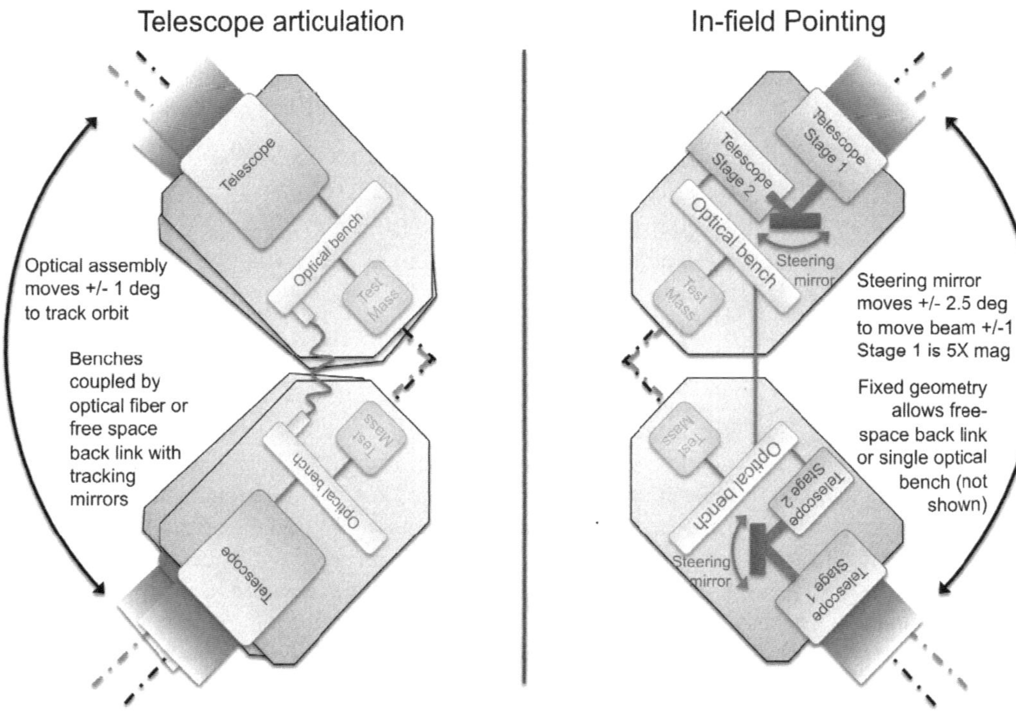

Figure 3: Proposed telescope pointing compensation solutions. Left: telescope articulation. Each telescope is mounted together with an optical bench and a gravitational reference sensor in an assembly that can be rotated as a unit to track the angular variations. Right: The in-field pointing alternative design has a steering mirror embedded into the telescope design so that the telescopes are mounted rigidly to the spacecraft. Specific values shown in the figure are representative of a TNO mirror design and assumed for the purposes of this study.

Both solutions perform the same function for the mission: they enable a continuous length measurement between widely separated proof masses and therefore operate with a transmitter and receiver simultaneously. An important requirement to accomplish this function is low coherent backscatter. This means that very little light from the transmitter must scatter back into the receiver. Since the transmitter is frequency stabilized and phase locked to the laser in the far spacecraft, any scattered light is coherent with the received light and with the local oscillator. Scattered light can therefore interfere with those beams and add noise to the main measurement.

The purpose of this report is to present the results of a preliminary study of the coherent backscatter performance of several representative in-field pointing designs. The study was motivated by concern over the stray light performance of the fixed mirror telescope designs. As will be shown below, the mirrors closest to the detector that collect the light and direct it onto the detector dominate the scattered light budget. An in-field pointing design will require a different optical design due to the practical limitations of the moveable mirror, and will typically have more of these small mirrors. If the scattered light performance scales with the number of mirrors, it may be difficult to meet the scattered light performance. We note that there are a number of other considerations that go into the design of these in-field pointing telescope systems [3,4,5,11].

2. Scattered Light Modeling

The scattered light modelling was done with FRED [6], a non-sequential ray tracing program. For the purposes of this study, two surface properties were considered for the modelling: surface roughness, and particulate contamination.

Surface roughness is specified with a Harvey-Shack model, and we considered two levels of roughness: 5 or 15 Angstroms rms. To some extent the achievable surface roughness depends on the material, the shape of the mirror, and the fabrication technique. State of the art for a flat mirror is < 1 Angstrom rms, so these specifications are not particularly challenging.

Particulate contamination is specified according to MILSTD-1246c [7], and two levels were considered: cleanliness level (CL) 200 or 300. Generally, the scattered light modelling results indicate that surface roughness can usually be made low enough that particulate contamination is the major contribution to the total backscatter.

The real requirement for the scattered light performance of either telescope is that the light that reaches the detector must not add noise to the measurement [8,9]. Since the scattered light is coherent with the transmitter, the local oscillator, and the received signal from the far spacecraft, the complete requirement must specify both the magnitude and the phase stability of the scattered light. To simplify the calculation for purposes of this study, we adopted the simplified specification that any light scattered by the transmitter into the solid angle subtended by the detector at the small beam exit pupils of the telescope would be counted toward the stray light. The solid angle is the instantaneous science field of view multiplied by the angular magnification, or approximately 8 μrad x 90 = 720 μrad. We required this scattered light to be < 10^{-10} of the transmitter power, or about 100 pW for a 1 W transmitter. This amplitude is consistent with other results [9], and should be considered to be a conservative specification.

3. Baseline Solution: Telescope Articulation

A telescope design with fixed mirrors that is representative of a possible solution for the baseline articulated telescope solutions is presented in the next subsection, along with a scattered light analysis, as a point of comparison for the in-field pointing designs.

3.1. Articulated Telescope Basic Design

The design for the baseline articulated telescope solution is a four mirror design with an off-axis Cassegrain/Gregorian primary and secondary, and a Schwarzschild-style pupil extender. An off-axis design is required for either telescope solution to avoid the "narcissus" reflection from the secondary mirror. A complete discussion of the design [10] is outside the scope of this report, but Table 1 shows the basic design assumptions and parameters.

Table 1: Telescope design assumptions.

Parameter	value	Comments
Primary	200 mm	NGO design
Small beam pupil	2.24 mm	real pupil
magnification	~ 90 X	calculated for information only
Instantaneous field of view	± 8 μradian	for science operations
Angular line of sight variation	± 1 degree	Assumed; may not be the final

3.2. Articulated telescope scattered light performance

The basic design described in the previous section along with the scattered light performance is shown in Figure 4. The design meets the simplified scattered light specification of < 100 pW, although with little margin, and of interest for this report is that most of the light comes from the M3/M4 mirror pair.

IOP Conf. Series: Journal of Physics: Conf. Series **840** (2017) 012015 doi:10.1088/1742-6596/840/1/012015

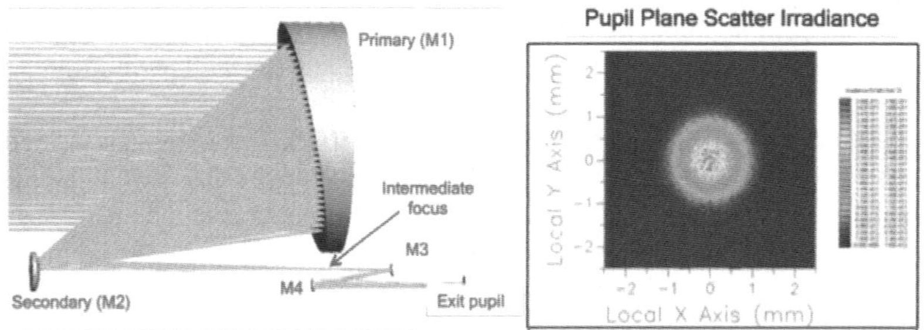

Mirror	RMS surface roughness (Å)	MIL-STD 1246D CL
M1	15	300
M2	15	200
M3	5	200
M4	5	200

- Source power = 1W
- Total power on the detector = 6.6×10^{-11} W → (barely) meets specification of less than 10^{-10}

	Path#	# Rays	Power %	Power	1st scatter surface
3	7	2291695	74.947	4.9421e-11	.20140417_elisa_baseline.M3.Front
4	3	2711030	23.053	1.5201e-11	.20140417_elisa_baseline.M4.Front
2	11	2565386	1.9733	1.3012e-12	.20140417_elisa_baseline.M2.Front
1	14	1399213	0.026184	1.7266e-14	.20140417_elisa_baseline.M1.Front

Figure 4. An example of scattered light modeling for a particular design for the articulated telescope assembly concept.

4. In-field Pointing Solution

For the purposes of this study, we assumed the performance of an in-field pointing (IFP) mirror similar to what has been demonstrated by TNO [12]. Specifically we assumed a mirror with a ± 2.5 degree angular rotation about one degree of freedom (giving ± 5 degree local beam angle change), designed for a 22.5 degree angle of incidence.

If we assume that the required line of sight variation is ± 1 degree at the output of the telescope, then the in-field pointing mirror must be placed inside the telescope where the maximum magnification is five, to accommodate the maximum angular deviation of the beam by reflection from the IFP mirror. This forces a 2 stage design for the telescope into a magnification five stage before the IFP mirror, and a stage with magnification 90/5 = ~ 18 after the IFP mirror.

Three representative designs were analyzed for their stray light performance. Figure 5 shows a design with the minimum mirror count (six) we could use to achieve the full magnification and get exit beam moving in the correct direction. The placement of the exit pupil is probably not practical, but the scattered light performance meets requirements with all mirrors at 5 angstroms rms surface roughness and CL300. The model indicates most of the light scattered into the reference solid angle is scattered from the final mirror.

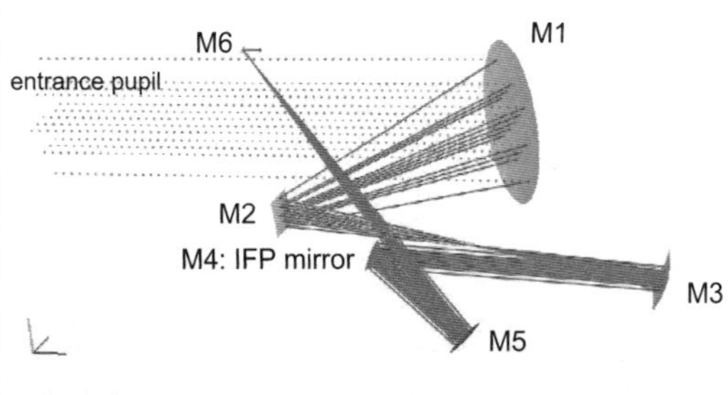

Scatterers	Rays	Power (W in 720 μrad)	RMS surf rough (Å)	CL
All	3004	2.0e-11	5	200
All	3004	1.0e-10	5	300
M5	3081	9.0e-11	5	300

Figure 5: Minimum mirror count (6 mirrors) in-field pointing design. This design meets the specification that the power scattered in the field of view of the detector is $\leq 10^{-10}$.

Figure 6 shows another design with better packaging and one additional mirror. The exit beam is going in the same direction as the beam coming off the primary mirror, which may make it difficult to use in a practical implementation. The scattered light performance is dominated by the final optical element. In this design, however, the scattered light from all of the mirrors is much higher in the minimal mirror design of Figure 5 with the addition of only one mirror. This result indicates that the details of the design and layout are important, not just the mirror count.

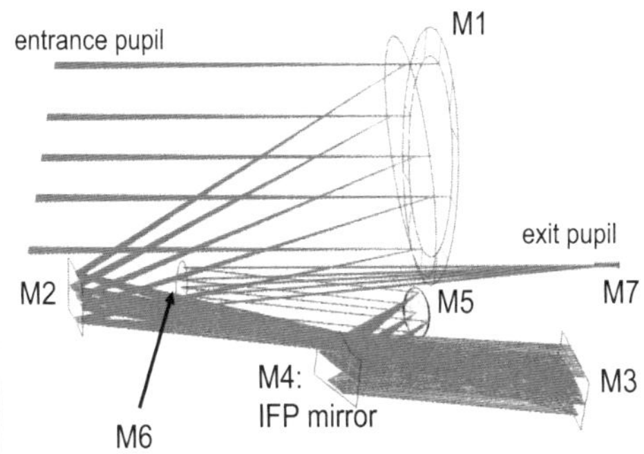

Scatterers	Rays	Power (W in 720 μrad)	RMS surf rough (Å)	CL
All	10046	1.6e-9	5	200
All except final mirror	23	1.5e-12	5	200

Figure 6: Repackaged in-field pointing design with 7 mirrors. This design meets the scattered light requirement only if there is no scattering from the final mirror.

The final design considered here, shown in Figure 7, includes some further repackaging plus a pair of relay lenses to try to push the exit pupil back toward the optical bench in a more realistic manner. The scattered light performance of the lenses is more complicated because in addition to scattering at the surface there is then generation of ghost beams from the reflections at the anti-reflection-coated interfaces. We found this design could meet the requirements only with unphysical near-perfect anti-reflective (AR) coatings and an unphysical non-scattering final lens. Further work would include a design with just mirrors, however it would be expected that those mirrors would contribute to the scatter given the proximity to the exit pupil.

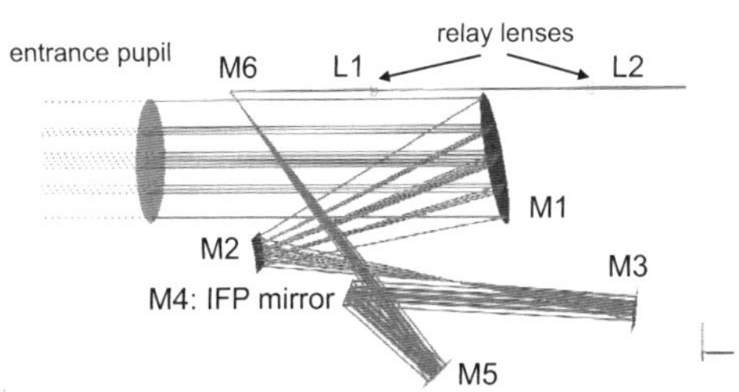

Scatterers	Rays	Power (W in 720 μrad)	RMS surf rough (Å)	CL
All, 10^{-4} AR	38,633	6.8e-7	5	200
All, perfect AR	1,084,665	1.2e-7	5	200
All, perfect AR, no L2	3004	2.0e-11	5	200

Figure 7: In-field pointing design with 6 mirrors plus a relay lens pair. This design meets requirements only if there is no scattering from the final lens.

5. Summary and Next Steps

The results are summarized below:

- Minimum mirror count (6) (but poor packaging)
  - Meets the requirement with 5Å rms surface roughness and CL 200
- Minimum mirror count plus one (7) (better packaging, but unrealistic ray paths and clearances)
  - Does not meet the requirement with 5Å rms surface roughness and CL 200
  - Meets the requirement with no scattering from the final mirror
- With 6 mirrors plus 2 relay lenses
  - Does not meet the requirement with realistic parameters

- Does not meet the requirements with perfect AR lens coatings
- Meets the requirement with perfect AR coatings and no-scatter lenses

Based on these results it does not seem that the scattered light increases in a simple way with the number of small mirrors in the back end of the design. It does suggest that the final optical elements tend to drive the scattered light budget. There has been no attempt to optimize the designs considered here, and that we have only shown that some simple in-field pointing designs do not meet the (simplified) scattered light requirements; we have not shown that it is impossible to do so. The results suggest that is important to consider the scattered light performance as part of the design process.

For the immediate next steps, it may be useful to extend the design with relay lenses to an all-reflective surfaces design to avoid the complication of AR coatings and ghost beams. Scattered light analysis of more realistic and optimized telescope designs than those considered here should be undertaken as well. It would also be useful to have a more realistic calculation for the stray light performance than the simplified metric considered for this study. Longer term, the modelling should be extended to include realistic light baffling and the telescope structure, as well as a better model of the receiver optics. Finally, we have neglected the contribution of thin films to the scattered light budget, such as those that might be formed from non-volatile residue in a space environment. We may need to consider this additional contribution to any scattered light analysis.

Acknowledgments
The authors would like to acknowledge support from the NASA Strategic Astrophysics Technology Program through grant 14-SAT14-0014, and fruitful discussions with Ulrich Johann and Dennis Weise (Airbus Defence and Space).

References

[1] Bender P, Danzmann K, and the LISA Study Team 1998 LISA for the Detection and Observation of Gravitational Waves *Max-Planck-Institut fur Quantenoptik, Garching* Technical Report No. MPQ233
[2] Hughes S Preliminary 2002 Optimal Orbit Design for the Laser Interferometer Space Antenna *25th Annual AAS Guidance and Control Conference* Breckinridge CO
[3] Brugger C An Experiment to Test In-Field Pointing for eLISA 2014 in *Proc. 10th International Conference on Space Optics*, Tenerife, Spain
[4] Brugger C System performance of In-Field Pointing for eLISA 2016 poster B2 *Proc. XIth International LISA Symposium* Zurich Switzerland
[5] Fitzsimons E Optical Bench Design for L3 2016 *poster B7 Proc. XIth International LISA Symposium* Zurich Switzerland
[6] Photon Engineering http://photoneng.com
[7] MILSTD 1246c http://www.phys.vt.edu/~kimballton/borex/ISIS/milstd1246c.pdf
[8] Spector A PhD Thesis University of Florida, 2015
[9] Spector A and Mueller G 2012 Back-reflection from a Cassegrain telescope for space-based interferometric gravitational-wave detectors *Classical and Quantum Gravity* **29** 205005
[10] Livas J, Arsenovic P, Crow J, Hill P, Howard J, Seals L, Shiri S Telescopes for space-based gravitational wave missions 2013 *Optical Engineering* **52** (9) 091811
[11] Kögel, H 2013 Interferometric characterization and modeling of pathlength errors resulting from beamwalk across mirror surfaces in LISA *Applied Optics* **52** 3516-25
[12] Witvoet G, Human J, Maniscalco M Realization and testing of an active mirror mechanism for In-Field Pointing in eLISA 2016 *Proc SPIE* **9912** 99126A

IOP Conf. Series: Journal of Physics: Conf. Series **840** (2017) 012016 doi:10.1088/1742-6596/840/1/012016

Suppressing ghost beams: Backlink options for LISA

K-S Isleif[1], O Gerberding[2], D Penkert[2], E Fitzsimons[3], H Ward[4], D Robertson[4], J Livas[5], G Mueller[6], J Reiche[2], G Heinzel[2] and K Danzmann[1,2]

[1] Leibniz Universität Hannover, Institute for Gravitational Physics, Callinstr. 38, 30167 Hannover, Germany

[2] Max Planck Institute for Gravitational Physics (Albert Einstein Institute), Callinstr. 38, 30167 Hannover, Germany

[3] UK Astronomy Technology Centre, Royal Observatory, Blackford Hill, Edinburgh EH9 3HJ

[4] University of Glasgow, R460 Level 4, IGR Physics & Astronomy, Kelvin Building, University Avenue Glasgow G12 8QQ

[5] NASA's Goddard Space Flight Center, 8800 Greenbelt Rd, Greenbelt, MD 20771, USA

[6] University of Florida, Department of Physics, Gainesville, Florida 32611, USA

E-mail: katharina-sophie.isleif@aei.mpg.de

Abstract. In this article we discuss possible design options for the optical phase reference system, the so called backlink, between two moving optical benches in a LISA satellite. The candidates are based on two approaches: Fiber backlinks, with additional features like mode cleaning cavities and Faraday isolators, and free beam backlinks with angle compensation techniques. We will indicate dedicated ghost beam mitigation strategies for the design options and we will point out critical aspects in case of an implementation in LISA.

1. Introduction

The laser interferometer space antenna (LISA) consists of three spacecraft (S/Cs) forming an equilateral triangle in space [1, 2]. Six laser links between the S/Cs establish the operation of three independent Michelson interferometers. Figure 1a shows the sketch of a possible LISA mission with 5 million km arm lengths and an angle of 60° between the laser links. Due to orbital motion the angle between the laser links will change by $\pm 1.5°$ and the arm length mismatch will reach a maximum deviation of the order of 1% over the period of one year. Accordingly to these effects the interferometric phase measurement will be influenced significantly: First, the arm length mismatch of 50000 km causes laser frequency noise dominantly couple into the phase measurement. Second, without compensation the angle variation will lead to a complete loss of the intra-S/C laser links.

1.1. LISA with Telescope Pointing and Backlink

The current baseline for LISA foresees two movable optical subassemblies per S/C, each consisting of an optical bench (OB) mounted to the back of a telescope along with a proof mass, and equipped with a local transmitted laser beam, so called TX beams, such that the complete optical subassemblies including the OBs can independently follow the incoming beam and compensate the angular variations. This technique, also referred to as Telescope Pointing, ensures the preservation of full contrast within the inter-S/C interferometry by actuating the

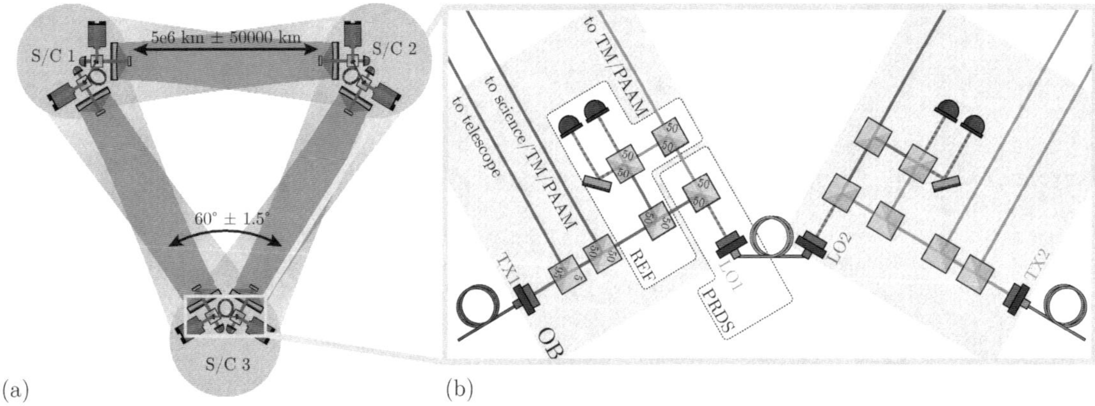

(a) (b)

Figure 1: (a) Sketch of the laser interferometer space antenna (LISA). (b) Magnified view of the two optical benches (OBs) within one S/C showing the phase reference distribution system (PRDS) including the injection of the local oscillator beam (LO) from the adjacent OB via backlink connection, the reference interferometer (REF) and the injection of the local laser light (TX beam). The interferometers per OB are namely the science, test mass (TM) and point ahead angle mechanism (PAAM) interferometer

OBs via two moving optical subassemblies. The data post-processing technique, Time Delay Interferometry (TDI), can deal with the laser frequency noise caused by the unequal arm length interferometer, but it requires, amongst other things, the knowledge about the phase behavior between the two TX beams. As consequence of implementing both, Telescope Pointing and TDI in LISA, two challenges need to be met:

Firstly a flexible optical connection is required that provides the TX beam as local oscillator (LO) beam for the other OB. Secondly, the exchange of the TX beam is required to perform a phase reference measurement. Together, this is the so called backlink, or phase reference distribution system (PRDS). The absolute path length of the PRDS is not crucial, but the path length difference between clockwise and counter-clockwise traveling light, the so called non-reciprocity, is required to be less than $1\,\mathrm{pm}/\sqrt{\mathrm{Hz}}$ such that a single Michelson link is sensitive to about 10^{-20} in strain [4]. Earlier experiments have shown that this phase stability can be reached by using quasi-monolithic interferometers in a thermally stable low-pressure environment [3]. For the actual PRDS the optics are not invariably stable. Due to the rotating OBs the backlink connection is either flexible itself, or actively controlled, or it is insensitive to beam tilts.

1.2. The LISA optical bench

The PRDS will be implemented in the LISA OB and thus depends on the current LISA design. A magnified view of two OBs within one S/C is shown in Figure 1b. Each OB has a local laser source that provides the TX beam. The TX beams are frequency shifted relative to each other by a few MHz to perform heterodyne interferometry. The part of the OB shown here can mainly be separated into three segments: Firstly, as laser preparation we denote the splitting of the TX beam into the interferometers, namely the science, test mass and point ahead angle mechanism interferometer and into the telescope that takes the major share of the laser light (95%). Via the telescope the light is sent to the remote S/C and received by it. Secondly, the PRDS that is located on the OB and provides the transfer of the TX beams between the OBs. The TX beam from the adjacent OB is used as local oscillator (LO) beam on the regarded OB. Thirdly, the reference interferometer interferes the LO and TX beams and therefore delivers the phase information between the two 5 million km Michelson links required for TDI.

IOP Conf. Series: Journal of Physics: Conf. Series **840** (2017) 012016 doi:10.1088/1742-6596/840/1/012016

Both, the engineering and the optical requirements, make the search for an optimal PRDS challenging. This article lists a selection of potential backlink candidates and we briefly estimate their complexity, their applicability for LISA (implementation and redundancy opportunity) and the parasitic noise influences, e.g. caused by ghost beams.

2. The classic fiber

The straight forward approach to achieve a flexible PRDS is the usage of a fiber that exchanges the TX beams between the benches. Figure 2a shows a layout of two OBs hosting a fiber backlink and the reference interferometer, but not the full LISA OB for simplification. Experiments, comparing a fiber backlink connection with an ultra stable quasi-monolithic optical connection implemented on one OB, have shown that a fiber-based PRDS is limited in its phase measurement by so called ghost beams that cause unwanted interferences. By applying designated correction and stabilization methods, like balanced detection for a ghost beam correction in post-processing and the usage of a fiber length stabilization to reduce the dynamic motion of ghost beams generated inside the fiber, the classic fiber backlink is able to achieve the desired LISA performance [4, 5]. But the balanced detection method requires photodiodes on both output ports of the recombining beam splitter and needs to provide about three orders of magnitude of suppression. For keeping the redundancy in LISA the number of photodiodes would increase by a factor of two. Furthermore, the active feedback control loop for the fiber length stabilization requires additional control electronics on board of the S/C. Most of the approaches that reduce ghost beams only increase the signal to ghost beam power ratio but do not fully remove the ghost beams from the set-up. At this point it is crucial to consider the complete LISA OB design including the telescope. Only the total number of ghost beams of each part of the LISA OB counteracting with the interferometers gives a valid estimation of the overall interferometer performance.

2.1. Fiber ghost beam

In general a ghost beam denotes a spurious beam traveling through the set-up in addition to the nominal beams that are used for the interference. There is a large variety of possible sources for ghost beams, like beams occurring due to secondary reflections at non-perfect anti-reflecting (AR) coatings, but this article will only focus on those ghost beams that are potentially limiting the performance in the classic fiber backlink experiment.

Due to the backlink path a direct connection between two fibers exists. Under the assumption that the fiber coupler and the fiber itself reflect a small amount of the light, a low finesse cavity is induced between the fibers. The reflection is caused by direct back reflections at the fiber coupler interface due to non-perfect AR coatings or by the fiber due to e.g. Rayleigh scattering or cladding modes. Both, nominal and ghost beams traveling collinearly inside this cavity, interfere among each other and spoil the phase performance. In the following we will distinguish between two ghost beam sources:

(i) **A ghost beam from the injection** denotes a beam coming from the fiber and fiber output coupler injecting the TX beam onto the optical bench.

(ii) **A ghost beam from the backlink** denotes a beam coming from the PRDS. Regarding a non-fiber PRDS the ghost beams due to the backlink fiber are eliminated.

2.2. Ghost beam coupling

When a spurious beam interferes with one of the nominal beams, or another ghost beam, the resulting signal might influence the phase detection. The level of the phase disturbance depends on the beam power, the polarization state, the beat frequency and the heterodyne efficiency of the ghost beam signal.

An optical signal can be described as a vector, decomposed in phase and amplitude in the complex plane. A signal contaminated with a ghost beam is described by the vector addition of the non-contaminated signal (amplitude a and phase φ) and the ghost beam signal, with an amplitude, a^{SL}, and a phase, φ^{SL}. The phase error φ^{err} caused by only one ghost beam signal can be calculated with simple trigonometry (here $\varphi = 0$ without loss of generality):

$$\tan(\varphi^{\mathrm{err}}) = \frac{a^{\mathrm{SL}}\sin(\varphi^{\mathrm{SL}})}{a + a_{\mathrm{SL}}\cos\varphi^{\mathrm{SL}}}. \tag{1}$$

Under the assumption that the ghost beam signal is a small vector noise its amplitude is negligible in comparison to the non-contaminated amplitude, $a^{\mathrm{SL}} \ll a$, and for small phase errors $\varphi^{\mathrm{err}} \ll 1$. Using these assumptions Equation 1 can be rewritten as

$$\varphi^{\mathrm{err}} = \frac{a^{\mathrm{SL}}}{a} \cdot \sin(\varphi^{\mathrm{SL}} - \varphi) \tag{2}$$

with $\varphi \neq 0$. By including n ghost beams, Equation 2 can be extended to

$$\varphi^{\mathrm{err}} = \frac{1}{a}\sum_{i}^{n} a_i^{\mathrm{SL}} \cdot \sin(\varphi_i^{\mathrm{SL}} - \varphi) \tag{3}$$

with $a = \sqrt{\eta P^{\mathrm{TX}} P^{\mathrm{LO}}}$ and $a_i^{SL} = \sqrt{\eta_i^{\mathrm{SL}} P_i^{\mathrm{SL}} P^x}$, while $P^x = \{P^{\mathrm{TX}}, P^{\mathrm{LO}}, P_i^{\mathrm{SL}}\}$. The decomposition of the amplitudes in heterodyne efficiencies, η, and laser powers, P, shows the coupling of ghost beam amplitudes into phase error in dependency on the mode overlap of ghost beam and local oscillator beam, and the contributing laser powers. However, this model neither includes non-clean polarization effects, which could indeed reduce the resulting coupling, nor does it include the small vector noise dynamics that couple non-linearly, leading to effects like frequency up-conversion of out-of-band signals.

3. Modifications of the fiber backlink

Regarding the classic fiber backlink a reduction of ghost beams can be achieved by improving the signal to ghost beam power ratio by installing attenuation beam splitters on the OB. While the nominal beam enters the interferometer only once the ghost beam passes the attenuation stages several times. A 5/95 beam splitter is envisaged to be implemented in LISA for the power splitting for the telescope. Due to the limited laser power consumption in the LISA telescope and the local interferometry the implementation of additional attenuation stages is not desirable.

Figure 2b shows an alternative design of a fiber-based backlink with a mode cleaning cavity implemented behind the injection fiber. The cavity has a narrow bandwidth and is matched to the laser frequency of the TX beam and thus filters the frequency shifted LO beam from the backlink. The generation of ghost beams from the injection is prevented. Due to the triangular cavity shape direct back reflections from the cavity itself can be avoided. However, an additional cavity would increase the complexity on the LISA OB and requires further locking electronics for stabilizing the cavity length. Furthermore, this solution only handles ghost beams coming from the injection fiber, not from the backlink.

Avoiding ghost beams from the injection can also be realized with a Faraday isolator that is inserted in the path from the injection fiber output coupler to the backlink as shown in Figure 2c. Back traveling light from the PRDS will be separated by the Faraday isolator and thus cannot be reflected at the injection fiber. Critical aspects in this design are the phase stability of the rotator and its magnetic field effecting the free floating test masses in the LISA S/C.

Also shown in Figure 2c is the installation of two acousto-optic modulators (AOMs) in front of the backlink fiber. The AOMs use the acousto-optic effect to shift the laser frequencies of the TX beams before they are coupled into the backlink fiber. The frequency shifts have a different

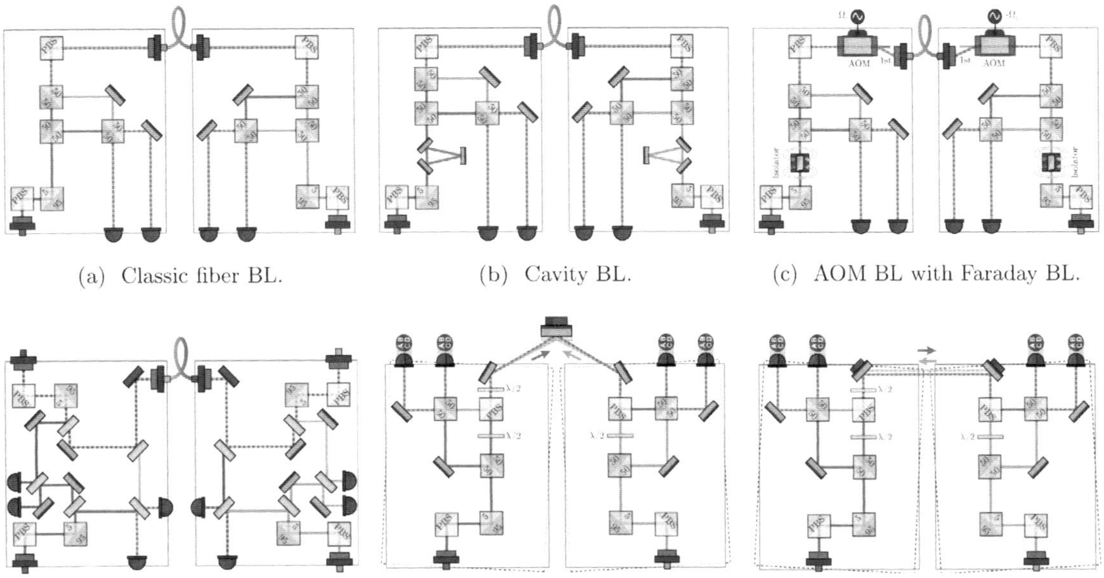

(a) Classic fiber BL. (b) Cavity BL. (c) AOM BL with Faraday BL.

(d) Frequency shift BL. (e) Mirror in the middle BL. (f) Free beam BL.

Figure 2: An overview of backlinks (BLs) for LISA.

sign such that the total frequency shift collected by the beams is kept in the detection bandwidth. A ghost beam passes the same AOM two times and thus collects the same frequency shift twice. Consequently ghost beams from the backlink are frequency shifted to non-detectable frequencies. Since the AOM cannot fully preserve the Gaussian beam quality both the heterodyne and the coupling efficiency will drop. Putting the AOM at a different position, such that the backlink fiber can be used as mode cleaner, can restore the interference quality. Unknown is the thermal behavior of the AOM, a significant heat source on the OB, and the electro magnetic crosstalk between two of these devices.

Another approach is shifting the resulting ghost beam beat note out-of-band by using additional laser sources, so called additional local oscillators (ALO). A possible layout is depicted in Figure 2d. The ALO beams are used for the light exchange between the two OBs. The four laser sources are all offset phase locked to each other such that they are all in the detection band. This ensures a minimal coupling of ghost beams from both sources, the injection and the backlink, into the interferometric measurement. However, the induced cavity between ALO and LO fiber generates ghost beams of second order, meaning laser beams that are reflected at both fibers. After two reflections the ghost beam power is significantly, but potentially not sufficiently, smaller. This alternative requires an additional single sideband, for example from another laser, and a third fiber output coupler per OB, further increasing the complexity of the LISA OB.

4. Free beam solutions

The exclusion of a fiber in the backlink connection eliminates the ghost beams from the backlink fiber itself and is therefore an attractive alternative.

The so called mirror in the middle backlink, shown in Figure 2e, demonstrates one thinkable design. To compensate the angular variations of the incoming beam due to the rotation of the two OBs a mirror is placed in the middle of the OBs such that the preservation of heterodyne efficiency in the local interferometry is ensured via the Law of Reflection. This alternative requires that the rotation axis of both OBs is at the laser reflection point on the mirror surface. The engineering complexity and the controlling of the moving optical subassemblies within the S/C would increase. Any significant deviation from the rotation axis would cause a parallel beam shift and the mirror needs to be controlled in its translational axis. Finally, with one mirror in the middle at one specific point the redundancy opportunity is not trivial and, additionally, the mirror needs to be installed outside of the OB at a fixed point. However, with the usage of polarizing optics in the backlink path, the backlink beam does not travel to the injection fiber, which makes this solution, theoretically, ghost beam free except for the uncertainties of the suppression properties of the polarizing optics. In earlier experiments the required $1\,\mathrm{pm}/\sqrt{\mathrm{Hz}}$ phase stability of polarizing optics had been demonstrated for the LISA OB and test mass readout [3].

A second, more likely, approach for a free beam backlink is the usage of two steering mirrors, one per OB, that will actively be actuated for compensating the angle of the incoming beam. A sketch is shown in Figure 2f. By controlling both, the horizontal and vertical degree of freedom, any bench motion and therefore beam rotation can be compensated. The challenge in this set-up is the implementation of two coupled active feedback control loops for steering the mirrors. The scattering due to the backlink fiber is eliminated and the reduction of back reflections from the LO beam at the injection fiber is feasible by using polarizing optics, as described before.

5. Conclusion

In this article we discussed the impact of the backlink on the LISA optical bench that is required when using Telescope Pointing. In this context ghost beams generated in the backlink and injection fiber were described as the critical noise source for the backlink and the phase reference measurement with a non-linear coupling depending on the ghost beam amplitude and phase.

While a previous experiment with a classic fiber backlink demonstrated $1\,\mathrm{pm}/\sqrt{\mathrm{Hz}}$ reciprocal phase stability, it required dedicated correction methods like balanced detection that are not desirable for LISA. Two approaches were pointed out as an alternative to a simple fiber backlink: One uses additional optics to reduce the relevant ghost beam power with a fiber solution. The other one relies on a steered free beam connection. Further studies and experiments are needed to determine the optimal implementation. To perform these studies under realistic conditions one critical aspect will be to measure the fiber backscatter after radiation, revealing the ghost beam power levels expected during the mission. It would be advisable to try to eliminate the possibility of cladding modes as a source of ghost beams with a direct measurement or by probing the influence of alignment on the ghost beams.

Acknowledgments

The authors would like to thank the DFG Sonderforschungsbereich (SFB) 1128 Relativistic Geodesy and Gravimetry with Quantum Sensors (geo-Q) for financial support. We also acknowledge support by the Deutsches Zentrum für Luft- und Raumfahrt (DLR) with funding from the Bundesministerium für Wirtschaft und Technologie (Project Ref. No. 50 OQ 0601). G. Mueller is supported by NASA grant NNX15AC48G.

References

[1] Danzmann K and others 2013 The gravitational universe: Whitepaper for the ESA L2/L3 selection
[2] Danzmann K and Rüdiger A 2003 *Classical and Quantum Gravity* **20** S1
[3] Dehne M, Tröbs M, Heinzel G and Danzmann K 2012 *Opt. Express* **20** 27273–27287
[4] Fleddermann R 2012 Interferometry for a space-based gravitational wave observatory - reciprocity of an optical fiber (Hannover: Leibniz Universität Hannover)
[5] Diekmann C 2013 Development of core elements for the LISA optical bench - Electro-optical measurement systems and test devices (Hannover: Leibniz Universität Hannover)

IOP Conf. Series: Journal of Physics: Conf. Series **840** (2017) 012017 doi:10.1088/1742-6596/840/1/012017

Enhanced Gravitational Wave Science with LISA and gLISA.

Massimo Tinto[1]

Jet Propulsion Laboratory,
California Institute of Technology,
4800 Oak Grove Drive, Pasadena, CA, 91109, U.S.A.

E-mail: `Massimo.Tinto@gmail.com`

Abstract. The geosynchronous Laser Interferometer Space Antenna (gLISA) is a space-based gravitational wave (GW) mission that, for the past five years, has been under joint study at the Jet Propulsion Laboratory, Stanford University, the National Institute for Space Research (I.N.P.E., Brazil), and Space Systems Loral. With an arm length of 73,000 km, gLISA will display optimal sensitivity over a frequency region that is exactly in between those accessible by LISA and LIGO. Such a GW frequency band is characterized by the presence of a very large ensemble of coalescing black-hole binaries (BHBs) similar to those first observed by LIGO and with masses that are 10 to 100 times the mass of the Sun. gLISA will detect thousands of such signals with good signal-to-noise ratio (SNR) and enhance the LIGO science by measuring with high precision the parameters characterizing such signals (source direction, chirp parameter, time to coalescence, etc.) well before they will enter the LIGO band. This valuable information will improve LIGO's ability to detect these signals and facilitate its study of the merger and ring-down phases not observable by space-based detectors. If flown at the same time as the LISA mission, the two arrays will deliver a joint sensitivity that accounts for the best performance of both missions in their respective parts of the milliHertz band. This simultaneous operation will result in an optimally combined sensitivity curve that is "white" from about 3×10^{-3} Hz to 1 Hz, making the two antennas capable of detecting, with high signal-to-noise ratios (SNRs), BHBs with masses in the range $(10 - 10^7) M_\odot$. Their ability of jointly tracking, with enhanced SNR, signals similar to that observed by the Advanced Laser Interferometer Gravitational Wave Observatory (aLIGO) on September 14, 2015 (the GW150914 event) will result in a larger number of observable small-mass binary black-holes and an improved precision of the parameters characterizing these sources. Together, LISA, gLISA and aLIGO will cover, with good sensitivity, the $(10^{-4} - 10^3)$ Hz frequency band.

1. Introduction

The first direct observation of a GW signal announced by the LIGO project [1] on February 11, 2016 [2], represents one of the most important achievements in experimental physics today and the beginning of GW astronomy [3]. By simultaneously measuring and recording strain data with two interferometers at Hanford (Washington) and Livingston (Louisiana), scientists were able to reach an extremely high level of detection confidence and infer unequivocally the GW source of the observed signal to be a coalescing binary system containing two black-holes

[1] Present address: Center for Astrophysics and Space Sciences, University of California San Diego, 9500 Gilman Drive, La Jolla, California 92093, U.S.A.

of masses $M_1 = 36^{+5}_{-4}\ M_\odot$ and $M_2 = 29^{+4}_{-4}\ M_\odot$ out to a luminosity distance of $410^{+160}_{-180}\ Mpc$ corresponding to a red-shift $z = 0.09^{+0.03}_{-0.04}$ (the above uncertainties being at the 90 percent confidence level). The network of the two LIGO interferometers could constrain the direction to the binary system only to a broad region of the sky because the French-Italian VIRGO detector [4] was not operational at the time of the detection and no electromagnetic counterparts could be uniquely associated with the observed signal.

Ground-based observations inherently require use of multiple detectors widely separated on Earth and operating in coincidence. This is because a network of GW interferometers operating at the same time can (i) very effectively discriminate a GW signal from random noise and (ii) provide enough information for reconstructing the parameters characterizing the wave's astrophysical source (such as its sky-location, luminosity distance, mass(es), dynamic time scale, etc.) [5, 6]. Space-based interferometers instead, with their six links along their three-arms, have enough data redundancy to validate their measurements and uniquely reconstruct an observed signal [7]. A mission such as LISA[2], with its three operating arms, will be able to assess its measurements' noise level and statistical properties over its entire observational frequency band. By relying on a time-delay interferometric (TDI) measurement [8] that is insensitive to GWs [9], LISA will assess its in-flight noise characteristics in the lower part of the band, i.e. at frequencies smaller than the inverse of the round-trip-light time. At higher frequencies instead, where it can synthesize three independent interferometric measurements, it will be able to perform a data consistency test by relying on the null-stream technique [6, 10, 11], i.e. a parametric non-linear combination of the TDI measurements that achieves a pronounced minimum at a unique point in the search parameter space when a signal is present. In addition, by taking advantage of the Doppler and amplitude modulations introduced by the motion of the array around the Sun on long-lived GW signals, LISA will measure the values of the parameters associated with the GW source of the observed signal [7].

Although a single space-based array such as LISA can synthesize the equivalent of four interferometric (TDI) combinations (such as the Sagnac TDI combinations $(\alpha, \beta, \gamma, \zeta)$) [8], its best sensitivity level is achieved only over a relatively narrow region of the mHz frequency band. At frequencies lower than the inverse of the round-trip light time, the sensitivity of a space-based GW interferometer is determined by the level of residual acceleration noise associated with the nearly free-floating proof-masses of the onboard gravitational reference sensor and the size of the arm-length. This is because in this region of the accessible frequency band the magnitude of a GW signal in the interferometer's data scales linearly with the arm-length. At frequencies higher than the inverse of the round-trip light time instead, the sensitivity is primarily determined by the photon-counting statistics [12]. Note that this noise grows linearly with the arm-length through its inverse proportionality to the square-root of the received optical power and because, in this region of the accessible frequency band, the GW signal no-longer scales with the arm-length. In summary, for a given performance of the onboard science instrument and optical configuration, the best sensitivity level and the corresponding wideness of the bandwidth over which it is reached are uniquely determined by the size of the array.

The best sensitivity level of a space-based interferometer and the corresponding bandwidth over which it is reached become particularly important when considering signals sweeping up-wards in frequency such as those emitted by coalescing binary systems containing black-holes. As recently pointed out by Sesana [13], it is theoretically expected that a very large ensemble of coalescing binary systems, with masses comparable to those of GW150914, will have characteristic wave's amplitudes that could be observable by LISA over an accessible frequency region from about 1.5×10^{-2} Hz to about 7.6×10^{-2} Hz. These frequency limits correspond to

[2] Over the years several LISA mission designs were considered, each resulting in a different mission sensitivity. In this article we will refer to LISA as being characterized by an arm-length of 5 Mkm, an acceleration noise of $3.0 \times 10^{-15}\ m/s^2\ (Hz)^{-1/2}$ and a high frequency noise of 12 $pm/\sqrt{Hz}$.

(i) the assumption of observing a GW150914-like signal for a period of five years (approximately equal to its coalescing time), and (ii) the value at which the signal's amplitude is equal to the LISA sensitivity. Although one could in principle increase the size of the LISA's optical telescopes and rely on more powerful lasers so as to increase the upper frequency cut-off to enlarge the observational bandwidth, in practice pointing accuracy and stability requirements together with the finiteness of the onboard available power would result in a negligible gain.

A natural way to broaden the mHz band, so as to fill the gap between the region accessible by LISA and that by ground interferometers, is to fly an additional interferometer with a smaller arm-length. The geosynchronous Laser Interferometer Space Antenna (gLISA) [12, 14, 15], which has been analyzed during the past five years by a collaboration of scientists and engineers at the Jet Propulsion Laboratory, Stanford University, the National Institute for Space Research (I.N.P.E., Brazil), and Space Systems Loral, would provide this capability[3]. With an arm-length of about 7.3×10^4 km, it could achieve a shot-noise limited sensitivity in the higher region of its accessible frequency band that is about a factor of 70 better than that of LISA[4]; gLISA will display a minimum of its sensitivity in a frequency region that perfectly complements those of LISA and aLIGO, resulting into an overall accessible GW frequency band equal to $(10^{-4} - 10^3)$ Hz (See Fig. 1 below).

Regarding the gLISA onboard science payload components (primarily the laser, the optical telescope, and the inertial reference sensor), we will assume them to have a similar noise performance as those of LISA. Other subsystems are regarded to have a noise performance that results in a high-frequency noise spectrum that is essentially determined by the photon-counting statistics. For further details we refer the reader to appendix A of Ref. [12].

2. LISA-gLISA joint sensitivity

To derive the expression of the joint LISA-gLISA sensitivity, we first note that the noises in the TDI measurements made by the two arrays will be independent. This means that the joint signal-to-noise ratio squared, averaged over an ensemble of signals randomly distributed over the celestial sphere and random polarization states, $\langle SNR^2_{L+gL} \rangle$, can be written in the following form

$$\begin{aligned} \langle SNR^2_{L+gL} \rangle &\equiv 4 \int_{f_1}^{f_2} \frac{|\tilde{h}(f)|^2}{S_h^{L+gL}(f)} \, df = \langle SNR^2_L \rangle + \langle SNR^2_{gL} \rangle \\ &= 4 \int_{f_1}^{f_2} \left\{ \frac{1}{S_h^L(f)} + \frac{1}{S_h^{gL}(f)} \right\} |\tilde{h}(f)|^2 \, df \ , \end{aligned} \tag{1}$$

where the functions $S_h^i(f)$, $i = L \ , \ gL$ correspond to the squared optimal sensitivities (as defined through the A, E, and T TDI combinations [18, 8]) of the LISA and gLISA arrays respectively. Note that the factor 4 in front of the integral in Eq.(1) reflects our adopted convention of using one-sided power spectral densities of the noises [19, 20].

From Eq. (1) it is straightforward to derive the following expression of the joint sensitivity

[3] Sean T. McWilliams independently proposed a mission concept very similar to gLISA called GADFLI. Both concepts were first (independently and simultaneously) proposed to NASA as part of its 2011 Request For Information (RFI) exercise for a LISA-like GW mission [16].

[4] The derivation of the gLISA sensitivity and the related discussion on the magnitude of the noises that define it have been presented in the main body and appendix of Ref. [12]. Although it was stated there that additional technology developments were needed to reduce the optical bench noise level below that of the photon-shot noise, the LISA Pathfinder experiment [17] has demonstrated this noise to be almost three orders of magnitude smaller than its anticipated value.

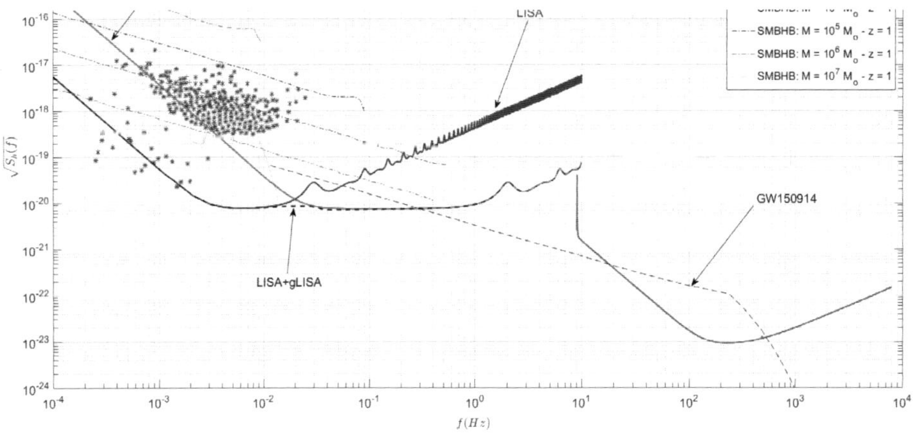

Figure 1. The optimal sensitivities of LISA (blue) and gLISA (red) averaged over sources randomly distributed over the sky and polarization states. Each of these curves is uniquely determined by the TDI A, E, and T combinations associated with the data from each mission. (See [18] for the derivation of the A, E, and T combinations and the resulting optimal sensitivity. Also refer to [8] for the revised expressions of their noise spectra.) The dashed-thick line (black) is the result of optimally combining the sensitivities of LISA and gLISA if they would fly at the same time [21]. As expected, in the lower part of the mHz band, LISA defines the network sensitivity; in the overlapping region the two sensitivities smoothly blend, resulting in a maximum sensitivity gain of $\sqrt{2}$. In the higher part of the accessible band gLISA defines the joint LISA-gLISA sensitivity. For completeness we have included the projected aLIGO sensitivity (magenta) together with the amplitude of GW150914 (dashed-thin black) as a function of the Fourier frequency f. Also included are the wave amplitudes of higher-mass black-hole binaries at $z = 1$, and a synthesized population of galactic stellar-mass binaries.

squared, $S_h^{L+gL}(f)$, in terms of the individual ones, $S_h^L(f)$ and $S_h^{gL}(f)$

$$S_h^{L+gL}(f) \equiv \frac{S_h^L(f)\ S_h^{gL}(f)}{S_h^L(f) + S_h^{gL}(f)} \ . \qquad (2)$$

In Fig. 1 we plot the optimal sensitivities of LISA (blue), gLISA (red) averaged over sources randomly distributed over the sky and polarization states, together with the joint LISA-gLISA (black) given by Eq. (2). For completeness, and to visually exemplify the scientific advantages of flying simultaneously two space-based missions of different arm-lengths, we have included the anticipated sensitivity of the third-generation aLIGO detector (green) together with the amplitude of the GW signal GW150914 (magenta) as functions of the Fourier frequency f. Shortly before the announcement made by aLIGO of the detection of a second signal emitted by another black-hole binary system, GW151226, with masses roughly half of those of GW150914 [2, 22], Sesana estimated that a large number of such systems [13] could be observed by LISA while they are still spiraling around each other for periods as long as the entire five years duration of the mission. Because of these compelling astrophysical reasons, to quantify the scientific advantages of flying gLISA jointly with LISA we will focus our attention on signals emitted by coalescing black-hole binaries with chirp masses in the range $(10 - 100)M_\odot$.

3. Results

In our analysis we will use the following expressions (valid for circular orbits) of the Fourier transform of the amplitude of the GW signal emitted by such systems, $\tilde{h}(f)$, and the time, t_c, it takes them to coalesce [23]

$$\tilde{h}(f) = \sqrt{\frac{5}{24}}\, \pi^{-2/3} \left[\frac{(1+z)GM_c}{c^3}\right]^{5/6} \frac{c}{D_L}\, f^{-7/6} , \quad (3)$$

$$t_c = \frac{5}{256} \left[\frac{(1+z)GM_c}{c^3}\right]^{-5/3} (\pi f_{gw})^{-8/3} . \quad (4)$$

In Eqs.(3, 4) G is the gravitational constant, c is the speed of light, z is the cosmological red-shift, D_L is the corresponding luminosity distance, $M_c \equiv (M_1 M_2)^{3/5}/(M_1 + M_2)^{1/5}$ is the chirp mass associated with the binary system whose components have masses M_1 and M_2, f is the Fourier frequency and f_{gw} the instantaneous frequency of the emitted GW.

By using the above expression for the signal amplitude and a time to coalescence of 5 years (which, in the case of the GW150914 signal, defines the lower-limit of integration in the integral of the SNR to be equal to 0.0151 Hz), we have estimated the SNRs achievable by the two interferometers when operating either as stand-alone or jointly. We find LISA can observe a GW150914-like signal with a SNR equal to 10.7, while gLISA with an SNR of 14.4 because of its better sensitivity over a larger part of its observable band (see Fig. 1). As expected, the joint LISA-gLISA network further improves upon the SNR of gLISA-alone by reaching a value of about 18.0. From these results we can further infer that gLISA will achieve a sufficiently high SNR to warrant the detection of a GW150914-like signal by integrating for a shorter time. We find that, by integrating for a period that corresponds to 135 days prior to the moment of coalescence of a GW150914-like system, gLISA can achieve a SNR of 10.7.

The level of SNR achievable by the LISA-gLISA network with an integration time of five years is about 80 percent higher than that of LISA alone. This implies, from the estimated parameter precisions derived by Sesana [13] and their dependence on the value of the SNR [24], that the LISA-gLISA network will estimate the parameters associated with the GW source of the observed signal with a precision that is 1.8 better than that obtainable by LISA alone. It should be said, however, that this is a lower bound on the improved precision by which the parameters can be estimated since it is based only on SNR considerations. As pointed out by McWilliams in an unpublished document [25], the Doppler frequency shift together with the larger diurnal amplitude modulation experienced by the GW signal in the gLISA TDI measurements will further improve the precision of the reconstructed parameters beyond that due to only the enhanced SNR. We will analyze and quantify this point in a follow-up article.

From the expressions of the SNR and of the Fourier amplitude of the GW signal, $\tilde{h}(f)$ (Eqs. 1, 3), and by fixing the SNR to a specific value for each operational configuration (stand-alone vs. network), it is possible to infer the corresponding average luminosity distance to a BHB in terms of its chirp mass parameter[5]. By assuming a $SNR = 10$, the corresponding three luminosity distances are monotonically increasing functions of the chirp-mass. In [21] it was

[5] To derive the redshift, z, we have assumed a fiducial ΛCDM flat cosmology with the matter density parameter $\Omega_M = 0.3$, the cosmological constant density parameter $\Omega_\Lambda = 0.7$, and the value of the Hubble parameter at the present time $H_0 = 70\,\mathrm{km\,s^{-1}\,Mpc^{-1}}$. To obtain the luminosity distance, D_L, in terms of the chirp mass, M_c, we have rewritten the SNR as a function of the redshift by expressing D_L in terms of z (see Eq. (13) in [12]). Then, for a given SNR, we have solved a transcendental equation to obtain z, and from it derived the luminosity distance.

shown that, for chirp-masses in the range $[10-36)\ M_\odot$, gLISA can see signals further away than LISA because of its better sensitivity at higher frequencies. Systems with chirp-mass larger that 36 $M_\odot$ instead can be seen by LISA at a larger luminosity distance than that achievable by gLISA alone. The LISA-gLISA network instead out-performs the stand-alone configurations by as much as 40 percent for BHBs with chirp-masses in the interval $(30-40)\ M_\odot$. This results into a number of observable events that is about 3 times larger than that detectable by each interferometer alone. Finally, a GW150914-like signal characterized by a chirp mass of 28.1 $M_\odot$ can be seen by LISA at an average luminosity distance of about 400 Mpc, by gLISA out to 600 Mpc, and by the LISA-gLISA system out to about 800 Mpc.

4. Conclusions

Significant scientific advantages can be achieved by jointly flying with the LISA mission an additional space-based interferometer such as gLISA. Because of its smaller arm-length, gLISA will be able to survey the region of the GW frequency band that is in between those accessible by LISA and aLIGO. By covering the entire mHz and kHz GW frequency band, LISA, gLISA and aLIGO will detect all known sources emitting in this broad frequency region, and observe signals requiring multi-band detection for understanding the physical nature of their sources.

References

[1] http://www.ligo.caltech.edu/ .
[2] Abbott B P, *et al.*, 2016 *Phys. Rev. Lett.,* **116**, 061102.
[3] Thorne K S, 1987, *Three Hundred Years of Gravitation*, Eds. S.W. Hawking, and W. Israel (Cambridge University Press, Cambridge), pp 330-458.
[4] http://www.virgo.infn.it/ .
[5] Schutz B F and Tinto M, 1987, *M.N.R.A.S.*, **224**, 131.
[6] Gürsel Y and Tinto M, 1989, *Phys. Rev. D,* **40**, 3884.
[7] P. Bender, K. Danzmann, & the LISA Study Team, July 1998, "Laser Interferometer Space Antenna for the Detection of Gravitational Waves", Pre-Phase A Report, MPQ233 (Max-Planck- Institüt für Quantenoptik, Garching).
[8] Tinto M and Dhurandhar S V, 2014, *Living Rev. Relativity,* **17**, 6.
[9] Tinto M, Armstrong J W and Estabrook F B, 2000 *Phys. Rev. D,* **63**, 021101(R).
[10] Tinto M and Larson S L, 2004, *Phys. Rev. D,* **70**, 062002.
[11] Searle A C, Sutton P J and Tinto M, 2009, *Class. Quantum. Grav.,* **26**, 15.
[12] Tinto M, de Araujo J C N, Aguiar O D and Alves M E S, 2013, *Astroparticle Physics*, **48**, 50.
[13] Sesana A, 2016, *Phys. Rev. Lett.,* **116**, 231102.
[14] Tinto M, DeBra D, Buchman S and Tilley S, 2015, *Review of Scientific Instruments*, **86**, 014501.
[15] Tinto M, de Araujo J C N, Kuga H K, Alves M E S and Aguiar O D, 2015, *Class. Quantum Grav.*, **32**, 185017.
[16] NASA solicitation # NNH11ZDA019L titled: "Concepts for the NASA gravitational-wave mission". http://nspires.nasaprs.com/external/
[17] Armano M *et al.*, 2016, *Phys. Rev. Lett.,* **116**, 231101.
[18] Prince T A, Tinto M, Larson S L and Armstrong J W, 2002, *Phys. Rev. D,* **66**, 122002.
[19] Flanagan E E and Hughes S A, 1998, *Phys. Rev. D,* **57**, 4535.
[20] Tinto M and Armstrong J W, 1999, *Phys. Rev. D,* **59**, 102003.
[21] Tinto M and de Araujo J C N, 2016, *Phys. Rev. D*, **94**, 081101 (R).
[22] Abbott B P, *et al.*, 2016, *Phys. Rev. Lett.,* **116**, 241103.
[23] M. Maggiore, *Gravitational Waves. Volume 1: Theory and Experiments - First Edition.*, 2008, (Oxford University Press, Oxford), pag. 174.
[24] Cutler C and Flanagan E E, 1994, *Phys. Rev. D,* **49**, 2658.
[25] S.T. McWilliams, http://arxiv.org/abs/1111.3708

IOP Conf. Series: Journal of Physics: Conf. Series **840** (2017) 012018 doi:10.1088/1742-6596/840/1/012018

Multi-band gravitational wave astronomy: science with joint space- and ground-based observations of black hole binaries

Alberto Sesana

School of Physics and Astronomy, University of Birmingham, Edgbaston, Birmingham B15 2TT, United Kingdom

E-mail: `asesana@star.sr.bham.ac.uk`

Abstract. Soon after the observation of the first black hole binary (BHB) by advanced LIGO (aLIGO), GW150914, it was realised that such a massive system would have been observable in the milli-Hz (mHz) band few years prior to coalescence. Operating in the frequency range 0.1-100 mHz, the Laser Interferometer Space Antenna (LISA) can potentially detect up to thousands inspiralling BHBs, based on the coalescence rates inferred from the aLIGO first observing run (O1). The vast majority of them (those emitting at $f < 10$ mHz) will experience only a minor frequency drift during LISA lifetime, resulting in signals similar to those emitted by galactic white dwarf binaries. At $f > 10$ mHz however, several of them will sweep through the LISA band, eventually producing loud coalescences in the audio-band probed by aLIGO. This contribution reviews the scientific potential of these new class of LISA sources which, in the past few months, has been investigated in several contexts, including multi-messenger and multi-band gravitational wave astronomy, BHB astrophysics, tests of alternative theories of gravity and cosmography.

1. Introduction

The detection by advanced LIGO (aLIGO) of a $36^{+5}_{-4}M_{\odot}$ and 29 $^{+4}_{-4}M_{\odot}$ coalescing black hole binary (BHB) at $z = 0.09^{+0.03}_{-0.04}$ on September 14, 2015 (GW150914) marked the beginning of the era of gravitational wave (GW) astronomy [1]. In the following few months, a second, lighter system (GW151226, [2]), plus a third candidate (LVT151012, [3]) were also observed.

Among the plethora of scientific consequences, those detections made possible for the first time to constrain observationally the coalescence rate of stellar BHBs in the Universe [4]. BHBs can in fact originate from a variety of astrophysical channels, including: standard evolution of field massive stellar binaries [5], dynamical formation in dense stellar clusters [6] or in dense compact remnant cusps surrounding nuclear massive black holes [7], evolution of chemically mixed massive binaries [8], population III remnants [9], primordial black holes [10]. Every channel, however, is subject to considerable uncertainties. Just to cite the mainstream ones as examples, the number of BHBs resulting from standard stellar evolution critically depends on the poorly understood physics driving common envelope evolution, supernova explosions and their subsequent kicks; similarly, the cluster dynamical formation channel is also affected by the unknown supernova kick velocities and by the retention fraction of BHs in the cluster. The net results is that BHB merger rate predictions vary wildly in the literature (see [11] for a review), including some models suggesting rates up to several hundreds events $\mathrm{yr}^{-1}\mathrm{Gpc}^{-3}$ [12, 13]. An extensive analysis of the aLIGO events places the 90% confidence range between 6 and 200 $\mathrm{yr}^{-1}\mathrm{Gpc}^{-3}$ (depending on the assumed BHB mass distribution [3]), close to the upper end of theoretical predictions. Moreover, aLIGO demonstrated that coalescing BHBs can be in excess of $30M_\odot$, whereas, based on mass estimates of BHs in X-ray binaries [14], it has been often customarily assumed those binaries would have masses around $10M_\odot$.

GW150914, in particular, prompted a burst of interest also within the low frequency GW community. Because of its high mass, it was immediately realised that GW150914 would have been visible by a spaceborne interferometer operating in the mHz band few years prior to coalescence [15]. Would the Laser Interferometer Space Antenna (LISA) be operating, we would have known few years in advance GW150914 would have coalesced in the LIGO band on September 14, 2015. Because of the poor constrains on the rates and the standard assumption of $10M_\odot$ objects, BHBs have never been considered a particularly interesting LISA sources, and multi-band GW observations have been only proposed in the context of seed [16] or intermediate mass [17] BHBs . On the contrary, analysis of the BHB population as constrained by aLIGO demonstrated that LISA might see up to thousands of these systems, with a variety of profound scientific consequences that are summarized in this contribution. The paper is organized as follows. Section 2 provides the recipe to construct the BHB population in the LISA band starting from the aLIGO rates and presents detection rates and parameter estimation precision of the observed LISA systems. Section 3 covers several scientific questions that can be tackled using those binaries, and the most relevant points are briefly summarized in Section 4.

2. Merger rates, observable sources and their properties

2.1. Merger rates

aLIGO O1 results place the 90% confidence BHB merger rate within the range 6 – 200 $\mathrm{yr}^{-1}\mathrm{Gpc}^{-3}$, with most of the uncertainty coming from the unknown shape of the underlying BHB mass function. If the BH mass function is biased-high, then aLIGO observations are consistent with lower intrinsic rates. In [4] therefore, rates' probability distributions were computed for two different mass functions: one in which both BH masses, $M_1 > M_2$, are drawn from a log-flat mass distribution (model *flat*), and one in which the primary BH of a binary is drawn from a Salpeter distribution and the secondary from a flat distribution (model *salp*). The *flat* model rate distribution peas at $\approx 35\mathrm{yr}^{-1}\mathrm{Gpc}^{-3}$, whereas the *salp* model favours relatively light BHBs, with a rate distribution peaking at ($\approx 100\mathrm{yr}^{-1}\mathrm{Gpc}^{-3}$).

as detailed in [15], having assumed a mass distribution, the differential merger rate density $d^2n/d\mathcal{M}_r dt_r$ can be converted into the number of emitting sources per unit chirp mass, redshift and frequency populating the sky *at any time* via

$$\frac{d^3N}{d\mathcal{M}_r dz df_r} = \frac{d^2n}{d\mathcal{M}_r dt_r}\frac{dV}{dz}\frac{dt_r}{df_r}, \tag{1}$$

where $\mathcal{M} = (M_1M_2)^{3/5}/(M_1+M_2)^{1/5}$ is the chirp mass, dV/dz is the standard volume shell per unit redshift in the fiducial ΛCDM cosmology ($h = 0.679$, $\Omega_M = 0.306$, $\Omega_\Lambda = 0.694$, [18]), and $dt_r/df_r \propto f_r^{11/3}$ is given by the standard quadrupole GW emission formula. Subscripts $_r$ refer to quantities evaluated in the source rest frame (as opposed to the detector frame): $t_r = t/(1+z)$, $f_r = (1+z)f$, $M_r = M/(1+z)$. Equation (1) can be used to generate Monte Carlo realizations of the BHB cosmic population emitting in the LISA band.

2.2. LISA sensitivity

The LISA concept has significantly evolved in the past few years. In 2014, the European Space Agency (ESA) appointed a gravitational observatory advisory committee (GOAT) to issue a

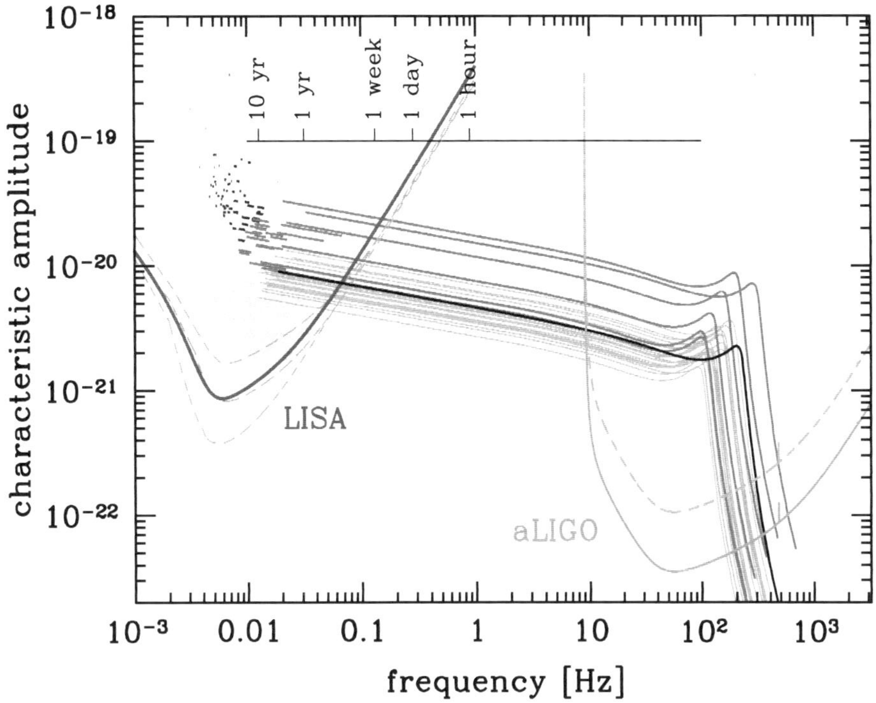

Figure 1. Visual representation of the multi-band GW astronomy concept. Violet dashed lines are, from top to bottom, the total sensitivity curves of LISA configurations N2A1, N2A2, N2A5 (from [19]). The thick solid purple line is the LISA baseline proposed by the LISA Consortium to address ESA's L3 call. Orange lines refer to current (dashed) and design (solid) aLIGO sensitivity curves. Lines in different shades of blue represent amplitude tracks of BHBs found in a selected Monte Carlo realization of the *flat* population model (see main text) seen with S/N> 1 in the new LISA configuration, integrated assuming a four year mission lifetime (baseline LISA4yr). Light and dark blue curves starting around 0.01Hz and extending to $\sim$ 100Hz are BHBs coalescing within the LIGO band during the LISA lifetime, and observable by LISA with S/N> 5 and S/N> 8 respectively; the dark blue ticks in the upper left corner are further sources with S/N> 8 by LISA but not crossing to the aLIGO band within the mission lifetime. Light turquoise lines clustering at the bottom are sources seen in LISA with S/N< 5 (for clarity those were down-sampled by a factor of 20 and sources extending to the aLIGO band were removed). The characteristic amplitude track completed by GW150914 is shown as a black solid line for comparison. The chart at the top of the figure indicates the frequency progression of this particular source in the last 10 years before coalescence. Adapted from [15].

recommendation for a space based GW observatory. During the study, different baselines for a spaceborne interferometer were considered. Full details can be found in [19]. In the following, six baselines featuring one, two or five million km arm-length (A1, A2, A5) and two possible low frequency noises – namely the LISA Pathfinder goal (N1) and the original LISA requirement (N2) – are considered. Those configurations, labelled N1A1, N1A2, N1A5, N2A1, N2A2, N2A5, all assume five years of observations with two equivalent Michelson interferometers (i.e. six active laser links). Building on the GOAT recommendation, the LISA Consortium proposed a somewhat different baseline [20], with slightly different technical specifications, including an armlength of 2.5M kilometres and a mission lifetime requirement of 4 years, and an extension goal to 10 years. The low frequency noise has been set to the level successfully demonstrated by the LISA Pathfinder [21]. We refer to these two latter designs as LISA4yr and LISA10yr. Numbers of detected BHBs are presented below for each of the eight configurations just described.

2.3. Observed systems and their properties

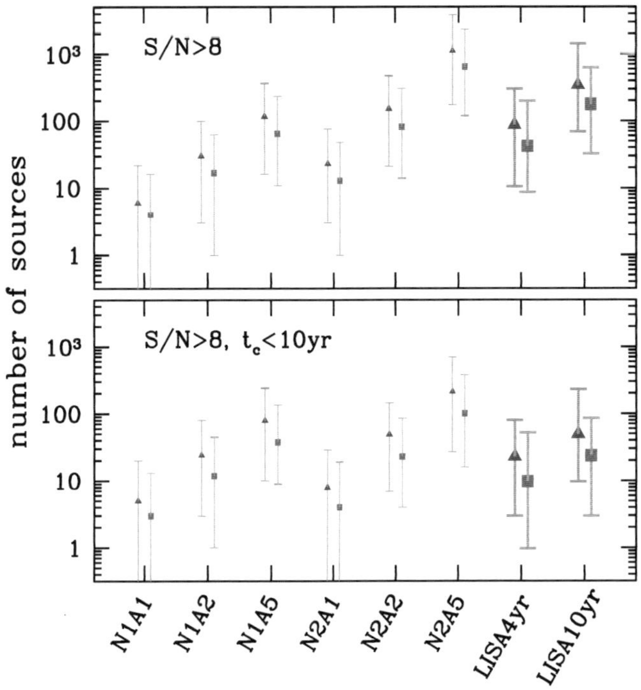

Figure 2. Number of resolved BHBs for different LISA baselines (as labelled on the x-axis). Orange triangles and blue squares are for models *flat* and *salp* respectively. Symbols and error-bars are the median and 95% confidence interval from 200 realizations of the BHB population. The top panel show the total number of sources with S/N> 8 across the whole LISA band, whereas the lower panel is restricted to sources that will eventually coalesce in the aLIGO band within 10 years from the start of the LISA mission. Results for the LISA configuration proposed to ESA, assuming a mission lifetime of either 4 or 10 years (LISA4yr, LISA10yr) are shown with thicker symbols.

Depending on the baseline, the number of sources that can be detected above the nominal signal-to-noise ratio (S/N) threshold of eight varies by more than two orders of magnitudes, from just a few in the N1A1 configuration to about a thousand in the N2A5 configuration. Perhaps counter-intuitively, the armlength appear to have a strong impact on the number of detections. This is because, besides shifting the low frequency sensitivity, armlength also severely affects the depth of the bucket, where most of the resolvable BHBs live (see sensitivity curves in figure 1). Therefore, even with equal high frequency noise (as it is the case for all NxAy configurations), the number of detectable sources increases by about two orders of magnitude going from one (A1) to five (A5) million km. LISA4yr performs rather similarly to N2A2, as expected from the similar mission specifications. The sensitivity in the bucket of the two configurations is essentially identical, resulting in a comparable number of total detections with S/N> 8 (upper panel). Numbers are actually slightly lower for the LISA4yr configuration, because the mission lifetime was assumed to be 5 years in the N2A2 case, and S/N$\propto T^{1/2}$. There is, however, a noticeable difference in the number of sources crossing to the aLIGO band in less then 10 years (lower panel). This is because those accumulate all their S/N at $f > 10$ mHz, where the LISA4yr sensitivity becomes about 30-40% worse than the N2A2 one. Since the cumulative number of sources is proportional to (S/N)^3, this results in a difference of a factor of more than two in detected systems. LISA would definitely benefit from reaching the 10 year lifetime goal. The number of observable sources is in fact boosted by almost a factor of four in the LISA10yr case, with typical detection numbers of several hundreds. This is again because S/N$\propto T^{1/2}$ and the the cumulative number of sources is proportional to (S/N)^3. In general the flat model, in

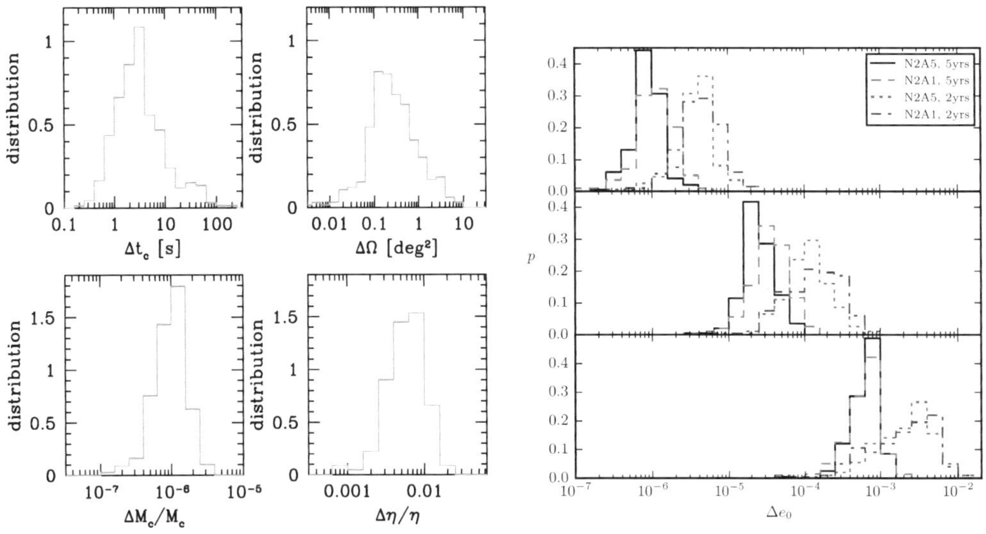

Figure 3. Left figure (from [15]): parameter estimation precision from LISA observations of circular BHBs. Top left: coalescence time; top right: sky localization; bottom left: relative error in the chirp mass $\mathcal{M}$; bottom right: relative error in the symmetric mass ratio $\eta = M_1 M_2/(M_1 + M_2)^2$. Histograms show normalized distributions obtained from a Monte Carlo realization of 1000 sources observed with S/N> 8 in the N2A5 configuration, assuming five years of mission operation (results for other configurations look similar). Estimates were obtained via Fisher Matrix analysis using 3.5 Post Newtonian non spinning waveforms [22] and the full time-dependent LISA response function. Right figure (from [23]): errors on the eccentricity measurement at frequency $f_0 = 10^{-2}$ Hz using eccentric non-spinning waveforms. From top to bottom, the three panels refer to systems with $e_0 = 0.1$, 0.01 and 0.001. Linestyles refer to different LISA baselines: N2A5 and $T_{\rm obs} = 5\,$yrs (solid black), N2A1 and $T_{\rm obs} = 5\,$yrs (dashed red), N2A5 and $T_{\rm obs} = 2\,$yrs (dotted green), N2A1 and $T_{\rm obs} = 2\,$yrs (dash-dotted blue).

virtue of its biased-heavy mass function, results in a factor of ≈ 2 more detections in all cases. Note that for each fixed baseline and BHB population, the 95% confidence region still spans more than one order of magnitude, because of the intrinsic uncertainties on the aLIGO inferred merger rates. In summary, the current proposed LISA baseline (in its LISA4yr and LISA10yr incarnations) has the potential to observe between few tens to a thousand stellar BHBs with S/N> 8, and about 10-to-20% of them will cross to the aLIGO band within 10 years of LISA operation kick-off, realizing the promise of multi-band GW astronomy.

Early calculations by [15] also showed that LISA can achieve exquisite accuracy in the measurement of selected BHB parameters. This is shown in figure 3 for systems that *cross to the aLIGO band* within the LISA lifetime. Time to coalescence can be typically established within less than 10 seconds and sky location accuracy is generally better than a square degree. The redshifted chirp mass and symmetric mass ratio can be determined to a 10^{-6} and 0.01 relative accuracy respectively (which means that individual redshifted masses can be measured to better than 1%). Note that for these sources, LISA and aLIGO measurement can be combined to reduce parameter estimation errors [24]. Considering mildly eccentric (up to 0.1) systems, [23] showed in a follow-up paper that eccentricities in excess of 10^{-3} can be confidently measured, as shown in the right panel of figure 3, which can provide useful information on the BHB formation channel (see below). Other parameters might not be determined with comparable precision. For example, the distance to the source can only be determined to $\Delta D_l/D_l \approx A/({\rm S/N})$, where in general $A > 1$; which gives rather poor results, since most BHBs have S/N< 20. Also spin determination might be problematic in general, as the two BHs are separated by thousands of Schwarzschild radii in the LISA band and the spin-orbit and spin-spin coupling corrections on the waveform are accordingly small. An extensive study on parameter estimation precision, including precessing spinning waveforms is currently ongoing (Klein et al. in preparation).

3. Scientific potential

Because of the relatively large number of observable systems in the LISA band, the exquisite precision to which some of the parameters will be measured, and the possibility of seeing few of them sweeping through the aLIGO band within few years of the initial LISA detection, massive stellar BHBs captured the interest of the wider GW community. This fairly 'unexpected' (in terms of typical mass and rates) class of sources can be exploited in several contexts, including BHB astrophysics, fundamental physics and cosmology. This section focuses on some of the main payouts, keeping in mind that this is relatively new investigation ground, and the list is inevitably incomplete.

3.1. Multi-band and multimessenger astronomy

One of the obvious benefits of sources like GW150914 is the realization of multi-band GW astronomy. Observations of the same signal in two different detectors provides an efficient independent way to cross check and validate the instruments (which is particularly valuable for a space-based detector). Moreover, the two instruments will observe at separate frequency bands, therefore covering different evolutionary stages of BHBs; while LISA will be sensitive to their adiabatic inspiral, most of the aLIGO S/N comes from the very late inspiral, merger and ringdown. Each stage of the evolution carries information about different physical aspects of the sources, that can be combined together to push the boundary of GW astronomy in several directions, as highlighted below.

Another obvious advantage of observing BHBs several months before coalescence, is the possibility of issuing early warnings for an upcoming coalescence. LISA will allow exquisite estimation of some key parameters of the BHBs crossing to the aLIGO band. In particular, the time to coalescence can be predicted within less then ten seconds (but see [25] for potential complications) and the sky localization known within less than a square degree few weeks *before* merger. Although it is not standard to associate electromagnetic (EM) counterparts to BHB mergers, several (sometimes exotic) EM production scenarios have been proposed (see e.g.[26, 27, 28]), triggered by a tentative association of a gamma-ray signal detected by Fermi to GW150914 [29]. The claim is controversial, and the signal has not been confirmed, among others, by simultaneous INTEGRAL observations [30]. In any case, since the Universe is certainly not short of surprises, it is wise keep an open mind and continue to point our telescopes, and this can be done better if we know in advance when and where to observe. Knowing the sky location will facilitate the early pointing of large field of view instruments at all wavelengths. In particular, deep searches can be performed *in coincidence* to the coalescence, testing any possible exotic scenario of emission on the BHB merger dynamical timescale.

3.2. BHB astrophysics

Combining high and low frequency GW detection, will also help in the identification of the astrophysical channel responsible of BHB formation. Different scenarios in fact result in different mass, mass ratio, spin and eccentricity distribution of the detected sources. Ground and space based detectors are conveniently sensitive to different parameters, providing complementary information. By measuring the last few inspiral cycles with high S/N, a network of second generation ground based detectors will allow to measure individual masses within 10% precision.

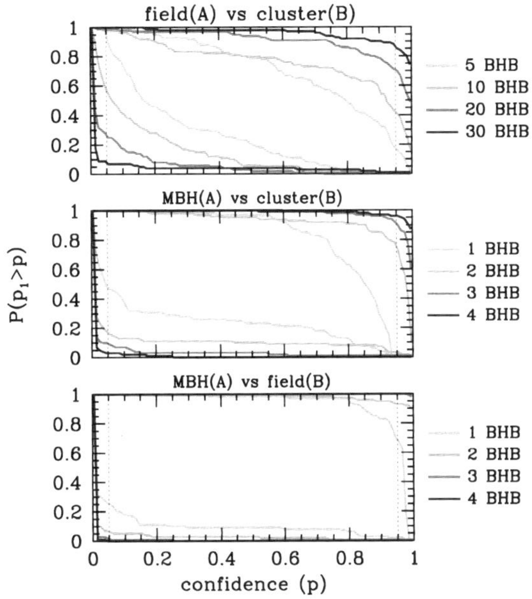

Figure 4. Cumulative distribution function (CDF) of the confidence in a given model over 100 Monte Carlo realizations of a specific number, $N_{\rm obs}$, of observed BHBs (several $N_{\rm obs}$ are shown in each panel as labelled on the right). The N2A2 configuration is assumed, but results are weakly dependent on the adopt LISA baseline. For each pair of curves, the top one marks the CDF of confidence in model A when model A is true, whereas the bottom one marks the CDF of confidence in model A when model B is true. For example, the upper panel compares the *field* (massive binary evolution) and *cluster* (dynamical capture) scenarios; observation of 30 binaries are sufficient to distinguish among them at a 95% confidence level in more than 90% of the realizations. From [31].

Information about spin magnitude and direction can be extracted by observing the effect of spin-orbit coupling induced precession, which is maximum in the late inspiral. However, because of GW circularization, BHBs will generally have a non-measurable eccentricity < 0.01 in the aLIGO band, regardless of their formation channel. LISA, on the other hand, by observing up to millions GW cycles, will allow the measurement of individual (redshifted) masses to better than 1% (and often better than 0.1%). Individual spins can be measured in a fraction of cases (Klein et al in preparation), but generally not to high precision. But perhaps most importantly, being sensitive to lower frequencies, LISA will catch GW signals from BHBs that did not have enough time to fully circularize, measuring any eccentricity in excess of $e \approx 10^{-3}$. As independently pointed out by [31] and [32], this will allow to differentiate between competing formation scenario, chiefly between dynamically formed systems and binary stellar evolution remnants. Figure 4, shows an example of how competitive formation scenarios can be distinguished after a given number of LISA detections, on the basis of eccentricity measurements only. The two lower panels show that a handful of measurements are enough to recognize merger induced by Kozai cycles triggered by the presence of a massive black hole in the proximity of the BHB (model *MBH* from [33]). Conversely, more than 30 detections are required to efficiently differentiate between dynamical formation (model *cluster* from [34]) and stellar evolution remnants (model *field* from [35]). The currently proposed LISA design will likely detect few hundred such systems. Eccentricity information can also be combined with mass and spin measurements (from both LISA and aLIGO) to optimize astrophysical inference.

3.3. Tests of alternative theories of gravity

Multi-band detections will also enhance the potential of gravity tests in the strong, dynamical field regime of merging BHBs. Massive systems will be observed by ground based detectors with high merger (and possibly ringdown) S/N, after being tracked for years by LISA in their adiabatic inspiral. The two portions of signal can be combined to push the search for deviations from General Relativity, due for example to dipolar radiation, as first discussed in [36]. Although timing of binary pulsars excludes the emission of dipole gravitational radiation by binaries [37], entire classes of theories predict this effect predominantly (or only) in binaries involving BHs (see [38] for a review). Joint observations of GW150914-like systems by aLIGO and LISA can potentially improve current bounds on dipole emission from BHBs by more than six orders of magnitude, as shown in figure 5. The figure highlights that even if observed by LISA alone, BHBs constrain these theories better than massive BHBs (MBHBs) and extreme mass ratio inspirals (EMRIs). This is because the signal has to be phased for millions of cycles to be recovered, being therefore sensitive to minuscule phasing deviations caused by dipole emission. When combined to LIGO observations of the same system, typical bounds improve by an additional factor of about five.

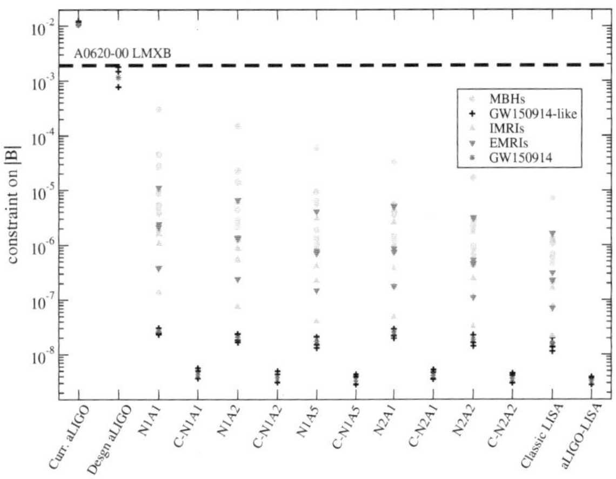

Figure 5. 1σ constraints on the BH dipole flux parameter B as a function of the assumed GW detector from different GW sources: GW150914-like systems (stars and pluses), massive BH binaries (filled circles) and EMRIs/IMRIs (filled triangles). Shown are the NxAy designs proposed in the GOAT study as well as Classic LISA; the prefix C- indicates observations in combination with aLIGO ad design sensitivity. The current best limit on vacuum dipole radiation from LMXB A0620-00 [39] is shown as a dashed horizontal line for comparison. From [36].

3.4. Cosmography and cosmology

Stellar BHBs observed in the LISA band potentially provide a new class of standard candles. In absence of a distinctive EM counterpart, [40] estimated the number of BHBs that can be localized with enough precision that only one galaxy falls in the GW measurement error cube. They concluded that LISA might measure H_0 to a few% precision, based on the analytical scaling of the average $\Delta\Omega$ and $\Delta D_l/D_l$ and the average number density of possible host galaxies. Although this is an useful result, it relies on average estimates of quantities ($\Delta\Omega$ and $\Delta D_l/D_l$) that vary by orders of magnitude from one source to another. Moreover, the outcome depends severely on the uncertain BHB merger rate and on the LISA design. A systematic investigation is currently ongoing (del Pozzo et al., in preparation). Preliminary results show in fact that $\Delta D_l/D_l$ errors in [40] are generally underestimated and association with an individual galaxy is not possible. One has therefore to rely on building a likelihood function based on the redshift distributions of plausible galaxy hosts in the GW measurement error cube. The technique has been first applied by [41] to EMRIs and subsequently by [42] to MBHBs. Note that LISA will provide excellent sky localization up to $\Delta\Omega < 0.1\text{deg}^2$ precision, it will not, however, provide a very good measurement of the luminosity distance, D_l, to the source. This latter might in fact be better measured by ground based detectors, that might see those sources with S/N> 100 at design sensitivity. Multi-band BHBs crossing between the two detectors' windows, might therefore be particularly valuable standard candles, combining LISA sky localization with aLIGO distance determination.

3.5. Milky way BHBs

As first pointed out in [43] and then further elaborated in [44], LISA might also detect few such BHBs within the Milky Way. This specific prospect is particularly appealing for a number of reasons. Considering the typical LISA error-cube, the location of those system within the

Milky Way will be rather accurate, shedding light on their formation mechanism. In case of few detections, a consistent presence of star clusters or globular clusters within the sky localization error would be a decisive element in support of the dynamical formation scenario (viceversa, the lack thereof would not discard this model, since BHBs can be efficiently ejected from their parent clusters, and the clusters themselves can evaporate by the time the BHBs merge). A frequent localization towards the Galactic centre might provide evidence of the role of dense BH cusps surrounding massive black holes in the formation of these systems, as predicted by some formation channels (see e.g. [45]). Their proximity will also allow to perform deep searches for associated EM signals. If BHBs are surrounded by debris disks left behind by partially failed supernovae, even a small amount of accretion leads to an observable photon flux within galactic distances. In the unlikely event that many such BHBs are detected, then their mass function can be linked to the metallicity content of the MW, to gather further clues about the physics driving their formation.

4. Conclusions

The first aLIGO detections showed that stellar BHBs with chirp masses in excess of $\approx 20 M_{\odot}$ are more abundant than previously thought, and the proposed LISA has the potential to see up to about a thousand of them, depending on the adopted mission design. About 10% of these systems, those caught emitting at $f > 10$ mHz, will sweep through the LISA window, eventually coalescing in the audio band probed by aLIGO. Multi-band and multi instrument observations of these sources will open a number of new possibilities for GW astronomy. Eccentricity measurement of BHB in the LISA band will allow to disentangle between different formation channels, and their exquisite sky localization will provide an independent local measurement of the Hubble constant. The phasing of millions of waveform cycles will constrain theories of gravity allowing dynamical scalarization with unprecedented precision, especially if combined with aLIGO observations of the merger phase. In general, joint and complementary LISA-aLIGO observations of the same source will enhance the scientific potential by combining measurements of the early inspiral (LISA) to observations of merger and ringdown (aLIGO). Moreover, S/N accumulated over years of observations in the LISA band will allow to issue early warnings (with few weeks notice) about when and where the BHB will coalesce in the LIGO band. Ground and space based probes across the EM spectrum can therefore be pre-pointed in the direction of the system, performing a deep search for possible EM precursors or for bursts coincident with the final coalescence. Multi-band GW observations of BHBs add a new exciting goal to the already rich LISA science case, opening new avenues in multi-messenger astronomy.

References

[1] Abbott B P, Abbott R, Abbott T D, Abernathy M R, Acernese F, Ackley K, Adams C, Adams T, Addesso P, Adhikari R X and et al 2016 *Physical Review Letters* **116** 061102 (*Preprint* 1602.03837)
[2] Abbott B P, Abbott R, Abbott T D, Abernathy M R, Acernese F, Ackley K, Adams C, Adams T, Addesso P, Adhikari R X and et al 2016 *Physical Review Letters* **116** 241103 (*Preprint* 1606.04855)
[3] Abbott B P, Abbott R, Abbott T D, Abernathy M R, Acernese F, Ackley K, Adams C, Adams T, Addesso P, Adhikari R X and et al 2016 *Physical Review X* **6** 041015 (*Preprint* 1606.04856)
[4] Abbott B P, Abbott R, Abbott T D, Abernathy M R, Acernese F, Ackley K, Adams C, Adams T, Addesso P, Adhikari R X and et al 2016 *ApJL* **833** L1 (*Preprint* 1602.03842)
[5] Postnov K A and Yungelson L R 2014 *Living Reviews in Relativity* **17** (*Preprint* 1403.4754)
[6] Benacquista M J and Downing J M B 2013 *Living Reviews in Relativity* **16** (*Preprint* 1110.4423)
[7] Antonini F and Rasio F A 2016 *ApJ* **831** 187 (*Preprint* 1606.04889)
[8] Mandel I and de Mink S E 2016 *MNRAS* **458** 2634–2647 (*Preprint* 1601.00007)
[9] Hartwig T, Volonteri M, Bromm V, Klessen R S, Barausse E, Magg M and Stacy A 2016 *MNRAS* **460** L74–L78 (*Preprint* 1603.05655)
[10] Bird S, Cholis I, Muñoz J B, Ali-Haïmoud Y, Kamionkowski M, Kovetz E D, Raccanelli A and Riess A G 2016 *Physical Review Letters* **116** 201301 (*Preprint* 1603.00464)

[11] Abadie J, Abbott B P, Abbott R, Abernathy M, Accadia T, Acernese F, Adams C, Adhikari R, Ajith P, Allen B and et al 2010 *Classical and Quantum Gravity* **27** 173001 (*Preprint* 1003.2480)
[12] Belczynski K, Dominik M, Bulik T, O'Shaughnessy R, Fryer C and Holz D E 2010 *ApJL* **715** L138–L141 (*Preprint* 1004.0386)
[13] Dominik M, Berti E, O'Shaughnessy R, Mandel I, Belczynski K, Fryer C, Holz D E, Bulik T and Pannarale F 2015 *ApJ* **806** 263 (*Preprint* 1405.7016)
[14] Farr W M, Sravan N, Cantrell A, Kreidberg L, Bailyn C D, Mandel I and Kalogera V 2011 *ApJ* **741** 103 (*Preprint* 1011.1459)
[15] Sesana A 2016 *Physical Review Letters* **116** 231102 (*Preprint* 1602.06951)
[16] Sesana A, Gair J, Mandel I and Vecchio A 2009 *ApJL* **698** L129–L132 (*Preprint* 0903.4177)
[17] Amaro-Seoane P and Santamaría L 2010 *ApJ* **722** 1197–1206 (*Preprint* 0910.0254)
[18] Planck Collaboration, Ade P A R, Aghanim N, Arnaud M, Ashdown M, Aumont J, Baccigalupi C, Banday A J, Barreiro R B, Bartlett J G and et al 2016 *A&A* **594** A13 (*Preprint* 1502.01589)
[19] Klein A, Barausse E, Sesana A, Petiteau A, Berti E, Babak S, Gair J, Aoudia S, Hinder I, Ohme F and Wardell B 2016 *Phys. Rev. D* **93** 024003 (*Preprint* 1511.05581)
[20] Audley H *et al.* 2017 *ArXiv e-prints* (*Preprint* 1702.00786)
[21] Armano M *et al.* 2016 *Physical Review Letters* **116** 231101
[22] Berti E, Buonanno A and Will C M 2005 *Phys. Rev. D* **71** 084025 (*Preprint* gr-qc/0411129)
[23] Nishizawa A, Berti E, Klein A and Sesana A 2016 *Phys. Rev. D* **94** 064020 (*Preprint* 1605.01341)
[24] Vitale S 2016 *Physical Review Letters* **117** 051102 (*Preprint* 1605.01037)
[25] Bonvin C, Caprini C, Sturani R and Tamanini N 2016 *ArXiv e-prints* (*Preprint* 1609.08093)
[26] Perna R, Lazzati D and Giacomazzo B 2016 *ApJL* **821** L18 (*Preprint* 1602.05140)
[27] Murase K, Kashiyama K, Mészáros P, Shoemaker I and Senno N 2016 *ApJL* **822** L9 (*Preprint* 1602.06938)
[28] Kotera K and Silk J 2016 *ApJL* **823** L29 (*Preprint* 1602.06961)
[29] Connaughton V *et al.* 2016 *ApJL* **826** L6 (*Preprint* 1602.03920)
[30] Savchenko V *et al.* 2016 *ApJL* **820** L36 (*Preprint* 1602.04180)
[31] Nishizawa A, Sesana A, Berti E and Klein A 2017 *MNRAS* **465** 4375–4380 (*Preprint* 1606.09295)
[32] Breivik K, Rodriguez C L, Larson S L, Kalogera V and Rasio F A 2016 *ApJL* **830** L18 (*Preprint* 1606.09558)
[33] Antonini F and Perets H B 2012 *Astrophys. J.* **757** 27 (*Preprint* 1203.2938)
[34] Rodriguez C L, Chatterjee S and Rasio F A 2016 *Phys. Rev.* **D93** 084029 (*Preprint* 1602.02444)
[35] Kowalska I, Bulik T, Belczynski K, Dominik M and Gondek-Rosinska D 2011 *Astron. Astrophys.* **527** A70 (*Preprint* 1010.0511)
[36] Barausse E, Yunes N and Chamberlain K 2016 *Physical Review Letters* **116** 241104 (*Preprint* 1603.04075)
[37] Yunes N and Hughes S A 2010 *Phys. Rev. D* **82** 082002 (*Preprint* 1007.1995)
[38] Berti E *et al.* 2015 *Classical and Quantum Gravity* **32** 243001 (*Preprint* 1501.07274)
[39] Yagi K 2012 *Phys. Rev. D* **86**(8) 081504 URL http://link.aps.org/doi/10.1103/PhysRevD.86.081504
[40] Kyutoku K and Seto N 2016 *ArXiv e-prints* (*Preprint* 1609.07142)
[41] MacLeod C L and Hogan C J 2008 *Phys. Rev. D* **77** 043512 (*Preprint* 0712.0618)
[42] Petiteau A, Babak S and Sesana A 2011 *ApJ* **732** 82 (*Preprint* 1102.0769)
[43] Seto N 2016 *MNRAS* **460** L1–L4 (*Preprint* 1602.04715)
[44] Christian P and Loeb A 2017 *ArXiv e-prints* (*Preprint* 1701.01736)
[45] Bartos I, Kocsis B, Haiman Z and Márka S 2017 *ApJ* **835** 165 (*Preprint* 1602.03831)

IOP Conf. Series: Journal of Physics: Conf. Series **840** (2017) 012019 doi:10.1088/1742-6596/840/1/012019

EMRIs and the relativistic loss-cone: The curious case of the fortunate coincidence

Tal Alexander

Dept. of Particle Physics and Astrophysics, Weizmann Institute of Science, Rehovot, Israel

E-mail: tal.alexander@weizmann.ac.il

Abstract. Extreme mass ratio inspiral (EMRI) events are vulnerable to perturbations by the stellar background, which can abort them prematurely by deflecting EMRI orbits to plunging ones that fall directly into the massive black hole (MBH), or to less eccentric ones that no longer interact strongly with the MBH. A coincidental hierarchy between the collective resonant Newtonian torques due to the stellar background, and the relative magnitudes of the leading-order post-Newtonian precessional and radiative terms of the general relativistic 2-body problem, allows EMRIs to decouple from the background and produce semi-periodic gravitational wave signals. I review the recent theoretical developments [1] that confirm this conjectured fortunate coincidence [2], and briefly discuss the implications for EMRI rates, and show how these dynamical effects can be probed locally by stars near the Galactic MBH.

1. Introduction

Extreme mass ratio inspiral (EMRI) gravitational wave (GW) emission events, where a stellar mass black hole (BH) of mass $M_\star$ gradually inspirals into a massive BH (MBH) of mass $M_\bullet$, are one of the main classes of anticipated extra-galactic low-frequency GW sources. Since $Q = M_\bullet/M_\star \gg 1$, such BHs probe spacetime near the MBH almost as test particles, and thereby offer an opportunity to test general relativity (GR) under theoretically favorable conditions.

However, the inspiral process is not simply a 2-body problem in the strong gravity regime. The MBH is surrounded by $\mathcal{O}(Q)$ background stars inside its radius of dynamical influence $r_h \sim GM_\bullet/\sigma^2$, where σ^2 is the stellar velocity dispersion in the host galaxy's spheroidal. The complex dynamics of the stellar cluster around the MBH, that supplies stars that fall into the MBH, whether directly on plunging orbits, or indirectly on inspiraling ones, also interferes with the idealized 2-body motion of the light BH relative to the MBH, and can even suppress inspiral altogether.

I show here that EMRIs are in fact possible by virtue of a three-way fortunate coincidence between the magnitude and timescale of collective Newtonian effects due to the background stars, which give rise to strong resonant gravitational torques, and those of the leading-order precessional and radiative post-Newtonian (PN) terms of GR [2]. This curious coincidence raises several questions: How fine-tuned is it? Is this a specific feature of Einstein's GR, or is it generic to a larger class of theories of strong gravity? Do still-viable alternatives to GR predict substantially different plunge / inspiral branching ratios? Could the very detection of an EMRI in the future rule out some of them? These questions remain open at this time.

Another consequence of the fortunate coincidence is that rapid relaxation by resonant torques in the symmetric potential near a MBH [3], which can dominate slow 2-body relaxation in some regions of phase space [4], is ultimately not very important in determining the steady-state loss-rates (e.g. by tidal disruption or GW inspiral). By a coincidence that can only be understood in the context of the relativistic loss-cone dynamics, early naive estimates that included only a partial treatment of Newtonian dynamics (2-body relaxation, but not resonant torques) and of GR dynamics (GW dissipation but not in-plane Schwarzschild precession), yielded correct order of magnitude loss-rates estimates [5].

Here I briefly review the recent developments in the understanding and modeling of the relativistic loss-cone that lead to these insights, focusing on the relevance for EMRI rates, and on the prospects of probing these dynamical mechanisms locally, near the Milky Way's MBH Sgr $A^{\star}$.

1.1. The loss-cone problem

Orbits whose periapse lies inside the tidal radius $r_t = R_\star Q^{1/3}$ (where $R_\star$ is the stellar radius) or inside the last stable orbit $r_\bullet \simeq 8r_g$ (where $r_g = GM_\bullet/c^2$) are classified as plunges. In terms of the normalized angular momentum $j = J/\sqrt{GM_\bullet a} = 1-e^2$ (where a is the orbital semimajor axis (sma) and e the eccentricity), the corresponding plunge conditions are $j < j_{lc} = j_t \simeq \sqrt{2r_t/a}$ or $j < j_{lc} = j_\bullet = \sqrt{16r_g/a}$. Since the set of velocity vectors that take a star from position $\boldsymbol{r}$ to the MBH span a cone, the set of such orbits are called the loss-cone. Stars in the loss-cone are destroyed in less than an orbital period. In steady state, new stars have to be deflected to loss-cone orbits, and therefore the loss-cone problem is the problem of how a stellar system around a MBH randomizes. In a spherical potential, where the angular momentum of individual orbits is conserved in collisionless motion, randomization is achieved by dynamical relaxation.

1.2. Loss-cone replenishment by dynamical relaxation

The classical treatments of the loss-cone problem considered only slow 2-body relaxation (denoted here also non-resonant relaxation, NR), which is inherent to any system composed of discrete interacting particles [6, 7, 8, 9]. A relaxed stellar system around a MBH settles into a high density powerlaw stellar cusp, $n_\star \propto r^{-\alpha}$. For a single mass population, which we consider here for simplicity, $\alpha = 7/4$ [7]. Near a MBH, the 2-body relaxation timescale can be expressed as $T_{NR}(a) \sim Q^2 P(a)/N_\star(a)\log Q$, where P is the orbital period, $N_\star(a) \sim Q(a/r_h)^{3-\alpha}$ is the number of stars with sma $< a$, and $\log Q$ is the Coulomb factor [1].

Generally, the 2-body relaxation timescale for changing the angular momentum J by order of itself, T_J, is related to the relaxation timescale for changing the energy E by order of itself, T_E, by $T_J \sim j^2 T_E$ [5]. It is therefore faster to reach the MBH by relaxation of angular momentum than by relaxation of energy (equivalently, relaxation of the sma, since $E = GM_\bullet/2a$, using the stellar dynamical convention $E > 0$ for bound orbits).

In the presence of orbital dissipation (e.g. by the emission of GWs, by tidal deformations of the star close to the MBH, or by hydrodynamical interactions with a massive accretion disk), the star can also fall into the MBH by gradually loosing orbital energy. Such orbits are classified as inspirals. Because inspiral, unlike a plunge, requires many consecutive periapse passages, such orbits are much more susceptible to perturbations by the stellar background, unless the inspiral already starts from a tight orbit around the MBH. phase space is therefore separated into two regimes (Fig. 1). Close to the MBH, below some critical sma a_{GW} (here we consider only dissipation by GWs), stars are statistically much more likely to reach the MBH by inspiral, whereas above a_{GW} and up to $a \sim r_h$, plunges are much more likely. The transition is sharp [10]. The plunge rate is $R_p \sim N_\star(r_h)/\log(1/j_{lc})T_E(r_h)$, while the inspiral rate is $R_i \sim N_\star(a_{GW})/\log(1/j_{lc})T_E(a_{GW})$ [11]. Since the relaxation timescale is typically not a strong function of distance, the branching ratio $R_i/R_p \sim N_\star(a_{GW})/N_\star(r_h)$ reflects the relative number of sources in the small phase space volume interior to a_{GW} and the much larger one

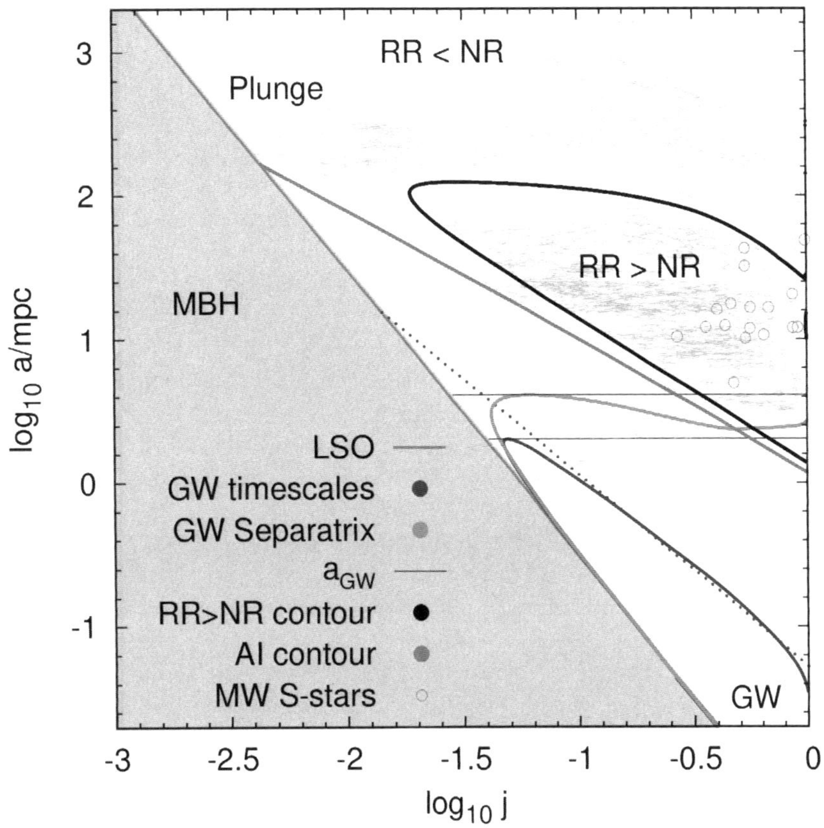

Figure 1. A schematic of the relativistic loss cone in $(\log j, \log a)$ phase space for a simplified model of the Milky Way (MW) nucleus with a $M_\bullet = 4 \times 10^6\,M_\odot$ MBH surrounded by a relaxed cluster of $10\,M_\odot$ stellar BHs [4]. The gray region to the left, below the last stable orbit (LSO) line demarcates the region of unstable orbits. Stars diffuse from the Galaxy beyond the radius of influence at the top ($a \gtrsim 2\,\mathrm{pc}$) and either wander back to the Galaxy, plunge directly into the MBH across the LSO (the light red track is a Monte Carlo generated track of a plunge event), or diffuse deeper into the cusp until the cross the GW separatrix (magenta curve) , where the streamlines of the probability density flux turn over to the region where GW dissipation is faster than 2-body relaxation (blue curve). The critical sma a_{GW} (black horizontal lines, at two possible extreme values) approximately separates phase space into a region $a > a_{GW}$ where the stellar BHs plunge directly into the MBH, and a small phase space volume at $a < a_{GW}$ where the BHs inspiral into the MBH. RR is quenched by fast precession below the AI line (gray line), and the trajectories appear to "bounce" against it due to the strong gradient in the effective RR diffusion coefficient (Fig. 3). RR is faster than NR only in a restricted region of phase space (yellow region), where motion along the j-direction is much faster than in the a-direction. Since the RR-dominated region is well-separated from the plunge and EMRI loss-lines, RR has a only a small effect on the steady state loss-rates (cf Fig. 5). However, it has a strong effect on the orbits of the observed S-stars, which happen to lie in this phase space region (Sec. 3.2). See on-line animation of the plunge and inspiral trajectories at `./EJ_animx08.mp4`.

inside r_h. For that reason inspiral events are generally much rarer than plunges, $R_i \sim \mathcal{O}(0.01)R_p$ [12].

Resonant relaxation [3, 2] is a form of rapid relaxation of angular momentum that occurs in potentials with a high degree of symmetry, which restrict orbital evolution (e.g. fixed Keplerian ellipses in the Newtonian potential of a point mass; fixed orbital planes in a spherical potential). The approximately spherical, point mass-dominated potential around a MBH is such a potential. Far enough from the MBH where GR effects are small, but close enough to it, where deviations from Keplerian motion due to the distributed stellar mass are also small, the orbits are nearly-Keplerian ellipses (for Sgr $A^\star$ this distance scale corresponds to $0.01 - 0.1$ pc [2]), which persist over a long coherence time $T_c \gg P$. On that timescale, the $N_\star$ background stars can be viewed as fixed elliptical mass "wires". A test star orbiting the MBH with sma a in the potential generated by the background, will conserve its energy (since the wires are stationary), but will be subject to a non-zero residual force of order $F_N \sim \mathcal{O}(\sqrt{N_\star(a)}GM_\star/a^2)$, which translates to a coherent torque $\tau_N \sim \mathcal{O}(\sqrt{N_\star(a)}GM_\star/a)$. This torque persists in magnitude and direction over time T_c, until the small deviations from perfect symmetry accumulate and randomize the background. The change in angular momentum over the coherence time, $(\Delta J)_c = \tau_N T_c$ then becomes the mean free path in J-space for a random walk on timescale $t \gg T_c$: $\Delta J = (\Delta J)_c\sqrt{t/T_c} \to \Delta j = \sqrt{t/T_{RR}}$, where the RR relaxation timescale is $T_{RR} \sim [Q^2/N_\star(a)]P^2(a)/T_c(a)$.

The ratio between the RR and NR relaxation times, $T_{RR}/T_{NR} \sim \log Q(P/T_c)$ reflects the fact that NR occurs by point-point interactions, and is boosted by the closest strong interactions, whereas RR occurs by orbit-orbit interactions, which as extended objects cannot approach each other arbitrarily close, but are boosted by the long coherence time. The relevance of RR to the loss-cone problem is due to the fact that near a MBH, it is possible to have $T_{RR}/T_{NR} \ll 1$. This implies very rapid angular momentum evolution, and specifically $j \to 0$, which allows strong interactions with the MBH.

1.3. The fortunate coincidence

The strong RR torques inside $a_{GW} \ll a \ll r_h$, if unquenched, would drive all stars directly into the MBH on plunge orbits and the EMRI rate would drop to zero. Hopman & Alexander conjectured in 2006 [2] that EMRIs will in fact occur because the PN1 $\mathcal{O}(\beta^2 j^{-2})$ GR precession becomes significant enough to quench RR before the PN2.5 $\mathcal{O}(\beta^5 j^{-7} Q^{-1})$ GW dissipation rate becomes fast (here $\beta = v/c$). This was later confirmed in N-body simulations [13, 14] (but see Sec. 1.4), and was also demonstrated by Monte Carlo (MC) simulations of diffusion in the relativistic loss-cone [4] (Sec. 2). Fig. (2) shows that GR Schwarzschild precession has a drastic effect on the steady state phase space densities and rates. EMRIs exist because of a fortunate coincidence in the ranking and phase space dependence of the magnitude of the RR torques, which are a collective Newtonian effect, and the magnitudes of GR's leading order precessional and radiative terms, which are post-Newtonian 2-body effects.

1.4. An unexpected result

The first direct N-body simulation with self-consistent post-Newtonian terms (up to PN2.5, the lowest-order radiative term) [13] yielded an unanticipated result. Not only did GR precession quench RR on eccentric orbits, as conjectured, but in fact it appeared that a barrier in phase space, the "Schwarzschild Barrier" (SB), prevented stars from reaching either the LSO or the GW line (cf Fig. 3). The phase space trajectories of stars seemed to linger near the SB for about a coherence time, while their orbital parameters oscillated at at the Schwarzschild precession frequency, before being reflected back to less eccentric orbits. Larger-scale N-body simulations subsequently confirmed the quenching of RR by GR precession near the SB [14]. Although an early analysis indicated that the SB is related to precession under the influence of a randomly changing uniform force [15, 13], the nature and implications of the SB phenomenon remained controversial.

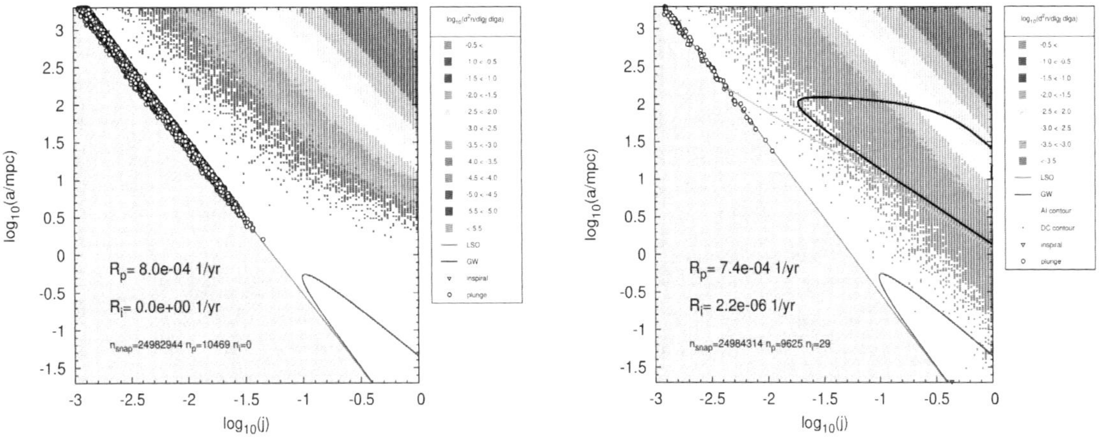

Figure 2. The steady state density in phase space as derived from Monte Carlo simulations of loss-cone dynamics for a simplified model of the Milky Way nucleus surrounded by stellar BHs (see Fig. 1), taking into account the dynamical effects of NR, RR (with the η-formalism) and GR (leading order precession and GW) [4]. The plunge and inspiral rates (R_p, R_i) are quoted on the plot. The end points of a sample of the trajectories are marked by small black circles. Top: GR Schwarzschild precession turned off. In the absence of quenching, the strong RR torques sweep all BHs to plunge orbits well above the GW loss-line, and the EMRI rate drops to zero. Bottom: All dynamical effects are included. The system reaches a steady state that is very close to thermal, with a finite inspiral rate $R_i \sim \mathcal{O}(10^{-6}\,\mathrm{yr}^{-1}) < 0.01R_p$.

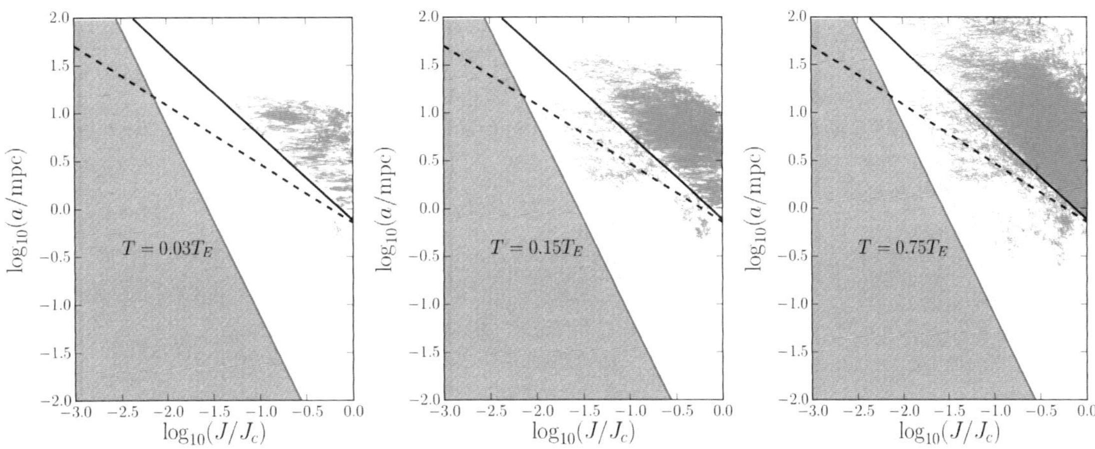

Figure 3. The nature of the SB / AI line. A time sequence (left to right) of three snapshots showing the phase space trajectories of 50 stars of mass 50 $M_\odot$ each, in a relaxed cusp ($\alpha = 7/4$) around a MBH of $10^6\ M_\odot$ [4]. The AI locus is denoted by a solid black line (the dashed line is the mis-identified locus of the SB [13]). At times much shorter than the NR energy relaxation timescale T_E, evolution by rapid RR dominates, but can occur only above the AI line. As time progresses, NR, which is unaffected by AI, populates phase space beyond the AI line, and ultimately establishes the maximal entropy (thermal) equilibrium on timescales $\gtrsim \mathcal{O}(T_E)$, irrespective of AI.

2. The η-formalism for relativistic loss-cone dynamics

Relativistic loss-cone dynamics, and in particular the interplay between secular precession, the coherent RR torques, and uncorrelated NR, lie in the difficult-to-treat interface between deterministic Hamiltonian dynamics and stochastic kinetic theory. The η-formalism [1] provides a formal framework for treating this regime by stochastic equations of motion (EOMs) that describe the dynamics, by identifying the relevant features of the stochastic effects of the stellar background on the test star, and importantly, by enabling the description of RR dynamics by effective diffusion coefficients, in spite of the fact that the long RR coherence time makes it a manifestly non-Markovian process that cannot be described, as is, by diffusion.

The key idea of the η-formalism is to describe the effect of the background stars in terms of a time-correlated noise model. A perturbative expansion of the phase-averaged post-Newtonian Hamiltonian to leading order shows that the symmetries of the noise are those of a vector in angular momentum space, $\boldsymbol{\eta}(t)$[1]. The dynamics of the system are then primarily determined by the temporal smoothness (differentiability) of the noise, and by its coherence time T_c. Neither of these properties are known at present from first principles or from N-body simulations. However, since the collective effect of the background is due to the superposition of many smooth orbital motions, it is likely that the noise is smooth (infinitely differentiable). A smooth noise must have an maximal frequency $\nu_{\max} \propto 1/T_c$, beyond which its power decays rapidly.

When the GR precession frequency, $\nu_{GR}(a,j) = 3(c/r_g)(r_g/a)^{5/2}j^{-2}$ (where $r_g = GM_\bullet/c^2$) is higher than $1/T_c$ (i.e. when the precession period is much shorter than the coherence time), the residual RR force is effectively constant over a precession period, so the RR torque on the precessing eccentric star is reversed every half cycle. Therefore, the net change in j per period is canceled to high precision—the rapidly precessing eccentric orbit is decoupled from the slowly varying stellar background by adiabatic invariance (AI). This happens along the phase space locus

$$j_{AI}(a) = \sqrt{T_c(a)\nu_{GR}(a,j=1)/2\pi}\,. \tag{1}$$

The stochastic EOMs allow to evolve in time the phase space trajectory of a test star, for a given random realization of the background noise. The fact that the noise is approximately a function of time only, makes it possible to formally derive an effective diffusion coefficient for RR, $D_{jj}(a,j)$, which is proportional to the power of the noise at $\nu_{GR}(a,j)$, and is therefore strongly suppressed below the AI locus. This then makes it possible to directly evolve the stellar distribution function by the Fokker-Planck (FP) equation. The great practical advantage of an effective RR diffusion formulation is that it is then possible to model the dynamics of the loss-cone in the realistic $N_\star \to \infty$ limit by an MC procedure that solves the FP equation statistically, and where it is easy to include also NR, the secular precessions due to GR and the enclosed stellar mass, and dissipation by GW (and by tidal heating, when relevant). The MC results are able to reproduce the SB phenomenology observed in the N-body simulations, as well as the plunge and inspiral rates [4]. Fig. (3) shows how the AI/SB phenomenon is reproduced by effective RR diffusion, and also how on long timescales the NR dominates the dynamics, and the system asymptotes to the maximal entropy solution, as it must irrespective of the randomization mechanism[2].

2.1. *How fine-tuned is the fortunate coincidence?*

A simple estimate for the degree of fine-tuning needed for the fortunate coincidence can be derived by artificially varying the Schwarzschild precession frequency by a constant factor,

[1] The noise can be approximated as independent of j because the mean free path in j due to RR is small.

[2] The fact the the system is not closed, and there is a small stellar flux into the less cone, makes possible small deviations from the maximal entropy solution.

IOP Conf. Series: Journal of Physics: Conf. Series **840** (2017) 012019 doi:10.1088/1742-6596/840/1/012019

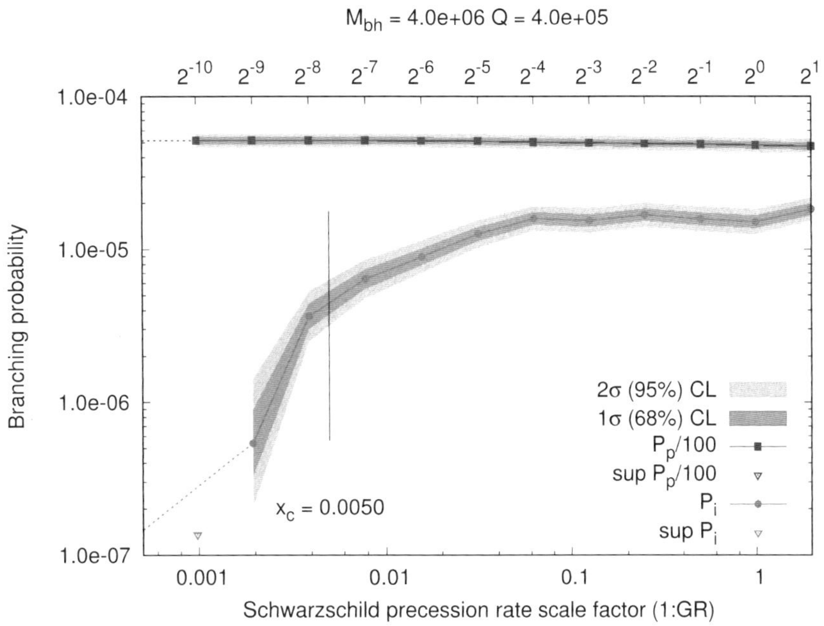

Figure 4. The run of the plunge and inspiral branching probabilities with the precession frequency scaling factor x ($x = 1$ corresponds to GR) for a simplified Milky Way model with $M_\bullet = 4 \times 10^6\, M_\odot$ and $M_\star = 10\, M_\odot$. The blue line is the plunge probability (divided by 100 to allow a more compact plot). A dotted blue line connects it to $P_p(x = 0)$. The red line is the inspiral (EMRI) probability. A dotted red line connects it to the result $P_i(x = 0) = 0$. The gray areas around the lines are the 1σ and 2σ regions, respectively. The vertical line is the theoretically derived critical value for x (see text).

$\nu_x = x\nu_{GR}$ ($x \geq 0$), and testing how this affects the branching ratio between plunges and inspirals. In terms of the phase space shown in Fig. (1), a scaling factor $x < 1$ shifts the AI line parallel to itself downward, and extends the region where RR is effective. A simple criterion for predicting the effect of the scale factor is the position of the intersection point of the AI locus and the LSO line relative to the the critical sma for EMRIs, a_{GW} (Figure 1). EMRIs are no longer possible when RR is effective at all values of $j_{lc} \leq j \leq 1$ down to the critical sma for EMRIs, a_{GW}, since then all stars are swept into the MBH on plunging trajectories before they can diffuse below a_{GW}. This occurs when the AI locus $j_0(a;x) = \sqrt{T_c(a)\nu_x(a, j = 1)/2\pi}$,, crosses the plunge loss cone $j_\bullet(a)$ below $a_{GW} \simeq 0.03(\log Q)^{-4/5} r_h$ [4], where a relaxed cusp with $\alpha = 7/4$ is assumed, and where the numeric prefactor depends on the approximation of the GW dissipation. The scaling of r_h with Q (assuming an $M_\bullet/\sigma$ relation $M_\bullet \propto \sigma^4$ and fixed $M_\star$) is $r_h \propto Q^{1/2}$, and so $a_{GW} \propto (\log Q)^{-4/5} Q^{1/2}$. The condition $j_0(a_{GW}; x_c)/j_{lc}(a_{GW}) = 1$ then defines a critical value x_c such that for $x < x_c$, RR remains unquenched everywhere above a_{GW}. Fig. (4) shows the run of the branching ratio with x for the Milky Way model of Fig. (1). The ratio changes in favor of plunges (cf. Fig. 2) for $x < x_c \sim \mathcal{O}(5 \times 10^{-3})$. This suggests that EMRIs are robust in the context of Einstein's GR.

3. Applications of the η-formalism

3.1. Cosmic loss-rates

MC simulations such as the one presented in Fig. (2) for $M_\bullet = 4 \times 10^6\, M_\odot$ can be carried out for any value of $M_\bullet$ once the properties of the galactic nucleus are related to the MBH mass via

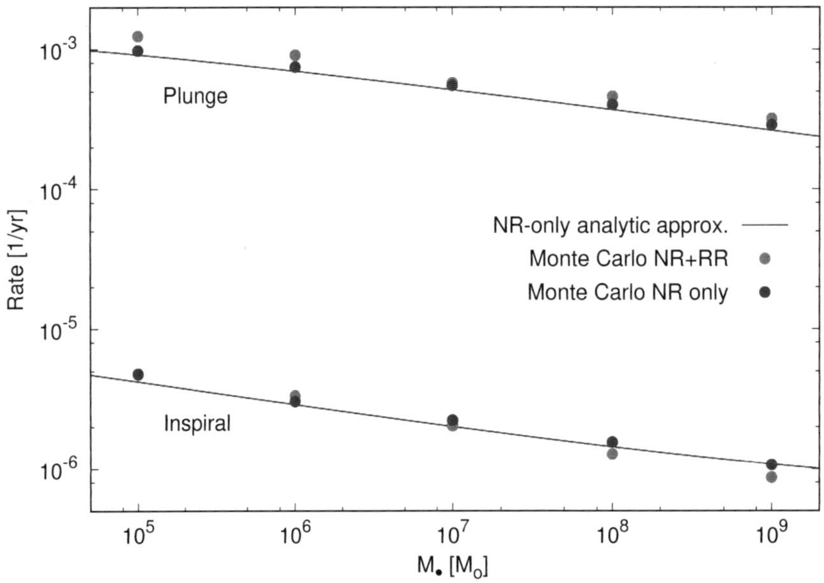

Figure 5. The plunge and inspiral rates, as function of the MBH mass $M_\bullet$, as derived from Monte Carlo simulations of the loss-cone in a sequence of simplified galactic nuclei models (adapted from [4]). The presence of RR makes only a small difference to the rates, since the loss lines (cf Figure 1) are well below the AI line (Eq. 1). The Inspiral rates are < 0.01 smaller than the plunge rate (Sec. 1.1), and both fall roughly as $M_\bullet^{-1/4}$ (for an $M_\bullet/\sigma$ relation with $\beta = 4$).

the $M_\bullet/\sigma$ relation. Fig. (5) Shows the run of the steady state plunge and EMRI inspiral rates as function of the MBH mass [4], and demonstrates the fact that RR plays only a small role in the rates, due to the very effective AI suppression of RR well away from the plunge and inspiral loss-lines (Fig. 1).

3.2. Loss-cone dynamics around the MBH of the Milky Way

One intriguing result that follows from the formal treatment of effective RR diffusion, and of the identification of the phase space region where RR dominates the dynamics, is that almost all the observed S-stars around Sgr $A^\star$ [16] (main sequence B stars, with masses $\sim 5 - 20\,M_\odot$ and lifespans $\sim 10^7 - 10^8$ yr) are found inside the strong RR region (Fig. 6). This suggests that whatever the unknown formation (or capture, or migration) process that is responsible for the puzzling presence of young massive stars so deep in the potential of a MBH, it is likely that they have undergone strong post-formation j-evolution. Fig. (6) shows a preliminary attempt to discriminate between the two leading models for the origin of the S-stars: Capture by a 3-body tidal interaction of the MBH with an incoming massive stellar binary [17, 18] (tidal binary capture, or the Hills mechanism [19]), or migration from the observed stellar disk around Sgr $A^\star$ [20]. In addition, such models also probe the unobserved population of faint stars and remnants around the MBH, which are necessary for generating the RR torques, but whose presence is observationally controversial [21, 22, 23] (but see recent detection of a cusp of faint low mass stars in the Galactic Center [24, 25]). The MC experiments indicate that the likeliest scenario is that the S-stars were captured in tidal binary separation events, and their orbits subsequently evolved in the presence of high density "dark" stellar cusp.

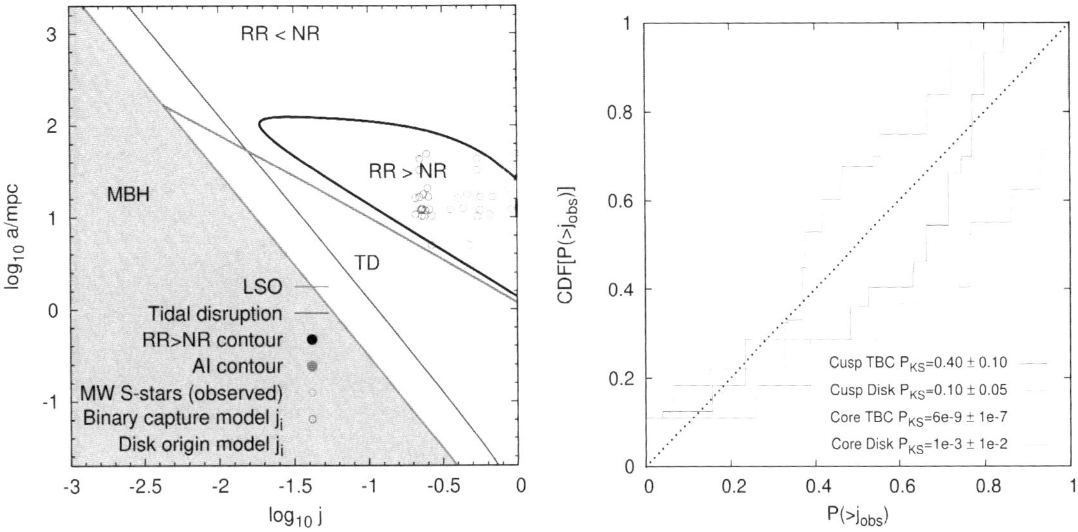

Figure 6. A joint test for the existence of a high density stellar cusp around Sgr $A^{\star}$ and for the origin of the S-stars that are observed to orbit the MBH in the central ~ 0.04 pc. Left: The S-stars [16] (red circles) lie in the phase space region where RR dominates the dynamics (the tidal disruption line for a typical S-star is shown in blue). If they originated from tidal binary captures (TBC, the Hills mechanism [19]), they start their existence near the MBH on highly eccentric orbits (blue circles), at an arbitrary phase of their lifespan. However, if the S-stars originated from the young ($\sim 6 \times 10^6$ yr [26]) stellar disk around observed around Sgr $A^{\star}$, they are only as old as the disk, and are expected to originate on nearly circular orbits (green circles). Right: A comparison of the final cumulative j-distribution of the S-stars for the two formation scenario, as derived from MC simulations for two possible stellar distributions around the MBH: a dense relaxed cusp, and an out-of-equilibrium lower density stellar core. The best fit scenario (highest K-S probability with distribution closest to a straight line) is the tidal binary capture in a high density stellar cusp (Sabsovich, Alexander & Bar-Or, in prep.).

4. Summary

The complex dynamics that ultimately lead stars and compact remnants to fall into the MBH can be modeled by MC simulations that introduce RR by effective diffusion coefficients that are derived from the η-formalism. Phase space is clearly separated into a restricted region where RR dominates the dynamics, and a region where AI strongly suppresses RR due to fast precession. Elsewhere, RR exists, but NR is faster. Importantly, both the the loss-lines for plunges (LSO) and for GW EMRIs lie below the AI locus, and so RR has only a small effect on the loss-rates. This situation appears to be a fortunate coincidence for the prospects of detecting EMRIs. It is a result of a three-way competition between RR torques that tend to push stars to plunge orbits, GR Schwarzschild precession, which suppresses RR, and GR GW dissipation, which when fast enough, can be completed before the stellar background interferes. This coincidence appears robust in the context of Einstein's GR. It is however still unclear how general it is in the wider context of theories of strong gravity.

The dynamical processes that play part in EMRI dynamics can be probed by the puzzling S-stars observed near the MBH of the Galactic Center, which are at the RR-dominated region of the phase space around Sgr $A^{\star}$.

Acknowledgments
This work was supported by the I-CORE Program of the Planning and Budgeting Committee and The Israel Science Foundation (grant No 1829/12).

References
[1] Bar-Or B and Alexander T 2014 *Classical and Quantum Gravity* **31** 244003 (*Preprint* 1404.0351)
[2] Hopman C and Alexander T 2006 *ApJ* **645** 1152–1163
[3] Rauch K P and Tremaine S 1996 *New Astronomy* **1** 149–170
[4] Bar-Or B and Alexander T 2016 *ApJ* **820** 129 (*Preprint* 1508.01390)
[5] Alexander T 2017 *ArXiv e-prints* (*Preprint* 1701.04762)
[6] Frank J and Rees M J 1976 *MNRAS* **176** 633–647
[7] Bahcall J N and Wolf R A 1976 *ApJ* **209** 214–232
[8] Shapiro S L and Marchant A B 1978 *ApJ* **225** 603–624
[9] Bahcall J N and Wolf R A 1977 *ApJ* **216** 883–907
[10] Hopman C and Alexander T 2005 *ApJ* **629** 362–372
[11] Lightman A P and Shapiro S L 1977 *ApJ* **211** 244–262
[12] Alexander T and Hopman C 2003 *ApJ* **590** L29–L32
[13] Merritt D, Alexander T, Mikkola S and Will C M 2011 *Phys. Rev. D* **84** 044024 (*Preprint* 1102.3180)
[14] Brem P, Amaro-Seoane P and Sopuerta C F 2014 *MNRAS* **437** 1259–1267 (*Preprint* 1211.5601)
[15] Alexander T 2010 *GW and EM signatures of MBH binaries and EMRIs, APC, Paris* ed Amaro-Seoane P and Porter E online http://www.aei.mpg.de/~pau/conf_vid4/Alexander.pdf URL http://www.aei.mpg.de/~pau/conf_vid4/Alexander.pdf
[16] Gillessen S, Eisenhauer F, Trippe S, Alexander T, Genzel R, Martins F and Ott T 2009 *ApJ* **692** 1075–1109
[17] Gould A and Quillen A C 2003 *ApJ* **592** 935–940
[18] Perets H B, Hopman C and Alexander T 2007 *ApJ* **656** 709–720
[19] Hills J G 1988 *Nature* **331** 687–689
[20] Levin Y 2007 *MNRAS* **374** 515–524 (*Preprint* astro-ph/0603583)
[21] Do T, Ghez A M, Morris M R, Lu J R, Matthews K, Yelda S and Larkin J 2009 *ApJ* **703** 1323–1337
[22] Buchholz R M, Schödel R and Eckart A 2009 *A&A* **499** 483–501
[23] Bartko H *et al.* 2010 *ApJ* **708** 834–840
[24] Gallego-Cano E, Schödel R, Dong H, Nogueras-Lara F, Gallego-Calvente A T, Amaro-Seoane P and Baumgardt H 2017 *ArXiv e-prints* (*Preprint* 1701.03816)
[25] Schödel R, Gallego-Cano E, Dong H, Nogueras-Lara F, Gallego-Calvente A T, Amaro-Seoane P and Baumgardt H 2017 *ArXiv e-prints* (*Preprint* 1701.03817)
[26] Bartko H *et al.* 2009 *ApJ* **697** 1741–1763

IOP Conf. Series: Journal of Physics: Conf. Series **840** (2017) 012020 doi:10.1088/1742-6596/840/1/012020

The stellar cusp around the Milky Way's central black hole

R. Schödel[1], E. Gallego-Cano[1] and P. Amaro-Seoane[2]

[1]Instituto de Astrofísica de Andalucía (CSIC), Glorieta de la Astronomía s/n, 18008 Granada, Spain

[2]Institut de Ciències de l'Espai (CSIC-IEEC) at Campus UAB, Carrer de Can Magrans s/n 08193 Barcelona, Spain

Institute of Applied Mathematics, Academy of Mathematics and Systems Science, CAS, Beijing 100190, China

Kavli Institute for Astronomy and Astrophysics, Beijing 100871, China

Zentrum für Astronomie und Astrophysik, TU Berlin, Hardenbergstraße 36, 10623 Berlin, Germany

E-mail: `rainer@iaa.es`

Abstract. The existence of stellar cusps in dense clusters around massive black holes is a fundamental, decades-old prediction of theoretical stellar dynamics. Yet, observational evidence has been difficult to obtain. With a new, improved analysis of high-angular resolution images of the central parsecs of the Galactic Center, we are finally able to provide the first solid evidence for the existence of a stellar cusp around the Milky Way's massive black hole. The existence of stellar cusps has a significant impact on predicted event rates of phenomena like tidal disruptions of stars and extreme mass ratio inspirals.

1. Introduction

It is a standard paradigm of modern astrophysics that the majority of galaxies – possibly with the exception of very low mass galaxies and/or dwarf irregulars – contains massive black holes (MBHs) at their centres (see, e.g., [1] and references therein). Over the past two decades, mainly thanks to studies with the Hubble Space Telescope and high-angular resolution observations with Adaptive Optics at major ground-based telescopes, we have also learned that the majority of galaxies contains so-called *nuclear star clusters* (NSCs) at their photometric and dynamical centres. These clusters are the densest and most massive clusters that can be found in the present-day Universe. They have masses between a few 10^5 to $10^8\,M_\odot$ and half light radii of a few to a few tens of parsecs. NSCs do not consist of single-age stellar populations, but rather show signs of repeated star formation along their host galaxies' lifetime, i.e. they are in their properties significantly different from globular clusters. Most intriguingly, NSCs have been found to coexist with massive black holes (for the properties of NSCs, see [2, 3, 4] and references therein).

The evolution of a star cluster containing an MBH is a decades-old problem of stellar dynamics that has been studied by a large number of authors with a wide range of techniques (analytical, Fokker-Planck, Nbody). Peebles (1972) [5] proved that, while the statistical thermal equilibrium must be violated close to a MBH in a galactic nucleus due to tidal disruptions, stellar collisions and gravitational captures, there exists a steady state with net inward flux of stars. This is a quasi-steady state solution, where the stellar density takes a power-law form $\rho \propto r^{-\gamma}$. This finding was corroborated for a single-mass population by Bahcall and Wolf [6] and subsequently for multi-mass stellar populations with realistic number fractions (e.g., [7, 8, 9, 10, 11, 12, 13, 14] and references therein). This so-called stellar density *cusp* will be fully developed after a so-called *relaxation time*, the time necessary for the randomisation of the cluster phase space via close encounters between stars (so-called two-body relaxation), which is typically below a Hubble time for a NSC in a Milky Way-like galaxy For a cluster composed only of stars of a single mass the predicted value is $\gamma = 1.75$. For realistic, multi-mass clusters, the lower mass stars have $\gamma = 1.5$, while the heavier stars (and in particular stellar mass black holes) will follow a steeper distribution ($\gamma_{BH} \approx 2$; see [13, 14]). The cusp will be well developed inside the radius of influence of the MBH, which is roughly the radius of a sphere that contains once to twice the black hole mass in form of stars (or stellar remnants) [11].

Although stellar cusps around MBHs are a robust theoretical prediction, they have been rather elusive observationally. That is mainly because of the great distances of MBHs: Their radii of influence are often resolved only by a few pixels in imaging or spectroscopy, which means that a handful of rare bright stars can bias the measured cluster structure significantly, as can be seen very well in the case of our own Milky Way (see discussion in [15]). When comparing theoretical predictions with observations in nature, it is also important to be aware that theoretical models are often highly simplified, e.g. they typically are setup with clusters of a single age stellar population, and do not suffer from observational effects such as strong and spatially variable interstellar extinction.

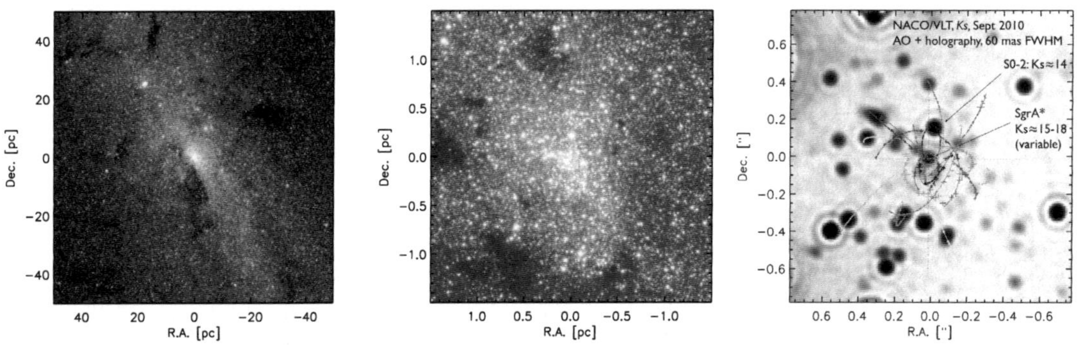

Figure 1. A progressive zoom into the Galactic Center. Left: Spitzer 4.5 μm image [16] of the central 100 pc × 100 pc of the Galaxy. Coordinates are given as offsets from the MBH Sgr A*. The NSC is the bright, compact structure in the centre of the image. Middle: NACO/VLT 2.2 μm image of the central 3 pc × 3 pc around Sgr A*. Right: NACO/VLT 2.2 μm image of the central 1.6" × 1.6" (0.06 pc × 0.06 pc) around Sgr A*. Stellar orbits determined by [17] are over-plotted onto the image. Sgr A* is visible as a near-infrared source at the image centre.

The best case for testing the stellar cusp hypothesis is the centre of the Milky Way. In the Galactic Centre (GC), an MBH of about 4×10^6 $M_\odot$, called Sagittarius A* (Sgr A*), is surrounded by an NSC of about 2.5×10^7 $M_\odot$, at a distance of ~8 kpc ([18, 19, 17, 20] and references therein). With high angular resolution observations we can therefore probe the cluster structure and dynamics on scales of milli-parsecs, which means that the GC is a prime laboratory for studying the interaction of a stellar cluster with an MBH [21]. Figure 1 provides an overview of the GC and a progressive zoom toward the MBH Sagittarius A* (Sgr A*).

It is important to keep in mind the limitations on observational studies of the GC: (1) Due to the extremely high interstellar extinction, stars can only be detected with reasonable sensitivity in the near-infrared. In this region, however, intrinsic stellar colours are small and in combination with the strong differential reddening caused by the small-scale variability of extinction it is therefore very hard to classify the stars. For example, it is not trivial to distinguish between an old, cold giant star and a young, hot massive star by means of imaging and stellar colours. (2) Due to the extreme density of the NSC, one needs to work with high angular resolution, typically at the diffraction limit of ground-based 8-10m-class telescopes, which can only be achieved with special techniques (Adaptive Optics and speckle imaging). Even then, most faint stars (main sequence stars of less than two solar masses) cannot be detected because of the source crowding. Hence, we can only observe the tip of the iceberg, on the order a few to 10% of all the stars suspected to exist at the GC.

Because of the observational limitations, evidence on the existence of a stellar cusp at the GC was elusive. Although the detection of a cusp was claimed by first high angular resolution observations [15, 22], the actual situation turned out to be more complex. Only stars older than the cluster relaxation time can serve as suitable tracers of a cusp, but there are many massive young stars present in the central parsec around Sgr A*. These are too young to be dynamically relaxed. Once the pollution from young stars was taken into account, it appeared that the stellar surface number density within a few 0.1 pc of the MBH at the GC was close to flat, or even decreasing [23, 24, 25]. The lack of late-type giants close to the MBH had been reported before (see, e.g., discussion fo this topic in the review by [18]), but these more detailed studies were now considered to be incompatible with the existence of a stellar cusp. This is the origin of the *missing cusp problem*.

The existence or not of a stellar cusp around the MBH at the GC has implications for other galaxies as well, assuming that the Milky Way's centre is representative for the nuclei of normal galaxies. Observational confirmation or rejection of a stellar cusp would not only have fundamental importance for our understanding of stellar dynamics. It is also of particular importance for gravitational wave astronomy. So called Extreme Mass Ratio Inspirals (EMRIs) are considered the most exquisite probes of General Relativity and of the related astrophysics of stellar remnant plus MBH systems [26]. The possibility of observing EMRIs with future gravitational wave observatories in space, such as LISA (Laser Interferometer Space Antenna) [27, 28] or Taiji [29] is directly linked to the question whether stellar cusps exist around MBHs or not [30, 14].

Motivated by the fundamental importance of this topic, we have recently undertaken new observational and theoretical studies. Observationally, we pushed the detection limit to fainter stars. The observations were then compared to new, more realistic Nbody simulations. We have found that a stellar cusp does indeed exist at the GC and that its properties agree with our expectations. The details of our work can be consulted in these three papers: 1. Gallego-Cano et al. (submitted to A&A, arXiv:1701.03816); 2. Schödel et al. (submitted to A&A, arXiv:1701.03817); 3. Baumgardt et al. (submitted to A&A, arXiv:1701.03818). We will briefly describe our findings in the following sections.

2. Surface density of stars around the MBH Sgr A*

Before interpreting the stellar surface and light densities, it is important to have an approximate understanding of the stellar population we expect to observe at the GC. While we are still far from having a detailed understanding of the stellar population and its formation history in the inner few parsecs of the Milky Way, the following assumptions appear to be robust, according to our current best knowledge: The NSC has undergone repeated star formation episodes throughout its lifetime, with the most recent events occurring a few 10^6 and a few 10 Myr ago. The majority of the stars are old, with an estimated 80% having formed more than

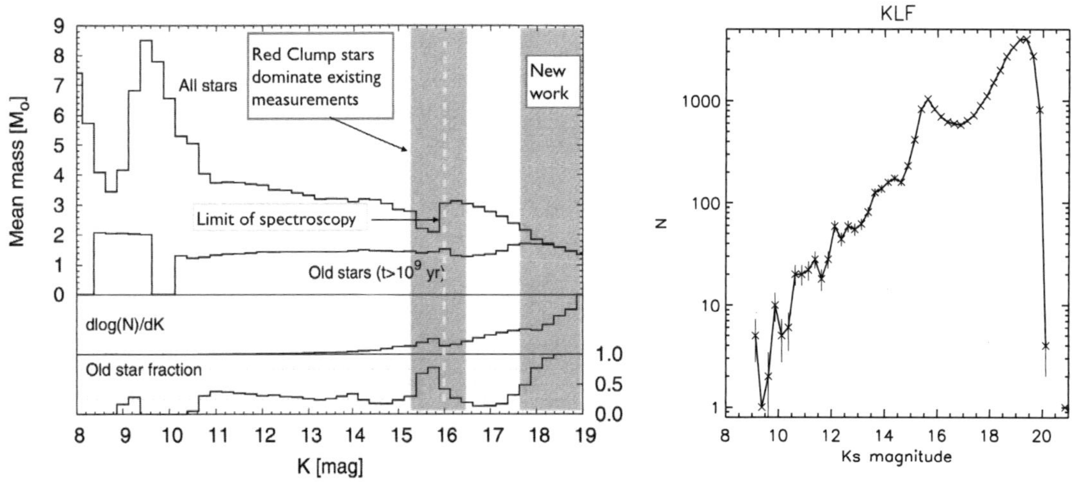

Figure 2. Properties of the observable stellar population at the GC. *Left*: Simple population synthesis model of the GC stellar population, assuming continuous star formation, modified from Fig. 16 in [15] (Reproduced with permission from Astronomy & Astrophysics, ©ESO). The distance modulus and extinction were added to the stellar magnitudes, which means that the magnitude values on the x-axis correspond to observed values. The upper panel shows the mean mass of all stars and of old stars (lifetime > 1 Gyr) only. The middle panel shows a luminosity function. The lower panel displays the fraction of old stars in each magnitude interval. The completeness limit due to source crowding of previous studies was limited to $K_s \lesssim 17.5$, while the sensitivity limit of spectroscopic studies is $K_s \approx 16$. At $K_s < 17.5$, number counts of old stars in the GC will therefore be dominated by horizontal branch/red clump (HB/RC) stars, core-helium burning giants. *Right*: K_s luminosity function of stars within a projected distance of $R \lesssim 10$" of Sgr A*, extracted from a deep NACO/VLT S13 image. The bump centred at $K_s \approx 15.75$ corresponds to the RC. The sharp drop in detected stars at $K_s > 18$ is due to extreme source crowding, which impedes the detection of fainter stars. While RC stars and brighter giants have dominated the stellar density measurements of all previous work, we use measurements of resolved and unresolved stars at $Ks \gtrsim 17.5$ in our new work.

5 Gyr ago [18, 31].

The left panel in Fig. 2 shows a simple model of the stellar population of the NSC assuming a continuous star formation history (taken from Fig. 16 in [15]). As mentioned above, it is very hard to classify stars at the GC because photometric studies do usually not cover a sufficient wavelength range and are usually not accurate and precise enough to break the degeneracy between intrinsic stellar colours and interstellar reddening and because spectroscopic studies are extremely time-consuming and limited to the brightest stars. Therefore, the diagram in the left panel in Fig. 2 is valuable because it can give us an approximate idea of the ages and masses of stars of a given observed brightness at the GC. If we are interested in identifying a stellar cusp, then it is important to focus on stars that are at least a few Gyr old. We can do this by selecting spectroscopically classified stars and/or by limiting the brightness range of the tracer population. Previous studies of the NSC structure ([23, 24, 25]) were dominated by stars of K-band magnitudes brighter than ~17.5, in particular by Red Clump (RC) stars, helium core burning giants with observed K-band magnitudes 15 − 16 at the distance and extinction of the GC.

RC stars are a convenient tracer population because they are, on average, a few Gyr old. We can also see in the left panel in Fig. 2 that stars of K-magnitudes 18 or fainter would also be suitable tracer populations because they have low mean masses and are therefore probably, on average, at least a few Gyr old.

In our new work, we focused explicitly on this faint, old, low mass stellar population. We used K-band high angular resolution observations from the camera NACO at the ESO VLT 8m telescope. Several epochs of high quality images were stacked in order to increase the sensitivity. Subsequently, correction factors for stellar crowding and interstellar extinction were determined and applied to the data. Systematic effects arising from different choices of key parameters in the software packages that identify the stars in images were explored. Finally, the stellar surface densities were determined as a function of distance from the central MBH. Spherical symmetry of the cluster was assumed, which is not strictly correct, but accurate on a 10-20% level. In parallel, we also explored the diffuse light density from unresolved stars that could be measured in the images after subtracting all identified, resolved stars and the emission from diffuse, ionised gas.

In order to constrain the structure of the entire NSC and to be able to deproject the observed quantities, these data were complemented, at projected distances $R \gtrsim 1\,\mathrm{pc}$ from Sgr A*, by star density/surface brightness measurements from other sources, with lower angular resolution and/or sensitivity, but taken over a larger field ([20, 32]). The latter were scaled to our data in the overlapping regions. In Fig. 3 we show the surface brightness from unresolved stars from 0.01 pc to 20 pc as well as the star counts for faint resolved stars (K-magnitudes ~18) and RC stars (K-magnitudes ~15-16). The diffuse flux from the unresolved stars traces sub-giants and main sequence stars of K-band magnitude 19 – 20, which have masses of about $0.8 - 1.8\,\mathrm{M}_\odot$. All these three tracer populations have similar masses and can be old enough to be dynamically relaxed.

3. Properties of the stellar cusp at the GC

As Fig. 3 shows, the three studied tracer populations show a very similar distribution. The projected densities can all be approximated well by single power-laws at projected distances $R < 1\,\mathrm{pc}$. Only the brightest population, the red clump giants, show some systematic deviations at around $R = 0.2\,\mathrm{pc}$ and a possible decline at $R < 0.05\,\mathrm{pc}$. The red line in Fig. 3 indicates a fit with a *Nuker* law [33]:

$$\rho(r) = \rho(r_b) 2^{(\beta-\gamma)/\alpha} \left(\frac{r}{r_b}\right)^{-\gamma} \left[1 + \left(\frac{r}{r_b}\right)^{\alpha}\right]^{(\gamma-\beta)/\alpha}, \tag{1}$$

where r is the 3D distance from the central MBH, r_b is the break radius, ρ is the 3D density, γ is the exponent of the inner and β the one of the outer power-law, and α defines the sharpness of the transition. We carried out fits with different parameters (e.g., for the value of α that was kept fixed at a value of 10 during the fits) and data (data at large R from [20] or [32]) and determined mean parameters and uncertainties from the resulting best-fit parameters of the different tries (the formal uncertainties were always very small). We obtain, for the unresolved stellar light $r_{b,diffuse} = 3.2 \pm 0.2\,\mathrm{pc}$, $\gamma_{diffuse} = 1.16 \pm 0.02$, $\beta_{diffuse} = 3.2 \pm 0.3$, and $\rho_{diffuse}(r_b) = 0.67 \pm 0.05\,\mathrm{mJy\,arcsec^{-3}}$, for the star counts of faint, resolved stars $r_{b,faint} = 3.0 \pm 0.4$ pc (77 ± 10"), $\gamma_{faint} = 1.29 \pm 0.02$, $\beta_{faint} = 2.1 \pm 0.1$, and a density at the break radius of $\rho_{faint}(r_b) = 3900 \pm 900\,\mathrm{pc^{-3}}$, and for the RC stars $r_{b,RC} = 3.4 \pm 0.5$ pc (110 ± 13"), $\gamma_{RC} = 1.28 \pm 0.07$, $\beta_{RC} = 2.26 \pm 0.04$, and $\rho_{RC}(r_b) = 1700 \pm 500\,\mathrm{pc^{-3}}$.

We cannot directly compare the values of the normalisation parameters ρ. As concerns β, it is not very well constrained. On the other hand, the break radius r_b is well constrained around 3 pc and the inner power-law index γ as well. The break radius corresponds roughly to the radius of influence of the Sgr A*, i.e. the radius inside which we expect the stellar cusp to appear [11]. Both the stellar number and light densities may suffer from not well constrained

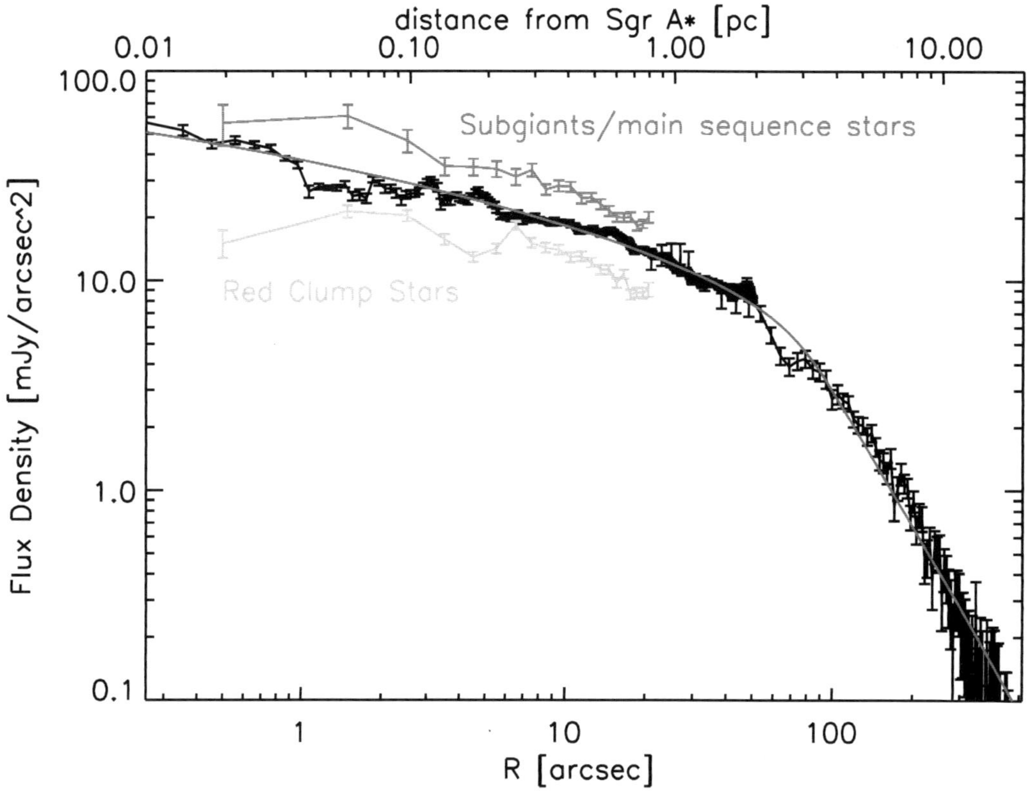

Figure 3. Surface light/stellar number density at the GC plotted against the projected distance from Sgr A*. The black line is the surface light density from the unresolved stars (Schödel et al., submitted to A&A, arXiv:1701.03817). In order to constrain the shape of the NSC at distances > 2 pc, we used the data from [32] and scaled them in the overlap region to our data. A constant flux of 6.4 mJy arcsec^{-2} was subtracted to remove the contamination by the surrounding nuclear bulge and Galactic bar, which can be approximated by a constant on these scales. The red line is a best-fit Nuker model. The blue and green lines are scaled stellar number densities from Gallego-Cano et al. (submitted to A&A, arXiv:1701.03816) for RC giants and faint subgiants/main sequence stars.

systematic biases (uncertainties in sky background subtraction or completeness corrections etc.). If we adopt conservative values from averaging the three values, we obtain $r_b = 3.2 \pm 0.2$ and $\gamma = 1.24 \pm 0.07$, taking the standard deviations as uncertainties. Obviously, the stellar density displays a cusp and a core-like (flat) density law can be excluded with very high confidence.

4. Discussion and conclusions

We find that the stars at the GC follow a cusp-like density distribution inside the radius of influence of the central MBH. Our findings resolve the ambiguity from previous work. The cusp is less steep than what is expected for a single age population that is older than the relaxation time. In the latter case one would expect to observe a so-called Bahcall-Wolf cusp with $\gamma \geq 1.5$. Numerical experiments were performed by Baumgardt et al. (arXiv:1701.03818) to interpret our data. They performed Nbody simulations of a more realistic NSC, assuming repeated star formation episodes every 1 Gyr. This means that most stars will not have had the time to fully relax dynamically, which will flatten the cusp. The results of the simulations agree very well with the data.

There appears to be a lack of giant stars at small projected distances of $R \lesssim 0.1$ pc from Sagittarius A*. The missing few dozens of giants could be explained if their envelopes were removed in the past, thus rendering them invisible to observations. While collisions between stars or between stellar remnants and stars are probably not effective enough, collisions with high-density clumps in a formerly existing star-forming gas disc around Sagittarius A* provide a viable mechanism [34]. We know that such a disc must have existed in the past because of the presence of many young, massive stars in this region that move in a coherent, disc-like pattern (e.g., (e.g., [35, 36, 18]).

We conclude that a stellar cusp exists around Sagittarius A* and that its properties agree with what we expect theoretically. Densities in excess of a few $10^7\,M_\odot\,pc^{-3}$ are reached at distances $r < 0.01$ pc from the central black hole (Schödel et al., submitted to A&A, arXiv:1701.03817). This has significant implications for gravitational wave astronomy. The existence of a stellar cusp in the Milky Way implies the existence of such structures in other galaxies with similar or smaller MBHs. Therefore we can expect to observe significant numbers of EMRIs with future space-based gravitational wave observatories [30, 37, 27, 28].

Acknowledgments
The research leading to these results has received funding from the European Research Council under the European Union's Seventh Framework Programme (FP7/2007-2013) / ERC grant agreement n° [614922]. PAS acknowledges support from the Ramón y Cajal Programme of the Spanish Ministerio de Economa, Industria y Competitividad.

[1] Gültekin K, Richstone D O, Gebhardt K, Lauer T R, Tremaine S, Aller M C, Bender R, Dressler A, Faber S M, Filippenko A V, Green R, Ho L C, Kormendy J, Magorrian J, Pinkney J and Siopis C 2009 *ApJ* **698** 198–221 (*Preprint* 0903.4897)
[2] Böker T 2010 Nuclear star clusters *IAU Symposium* (*IAU Symposium* vol 266) ed R de Grijs & J R D Lépine pp 58–63
[3] Georgiev I Y and Böker T 2014 *MNRAS* **441** 3570–3590 (*Preprint* 1404.5956)
[4] Georgiev I Y, Böker T, Leigh N, Lützgendorf N and Neumayer N 2016 *MNRAS* **457** 2122–2138 (*Preprint* 1601.02613)
[5] Peebles P J E 1972 *ApJ* **178** 371–376
[6] Bahcall J N and Wolf R A 1976 *ApJ* **209** 214–232
[7] Bahcall J N and Wolf R A 1977 *ApJ* **216** 883–907
[8] Lightman A P and Shapiro S L 1977 *ApJ* **211** 244–262
[9] Murphy B W, Cohn H N and Durisen R H 1991 *ApJ* **370** 60–77
[10] Amaro-Seoane P, Freitag M and Spurzem R 2004 *MNRAS* **352** 655–672 (*Preprint* astro-ph/0401163)
[11] Alexander T 2005 *Phys. Rep.* **419** 65–142 (*Preprint* arXiv:astro-ph/0508106)
[12] Merritt D 2006 *Reports on Progress in Physics* **69** 2513–2579 (*Preprint* arXiv:astro-ph/0605070)
[13] Alexander T and Hopman C 2009 *ApJ* **697** 1861–1869 (*Preprint* 0808.3150)
[14] Amaro-Seoane P and Preto M 2011 *Classical and Quantum Gravity* **28** 094017 (*Preprint* 1010.5781)
[15] Schödel R, Eckart A, Alexander T, Merritt D, Genzel R, Sternberg A, Meyer L, Kul F, Moultaka J, Ott T and Straubmeier C 2007 *A&A* **469** 125–146 (*Preprint* arXiv:astro-ph/0703178)
[16] Stolovy S, Ramirez S, Arendt R G, Cotera A, Yusef-Zadeh F, Law C, Gezari D, Sellgren K, Karr J, Moseley H and Smith H A 2006 *Journal of Physics Conference Series* **54** 176–182
[17] Gillessen S, Plewa P, Eisenhauer F, Sari R, Waisberg I, Habibi M, Pfuhl O, George E, Dexter J, von Fellenberg S, Ott T and Genzel R 2016 *ArXiv e-prints* (*Preprint* 1611.09144)
[18] Genzel R, Eisenhauer F and Gillessen S 2010 *Reviews of Modern Physics* **82** 3121–3195
[19] Boehle A, Ghez A M, Schödel R, Meyer L, Yelda S, Albers S, Martinez G D, Becklin E E, Do T, Lu J R, Matthews K, Morris M R, Sitarski B and Witzel G 2016 *ApJ* **830** 17 (*Preprint* 1607.05726)
[20] Schödel R, Feldmeier A, Kunneriath D, Stolovy S, Neumayer N, Amaro-Seoane P and Nishiyama S 2014 *A&A* **566** A47 (*Preprint* 1403.6657)
[21] Schödel R, Feldmeier A, Neumayer N, Meyer L and Yelda S 2014 *Classical and Quantum Gravity* **31** 244007 (*Preprint* 1411.4504)

[22] Genzel R, Schödel R, Ott T, Eisenhauer F, Hofmann R, Lehnert M, Eckart A, Alexander T, Sternberg A, Lenzen R, Clénet Y, Lacombe F, Rouan D, Renzini A and Tacconi-Garman L E 2003 *ApJ* **594** 812–832 (*Preprint* arXiv:astro-ph/0305423)
[23] Buchholz R M, Schödel R and Eckart A 2009 *A&A* **499** 483–501 (*Preprint* 0903.2135)
[24] Do T, Ghez A M, Morris M R, Lu J R, Matthews K, Yelda S and Larkin J 2009 *ApJ* **703** 1323–1337 (*Preprint* 0908.0311)
[25] Bartko H, Martins F, Trippe S, Fritz T K, Genzel R, Ott T, Eisenhauer F, Gillessen S, Paumard T, Alexander T, Dodds-Eden K, Gerhard O, Levin Y, Mascetti L, Nayakshin S, Perets H B, Perrin G, Pfuhl O, Reid M J, Rouan D, Zilka M and Sternberg A 2010 *ApJ* **708** 834–840 (*Preprint* 0908.2177)
[26] Amaro-Seoane P, Gair J R, Freitag M, Miller M C, Mandel I, Cutler C J and Babak S 2007 *Classical and Quantum Gravity* **24** 113 (*Preprint* astro-ph/0703495)
[27] Amaro-Seoane P, Aoudia S, Babak S, Binétruy P, Berti E, Bohé A, Caprini C, Colpi M, Cornish N J, Danzmann K, Dufaux J F, Gair J, Jennrich O, Jetzer P, Klein A, Lang R N, Lobo A, Littenberg T, McWilliams S T, Nelemans G, Petiteau A, Porter E K, Schutz B F, Sesana A, Stebbins R, Sumner T, Vallisneri M, Vitale S, Volonteri M and Ward H 2012 *Classical and Quantum Gravity* **29** 124016 (*Preprint* 1202.0839)
[28] Amaro-Seoane P, Aoudia S, Babak S, Binétruy P, Berti E, Bohé A, Caprini C, Colpi M, Cornish N J, Danzmann K, Dufaux J F, Gair J, Hinder I, Jennrich O, Jetzer P, Klein A, Lang R N, Lobo A, Littenberg T, McWilliams S T, Nelemans G, Petiteau A, Porter E K, Schutz B F, Sesana A, Stebbins R, Sumner T, Vallisneri M, Vitale S, Volonteri M, Ward H and Wardell B 2013 *GW Notes, Vol. 6, p. 4-110* **6** 4–110 (*Preprint* 1201.3621)
[29] Gong X, Lau Y K, Xu S, Amaro-Seoane P, Bai S, Bian X, Cao Z, Chen G, Chen X, Ding Y, Dong P, Gao W, Heinzel G, Li M, Li S, Liu F, Luo Z, Shao M, Spurzem R, Sun B, Tang W, Wang Y, Xu P, Yu P, Yuan Y, Zhang X and Zhou Z 2015 Descope of the ALIA mission *Journal of Physics Conference Series* (*Journal of Physics Conference Series* vol 610) p 012011 (*Preprint* 1410.7296)
[30] Hopman C and Alexander T 2005 *ApJ* **629** 362–372 (*Preprint* astro-ph/0503672)
[31] Pfuhl O, Fritz T K, Zilka M, Maness H, Eisenhauer F, Genzel R, Gillessen S, Ott T, Dodds-Eden K and Sternberg A 2011 *ApJ* **741** 108 (*Preprint* 1110.1633)
[32] Fritz T K, Chatzopoulos S, Gerhard O, Gillessen S, Genzel R, Pfuhl O, Tacchella S, Eisenhauer F and Ott T 2016 *ApJ* **821** 44
[33] Lauer T R, Ajhar E A, Byun Y I, Dressler A, Faber S M, Grillmair C, Kormendy J, Richstone D and Tremaine S 1995 *AJ* **110** 2622
[34] Amaro-Seoane P and Chen X 2014 *ApJL* **781** L18 (*Preprint* 1310.0458)
[35] Beloborodov A M, Levin Y, Eisenhauer F, Genzel R, Paumard T, Gillessen S and Ott T 2006 *ApJ* **648** 405–410 (*Preprint* astro-ph/0601273)
[36] Lu J R, Ghez A M, Hornstein S D, Morris M R, Becklin E E and Matthews K 2009 *ApJ* **690** 1463–1487 (*Preprint* 0808.3818)
[37] Amaro-Seoane P 2012 *ArXiv e-prints* (*Preprint* 1205.5240)

IOP Conf. Series: Journal of Physics: Conf. Series **840** (2017) 012021 doi:10.1088/1742-6596/840/1/012021

Prospects for observing extreme-mass-ratio inspirals with LISA

Jonathan R Gair[1,†], Stanislav Babak[2], Alberto Sesana[3], Pau Amaro-Seoane[4,5,6,7], Enrico Barausse[8,9], Christopher P L Berry[3], Emanuele Berti[10,11] and Carlos Sopuerta[4]

[1]School of Mathematics, University of Edinburgh, The King's Buildings, Peter Guthrie Tait Road, Edinburgh, EH9 3FD, UK
[2]Max Planck Institute for Gravitational Physics, Albert Einstein Institute, Am Mühlenberg 1, 14476 Golm, Germany
[3]School of Physics and Astronomy, University of Birmingham, Edgbaston, Birmingham B15 2TT, UK
[4]Institut de Ciències de l'Espai (CSIC-IEEC), Campus UAB, Carrer de Can Magrans s/n, 08193 Cerdanyola del Vallès, Spain
[5]Kavli Institute for Astronomy and Astrophysics, Beijing 100871, China
[6]Institute of Applied Mathematics, Academy of Mathematics and Systems Science, CAS, Beijing 100190, China
[7]Zentrum für Astronomie und Astrophysik, TU Berlin, Hardenbergstraße 36, 10623 Berlin, Germany
[8]Sorbonne Universités, UPMC Univesité Paris 6, UMR 7095, Institut d'Astrophysique de Paris, 98 bis Bd Arago, 75014 Paris, France
[9]CNRS, UMR 7095, Institut d'Astrophysique de Paris, 98 bis Bd Arago, 75014 Paris, France
[10]Department of Physics and Astronomy, The University of Mississippi, University, MS 38677, USA
[11]CENTRA, Departamento de Física, Instituto SuperiorTécnico, Universidade de Lisboa, Avenida Rovisco Pais 1, 1049 Lisboa, Portugal

E-mail: [†]j.gair@ed.ac.uk

Abstract. One of the key astrophysical sources for the Laser Interferometer Space Antenna (LISA) are the inspirals of stellar-origin compact objects into massive black holes in the centres of galaxies. These extreme-mass-ratio inspirals (EMRIs) have great potential for astrophysics, cosmology and fundamental physics. In this paper we describe the likely numbers and properties of EMRI events that LISA will observe. We present the first results computed for the 2.5 Gm interferometer that was the new baseline mission submitted in January 2017 in response to the ESA L3 mission call. In addition, we attempt to quantify the astrophysical uncertainties in EMRI event rate estimates by considering a range of different models for the astrophysical population. We present both likely event rates and estimates for the precision with which the parameters of the observed sources could be measured. We finish by discussing the implications of these results for science using EMRIs.

1. Introduction

The first gravitational wave (GW) detections by LIGO [1, 2] in 2015 began the new field of GW astronomy. These events indicated that the ground-based interferometers will see many more

IOP Conf. Series: Journal of Physics: Conf. Series **840** (2017) 012021 doi:10.1088/1742-6596/840/1/012021

systems over the coming decades [3]. However, seismic noise limits the sensitivity of ground-based detectors to frequencies above ~ 1 Hz and hence to systems with total mass no more than a few hundred solar masses. Observational evidence indicates that the centres of most galaxies contain massive black holes (MBHs), with masses between a few tens of thousands and a few billion solar masses [4]. Mergers involving MBHs are powerful sources of GWs, but these GWs can only be observed from space. ESA selected GW detection from space as the theme to be addressed by the L3 mission in the Cosmic Vision programme, scheduled for launch in 2034 [5]. The technology for a space-based interferometer was successfully demonstrated in 2016 by the LISA Pathfinder satellite [6], which paved the way for an ESA call for L3 mission proposals that closed in early 2017. Given the success of LISA Pathfinder, it is highly likely that the successful proposal will be the Laser Interferometer Space Antenna (LISA) [7].

LISA will have sensitivity to GWs in the millihertz band, which are generated by merging systems having total mass in the range $\sim 10^4$–$10^7 M_\odot$. LISA is expected to observe a variety of sources, including hour-period compact binaries (primarily white dwarf–white dwarf binaries) in the Milky Way, the early inspiral phase of some of the heavier stellar-origin black hole binaries that LIGO has shown to exist [8], and possibly cosmological sources such as GWs from cosmic string cusps or a stochastic background generated by phase transitions in the early Universe. However, the primary source for LISA are the MBHs in the centres of galaxies. These MBHs will generate GWs either when they merge with other MBHs, which is expected to occur following mergers between their host galaxies, or when they merge with much smaller compact objects formed from stars. These latter systems are called extreme-mass-ratio inspirals (EMRIs), because of the large difference in mass between the two objects involved. Galactic MBHs are typically surrounded by a stellar cluster, and EMRIs occur when compact objects formed as the end point of the evolution of stars in that cluster are captured and gradually fall into the central MBH. EMRIs occur over long timescales, meaning that $\sim 10^4$–10^5 cycles of gravitational radiation are generated while the compact object is in the strong gravitational field region close to the central MBH. In addition, the process by which EMRIs begin, through a scattering capture of the compact object by the MBH, tends to lead to high initial eccentricities, with the result that EMRI orbits are expected to retain significant eccentricity and non-equatorial inclinations when they are radiating in the LISA band. These properties lead to a great richness in EMRI gravitational waveforms, created by a superposition of the three fundamental frequencies — the orbital frequency, the perihelion precession frequency and the frequency of precession of the orbital plane. This complexity contains a great deal of information that can be used for science. EMRI GWs can be used both to measure the parameters of the system to high precision and to detect small deviations in the waveforms from the predictions of general relativity that are indicative of a breakdown of the theory. EMRIs thus have tremendous potential for astrophysics, cosmology and fundamental physics; this will be discussed again at the end of this article, but we refer the reader to [9, 10] for a more comprehensive discussion.

In this article we will compute the likely numbers of EMRI events and the precision with which LISA will be able to measure their parameters. There have been previous studies computing expected numbers of EMRI events [11]. However, the astrophysical model employed in those calculations was a combination of simple power laws and no attempt was made to quantify the uncertainties in that model. EMRI parameter-estimation studies have also been carried out [12, 13], but only for a small sample of representative cases and not for a full astrophysical population. We address both of these shortcomings in the current article. We compute event rates for several different astrophysical models that cover the range that is plausible given current observational constraints and compute estimates of the parameter-estimation accuracies for all the events in each population. In addition, we compute these results for the first time for a 2.5 Gm LISA detector with six laser links, which was proposed as the new mission baseline in the response to the ESA call in January 2017 [7].

This article is a summary of a paper that will be submitted shortly and we refer the interested reader to that paper for more details of the calculations and final results. This paper is organised as follows. In Section 2, we describe the models we employ for the detector sensitivity, the EMRI waveforms and the astrophysical population of EMRI sources. In Section 3 we present a comparison of the likely numbers of EMRI events that will be observed under each model and the parameter-estimation precision. In Section 4 we describe the implications of these results for science with EMRIs, before finishing in Section 5 with a short summary of conclusions.

2. Model

To estimate the detectability of EMRIs by LISA, we must model three different things — the intrinsic astrophysical population of EMRI events; the GWs generated by EMRI events; and the sensitivity of our detector to those GWs. We describe the models we use for each of these in this section.

2.1. Intrinsic astrophysical population

Our astrophysical population model includes models for the MBH mass and spin distribution, the intrinsic rate of EMRIs occurring in systems containing black holes with particular properties and the properties of the compact object in the EMRI.

MBH mass distribution Previous estimates [11] of EMRI rates have used a simple power law MBH mass function of the form

$$\frac{\mathrm{d}n}{\mathrm{d}\log M} = A\left(\frac{M}{3\times 10^6 M_\odot}\right)^{\alpha}\mathrm{Mpc}^{-3} \tag{1}$$

which is known to be a good fit to the observed MBH population in the relevant range [14]. Values of $A = 0.002$ and $-0.3 \leq \alpha \leq 0.3$ are usually assumed. More sophisticated models that self-consistently evolve the MBH population are now available [15], which have been used to predict the expected number of MBH binary mergers that will be observed by LISA [16]. We will denote this self-consistent MBH model as "B12". These models generate mass functions that are roughly consistent with the above power-law form, with $A = 0.005$ and $\alpha = -0.3$. To obtain a lower bound on the EMRI rate we will also consider a strict power-law MBH mass function with $A = 0.002$ and $\alpha = 0.3$. We will denote this model as "G10".

During a major merger, we expect the stellar cusp around the MBH to be disrupted. The cusp will regrow, but this takes a quarter of the relaxation time for a Milky Way-like galaxy [17], during which the standard EMRI picture does not apply. We include this effect in our model by assuming an MBH is unavailable as a host for EMRIs for a time

$$t_{\mathrm{cusp}} = T_0 (M/10^6 M_\odot)^{1.19} q^{0.35}\,\mathrm{Gyr} \tag{2}$$

following a merger with total binary mass M and mass ratio q. The pre-factor T_0 depends on the assumed M–σ relation, which is discussed below. We cannot apply this correction for the "G10" MBH population, as we do not track mergers in that case.

MBH spin distribution The "B12" model also tracks MBH spins. These are driven partially by mergers but primarily by accretion. We denote the MBH spin distribution computed in "B12" by "a98" as it predicts [18] that most MBHs will be spinning close to the maximal imposed limit, i.e., with $a \approx 0.98$. To understand the significance of this assumption we also consider a "flat" model with spins uniformly distributed in the range $a \in [0, 0.98]$, and a conservative "a0" model in which all spins are set to $a = 0$.

EMRI rate per black hole We base our EMRI rate per MBH on [19]

$$R_0 = 300\left(\frac{M}{10^6 M_\odot}\right)^{-0.19}\mathrm{Gyr}^{-1}. \tag{3}$$

ID	MF	MBH spin	Erosion	M–σ relation	N_p	CO mass $[M_\odot]$	EMRI rate $[yr^{-1}]$
M1	B12	a98	yes	Gultekin09	10	10	1600
M2	B12	a98	yes	KormendyHo13	10	10	1400
M3	B12	a98	yes	GrahamScott13	10	10	2770
M4	B12	a98	yes	Gultekin09	10	30	520
M5	G10	a98	no	Gultekin09	10	10	140
M6	B12	a98	no	Gultekin09	10	10	2080
M7	B12	a98	yes	Gultekin09	10	10	15800
M8	B12	a98	yes	Gultekin09	100	10	180
M9	B12	flat	yes	Gultekin09	10	10	1530
M10	B12	a0	yes	Gultekin09	10	10	1520
M11	G10	a0	no	Gultekin09	100	10	13
M12	B12	a98	no	Gultekin09	10	10	20000

Table 1. Summary of the twelve models used in this study. The columns indicate, respectively, the model identifier, the mass function prescription, the model for MBH spins, whether cusp erosion following mergers is included or not, the model used for the M-σ relation, the number of direct plunges per EMRI, the assumed mass of the compact object in all EMRIs, and the resulting astrophysical rate of EMRI events in the Universe at redshift $z < 4.5$.

If this rate was maintained over the age of the Universe, it would overgrow lower mass black holes, particularly when we account for the fact that for every EMRI there will be $N_p \sim 1$–100 "failed EMRIs" that directly plunge [20]. Therefore, we add two corrections. First, we compute $\Gamma = \min\{t_d/t_{relax}, 1\}$, where t_{relax} is the relaxation time in the MBH stellar cusp and $t_d = 0.06M/[mR_0(1+N_p)]$ is the time it would take to deplete the stellar cusp of compact objects through EMRIs and direct plunges. Second, we compute $\kappa = \min\{\exp(-1)M/(\Gamma\Delta M), 1\}$, where ΔM is the total mass that would be accreted by the MBH over its evolution history according to (3). The final EMRI rate is taken to be $\kappa\Gamma R_0$. The Γ factor corrects for the time taken for the stellar loss cone to be repopulated by diffusion, while the κ factor imposes a limit (of one e-fold) on the total mass that an MBH can gain by compact object accretion.

Black hole M–σ relation The factor T_0 in (2) and the stellar relaxation time depend on the velocity dispersion σ in the stellar cluster. This can be determined from the black hole mass via the M–σ relation. We use the model of [21] as our reference model "Gultekin09", which predicts $T_0 \sim 6$, plus an optimistic case "GrahamScott13" [22], which predicts $T_0 \sim 2$, and a pessimistic case "KormendyHo13" [23], which predicts $T_0 \sim 10$.

Compact object properties EMRI rates are expected to be dominated by inspirals of black holes, as these are visible to much greater distances and are intrinsically enhanced by mass segregation [9, 11]. Typically the mass, m, of the inspiralling object has been taken to be fixed at $m = 10M_\odot$. Given recent LIGO results [1, 3] we will consider both this standard case and one in which all EMRIs are assumed to have $m = 30M_\odot$. We take the eccentricity of the EMRI orbit at plunge to be uniformly distributed in the range $[0, 0.2]$. EMRIs on prograde orbits into more rapidly spinning MBHs can be more easily captured as inspirals rather than direct plunges. We account for this by enhancing the EMRI rate by a factor $W(a)$ that is a function of the MBH spin a, and by using the inclination distribution $p(\iota) \propto \sin(\iota)[W(a,\iota)]^{-0.83}$ [24].

Overall, we use 12 different models that are constructed using different combinations of the above ingredients. The models are summarised in Table 1. M1 is our reference model. M10 and M11 will be used only to estimate event rates, as degeneracies present for non-spinning EMRIs make parameter-estimation calculations based on the Fisher matrix more challenging. Full parameter-estimation results will appear in the longer journal article based on this work.

2.2. EMRI gravitational waveforms

The extreme-mass-ratio in an EMRI system means that the emitted GWs can be computed accurately using black hole perturbation theory. This involves calculating the self-force acting on the inspiralling object. We refer the reader to [25] for details on this approach. Self-force calculations have not yet been completed at the necessary order for objects on arbitrary orbits about spinning black holes, and those calculations that do exist for simpler cases are computationally intensive and therefore not well suited to studies of populations such as this one. As an alternative, there are two different families of kludge waveforms available. These are approximate models that have been constructed to include the most important qualitative features of EMRI waveforms, while not necessarily being quantitatively accurate. The analytic kludge (AK) model [12] constructs an EMRI waveform starting from the simple GW emission from a Keplerian orbit, and then imposing precession and evolution of the orbit onto the system. The precessional and evolution effects are governed by low-order post-Newtonian equations, which are accurate in the weak-field, low-velocity regime. The AK model is cheap to generate, but it rapidly goes out of phase with more accurate waveforms. The numerical kludge (NK) model [26, 27] uses Kerr geodesics as the basis for constructing EMRI orbits, adding inspiral determined by evolution equations that are based on post-Newtonian expansions, but augmented by fits to perturbation theory results. The NK waveform is generated from the trajectory using a flat-space emission formula [27]. NK waveforms are much more faithful to accurate inspirals than AK waveforms, staying in phase down to the last stages of inspiral [27]; however, they are more computationally expensive.

For the purposes of this study we need to be able to generate large numbers of gravitational waveforms to determine the detectability and parameter-estimation precision for astrophysical populations of EMRI events. We therefore use the computationally cheapest of the three models, the AK. While these waveforms are not faithful to the true EMRI signals produced in nature, it is believed that they capture the richness of real EMRI gravitational waveforms and therefore will provide an accurate guide to signal to noise ratios and parameter-estimation precisions. To attempt to characterise the uncertainty from the waveform choice, we consider two different versions of the AK. In the original AK model [12], the waveforms are terminated when the object reaches the Schwazrschild innermost stable orbit (ISO). This is inaccurate for prograde inspirals into spinning black holes, which can get much closer to the central black hole. As the spins of the black holes in our population tend to be quite high, this is likely to significantly underestimate the detectability of many events. A simple fix is to continue the inspiral until the EMRI reaches the Kerr ISO. As the post-Newtonian expressions on which the AK model is built break down in the strong-field regime at the end of the inspiral, this continuation is not going to be accurate and will most likely lead to an overestimate of detectability. Therefore, we expect the two sets of results to bracket the true answer.

2.3. LISA sensitivity to EMRIs

An EMRI will be detectable by LISA if it is sufficiently loud. This can be characterised by a requirement on the matched filtering signal-to-noise ratio (SNR) ρ, which the source has in the detector data stream

$$\rho^2 = \langle h|h\rangle, \quad \text{where } \langle a|b\rangle = 2\int_0^\infty \frac{\tilde{a}(f)\tilde{b}^*(f) + \tilde{a}^*(f)\tilde{b}(f)}{S_n(f)}\mathrm{d}f. \tag{4}$$

This expression depends on the final detector configuration through $S_n(f)$, the power spectral density of noise in the dctector. We use the LISA noise model described in [7]

$$S_n(f) = \frac{20}{3}\left[L^2 + \left(\frac{2f}{0.41c}\right)^2\right]\left(4S_n^{\mathrm{acc}}(f) + 2S_n^{\mathrm{loc}}(f) + S_n^{\mathrm{sn}}(f) + S_n^{\mathrm{omn}}(f)\right), \tag{5}$$

where $S_n^{\rm acc}(f) = \{9 \times 10^{-30} + 3.24 \times 10^{-28} \left[(3 \times 10^{-5}{\rm Hz}/f)^{10} + (10^{-4}{\rm Hz}/f)^2\right]\} / (2\pi f)^4 \, {\rm m}^2\,{\rm Hz}^{-1}$, $S_n^{\rm loc}(f) = 2.89 \times 10^{-24} \, {\rm m}^2\,{\rm Hz}^{-1}$, $S_n^{\rm sn}(f) = 7.92 \times 10^{-23} \, {\rm m}^2\,{\rm Hz}^{-1}$ and $S_n^{\rm omn}(f) = 4 \times 10^{-24} \, {\rm m}^2\,{\rm Hz}^{-1}$. We set the detector arm length to $L = 2.5$Gm. The complexities of EMRI data analysis mean that a relatively high SNR will be required for confident detection. The value $\rho_{\rm thresh} = 30$ was historically assumed [28], but results from the Mock LISA Data Challenge suggest that EMRIs with SNR as low as $\rho_{\rm thresh} = 20$ could be identified [29], albeit under somewhat idealised conditions. We will present results for both thresholds.

We assume conservatively that the duration of the LISA mission is two years and that any EMRIs that are detected must plunge during the mission lifetime. EMRIs accumulate SNR gradually over the whole inspiral, so LISA will also observe systems that are in the final stages of inspiral but do not reach plunge over the mission duration. The event rates reported here are therefore conservative in this regard. To extrapolate to longer mission durations we can multiply by the ratio of the mission duration to two years. Again, this will be conservative, as some EMRI signals that do not accumulate enough SNR over two years will pass the SNR threshold when integrated over a longer observation.

To evaluate parameter-estimation precision we use the Fisher matrix, defined by

$$\Gamma_{ij} = \left\langle \frac{\partial h}{\partial \lambda_i} \middle| \frac{\partial h}{\partial \lambda_j} \right\rangle \tag{6}$$

where λ^i denotes the waveform model parameters. An estimate of the precision of measurement is given by $\Delta\lambda_i^2 = (\Gamma^{-1})_{ii}$ (no sum over i).

3. Results

3.1. Event rates

Table 2 indicates the number of EMRI events that would be detected over the two year mission lifetime for each model, using SNRs computed for AK waveforms with the standard Schwarzschild plunge condition. For the reference model we would expect to observe several hundred events with SNR above 20, of which approximately one third would have SNR above 30. The number of observed EMRIs can be no more than a factor of 10 larger (M12), if we assume that MBHs are always available as EMRI hosts and MBHs gain no mass from direct plunges of compact objects. The EMRI rate could be as low as 0, if we assume a mass function that falls steeply toward lower MBH masses and that MBHs have small spins (M11). That model is overly pessimistic, and M5 and M8 probably give more realistic lower bounds, which are a factor of 10 smaller. The majority of the models predict several hundred EMRI events, so this figure appears to be fairly robust to astrophysical uncertainties.

If the inspiralling compact objects in EMRIs tend to be more massive (M4), we see a comparable number of events. While such EMRIs can be seen to larger redshift, this is compensated by a decrease in the intrinsic rate to prevent MBH overgrowth. Table 3 shows the corresponding results computed using the Kerr plunge criterion to terminate the EMRI. These rates are somewhat higher, as expected, but only by a factor $\sim$ 1–2.

In our model we have prevented MBH overgrowth through compact object accretion by imposing a constraint on the EMRI rate. An alternative solution is that the number of light MBHs is significantly reduced, because these MBHs grow rapidly from EMRI consumption. Results in the table are divided into mass bins and we see that between 20% and 60% of the events have MBH mass greater than $10^{5.5} M_\odot$, so even if no lighter MBHs exist we should see approximately one hundred events in the reference model. The rate drops significantly, by a factor of 40 or more, if we impose a cut off at $10^6 M_\odot$, when using the Schwarzschild plunge condition. However, a higher proportion of the events are heavier mass when we use the Kerr plunge condition. This is because in that model the EMRI can get closer to the MBH, shifting

Model	$M_{10} < 5$	Number of events in mass range $5 < M_{10} < 5.5$	$5.5 < M_{10} < 6$	$6 < M_{10}$	Total
M1	20 (10)	240 (60)	110 (50)	10 (0)	380 (130)
M2	30 (10)	190 (50)	70 (30)	0 (0)	290 (90)
M3	20 (0)	310 (90)	510 (220)	40 (20)	880 (340)
M4	70 (20)	280 (130)	80 (50)	0 (0)	440 (200)
M5	0 (0)	10 (0)	20 (10)	0 (0)	30 (10)
M6	20 (0)	270 (70)	210 (90)	20 (10)	520 (180)
M7	230 (50)	2190 (600)	1040 (480)	60 (40)	3530 (1170)
M8	0 (0)	30 (10)	10 (10)	0 (0)	50 (10)
M9	20 (10)	210 (60)	110 (50)	10 (10)	350 (130)
M10	30 (10)	240 (70)	100 (40)	10 (10)	370 (130)
M11	0 (0)	0 (0)	1 (0)	0 (0)	1 (0)
M12	230 (50)	2420 (670)	1730 (730)	180 (110)	4560 (1560)

Table 2. Number of events detected in each mass range, and total number of events, for each model. Mass ranges are indicated in terms of $M_{10} = \log_{10}(M/M_\odot)$. The primary values in each cell assume an SNR threshold of 20 is required for detection, while the bracketed numbers given the corresponding results for an SNR threshold of 30. SNRs are computed using the AK model with the Schwarzschild plunge condition. All numbers are rounded to the nearest 10 apart from the M11 results which are rounded to the nearest 1.

Model	$M_{10} < 5$	Number of events in mass range $5 < M_{10} < 5.5$	$5.5 < M_{10} < 6$	$6 < M_{10}$	Total
M1	20 (0)	260 (60)	230 (100)	80 (60)	590 (230)
M2	20 (0)	210 (50)	160 (70)	50 (40)	440 (160)
M3	10 (0)	360 (90)	1000 (470)	240 (180)	1620 (750)
M4	50 (10)	300 (150)	140 (100)	30 (30)	520 (280)
M5	0 (0)	10 (0)	40 (20)	40 (30)	90 (50)
M6	20 (0)	300 (80)	430 (200)	200 (150)	960 (440)
M7	190 (40)	2390 (600)	2110 (930)	730 (510)	5420 (2090)
M8	0 (0)	30 (10)	30 (10)	10 (10)	70 (30)
M9	20 (0)	230 (60)	160 (70)	30 (20)	430 (160)
M10	30 (10)	240 (70)	100 (40)	10 (10)	370 (130)
M11	0 (0)	0 (0)	1 (0)	0 (0)	1 (0)
M12	190 (40)	2700 (680)	3710 (1690)	1830 (1380)	8440 (3790)

Table 3. As Table 2, but now with SNRs computed using the AK model with the Kerr plunge condition.

the GW emission to higher frequencies and hence providing sensitivity to heavier MBHs. The prospects for significant numbers of EMRI detections are therefore good even if the number of lower mass MBHs is significantly depleted.

Table 4 shows how the number of events detected depends on the configuration of the detector for the reference model M1. If the final LISA configuration is similar to the 1 Gm, four-link NGO model used in [5], the number of events would be about a factor of 10 smaller. If a more sensitive 5 Gm configuration was launched, then the event rate could be increased by a factor of about three. The change in the number of detected events is similar for the other astrophysical models, except for the high mass compact object model (M4), for which the decrease in number

Armlength					
1 Gm		2.5 Gm		5 Gm	
4 links	6 links	4 links	6 links	4 links	6 links
0.1	0.2	0.4	1	2	3

Table 4. Variation in the number of observed EMRI events with detector configuration. Numbers are computed for the reference model, M1, and are expressed relative to the number detected for the new baseline 2.5 Gm, six-link configuration.

of events going to the 1 Gm configuration is only a factor of ~ 5, and the increase going to the 5 Gm configuration is only a factor of ~ 1.5. This difference arises because the EMRIs in M4 are intrinsically louder and so all detector configurations can detect most of the EMRI events out to moderate redshift.

3.2. Parameter-estimation accuracy

Figure 1 shows the distribution of parameter-estimation precisions that we would expect for the set of EMRIs observed in each population model. We present results for the redshifted masses of the MBH and compact object, for the spin of the MBH and for the eccentricity at plunge. We have normalised results to a fixed SNR of 20 (accuracies scale as $\Delta X \propto 1/\rho$) and we see that after doing that the precision is essentially independent of the population model. The distribution of SNRs is determined by the spatial distribution of sources, which is uniform in comoving volume and hence also similar in each model. However, the models that predict larger numbers of events will have some sources at high SNRs which will have correspondingly more precise parameter determinations. We see that we would expect to measure the redshifted masses of the components and the orbital characteristics such as eccentricity to precisions of $\sim 10^{-6}$–10^{-3}, while also measuring MBH spins to $\sim 10^{-5}$–10^{-3}. Figure 2 shows corresponding results for EMRI measurements of source distances and sky positions. Luminosity distance should typically be determined to a few to a few tens of percent, and sky location should be measured to about a thousandth of a steradian, which is a few square degrees. Sky position will be slightly less well measured for EMRIs with larger mass inspiraling objects, since such sources are observable for a smaller fraction of a LISA orbit. The luminosity distance accuracy is also approximately the accuracy with which the intrinsic masses of the sources will be determined, since the redshift will be inferred from the luminosity distance. We would therefore expect to determine the intrinsic MBH mass to better than 10% in the majority of cases.

These results were computed using the Schwarzschild plunge condition. Results computed with the Kerr plunge condition are typically a factor of ~ 10 better for the intrinsic parameters (and comparable for the extrinsic parameters), so these results are most likely conservative.

4. Implications for EMRI science

4.1. Astrophysics

In [14] it was shown that the classic 5 Gm LISA configuration would be able to measure the slope of the mass function of MBHs in the LISA range to a precision of $\sim \pm 0.3$ if 10 EMRI events were observed. This is the precision with which the mass function is currently known. This conclusion is robust to assumptions about the configuration of the detector. In the astrophysical models presented here, we expect more than 10 EMRI events in two years for all but the most pessimistic model. In most cases, the number of EMRIs expected is several hundreds. The precision scales as the square root of the number of events, so with a factor of 36 more EMRIs we would measure the slope of the mass function to a precision of ± 0.05. This would provide constraints on the low mass end of the MBH mass function that can not be determined in any other way. We note,

IOP Conf. Series: Journal of Physics: Conf. Series **840** (2017) 012021 doi:10.1088/1742-6596/840/1/012021

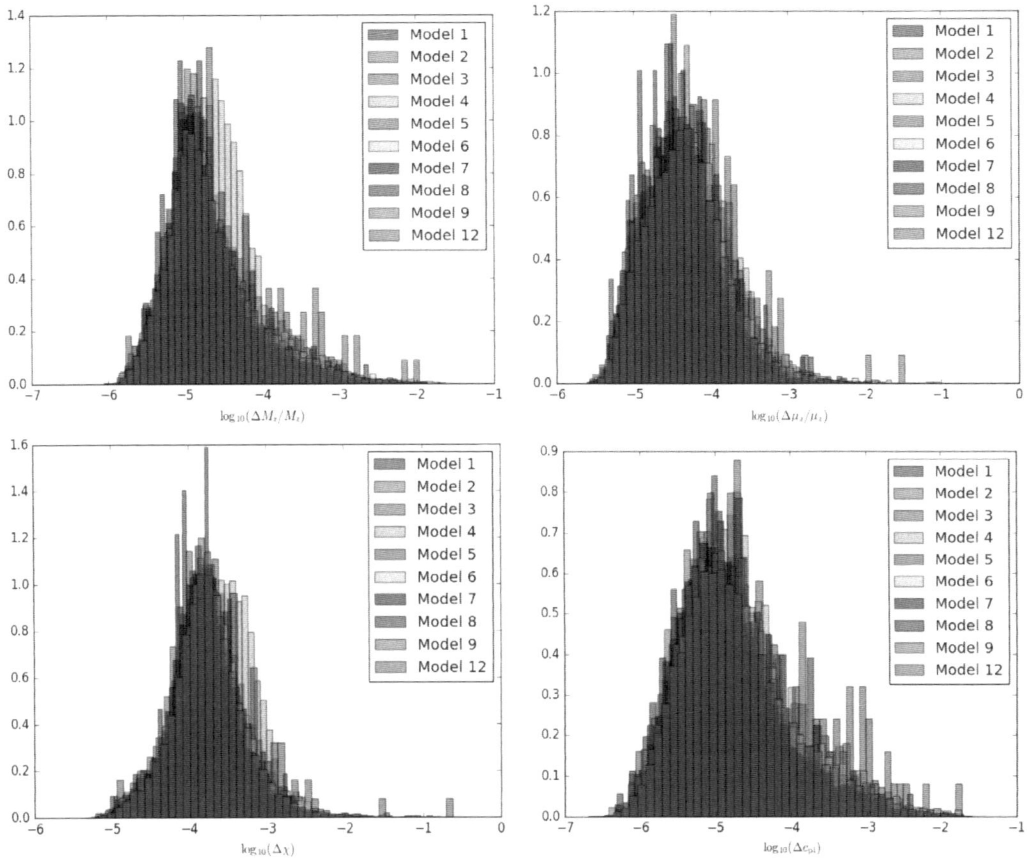

Figure 1. Expected parameter-estimation accuracies for redshifted MBH mass (top left), redshifted compact object mass (top right), MBH spin (bottom left) and eccentricity at plunge (bottom right).

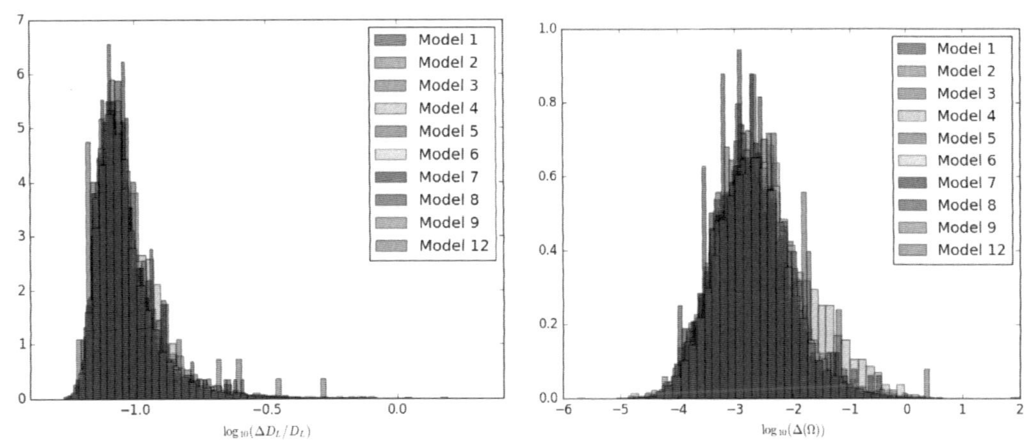

Figure 2. Expected parameter-estimation accuracies for luminosity distance (left) and sky position (right).

	Schwarzschild plunge condition		Kerr plunge condition	
Model	$N(z < 0.5)$	$N(z < 0.5$; small error)	$N(z < 0.5)$	$N(z < 0.5$; small error)
M1	30	5	29	7
M2	23	4	22	4
M3	62	15	60	16
M4	11	4	11	4
M5	2	0	3	1
M6	35	6	35	8
M7	298	48	285	52
M8	4	0	4	1
M9	25	3	25	5
M10	24	0	24	0
M11	0	0	0	0
M12	354	60	354	74

Table 5. Number of EMRIs detected at redshift $z < 0.5$ for each model, computed using each of the two waveform plunge conditions. The first column in each case gives all EMRIs at $z < 0.5$, while the second gives those EMRIs at $z < 0.5$ that also satisfy the error conditions that were assumed in [30], i.e., $\Delta(\ln D_L) < 0.07z$ and $\Delta\Omega < 16z^2$.

however, that [14] assumed the dependence of the EMRI rate on MBH mass was known, which is not the case due to the various complications discussed here. LISA will actually determine the product of the MBH mass function with the rate per MBH to the aforementioned precision. Combining EMRI observations with those of MBH mergers could break this degeneracy, but more work is required to properly understand this.

4.2. Cosmology

EMRI observations can be used to probe cosmological parameters. A GW observation cannot determine the redshift of an EMRI source and electromagnetic counterparts are unlikely to be observed. However, constraints can also be determined statistically. McLeod and Hogan [30] showed that by using the electromagnetically determined redshifts of all galaxies within any given LISA EMRI error box, a statistical constraint on the Hubble constant could be determined. A determination of H_0 to 1% precision would be possible provided LISA observes ~ 20 EMRIs at redshift $z < 0.5$. This calculation was done for the classic 5 Gm LISA configuration and assumed that LISA could determine luminosity distance and sky position to precisions $\Delta(\ln D_L) = 0.07z$ and $\Delta\Omega = 16z^2$. The new 2.5 Gm baseline will not be able to measure parameters to the same precision. However, these precisions will be reached for some events. In Table 5 we show how many EMRI events will be observed at redshift $z < 0.5$ in each model and how many will be observed with the assumed parameter-estimation accuracy. We see that most models predict the requisite number of EMRIs, but we would only expect ~ 5 events with the assumed precision. We would expect a factor of 2 worse precision with a factor of 4 fewer events, and so it is clear that a 2% measurement of H_0 will be possible in most cases. The LISA mission is likely to be longer than the two years assumed here, which will increase the number of accurately determined EMRIs. Moreover, all of the EMRIs will contribute something to the H_0 measurement, so assuming only the well localised EMRIs are useful is clearly conservative. Thus it is likely EMRI observations with LISA will provide interesting cosmological constraints, unless the total number of observed EMRIs is toward the low end.

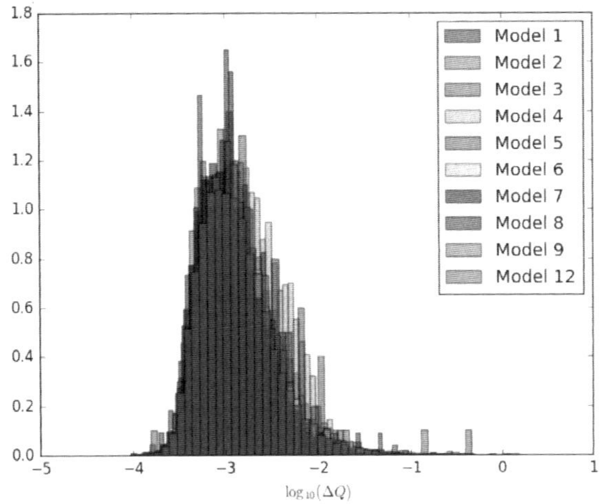

Figure 3. Precision with which LISA could detect a deviation in the quadrupole moment of the MBH spacetime away from the Kerr value.

4.3. Fundamental physics

EMRIs are excellent probes of fundamental physics, as they generate many cycles in the strong field regime. In particular, EMRIs can be used to map out the space-time structure outside the central MBH, and identify any deviations from the Kerr metric structure predicted by general relativity. There has been a lot of research in this area and a full discussion of tests of fundamental physics with LISA can be found in [10]. An example of the application of EMRIs to fundamental physics is to ask with what precision an EMRI could measure a deviation in the quadrupole moment of a black hole away from the Kerr metric value. Such a study was performed in [31], using a gravitational waveform model constructed by adding the leading order effect of a quadruple deviation into the AK model [12]. We have repeated the same calculation for the current EMRI populations. We assume that EMRIs are occurring in a Kerr MBH spacetime, i.e., the true $\Delta Q = 0$, but compute a Fisher matrix including this quadrupole deviation parameter. The corresponding element of the inverse Fisher Matrix is then a measure of how large a quadrupole deviation could be present in an observed EMRI before we would be able to identify that it was there. These results are shown in Figure 3, again normalised to a fixed SNR $\rho = 20$. We see that every EMRI should provide a constraint on deviations from the Kerr metric at a level of 0.01–1%, irrespective of our assumptions about the underlying astrophysical model.

5. Summary

We have described a comprehensive study of the prospects for detection of EMRIs with LISA. Our study has attempted to quantify, for the first time, the uncertainties in EMRI rate predictions arising from astrophysical uncertainties, as well as updating predictions for the new LISA baseline configuration used in [7]. We find that LISA should observe several hundred EMRI events over two years, with uncertainties of about one order of magnitude in each direction. These predictions are robust to the distribution of MBH spins and the possible depletion of MBHs at low masses. For all of these events LISA will determine the intrinsic parameters to high precision (sub-percent accuracy), determine sky location to a few square degrees and determine luminosity distance to $O(10\%)$. The parameter-estimation results are largely independent of the astrophysical model assumptions.

These EMRI detections have tremendous scientific potential. EMRIs will provide new constraints on the mass function and spin distribution of MBHs in the LISA range. EMRIs will provide constraints on cosmological parameters that are competitive to existing constraints, but with completely independent systematics. Finally, EMRIs will provide strong tests of aspects of the theory of general relativity. We have shown that all of these scientific objectives will be realised irrespective of the true nature of the astrophysical population. The science will be enhanced if the number of observed EMRIs is at the high end of the plausible range, but these goals will still be met even if the rates are nearer to the low end.

Acknowledgments

AS is supported by the Royal Society. E. Barausse and E. Berti acknowledge support from the H2020-MSCA-RISE-2015 Grant No. StronGrHEP-690904. This work has made use of the Horizon Cluster, hosted by the Institut d'Astrophysique de Paris. We thank Stephane Rouberol for running smoothly this cluster for us. E. Berti was supported by NSF Grant No. PHY-1607130 and by FCT contract IF/00797/2014/CP1214/CT0012 under the IF2014 Programme. PAS acknowledges support from the Ramón y Cajal Programme of the Ministry of Economy, Industry and Competitiveness of Spain. PAS's work has been partially supported by the CAS President's International Fellowship Initiative.

References

[1] Abbott, B P et al. 2016 *Phys. Rev. Lett.* **116** 061102
[2] Abbott, B P et al. 2016 *Phys. Rev. Lett.* **116** 241103
[3] Abbott, B P et al. 2016 *Phys. Rev.* X **6** 041015
[4] Croton D J et al. 2006 *Mon. Not. Roy. Astron. Soc.* **365** 11
[5] The eLISA Consortium 2013 *The Gravitational Universe* arxiv:1375.5720
[6] Armano M et al. 2016 *Phys. Rev. Lett.* **116** 231101
[7] Amaro-Seoane P et al. 2017 *Laser Interferometer Space Antenna* arxiv:1702.00786
[8] Sesana A 2016 *Phys. Rev. Lett.* **116** 2311021
[9] Amaro-Seoane P, Gair J R, Freitag M, Miller M C, Mandel I, Cutler C and Babak S 2007 *Class. Quantum Grav.* **24** R113
[10] Gair J R, Vallisneri M, Larson S L and Baker J G 2013 *Liv. Rev. Rel.* **16** 7 (cited on February 28th 2017)
[11] Gair J R 2009 *Class. Quantum Grav.* **26** 094034
[12] Barack L and Cutler C J 2004 *Phys. Rev.* D **69** 082005
[13] Huerta E A and Gair J R 2009 *Phys. Rev.* D **79** 084021
[14] Gair J R, Tang C and Volonteri M 2010 *Phys. Rev.* D **81** 104014
[15] Barausse E 2012 *Mon. Not. Roy. Astron. Soc.* **423** 2533
[16] Klein A et al. 2016 *Phys. Rev.* D **93** 024003
[17] Preto M and Amaro-Seoane 2010 *Astrophys. J.* **708** L42
[18] Sesana A, Barausse E, Dotti M and Rossi E M 2014 *Astrophys. J.* **794** 104
[19] Amaro-Seoane P and Preto M 2011 *Class. Quantum Grav.* **28** 094017
[20] Merritt D 2015 *Astrophys. J.* **814** 57
[21] Gültekin K et al. 2009 *Astrophys. J.* **698** 198
[22] Graham A W and Scott N 2013 *Astrophys. J.* **764** 151
[23] Kormendy J and Ho L C 2013 *Ann. Rev. Astron. & Astrophys.* **51** 511
[24] Amaro-Seoane P, Sopuerta C F and Freitag M D 2013 *Mon. Not. Roy. Astron. Soc.* **429** 3155
[25] Poisson E, Pound A and Vega I 2011 *Liv. Rev. Rel.* **14** 7 (cited on February 28th 2017)
[26] Gair J R and Glampedakis K 2006 *Phys. Rev.* D **73** 064037
[27] Babak S, Fang H, Gair J R, Glampedakis K and Hughes S A 2007 *Phys. Rev.* D **75** 024005
[28] Gair J R et al. 2004 *Class. Quantum Grav.* **21** S1595
[29] Babak S et al. 2010 *Class. Quantum Grav.* **27** 084009
[30] MacLeod C L and Hogan C J 2008 *Phys. Rev.* D **77** 3512
[31] Barack L and Cutler C 2007 *Phys. Rev.* D **75** 042003

Modelling EMRIs with gravitational self-force: a status report

M van de Meent

Mathematical Sciences, University of Southampton, United Kingdom

E-mail: M.vandeMeent@soton.ac.uk

Abstract. We give a status report on the progress in modelling extreme mass-ratio inspiral (EMRI) evolution using the gravitational self-force formalism.

1. Introduction

Extreme mass ratio inspirals (EMRIs) are black hole binaries where there primary object is a supermassive black hole of 10^5–10^6 Solar masses and the secondary object a stellar massive black hole of 10–60 Solar masses. EMRIs have been identified as key potential source for the LISA mission[1, 2]. In contrast to comparable mass black hole binaries, EMRIs evolve very slowly spending a large number 10^3–10^5 orbital cycles in the LISA sensitivity band before finally merging. As a consequence matched filtering can be used to make unprecedented accurate measurements of the orbital parameters of the systems obtaining typical relative errors of 1 part in 10^3–10^4 [3]. Beside the masses, spins, eccentricity, and inclination, it will also be possible to measure some of the higher gravitational multipoles of the primary object allowing tests of the no-hair theorem of general relativity [4].

In order to realize such measurements we need accurate theoretical models for the evolution of EMRIs. However the modelling of EMRIs introduces some novel complications. On the one hand, the large disparity in length and time scales (essentially of order the mass-ratio) makes the application of full numerical relativity unrealistic. On the other hand, EMRIs are highly relativistic making post-Newtonian models unreliably. Instead of these familiar cornerstones black hole binary modelling, the modelling of EMRIs requires the application of multi time and length scale perturbation methods to obtain a systematic expansion of the binary motion in terms of the ratio of the two masses. At zeroth order, the secondary object follows a free falling geodesic. Higher order corrections then take the form of an effective force generated by the secondary's own gravitational field, the *gravitational self-force* (GSF).

In this note we summarize the current progress in modelling EMRIs using gravitational self-force.

2. Modelling requirements

Population models suggest that EMRIs, unlike comparable mass binaries, can enter the LISA sensitivity band with significant eccentricity $e \lesssim 0.8$, and almost any spin configuration. Consequently, to provide accurate waveforms for matched filtering we need the ability to model the evolution of an EMRI with generic eccentricity, inclination and spins up to an phase accuracy of less than ~ 0.1 radian over the course of the full evolution covering 10^4–10^6 radians. How far do we need to go in the systematic expansion of the equations of motion to achieve this? This can be answered using a two time scale expansion [5].

For this purpose it has proven useful to adopt coordinates on the binary phase space using an osculating elements scheme. At zeroth order, the spacetime can be approximated by the background (Kerr) spacetime generated by the primary object. Each point in the binary phase space can then be identified by a geodesic tangent to that point. To identify each geodesic we need a set of invariants $\vec{P}$ (e.g. semi-latus rectum, eccentricity, and inclination). A further three phases $\vec{q}$ are then needed to identify the (radial, polar, azimuthal) position along the geodesic. In first order form, the formal expansion of the equations of motions then become,

$$\dot{\vec{q}}(\tau) = \vec{\omega}(\vec{P}) + \epsilon\vec{g}_1(\vec{P},\vec{q}) + \epsilon^2\vec{g}_2(\vec{P},\vec{q}) + O(\epsilon^3) \tag{1}$$

$$\dot{\vec{P}}(\tau) = \epsilon\vec{G}_1(\vec{P},\vec{q}) + \epsilon^2\vec{G}_2(\vec{P},\vec{q}) + O(\epsilon^3), \tag{2}$$

where ϵ is the mass-ratio and the $\vec{g}_i$ and $\vec{G}_i$ encode the self-force corrections to the motion. Hinderer and Flanagan [5] performed a multi-timescale analysis of the solutions of this system leading to a systematic post-Adiabatic expansion of the phase evolution.

2.1. Adiabatic approximation

The leading term in the expansion, is the adiabatic approximation, where we only include the orbit averaged changes $\langle\vec{G}_1\rangle(\vec{P})$ to the orbital invariants $\vec{P}$, which can be calculated using the gravitational wave flux to infinity and down the horizon. This approximation contains all the terms in the phase evolution up to $O(\epsilon^{-1})$. Consequently, we expect a phase error of at least $O(1)$ over a full inspiral. In fact, orbital resonances (see below) increase this error to $O(\epsilon^{-1/2})$. This might be accurate enough for the detection of EMRI signals in the LISA data, but will certainly not be enough for accurate parameter estimation.

2.2. 1-Post Adiabatic correction

The next (full) step in the expansion of the phase evolution is the 1-post-adiabatic (1PA) approximation, which includes all terms in the phase evolution up to $O(\epsilon^0)$, leading a phase error of at most $O(\epsilon^{1/2})$, which should be enough for accurate parameter estimation. Hinderer and Flanagan [5] showed that to obtain the 1PA approximations one needs the full oscillatory first order corrections, $\vec{G}_1(\vec{P},\vec{q})$ and $\vec{g}_1(\vec{P},\vec{q})$, and the orbit average second order change to $\vec{P}$, $\langle\vec{G}_2\rangle(\vec{P})$. The former can be obtained from a local first order metric perturbation, whereas the later requires second order perturbation theory.

2.3. 1/2-PA correction: orbital resonances

The above picture breaks down in the case of a 'resonance' between the evolution of the different orbital phases, i.e. the situation where two or more phases evolve in tandem. A resonance allows normally oscillatory terms in the equations of motion to build up coherently over many orbits leading to an $O(\epsilon^{1/2})$ amplification of certain oscillatory effects [6], leading to the introduction of a 1/2-PA term in the phase evolution. The evolution of certain extrinsic parameters that miss a leading order average term such as the center-of-mass velocity, is in fact dominated by this correction [7, 8].

3. State-of-the-art

A dedicated community—informally knwown as the "Capra community"—has formed around the challenge of modelling EMRIs using the gravitational self-force formalism. The first major breakthrough in the formalism came with the introduction of the MiSaTaQuWa [9][10] formula for the calculation of the first order self-force. In the two decades since, a solid rigorous theoretical framework has arisen for the systematic definition of the perturbative corrections to the equations of motion of EMRIs [11][12].

3.1. Flux calculations
The orbital averaged changes to the constants of motion, $\langle \vec{G}_1 \rangle(\vec{P})$, can extracted directly from the average gravitational wave flux to infinity and down the horizon [13, 14, 15]. The fluxes of energy and angular momentum can easily be extracted from solutions of the Teukolsky equation [16]. With some more effort changes to the Carter can be obtained as well [17]. Techniques for obtaining numerical solutions to the Teukolsky equation have been developed [18, 19, 20] that allow for the calculation of the averaged fluxes sourced by arbitrary geodesics, and thereby $\langle \vec{G}_1 \rangle(\vec{P})$.

3.2. First-order
The first order corrections $\vec{G}_1(\vec{P}, \vec{q})$ and $\vec{g}_1(\vec{P}, \vec{q})$ to the equations of motion split into two distinct parts. The first is the so called 'spin force' and depends only on the spin of the secondary and is independent of the metric perturbation it generates. The analytic expression for this force was formulated already in 1951 by Papapetrou [21]. The second is the gravitational self-force proper, which depends on the metric perturbation generated by the secondary, but is insensitive to its spin.

Various practical techniques have been developed for calculating the gravitational self-force [22]. In the case of a Schwarzschild primary, the first calculations for circular [23] and eccentric [24] orbits were published in 2007 and 2011, respectively. We now have a wide range of techniques [23][24][25, 26, 27, 28, 29] using both time domain and frequency domain methods and a variety of regularization methods for calculating the gravitational self-force on generic geodesics in Schwarzschild spacetime.

Calculation of the self-force in Kerr spacetime for a long time was considered a major problem of the program, because the linearized Einstein equation is not separable on a Kerr background. This issue can be circumvented by reconstructing the metric perturbation from solutions of the Teukolsky equation. We now have a rigorous understanding [30, 31] of how the self-force can be obtained from such a reconstructed metric. After much preliminary work [32, 33, 34, 35, 36], we can now calculate the gravitational self-force for any equatorial orbit around a Kerr black hole [37, 38], with no obvious obstructions for extending the calculations to generic (inclined) orbits.

3.3. Second-order GSF
Calculation of the second order self-force remains one of the major outstanding challenges of the self-force program, but much progress has been made. The formalism for the calculation of the second order self-force is now in place [39, 40, 41, 42, 12, 43], and major steps are being made towards solving practical obstructions [44, 45] for the first numerical calculations.

3.4. Evolution
Although a fully consistent 1PA evolution of an EMRI requires the orbit averaged second order self-force, it is of practical use to study the implementation of an osculating elements scheme sourced by only the first order force. In the case of a non-spinning Schwarzschild primary, the entire EMRI parameter space has now been covered with first order self-force data, allowing the evolution of inspirals [46, 47]. For example, this allows us to gauge the error made by using just using the adiabatic approximation to several tens of radians over a full inspiral.

3.4.1. Resonances The occurrence of orbital resonances acts as a further complication to the evolution of EMRIs. From symmetry considerations, it is clear that resonances can only impact the evolution for inspirals that are both inclined and eccentric. The general impact of a resonance of the EMRI evolution has been studied [6, 48, 49], and can be characterized by a jump in the evolution of the orbital invariants at the resonance, the size of which can be obtained from the orbit averaged quantities $\langle \vec{G}_1 \rangle$ [49].

A potential further complication would occur if the self-force dynamics of an EMRI could get locked in a sustained resonance, leading to major qualitative change to the evolution of the orbit over the inspiral time scale. The conditions necessary for such resonant locking to occur were studied [49], and based on the (limited) available data [20] it appears unlikely that they are satisfied anywhere in the parameter space. Moreover, if they could one would expect resonant locking to occur only in a small fraction (proportional to $\epsilon^{1/2}$) of all EMRIs [49].

4. Interaction with other modelling approaches

In the absence of fully modelled gravitational waveforms produced by EMRIs, it is important to be able to compare intermediate results between different calculations; both between different self-force calculations and between completely different modelling approaches such as post-Newtonian (PN) theory, Numerical relativity (NR), and effective one body (EOB) theory. This requires the identification and calculation of coordinate invariant (physical) observables that are completely independent of the used approximation scheme.

The gravitational wave fluxes are an example of such quantities and have long been used compared between different approaches. However, they are completely insensitive to the subtleties of the conserved dynamics. This requires different invariants such as the Detweiler-Sago-Barack redshift invariant [25, 24], the frequency shift of the inner most stable orbit [50, 51, 52], the periapsis advance of (nearly) circular orbits [53, 24, 52], and corrections to the precession of the secondary spin [54, 55].

Calculation of such quantities has allowed valuable validation of the self-force formalism. Successful comparison between self-force and numerical relativity results has, for example, taught use that even at first order some self-force results remain fairly accurate at even comparable mass-ratios [56, 57, 58, 59].

In recent years, self-force results have become increasingly useful in aiding the development of other modelling approaches. Self-force calculations have allow the calculation of the linear in mass-ration part of the post-Newtonian expansionist to very high (22.5) PN order, both through numerical [60, 61, 62, 63, 64, 65, 37, 52] and analytical [66, 67, 68, 69, 70, 71, 72, 73, 74, 75, 76] techniques. Moreover, highly accurate self-force calculations of periapsis advance [52] have allowed to discriminate between competing derivations of the 4PN equations of motion.

Finally, self-force calculations of various invariants have allowed calibration of various EOB potentials [77, 78, 69, 71, 79, 80].

5. Outlook

Although much progress has been made in the modelling of EMRIs over the last two decades, a lot of work remains to be done. We should soon have the calculation tools to calculate the first-order gravitational self-force on arbitrary geodesic orbits. However, it will still be a challenge to fill the entire four dimensional EMRI parameter space with data, since high eccentricity/high inclination data points currently take up to 1000s of CPU hours. We will need to both ncreases in efficiency of the calculation, and optimize the strategy for covering the entire parameter space.

At second order, calculational methods are still in its infancy. Once the simplest case of the calculation of the second order self-force on (quasi-)circular orbits on a Schwarzschild background has been achieved, this will need to be extended to general orbits on Kerr backgrounds.

Achieving high enough computational efficiency to fill the parameter space with second order data will be an even greater challenge than at first order.

Having covered the EMRI parameter space with self-force data, we will be in a position to model the evolution of EMRIs. The basic techniques for this have been implemented and tested. A remaining open question concerns the fact that formally the self-force is not just a local function of the parameter space, but depends on the entire past history of the EMRI system. By approximating the local self-force by that generated by a tangent geodesic, we are making an error of currently unknown size. This needs to be validated by comparison with time domain self-consistent evolution of an EMRI. Techniques for this are being developed. Initial investigations with scalar toy models suggest this error may be negligible.

Finally, once inspiral evolutions can be computed, waveforms need to be generated. Time domain integration of the Teukolsky equation over an entire inspiral will be costly and slow. In principle, it should be possible to synthesize the waveform from frequency domain data using a two timescale expansion. It remains to be seen if this can be done fast enough to generate waveforms on-the-fly for LISA data analysis. It is likely that effective surrogate (kludge) models will need to be developed for data analysis purposes.

Acknowledgments

This work was supported by the European Research Council under the European Union's Seventh Framework Programme (FP7/2007-2013) ERC grant agreement no. 304978.

References

[1] Amaro-Seoane P, Gair J R, Freitag M, Coleman Miller M, Mandel I, Cutler C J and Babak S 2007 *Class. Quant. Grav.* **24** R113–R169 (*Preprint* astro-ph/0703495)
[2] LISA Consortium 2017 LISA: Laser interferometer space antenna URL https://www.lisamission.org/proposal/LISA.pdf
[3] Barack L and Cutler C 2004 *Phys. Rev. D* **69** 082005 (*Preprint* gr-qc/0310125)
[4] Barack L and Cutler C 2007 *Phys. Rev. D* **75** 042003 (*Preprint* gr-qc/0612029)
[5] Hinderer T and Flanagan E E 2008 *Phys. Rev. D* **D78** 064028 (*Preprint* 0805.3337)
[6] Flanagan E E and Hinderer T 2012 *Phys. Rev. Lett.* **109** 071102 (*Preprint* 1009.4923)
[7] Hirata C M 2011 *Phys. Rev. D* **83** 104024 (*Preprint* 1011.4987)
[8] van de Meent M 2014 *Phys. Rev. D* **90** 044027 (*Preprint* 1406.2594)
[9] Mino Y, Sasaki M and Tanaka T 1997 *Phys. Rev. D* **55** 3457–3476 (*Preprint* gr-qc/9606018)
[10] Quinn T C and Wald R M 1997 *Phys. Rev. D* **56** 3381–3394 (*Preprint* gr-qc/9610053)
[11] Poisson E, Pound A and Vega I 2011 *Living Rev. Rel.* **14** 7 (*Preprint* 1102.0529)
[12] Pound A 2015 *Motion of small objects in curved spacetimes: An introduction to gravitational self-force* (*Fund. Theor. Phys.* vol 179) pp 399–486 (*Preprint* 1506.06245)
[13] Mino Y 2003 *Phys. Rev. D* **67** 084027 (*Preprint* gr-qc/0302075)
[14] Mino Y 2005 *Prog. Theor. Phys.* **113** 733–761 (*Preprint* gr-qc/0506003)
[15] Mino Y 2005 *Classical Quant. Grav.* **22** S717–S738 (*Preprint* gr-qc/0506002)
[16] Teukolsky S A and Press W H 1974 *Astrophys. J.* **193** 443–461
[17] Sago N, Tanaka T, Hikida W, Ganz K and Nakano H 2006 *Prog. Theor. Phys.* **115** 873–907 (*Preprint* gr-qc/0511151)
[18] Drasco S and Hughes S A 2006 *Phys. Rev. D* **73** 024027 (*Preprint* gr-qc/0509101)
[19] Fujita R and Hikida W 2009 *Classical Quant. Grav.* **26** 135002 (*Preprint* 0906.1420)
[20] Flanagan E E, Hughes S A and Ruangsri U 2014 *Phys. Rev. D* **89** 084028 (*Preprint* 1208.3906)
[21] Papapetrou A 1951 *Proc. Roy. Soc. Lond.* **A209** 248–258
[22] Wardell B and Gopakumar A 2015 **179** 487–522 (*Preprint* 1501.07322)
[23] Barack L and Sago N 2007 *Phys. Rev. D* **75** 064021 (*Preprint* gr-qc/0701069)
[24] Barack L and Sago N 2011 *Phys. Rev. D* **83** 084023 (*Preprint* 1101.3331)
[25] Detweiler S L 2008 *Phys. Rev. D* **77** 124026 (*Preprint* 0804.3529)
[26] Hopper S and Evans C R 2010 *Phys. Rev. D* **82** 084010 (*Preprint* 1006.4907)
[27] Akcay S 2011 *Phys. Rev. D* **83** 124026 (*Preprint* 1012.5860)
[28] Akcay S, Warburton N and Barack L 2013 *Phys. Rev. D* **88** 104009 (*Preprint* 1308.5223)
[29] Osburn T, Forseth E, Evans C R and Hopper S 2014 *Phys. Rev. D* **90** 104031 (*Preprint* 1409.4419)

[30] Pound A, Merlin C and Barack L 2014 *Phys. Rev. D* **89** 024009 (*Preprint* 1310.1513)
[31] Merlin C, Ori A, Barack L, Pound A and van de Meent M 2016 *Phys. Rev. D* **94** 104066 (*Preprint* 1609.01227)
[32] Ori A 2003 *Phys. Rev. D* **67** 124010 (*Preprint* gr-qc/0207045)
[33] Lousto C O and Whiting B F 2002 *Phys. Rev. D* **66** 024026 (*Preprint* gr-qc/0203061)
[34] Keidl T S, Friedman J L and Wiseman A G 2007 *Phys. Rev. D* **75** 124009 (*Preprint* gr-qc/0611072)
[35] Keidl T S, Shah A G, Friedman J L, Kim D H and Price L R 2010 *Phys. Rev. D* **82** 124012 (*Preprint* 1004.2276)
[36] Shah A G, Keidl T S, Friedman J L, Kim D H and Price L R 2011 *Phys. Rev. D* **83** 064018 (*Preprint* 1009.4876)
[37] van de Meent M and Shah A G 2015 *Phys. Rev. D* **92** 064025 (*Preprint* 1506.04755)
[38] van de Meent M 2016 *Phys. Rev. D* **94** 044034 (*Preprint* 1606.06297)
[39] Pound A 2012 *Phys. Rev. Lett.* **109** 051101 (*Preprint* 1201.5089)
[40] Gralla S E 2012 *Phys. Rev. D* **85** 124011 (*Preprint* 1203.3189)
[41] Detweiler S 2012 *Phys. Rev. D* **85** 044048 (*Preprint* 1107.2098)
[42] Pound A and Miller J 2014 *Phys. Rev. D* **89** 104020 (*Preprint* 1403.1843)
[43] Pound A 2015 *Phys. Rev. D* **92** 044021 (*Preprint* 1506.02894)
[44] Miller J, Wardell B and Pound A 2016 *Phys. Rev. D* **94** 104018 (*Preprint* 1608.06783)
[45] Pound A 2015 *Phys. Rev.* **D92** 104047 (*Preprint* 1510.05172)
[46] Warburton N, Akcay S, Barack L, Gair J R and Sago N 2012 *Phys. Rev. D* **85** 061501 (*Preprint* 1111.6908)
[47] Osburn T, Warburton N and Evans C R 2016 *Phys. Rev.* **D93** 064024 (*Preprint* 1511.01498)
[48] Gair J, Yunes N and Bender C M 2012 *J. Math. Phys.* **53** 032503 (*Preprint* 1111.3605)
[49] van de Meent M 2014 *Phys. Rev. D* **89** 084033 (*Preprint* 1311.4457)
[50] Barack L and Sago N 2009 *Phys. Rev. Lett.* **102** 191101 (*Preprint* 0902.0573)
[51] Isoyama S, Barack L, Dolan S R, Le Tiec A, Nakano H *et al.* 2014 *Phys. Rev. Lett.* **113** 161101 (*Preprint* 1404.6133)
[52] van de Meent M 2017 *Phys. Rev. Lett.* **118** 011101 (*Preprint* 1610.03497)
[53] Barack L, Damour T and Sago N 2010 *Phys. Rev. D* **82** 084036 (*Preprint* 1008.0935)
[54] Dolan S R, Warburton N, Harte A I, Le Tiec A, Wardell B *et al.* 2014 *Phys. Rev. D* **89** 064011 (*Preprint* 1312.0775)
[55] Dolan S R, Nolan P, Ottewill A C, Warburton N and Wardell B 2015 *Phys. Rev. D* **91** 023009 (*Preprint* 1406.4890)
[56] Le Tiec A, Mrou A H, Barack L, Buonanno A, Pfeiffer H P, Sago N and Taracchini A 2011 *Phys. Rev. Lett.* **107** 141101 (*Preprint* 1106.3278)
[57] Le Tiec A, Buonanno A, Mroué A H, Pfeiffer H P, Hemberger D A, Lovelace G, Kidder L E, Scheel M A, Szilágyi B, Taylor N W and Teukolsky S A 2013 *Phys. Rev. D* **88** 124027 (*Preprint* 1309.0541)
[58] Zimmerman A, Lewis A G M and Pfeiffer H P 2016 *Phys. Rev. Lett.* **117** 191101 (*Preprint* 1606.08056)
[59] Lewis A G M, Zimmerman A and Pfeiffer H P 2016 (*Preprint* 1611.03418)
[60] Akcay S, Barack L, Damour T and Sago N 2012 *Phys. Rev. D* **86** 104041 (*Preprint* 1209.0964)
[61] Akcay S and van de Meent M 2016 *Phys. Rev. D* **93** 064063 (*Preprint* 1512.03392)
[62] Shah A G, Friedman J L and Whiting B F 2014 *Phys. Rev. D* **89** 064042 (*Preprint* 1312.1952)
[63] Shah A G 2014 *Phys. Rev.* **D90** 044025 (*Preprint* 1403.2697)
[64] Shah A G and Pound A 2015 *Phys. Rev. D* **91**(12) 124022 (*Preprint* 1503.02414)
[65] Johnson-McDaniel N K, Shah A G and Whiting B F 2015 (*Preprint* 1503.02638)
[66] Bini D and Damour T 2013 *Phys. Rev.* **D87** 121501 (*Preprint* 1305.4884)
[67] Bini D and Damour T 2014 *Phys. Rev. D* **89** 104047 (*Preprint* 1403.2366)
[68] Bini D and Damour T 2014 *Phys. Rev. D* **90** 024039 (*Preprint* 1404.2747)
[69] Bini D and Damour T 2014 *Phys. Rev. D* **90** 124037 (*Preprint* 1409.6933)
[70] Bini D and Damour T 2015 *Phys. Rev. D* **91** 064050 (*Preprint* 1502.02450)
[71] Bini D, Damour T and Geralico A 2016 *Phys. Rev. D* **93** 064023 (*Preprint* 1511.04533)
[72] Bini D and Damour T 2015 *Phys. Rev. D* **91** 064064 (*Preprint* 1503.01272)
[73] Bini D, Damour T and Geralico A 2016 *Phys. Rev. D* **93** 104017 (*Preprint* 1601.02988)
[74] Kavanagh C, Ottewill A C and Wardell B 2015 *Phys. Rev. D* **92** 084025 (*Preprint* 1503.02334)
[75] Hopper S, Forseth E, Osburn T and Evans C R 2015 *Phys. Rev. D* **92** 044048 (*Preprint* 1506.04742)
[76] Kavanagh C, Ottewill A C and Wardell B 2016 *Phys. Rev. D* **93** 124038 (*Preprint* 1601.03394)
[77] Damour T 2010 *Phys. Rev. D* **81** 024017 (*Preprint* 0910.5533)
[78] Akcay S, Le Tiec A, Barack L, Sago N and Warburton N 2015 *Phys. Rev. D* **91** 124014 (*Preprint* 1503.01374)
[79] Barausse E, Buonanno A and Le Tiec A 2012 *Phys. Rev. D* **85** 064010 (*Preprint* 1111.5610)

[80] Le Tiec A, Barausse E and Buonanno A 2012 *Phys. Rev. Lett.* **108** 131103 (*Preprint* 1111.5609)

IOP Conf. Series: Journal of Physics: Conf. Series **840** (2017) 012023 doi:10.1088/1742-6596/840/1/012023

The Future of Black Hole Astrophysics in the LIGO-VIRGO-LPF Era

Roger Blandford
KIPAC, Stanford, CA 94305
E-mail: rdb3@stanford.edu

Richard Anantua
Dept. Astronomy, UC Berkeley, CA 94720
E-mail: ranantua@berkeley.edu

Abstract. There is a resurgence of interest in black holes sparked by the LIGO-VIRGO detection of stellar black hole mergers and recent astronomical investigations of jets and accretion disks which probe the spacetime geometry of black holes with masses ranging from a few times the mass of the sun to tens of billions of solar masses. Many of these black holes appear to be spinning rapidly. Some new approaches are described to studying how accreting black holes function as cosmic machines paying special attention to observations of AGN jets, especially with VLBI and γ-ray telescopes. It is assumed that these jets are powered by the electromagnetic extraction of the spin energy of their associated black holes, which are described by the Kerr metric, and that they become simpler and more electromagnetically dominated as the event horizon is approached. The major uncertainty in these models is in describing acceleration and transport of relativistic electrons and positrons and simple phenomenological prescriptions are proposed. The application of these ideas to M87 and 3C279 is outlined and the prospects for learning more, especially from the Event Horizon Telescope and the Cerenkov Telescope Array, are discussed. The main benefit of a better understanding of black hole astrophysics to the LISA mission should be a firmer understanding of the source demographics.

1. Introduction

The last time one of us (RB) attended a LISA meeting was a decade ago and we would like to start by summarizing what was said then just to give give some context to how things have changed in studying both gravitational radiation and black holes. In 2006, there was an expectation that LIGO would make the first direct detection of gravitational radiation, as has happened, but that LISA sources (white dwarf binaries, EMRIs, MBH binaries, stochastic background and, of course, the unanticipated) would be more diverse, could be studied in greater detail and would directly impact more astrophysical questions. RB highlighted what he called "harbingers" – advanced notice of incipient mergers in distant AGN with few degree positions from LISA waveforms – and the use of LSST images to seek an electromagnetic signal in phase with the waveform ahead of the merger and thereby make an identification. It now looks like this will not be practical because the anticipated accuracy of the LISA positions has been downgraded. However the search for "afterglows" will still be practical and optimism remains high that these will be seen. In this connection RB resurrected an old idea that merging black holes might detonate enormous electromagnetic explosions – a tiny fraction of the gravitational radiation energy release would suffice – by acting as two-sphere or three-sphere dynamos. Although he expressed some skepticism, this might allow a seed electromagnetic field to grow exponentially, just prior to the merger, provided that the effective resistive decay did not quench the growth. Recent simulations (East and RB, in preparation) demonstrate that there is no such growth. This result is supported by observation! RB also discussed resonant excitation of stellar oscillations when white dwarfs follow elliptical orbits around more massive black holes [13]. The resonances are inevitably crossed as gravitational radiation shrinks the orbit and the energy transfer could be large enough to explode the white dwarf, especially when the mass ratio is large. We still think this is an interesting possibility and could lead to a new class of supernova explosion.

Observationally, RB described Very Long Baseline Interferometry (VLBI) with its improving resolution and sensitivity. Here the big development has been great strides in mm interferometry. It has already been found that the M87 jet appears to be collimated on scales less than of order ten gravitational radii and that the Sgr A* polarization is high, implying an organized magnetic field close to the black hole[2][?]. We should soon have even more impressive images and model fitting for these and other sources especially under the aegis of the Event Horizon Telescope (EHT). JWST, which then had a scheduled 2013 launch is now expected to start observing in 2019 but may have little overlap with LISA, as currently planned. The wide field OIR space telescope JDEM has become Euclid, to be launched at the start of the next decade and WFIRST, scheduled for mid-decade. The next generation X-ray proposals, Con-X and XEUS have evolved into Athena, which is now seen as being on a similar timescale to LISA, late decade. GLAST was launched in 2008 as Fermi Gamma-ray Space Telescope (FGST) and opened up the GeV sky. As a widefield instrument, it sets important limits on afterglows but is now unlikely to overlap with LISA. TeV astronomy with Atmospheric and Water Cerenkov detectors was also discussed and there have been great advances with the discovery of many new blazars as well as many intriguing Galactic sources. Here, the future lies with the Cerenkov Telescope Array (CTA), two observatories that will soon produce their first data and which should be fully operational in the early part of the next decade. The optimism for neutrino astronomy has been vindicated, especially with the successful deployment of Ice Cube and its detection of (still unidentified) cosmic sources.

The bad news, though, is that the launch date of LISA has been pushed out from the (2006) nominal date of 2016 by at least a decade. Others here discussed the prospects for accelerating and funding the project, especially following the great success of LPF and the scientific impetus of GW150914. The era of gravitational wave astronomy, with its associated physics program, has begun and there is a a whole spectrum extending from the LIGO band through the LISA, IPTA and CMB bands to explore.

2. Black Hole Astrophysics

2.1. Generalities

Although once regarded with suspicion by some theoretical physics intimidated by the perplexing questions they raised and dismissed by the more conservative observational astronomers, black holes are now part of the astrophysics mainstream. We are pretty sure that we are witnessing the births of stellar holes on a daily basis as γ-ray bursts and, even more frequently, as a minority of core collapse supernovae. The evidence that normal galaxy nuclei contain massive black holes is just as compelling. The few GW events provide an even stronger existence proof and, more importantly, they confirm that general relativity can be used to describe them. The masses that we observe are in the range $\sim 4\mathrm{M}_\odot - 2\times10^{10}\mathrm{M}_\odot$. Spin measurements are model-dependent and probably less accurate but do suggest that many holes have large spins, $a \gtrsim 0.9m, \Omega_H \gtrsim 0.3m^{-1}$. Even less secure are the basic principles of disk accretion despite considerable, thought, analysis and simulation. We still do not have a description of the physics at the level of our description of main sequence stars.

Most instances of black hole accretion are believed to have enough angular momentum to form an extended disk. There appear to be three main accretion modes, mostly distinguished by the mass supply rate in units of the "critical" or "Eddington" or "Salpeter" rate, $4\pi GMm_p/\sigma_T c^2 \equiv \dot{m}$. (Less important factors include the hole mass and spin, the outer disk radius in units of the gravitational radius $GM/c^2 \equiv m$, the composition of the accreting gas and environmental factors, especially in for stellar holes.) In the case of AGN, it is clear that overall disk accretion has to be very efficient - $\sim 0.1 - 0.3c^2 \equiv 10 - 30$ PJ kg^{-1}, because we observe the radiant energy density and measure the local hole mass density [15]. However, there is now pretty strong evidence that accretion disks are only thin down to their inner radius and are capable of radiating their released gravitational binding energy on the inflow timescale when the mass supply is intermediate. $0.01 \lesssim \dot{m} \lesssim 1$, say. At lower mass supply, the ion-electron energy exchange time is longer than the inflow time and a thick accretion torus seems to form. At higher mass supply, radiation is efficient but escape is slow on account of the high Thomson optical depth and the radiation is advected with the flow. So, at high and low $\dot{m}$, the accretion flow is roughly adiabatic. Under these circumstances, the torque that drives the accretion must transport energy outward and the release of this energy is enough to unbind the gas in the disk[7]. There has to be an outflow to carry away the accretion energy when radiation fails to do the job. In the low $\dot{m}$ case, this makes cooling even harder and the mass accretion rate can be many orders of magnitude smaller than the supply rate. This appears to be the case in Sgr A* and M87 where there is no conclusive observational evidence for an accretion disk close to the hole.

It is also clear that the internal disk torque is hydromagnetic because weakly magnetized disks are generically subject to the magnetorotational instability which grows the field on a rotational timescale[5]. It is also quite likely that this magnetic field is not confined to the disk and, instead is responsible for an external torque and an outflowing wind. It is difficult to be quantitative. Magnetic flux may also be trapped within the inner disk radius and can thread the event horizon of the black hole. When the hole spins, electromagnetic power is extracted and this appears to be responsible for relativistic jets[6]. The bulk Lorentz factors in the jets associated with AGN and of most interest to the LISA community are $\Gamma \sim 10$; those associated with X-ray binaries have $\Gamma \sim 3$ while$\Gamma \sim 300$ for γ-ray bursts. The jets are thought to be confined by a thick accretion torus and magnetized outflows. In this interpretation, the high latitude magnetosphere at the base of a black hole jet is likely to be electromagnetically dominated as low altitude plasma will drain across the horizon and high altitude plasma will fly away, being replenished by electron-positron pair production or cross-field transport of ionic plasma within the black hole magnetosphere. In addition to the observational evidence that relativistic jets involved strong magnetic field we are starting to see evidence that pair plasma is also present[14].

2.2. Electromagnetic Jets

When a black hole with angular frequency Ω_H is threaded by magnetic flux Φ_H sourced by external toroidal current, an EMF $V \sim \Omega_H\Phi_H$ will be generated. Under electromagnetic conditions, a current $I \sim V/Z_0$, where Z_0 is the impedence of free space will flow and the extracted power, derived from the hole's rotational energy, is $\sim V^2/Z_0$. A comparable power will be dissipated within the event horizon, increasing the mass of the hole. AGN jets typically have $\Phi_H \sim 10^{25-28}$ Wb and $V \sim 1 - 100$ EV, $I \sim 10$ PA $- 1$ EA. Particle acceleration is not a challenge given these large voltages. Indeed the most powerful AGN jets remain a candidate site for the acceleration of Ultra High Energy Cosmic Rays. Another feature of this mechanism is that the angular velocity associated with the field lines (that of a reference frame in which the local electric field vanishes) will be $\sim 0.5\Omega_H$, being larger for polar than low latitude field lines. Three (space) dimensional general relativistic MHD simulations bring out and quantify these general features[12]. In particular they confirm that fairly close to the hole, within $\lesssim 100m$, say, the jet structure is quite simple and electromagnetically-dominant. In addition the jets are surpringly robust. This will quite likely be a boon when the EHT investigates jets closer to their originating black holes. In fact these inner jets can be described approximately by a simple analytical solution with a paraboloidal profile that brings out the main features of the simulations. In addition, it is found the return current from the jet is concentrated close to the jet wall. In other words, the electromagnetic stress within the jet is balanced by the external gas pressure.

There are three consequences of the presumed electromagnetic dominance close to the event horizon. The first is that the commonly invoked diffusive shock acceleration is irrelevant. Secondly, while the electromagnetic field and current structure may be fairly simple and well-defined, the fluid (plasma) component is quite unknown. There is no prescription for specifying where and how this plasma is created though the amount needed to supply electrical current and provide space charge is tiny. However, we can say something about the velocity of this plasma, presuming that it can be treated as a single fluid. The component of velocity measured in a local inertial frame perpendicular to the presumably dominant magnetic field will be given by $\vec{E} \times \vec{B}/B^2$. The component along the magnetic field may also be fairly well-defined except close to the black hole. In the perfect MHD limit, it will be essentially the mechanical speed of a bead on a rotating wire with the same shape as the magnetic field. Of course, this depends upon the initial conditions and instabilities, radiation reaction or particle drift may vitiate MHD but it is a good starting assumption that the velocity field, as furnished by a simulation is a fair representation of that found in an actual relativistic jet.

The third consequence that emerges from these models is that at larger distances from the hole, $\gtrsim 1000m$, say, the jets build up internally created pairs, interact strongly with their environment and are subject to instabilities, which, while not completely disrupting them, do lead to much entrainment. We expect the jets to accelerate first then to decelerate. As the jet inertia builds up, more of the jet current closes within the jet, thereby accelerating the plasma. Note that the jet speed is quite variable in space and time in contrast to what is commonly assumed.

3. "Observing" Numerical Simulations

We have argued that RMHD simulations now provide a good foundation for modeling a jet close to the hole where the gas inflow and the electromagnetic outflow might be relatively simple. In order to compare the model with actual observations – to "observe" the simulation – it is necessary to add a prescription for particle acceleration[3]. There are several such prescriptions that can be used.

3.1. Particle Acceleration

Let us consider the synchrotron emission and suppose that the the particle distribution function is isotropic. It is helpful to measure the particle density by the partial pressure $\tilde{P}(\nu)$ of the electrons and positron with energy such that emit syncrotron radiation in the local magnetic field within an octave of frequency of the actual observed frequency. This focuses attention on the electrons that have to be present and must have been accelerated instead of a much more extensive distribution with density that we cannot measure. A variety of prescriptions have been entertained. The simplest is that $\tilde{P}(\nu)$ is a simple fraction of the magnetic pressure P_M. (The further assumption that $\tilde{P}(\nu)$ is independent of ν is essentially equivalewnt to saying that the observed spectral index is 0.5.) This is a natural generalization of the "equipartition hypothesis". A further extension is to suppose that $\tilde{P}(\nu) \propto P_M^n$.

Other prescriptions are based upon electromagnetic generalizations of fluid dissipation prescriptions. For example, following the practice with accretion disks, we can suppose that the Maxwell shear stress is a fixed fraction – α – of P_M. The rate of dissipation can then be estimated by the product of this stress and the transverse velocity gradient. Alternatively, we can introduce a coefficient of shear viscosity μ as some multiple of cL where L is the characteristic length scale associated and use the full Maxwell stress tensor. An electromagnetic prescription introduces a resistivity $\eta \sim \mu_0 cL$ and evaluates the rate of dissipation using ηj^2. These estimates of the rate of dissipation can be converted into an estimate of $\tilde{P}(\nu)$ using an acceleration time scale which is most simply chosen to be the shorter of the expansion time scale and the radiative cooling timescale.

These essentially phenomenological models of particle acceleration lead to quite varied jet behavior when they are "observed". It is hoped that with future observations, they will help us understand the microphysical processes at work when particles are energized.

3.2. Radiative Transfer

The treatment of radiative transfer should be straightforward but new approaches have been developed to provide an efficient interface with a relativistic hydromagnetic simulation. The equation of polarized radiative must be solved in the frame in which the simulation is performed, essentially that of the host galaxy. However, the emissivity and absorption coefficient are specified in the local rest frame – the comoving frame – inside the jet, and have to be transformed allowing for the Doppler shift, aberration and rotation of the plane of polarization. They must be expressed in terms of electromagnetic variables evaluated in the comoving frame.

3.3. mm Observations

The accreting, massive black holes in the nuclei of active galaxies create antiparallel jets, most commonly observed at radio frequencies. The smallest scale structure at the base of these jets is resolved using VLBI and it is now possible to use this technique at mm wavelengths with intercontinental baselines thereby exploring the mechanisms at work in powering and collimating them. We already know that the jets are collimated quite close to the black hole and the jet emission can be limb-brightened.

Imaging polarimetry at mm wavelengths promises to be very prescriptive. The linear polarization should help us to determine if the magnetic field has the basic structure predicted, becoming increasingly toroidal with increasing cylindrical radius. In particular, as the plasma near the hole is moving slowly or even inwards. we may be able to observe the counterjet[9], gravitationally lensed by the black hole (as well as the torus). This should also allow us to see if the magnetic field geometry is basically dipolar (as generally expected) or quadrupolar. The inferred sense of rotation of the hole is expected to be the same as that of the gas observed to be orbiting the hole at large radius. If circular polarization can be measured, then it may distinguish an electron-positron plasma from an electron-ion plasma especially if the two frequencies can be observed.

3.4. γ-ray Observations

The small minority of jets that are directed towards us are called blazars. (However, Doppler beaming allows them to dominate flux-limited samples.) Blazars were originally studied using radio and optical telescopes. However, many of them are most prominent inγ-rays both at GeV energies, as observed by FGST and at TeV energies from the ground. One of the big surprises was how many of the sources are extremely variable with time scales as short as a few minutes. This suggests to many, though not all, astronomers that the emission comes from close to the black hole. However, it must originate sufficiently far out to avoid pair production on the photon background. Now most blazars have "Bactrian" (two humped) spectra when plotted as $\log(\nu F_\nu)$

as a function of $\log(\nu)$ ranging from ~ 8 to ~ 25. It is conventional to associate the low frequency hump with synchrotron radiation and the high frequency hump with Compton scattering. It is also generally supposed that the soft photons that are being scattered mostly originate outside the jet, originally as disk photons – the External Compton representation – in the higher power "Flat Spectrum Radio Quasars" (FSRQ) and that they mostly originate inside the jet – the Synchrotron Self Compton representation – in the lower power "BL Lac" sources.

4. M87

The M87 jet has been studied carefully for a century. The black hole mass is measured to be $M \sim 6.6 \times 10^9 M_\odot \equiv m \sim 10^{13}$m $\equiv 4\mu$as[10] and is now well-imaged down to projected radii $\sim 100m$ and modeled down to $\sim 10m$. Fitting a stationary analytical approximation to a numerical simulation suggests $\Omega_H \sim 10^{-5}$rad s^{-1}, $\Phi_H \sim 10^{26}$Wb, $V \sim 50$EV, $I \sim 1$EA. The outflow speeds at jet radii $\sim 100m$ are $\sim 0.7 - 0.95$c. The M87 jet and, by inference, the hole spin axis is inclined at an angle $\sim 20°$ to the line of sight. This means that Doppler boosting and polarization rotation will be important. Nevertheless, if the magnetic field is organized, as we conjecture, then the polarization will be large and we should be able to learn details of its geometry.

One feature of the models that is both striking and robust is the presence of a lateral shift of the "spine" of the jets due to the dependence of the single particle emissivity on the pitch angle which has to be different on the two sides of the jet. (See Fig. 1.) The observations should also provide indications of entrainment. The innermost parts of the jet are self-absorbed and so one is really viewing a radio photosphere. as this will move inward with increasing radio frequency, the centroid of the jet "core" emission will move closer to the hole as the frequency of observation is increased.

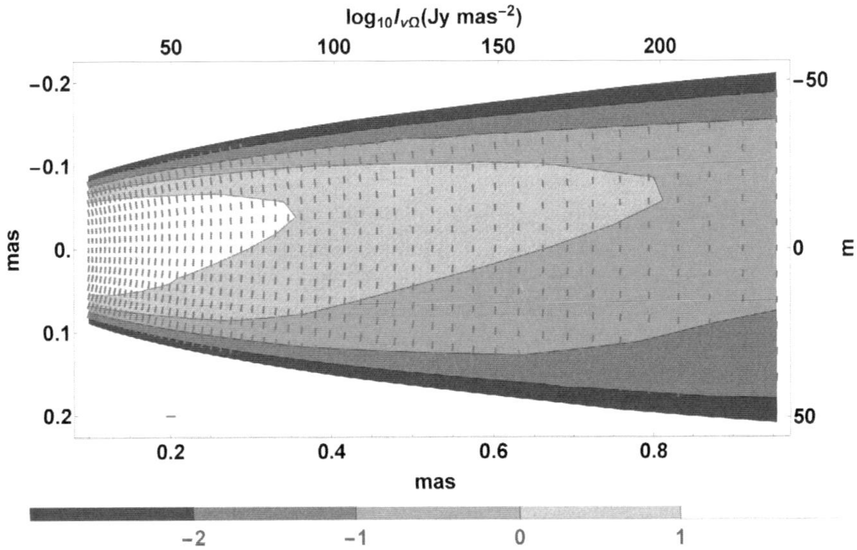

Figure 1. Simple model of the M87 jet that shows a vertical shift in the spine of the jet relative to the jet axis due to the dependence of the synchrotron emisssivity on pitch angle. The red bars represent linear polarization, with the isolated bar at the bottom corresponding to 100% polarization.

IOP Conf. Series: Journal of Physics: Conf. Series **840** (2017) 012023 doi:10.1088/1742-6596/840/1/012023

5. 3C279

3C279[1] is a good example of a γ-ray very luminous FSRQ. It has an optical luminosity of $\sim 3\times10^{38}$ W and an isotropic γ-ray flux that varies between $\sim 10^{40}$ W and $\sim 10^{42}$ W , consistent with strong relativistic beaming. The mass of the black hole is estimated to be $\sim 10^9$ $M_\odot$ and a bulk Lorentz factor $\Gamma \sim 30$ is indicated during major flares. In the simplest models all the jet components are produced by a single moving source. Consequently, the magnetic energy density would have to be much lower than the soft photon and the particle energy density. A recent discovery is that occasional γ-ray flares show variability on time scales (after correcting for cosmological redshift) as short as $t_{\rm var} \sim 3$ min $\equiv m/30$. Relativistic kinematics allows this rapid a variation to arise at a radius $r \sim \Gamma^2 ct_{\rm var}$. The conclusion from more detailed calculations is that this can be just compatible with a minimalist model of the soft photon background.

In order to reconcile these observations with the electromagnetic jet model, it is nexcessary to consider alternative models for the emission. The first possibility to consider is that the "synchrotron hump" originates from much larger radii than the γ-rays. There is no evidence for minute-scale variation at these much lower frequencies. The second modification is to suppose that the γ-ray emission is synchrotron radiation. There is a precedent for this as the variable ~ 300 MeV emission from the Crab Nebula is almost certainly made by the synchrotron process. However, it is commonly claimed that synchrotron photons have a maximum energy $\sim \alpha_F^{-1} m_e c^2 \sim 70$ MeV which may be hard to reconcile with $\sim$ TeV emission, even after allowing for a Doppler-boost of the photon energy by a factor $\sim \Gamma$. This conclusion follows from noting that the radiation reaction force on an electron emitting photons with characteristic frequency $\mathcal{E}$ can be expressed as $\sim (\mathcal{E}\alpha_F/m_e c^2)ecB$. This must be balanced by an accelerating force $\sim eE$ and the expectation that $E \lesssim cB$ delivers the result.

Before examining this argument more carefully, let us illustrate it with some quite model-dependent numbers. Suppose that the ~ 300 MeV flare emission comes from $r \sim 100m$. A jet that is still electromagnetic will have a a comoving field strength ~ 3 mT and be emitted by electrons with comoving energy ~ 3 TeV. The comoving radiative cooling time is ~ 0.1 s. Short, sporadic electric fields with strength $\sim 0.1cB$ are sufficient to accelerate the particles to these energies and dissipate the electromagnetic energy density, a generic process called "magnetoluminescence"[8]. Specific mechanisms that achieve this can be exhibited using "Particle in Cell" simulations[16] for example. The most extreme acceleration involves the creation of occasional linear magnetic-field parallel electric field sites. Electrons are rapidly accelerated and radiate when they escape these sites and are scattered into large pitch angle gyrastions about the magnetic field. In our example, freshly accelerated electrons should continue to cool down to energies ~ 20 MeV, creating a spectrum $\nu F_\nu \propto \nu^{0.5}$ in the simplest case, when expansion losses will be more important than radiative losses.

Powerful blazars like 3C279 emit γ-rays up to energies ~ 1 TeV. These are not known to be so variable and can originate from larger jet radii. Can these be synchrotron photons too? Two variations on the scheme just sketched can be mentioned. The first is that the electrons are impulsively accelerated by electric fields directed along the comoving magnetic field and so develop small pitch angles alllowing them to achieve higher energy and emit higher energy photons before radiation reaction becomes significant. They can be scattered in pitch angle by plasma instability and then start to radiate very efficiently. The second possibility is that it is protons that are accelerated so that direct radiation reaction is unimportant. They could be accelerated to vey high energy, even EeV. The major loss may arise from electron-positron pair production by soft photons interacting with the proton Coulomb fields. This would initiate electromagnetic showers. In general electromagnetic dissipation at small radius is likely to create showers until the γ-rays can escape from the jet. More detailed simulations are clearly needed to determine if models like these are viable[4], for example.

6. Summary and Outlook

Although, in the popular imagination, black holes are associated with closure, astronomers regard them as vital entities. From their proud birth announcements as γ-ray bursts and GW events, through the brilliant lives that some of them undergo to the unlikely, though not impossible, new contributions that they could yet make to fundamental physics as primordial, cosmological agents, firewalls and even macroscopic wormholes, there is a renaissance in black hole astrophysics. The energy released by the accreting black hole in the nucleus of a typical galaxy is comparable with that released by the stars (and ten times the gravitational binding energy of the galaxy). Black holes are now recognized as important, alongside stars, in bringing about reionization around $z \sim 10$ and their long range, radiative influence, which can affect, the subtle business of galaxy formation and evolution may limit the accuracy with which quantitative analyses of large scale structure can be conducted. On the stellar scale, we are about to learn much from general, transient astronomy, now including gravitational radiation, which may transform our understanding of advanced stellar evolution and X-ray binaries.

However, it is with spinning black holes, interpreted as cosmic machines, where most progress seems imminent. It is a very good working hypothesis that they are described by the Kerr metric and that strong field relativity has no surprises in store. LIGO-VIRGO has already affirmed this belief and has the opportunity to challenge it in much more detail in the near future. In order to understand accretion disks and relativistic jets, we have to deal with the classical physics of fluid mechanics, radiative transfer and electromagnetism. The capability of modern simulations is impressive but they are now limited by our ignorance of the principles that control where and how dissipation takes place, especially leading to the acceleration of non-thermal electrons and positrons. However many of the issues are being forced by observations of rapid γ-ray variability and imaging polarimetry of the innermost segments of relativistic jets at mm and eventually submm wavelengths. The growing scope of PIC simulations will also help us to determine the rules that magnetized, relativistic plasma obeys when it dissipates and radiates.

The major impact of these investigations on LISA will be to give a more quantitative understanding of black hole demographics and its redshift dependence. We hope to attend this symposium in a decade when LISA is just about to launch!

Acknowledgments

We thank Greg Madejski, Jonathan McKinney, Tony Readhead, Sasha Tchekhovskoy, and Craig Walker for advice and collaboration. RA acknowledges the Stanford DARE program and the CA Alliance for Graduate Education and the Professoriate for fellowships.

References

[1] Ackermann, M. *et al.*2016 ApJ 824 L20
[2] Akiyama, K. *et al.* 2015 ApJ 807 150
[3] Anantua, R. 2016 Unpublished thesis, Stanford
[4] Anantua, R. *et al.* 2017 in preparation
[5] Balbus, S. & Hawley, J . 1991 ApJ 376 214
[6] Blandford, R & Znajek, R. 1977 MNRAS 179 433
[7] Blandford, R. & Begelman, M. 1999 MNRAS 303 L1
[8] Blandford, R. *et al.* 2014 Nucl. Phys. B. 256 9
[9] Broderick, A. *et al.* 2016 ApJ 820 137
[10] Gebhardt, K. *et al.* 2011 ApJ 729 119
[11] Johnson, M. *et al.* 2015 Science 350 1242
[12] McKinney, J. *et al.* 2012 MNRAS 423 3083
[13] Rathore, Y. 2005 Unpublished thesis, Caltech
[14] Siegert, T. *et al.* 2016 Nature 531 341
[15] Soltan, A. 1982 MNRAS 200 115
[16] Yuan, Y. 2016 ApJ 828 92

IOP Conf. Series: Journal of Physics: Conf. Series **840** (2017) 012024 doi:10.1088/1742-6596/840/1/012024

Galactic binary science with the new LISA design

Neil Cornish and Travis Robson

eXtreme Gravity Institute, Department of Physics, Montana State University, Bozeman, MT 59717

E-mail: ncornish@montana.edu

E-mail: travis.robson@montana.edu

Abstract. Building on the great success of the LISA Pathfinder mission, the outlines of a new LISA mission design were laid out at the $11^{\rm th}$ International LISA Symposium in Zurich. The revised design calls for three identical spacecraft forming an equilateral triangle with 2.5 million kilometer sides, and two laser links per side delivering full polarization sensitivity. With the demonstrated Pathfinder performance for the disturbance reduction system, and a well studied design for the laser metrology, it is anticipated that the new mission will have a sensitivity very close to the original LISA design. This implies that the mid-band performance, between 0.5 mHz and 3 mHz, will be limited by unresolved signals from compact binaries in our galaxy. Here we use the new LISA design to compute updated estimates for the galactic confusion noise, the number of resolvable galactic binaries, and the accuracy to which key parameters of these systems can be measured.

1. Introduction

The first direct detection of gravitational waves by the advanced LIGO [?] and the spectacular success of the LISA Pathfinder mission [?] added a renewed energy to the $11^{\rm th}$ International LISA Symposium in Zurich. During the symposium, the LISA Consortium held a half-day meeting to begin the preparation of a formal mission proposal for the L3 launch opportunity in ESA's Cosmic Vision Program. The many decades of experience with LISA technologies and designs meant that the broad outlines of the mission quickly fell into place. The new LISA mission [?] is envisioned to comprise of three identical spacecraft in a triangular formation separated by 2.5 million km in an Earth trailing orbit, with six continuously operating laser links providing heterodyne laser interferometry with pm/$\sqrt{\rm Hz}$ sensitivity along each arm. The nominal design calls for 2 W of laser power being delivered to the optical system, with 30 cm telescopes transmitting and receiving the laser light between the spacecraft. The six laser links allow the synthesis of two Michelson-like channels that provide instantaneous measurements of the two gravitational wave polarization states, and a third Sagnac-like channel that is relatively insensitive to gravitational waves that can be used to monitor the average noise level in the detector.

Figure 1 compares the predicted Michelson-equivalent sensitivity of the LISA design from the 1998 pre-phase A report [?] and the new LISA design [?]. Also shown is a simple approximation to the new sensitivity curve [?] given by

$$S_n(f) = \frac{20}{3}\frac{1}{L^2}\left(1 + \left(\frac{f}{1.29 f_*}\right)^2\right)\left(S_p(f) + \frac{S_a(f)}{(2\pi f)^4}\right). \tag{1}$$

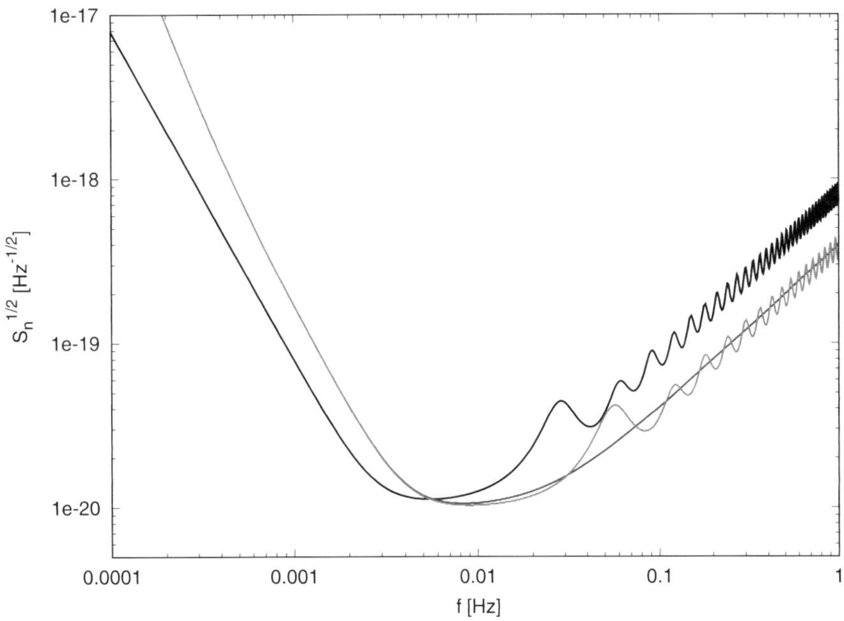

Figure 1. Sensitivity curves for the 1998 LISA design (black) and the new LISA design (red). Also shown is a simplified model for the sensitivity (blue), given in equation (1) of the text. The differences in the old and new sensitivity curves are almost entirely due to the reduction in arm length from 5 million km to 2.5 million km, which has the effect of reducing the low frequency sensitivity while enhancing the high frequency sensitivity.

Here $f_* = c/(2\pi L)$ is the transfer frequency, $L = 2.5$ million km is the arm length, $S_p(f) = 8.9 \times 10^{-23}\ \mathrm{m^2\,Hz^{-1}}$ is the white position noise, and

$$S_a(f) = 9 \times 10^{-30} \left(1 + \left(\frac{10^{-4}\,\mathrm{Hz}}{f}\right)^2 + 16\left(\frac{2 \times 10^{-5}\,\mathrm{Hz}}{f}\right)^{10}\right)\ \mathrm{m^2\,s^{-4}\,Hz^{-1}} \tag{2}$$

is the colored acceleration noise. The acceleration noise model is based on a fit to the measured LISA Pathfinder performance.

The new LISA design balances technical and budgetary constraints against a set of science requirements. A key component in this iterative design process is an assessment of the science reach of the instrument. The first step in such an assessment is to consider compact galactic binaries as these are expected to be the most numerous of the sources; so numerous in-fact that the unresolved component of the galactic population produces an effectively stochastic "confusion noise". Here we provide estimates for the confusion noise level as it evolves over the lifetime of the mission, as well as estimates for the number of resolvable galactic binaries and how well key parameters, such as the orbital periods and sky location, can be determined. Our work is updates the confusion noise estimates of Nissanke *et al* [?].

2. Galactic Confusion Noise

The starting point for our analysis is a model for the population of compact binaries in our galaxy. We use an updated version [?] of the Nelemans *et al* [?] model provided to us by Valeriya Korol and Gijs Nelemans . The space density of interacting white dwarf binaries is reduced by a factor of ten relative to earlier models in response to the findings of recent observational studies[?, ?].

The model predicts there are $\sim$ 26 million galactic binaries emitting in the frequency band 0.1-10 mHz. The confusion noise is estimated using the iterative subtraction procedure described in Timpano *et al* [?] using a fast algorithm to generate the signals discussed in Cornish and Littenberg [?]. The first step is to generate a realization of the instrument noise in each data channel, then co-add the signals from the population of galactic binaries. A smooth fit to the the power spectral density of the signal plus noise is found using a variant of the BayesLine algorithm [?]. This fit is used as an initial estimate of the effective noise spectral density, against which the signal-to-noise ratio (SNR) of each binary is computed. All binaries with SNR > 7 are then subtracted from the data, and the estimate for the power spectral density of the remaining signal plus noise is computed. The SNRs of the remaining systems are re-computed using the updated (lowered) noise model, and those above threshold are subtracted. The subtraction procedure is iterated several times until few additional systems are flagged as loud. The end product is an estimate for the confusion noise and a list of resolvable sources. The mission duration has a significant impact on the level of the confusion noise. The longer the mission continues the more sources are resolved and the lower the confusion noise.

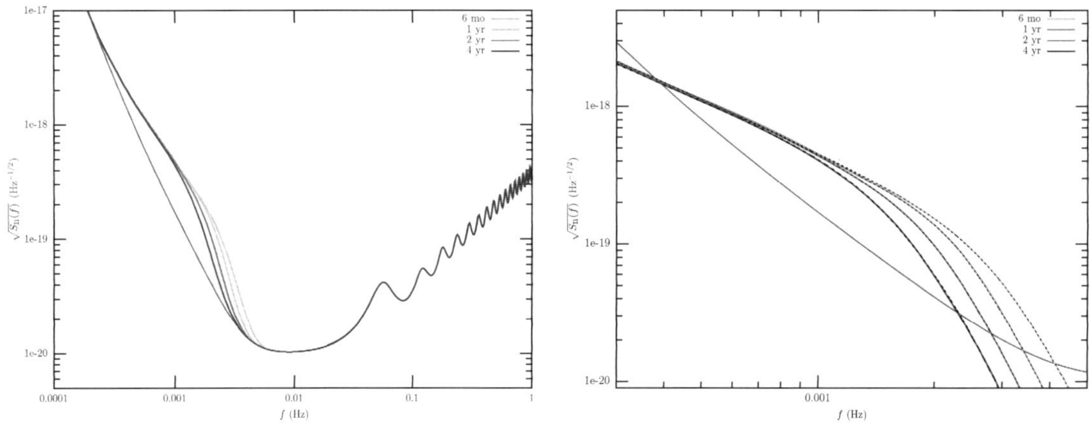

Figure 2. The panel on the left shows the sensitivity curve for the new LISA design including the estimated confusion noise level for a range of mission durations. The panel on the right focuses on the confusion noise impact on the sensitivity, and compares our analytic fit (black dashed lines) to the output of the simulation.

Shown in the first panel of Figure 2 are sensitivity curves that include the galactic confusion noise after 6 months, 1 year, 2 years and 4 years of data collection. The second panel in Figure 2 focuses on the confusion noise contribution, and compares our analytic fits to the results of the simulation. The analytic fits to the confusion noise have the functional form

$$S_c(f) = A\, f^{-7/3}\, e^{-f^{\alpha}+\beta f \sin(\kappa f)}\, [1+\tanh(\gamma(f_k - f))]\ \mathrm{Hz}^{-1} \tag{3}$$

The overall $f^{-7/3}$ slope is the theoretical prediction for the power spectrum of a population of quasi-circular binaries evolving due to gravitational wave emission. The other multiplicative factors were chosen to match the departures from the power law spectrum. The form of the fit was originally suggested by Stas Babak. The parameters for the fits are given in Table 1. The overall amplitude and slope of there fit remain roughly constant with increasing observing time. The main change is in the knee frequency f_k, which gets steadily smaller with time.

	6 mo	1 yr	2 yr	4 yr
α	0.133	0.171	0.165	0.138
β	243	292	299	-221
κ	482	1020	611	521
γ	917	1680	1340	1680
f_k	0.00258	0.00215	0.00173	0.00113

Table 1. Parameters of the analytic fit the Galactic confusion noise as described by equation (1). The amplitude A has been fixed to $1.80e-44$. Note that the knee frequency f_k decreases with observation time as expected from Figure 2. Similarly, γ increase with observation time to leading to steeper drop off in confusion noise with increasing frequency.

3. Galactic Binary Science

While the unresolved galactic binaries are a source of noise that negatively impacts the science that can be done with other sources, the resolvable galactic binaries tell a rich story about stellar evolution and the distribution of stars in the galaxy. The distribution of resolved signals as a function of orbital period can be used to constrain population synthesis models. The orbital period is very well measured even for systems at the detection threshold. Figure 3 shows the number density of resolved signals per $1/T_{\rm obs}$ frequency bin for a 4 year mission using both Michelson-like data channels.

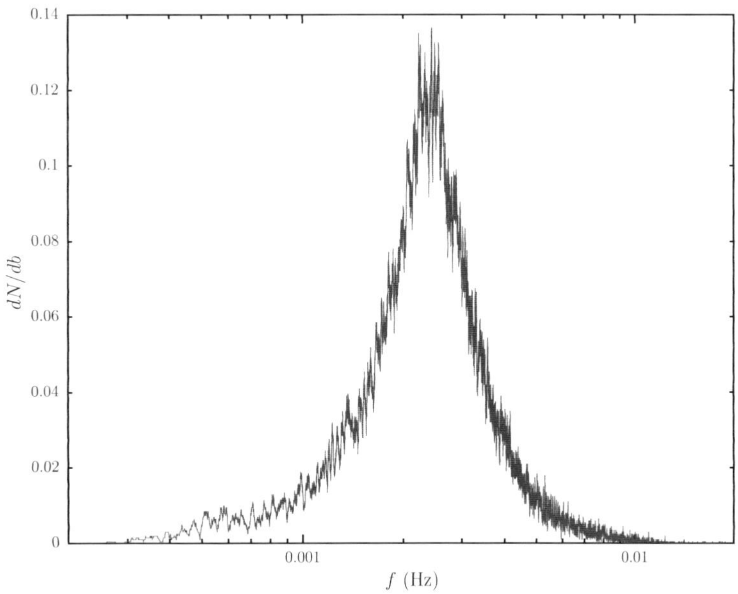

Figure 3. The number density of resolved signals per frequency bin for a 4 year mission using both Michelson-like data channels. Note that the density of resolved signals is largest between 2 - 3 mHz, which is where the Galactic confusion noise changes the most over time as shown in Figure 2.

A sub-set of the resolved signals will be well localized on the sky, and a smaller sub-set will undergo sufficient frequency evolution to allow measurement of the chirp mass and distance. Table 2 lists the predicted number of detected systems, along with the number that are well

IOP Conf. Series: Journal of Physics: Conf. Series **840** (2017) 012024 doi:10.1088/1742-6596/840/1/012024

	6 mo	1 yr	2 yr	4 yr
# detected	6,590	11,142	18,281	29,059
2D mapped	104	1,065	4,138	6,304
3D mapped	19	129	1,010	2,373
$\mathcal{M}$ measured	233	737	4,432	10,770

Table 2. The first row indicates the number of sources detected using the two Michelson-like data channels for 6 months, 1 year, 2 year, and 4 year observation periods. If a source is localized to better than a square degree we deemed it to be mapped on the sky (2D mapped). Additionally, if its distance has also been obtained to better than 10% we deemed to have been mapped in 3D. The last row indicates the number of sources whose chirp mass $\mathcal{M}$ have been determined to better than 20%.

localized in sky position and distance, and the number for which we can measure the chirp mass. The parameter measurement accuracy is estimated using the Fisher information matrix. The limitations of this approach are well known, but we expect the estimates given in Table 2 would be close to those found using a full Bayesian analysis since the results are for the best measured systems, which typically have high SNR.

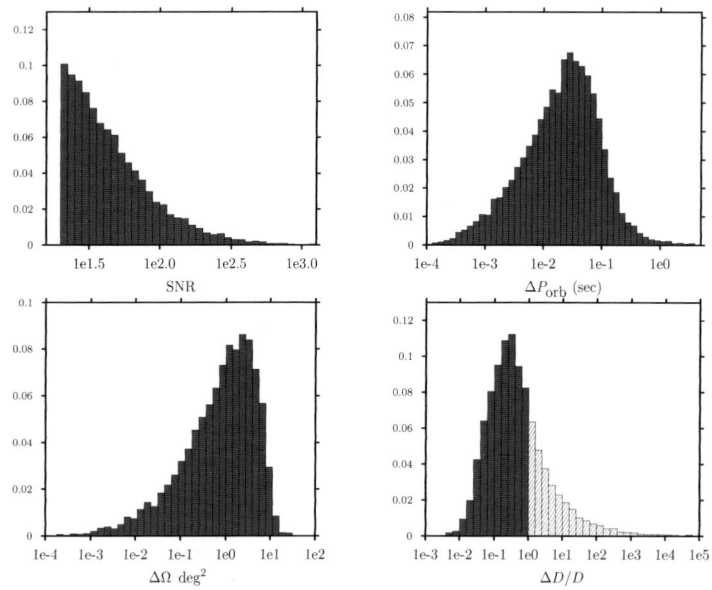

Figure 4. Histograms for the SNR and parameter estimation errors for sources with SNR > 20 using the two Michelson-like data channels for 4 year observation period. The orbital period (upper right) is determined to better than a second accuracy, reflecting the precision to which we expect to measure the galactic binary frequencies. In the lower left panel we see that the angular resolution of many sources is determined to within a square degree. Distance is the hardest parameter to extract, as depicted by the lower right panel. Results which are nonphysical (i.e. $\Delta D/D > 1$) are dashed.

Figure 4 shows histograms of the SNR, and the expected accuracy in the measurement of orbital period, sky location and distance for systems with SNR > 20, assuming a 4 year mission with 6 links. The limitations of the Fisher matrix approximation for estimating parameter estimation motivated the higher SNR cut.

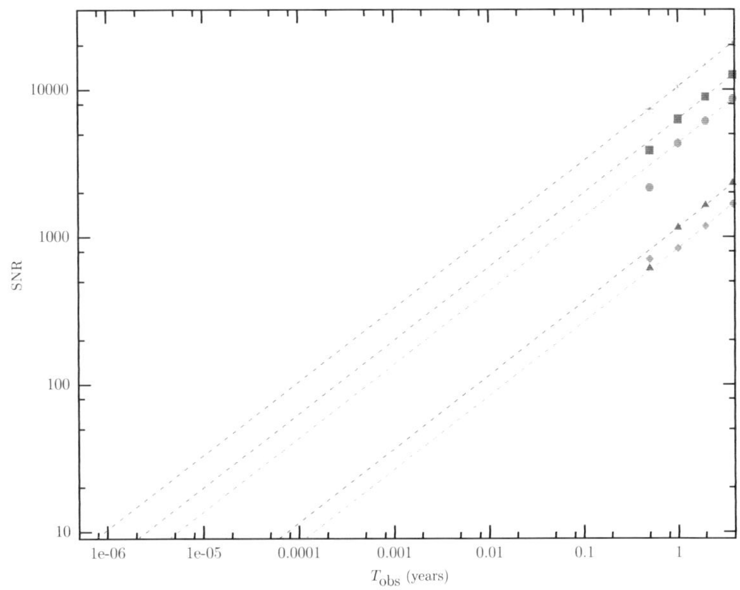

Figure 5. The signal-to-noise ratios for the five loudest galactic binaries as a function of observation time. The dashed lines show the $T_{obs}^{1/2}$ growth in SNR that we expect for a fixed noise level. At early times the confusion noise level is dropping rapidly form many of the brightest sources so the SNR growth is faster.

The SNR distribution has a long tail that extends to very high SNR. These very high SNR systems are of particular interest as they will typically be very well localized (especially the higher frequency systems), making them good candidates for joint electromagnetic-gravitational observation [?], and they will also be among the first sources detected by LISA. Figure 6 plots the SNR of the five brightest galactic binaries as a function of mission duration. The dashed lines indicate the $T_{obs}^{1/2}$ growth in SNR that we expect if the noise level is constant. For most galactic binaries the SNR grows faster than this since the effective noise level drops as weaker signals are resolved and removed. Extrapolating the curves back below 6 months is not entirely justified since the confusion noise is non-stationary and varies on a 6 month period, and the SNR does not precisely follow the $T_{obs}^{1/2}$ scaling, but even with the caveats, we see that the loudest signals will be detectable within hours of starting data collection.

4. Discussion

Using the current best estimates for the galactic binary population we computed the galactic binary confusion noise for the new LISA design and found that is the dominating source of noise in the frequency band $0.5 - 3.0$ mHz. Fits for the confusion noise curves are been provided. Additionally we provided estimates for how many galactic binary sources will be detected, and how well we expect to be able to estimate their parameters. Determining the 3D position will be the most challenging; many sources will be located to better than a square degree on the sky, but breaking the degeneracy between chirp mass and distance hurts distance estimation. We found that most sources will be resolved above 3 mHz, with orbital periods extracted to better than one second accuracy.

5. Acknowledgments
We wish to thank the organizers of the 11th International LISA Symposium for putting together a memorable meeting, and our colleagues in the LISA Consortium for their work on the mission design. We are grateful for the support provided by NASA grant NNX16AB98G.

6. References
[1] Abbott B P *et al* (LIGO Scientific Collaboration and Virgo Collaboration) 2016 *Phys. Rev. Lett.* **116**(6) 061102 URL `http://link.aps.org/doi/10.1103/PhysRevLett.116.061102`
[2] Armano M *et al* 2016 *Phys. Rev. Lett.* **116**(23) 231101 URL `http://link.aps.org/doi/10.1103/PhysRevLett.116.231101`
[3] Danzmann K *et al* 2016 Laser interferometer space antenna Tech. rep. LISA Consortium URL `https://www.elisascience.org/files/publications/LISA_L3_20170120.pdf`
[4] Bender P, Brillet A, Ciufolini I, Cruise A, Cutler C, Danzmann K, Fidecaro F, Folkner W, Hough J, McNamara P, Peterseim M, Robertson D, Rodrigues M, Rüdiger A, Sandford M, Schäfer G, Schilling R, Schutz B, Speake C, Stebbins R, Sumner T, Touboul P, Vinet J, Vitale S, Ward H and Winkler W 1998 Laser interferometer space antenna for the detection and observation of gravitational waves Tech. rep. NASA mPQ 233, Second Edition URL `https://lisa.gsfc.nasa.gov/Documentation/ppa2.08.pdf`
[5] Petiteau A, Hewitson M, Heinzel G, Fitzsimons E and Halloin H 2016 Lisa noise budget Tech. rep. LISA Consortium lISA-CST-TN-0001
[6] Nissanke S, Vallisneri M, Nelemans G and Prince T A 2012 *Astrophys. J.* **758** 131 (*Preprint* `1201.4613`)
[7] Toonen S, Nelemans G and Portegies Zwart S 2012 A&A **546** A70 (*Preprint* `1208.6446`)
[8] Nelemans G and Tout C A 2005 MNRAS **356** 753–764 (*Preprint* `astro-ph/0410301`)
[9] Roelofs G H A, Nelemans G and Groot P J 2007 MNRAS **382** 685–692 (*Preprint* `0709.2951`)
[10] Carter P J, Marsh T R, Steeghs D, Groot P J, Nelemans G, Levitan D, Rau A, Copperwheat C M, Kupfer T and Roelofs G H A 2013 MNRAS **429** 2143–2160 (*Preprint* `1211.6439`)
[11] Timpano S E, Rubbo L J and Cornish N J 2006 *Phys. Rev. D* **73**(12) 122001 URL `http://link.aps.org/doi/10.1103/PhysRevD.73.122001`
[12] Cornish N J and Littenberg T B 2007 *Phys. Rev. D* **76**(8) 083006 URL `http://link.aps.org/doi/10.1103/PhysRevD.76.083006`
[13] Littenberg T B and Cornish N J 2015 *Phys. Rev. D* **91**(8) 084034 URL `http://link.aps.org/doi/10.1103/PhysRevD.91.084034`
[14] Littenberg T B, Larson S L, Nelemans G and Cornish N J 2013 *Mon. Not. Roy. Astron. Soc.* **429** 2361 (*Preprint* `1207.4848`)

IOP Conf. Series: Journal of Physics: Conf. Series **840** (2017) 012025 doi:10.1088/1742-6596/840/1/012025

Multiple regimes and coalescence timescales for massive black hole pairs ; the critical role of galaxy formation physics

Lucio Mayer

Center for Theoretical Astrophysics and Cosmology, Institute for Computational Science, University of Zürich, Winterthurestrasse 190, 8057, Zürich, Switzerland

E-mail: lmayer@physik.uzh.ch

Abstract. We discuss the latest results of numerical simulations following the orbital decay of massive black hole pairs in galaxy mergers. We highlight important differences between gas-poor and gas-rich hosts, and between orbital evolution taking place at high redshift as opposed to low redshift. Two effects have a huge impact and are rather novel in the context of massive black hole binaries. The first is the increase in characteristic density of galactic nuclei of merger remnants as galaxies are more compact at high redshift due to the way dark halo collapse depends on redshift. This leads naturally to hardening timescales due to 3-body encounters that should decrease by two orders of magnitude up to $z = 4$. It explains naturally the short binary coalescence timescale, $\sim$ 10 Myr, found in novel cosmological simulations that follow binary evolution from galactic to milliparsec scales. The second one is the inhomogeneity of the interstellar medium in massive gas-rich disks at high redshift. In the latter star forming clumps 1-2 orders of magnitude more massive than local Giant Molecular Clouds (GMCs) can scatter massive black holes out of the disk plane via gravitational perturbations and direct encounters. This renders the character of orbital decay inherently stochastic, often increasing orbital decay timescales by as much as a Gyr. At low redshift a similar regime is present at scales of $1 - 10$ pc inside Circumnuclear Gas Disks (CNDs). In CNDs only massive black holes with masses below $10^7 M_\odot$ can be significantly perturbed. They decay to sub-pc separations in up to $\sim 10^8$ yr rather than the in just a few million years as in a smooth CND. Finally implications for building robust forecasts of LISA event rates are discussed.

1. Introduction

The orbital decay of massive black hole pairs from galactic scale separations to distances at which gravitational wave (GW) emission takes over is governed by a number of different physical mechanisms. In particular the decay has a different nature in a predominantly gaseous background as opposed to a predominantly stellar background. These two regimes are usually studied separately, although galactic nuclei in a galaxy host might transition from one regime to the other even multiple times in their evolution as a result of galaxy mergers, starvation of cosmological gas accretion, disk instabilities, and any process that can significantly affect the gas supply and star formation rate in galactic nuclei. The last decade has seen significant improvements in the realism of the numerical models dedicated to tackle one or the other regime, and even the first attempts to model both regimes in the same calculation (Khan et al. 2013a).

Two important general results can be drawn from the studies of the last few years. First, it is now largely agreed that the last parsec problem that once plagued the stellar background calculations does not exist in a realistic galaxy host, namely one with a triaxial, rotating stellar distribution (Berczik et al. 2006;Khan et al. 2011;2012;2013b; Vasiliev, Antonini & Merritt 2015). One of the most convincing demonstrations comes from a recent analysis of orbital diffusion in the loss cone using different types of potentials, including triaxial ones, with increasing number of particles in collisionless simulations (Gualandris et al. 2016). Loss cone refilling thus appears to be effective enough in realistic galaxy hosts, allowing a successful hardening via 3-body encounters down to separations at which gravitational waves take over. Recently cosmological hydrodynamical simulations have confirmed that realistic merger remnants, in which gas dissipation and star formation takes place, do possess the required degree of triaxiality to keep the loss cone nearly full and drive coalescence (Khan et al. 2016).

The second result is that the decay in a gaseous background, once believed to be much faster than in a stellar background (eg Escala et al. 2005; Dotti, Colpi & Haardt 2006; Mayer et al. 2007), turned out to be a much more complex problem than initially suspected, with the consequence that there is no simple general scenario at the moment. At variance with the stellar background case, in which dynamical fiction and 3-body encounters are well defined and well understood sources of drag, in gaseous backgrounds different types of torques can dominate at different scales and in different stages of the orbital evolution, such as dynamical friction, linear and nonlinear torques from a background circumnuclear rotating disk (CND), stochastic torques from massive clouds/clumps and large-amplitude spiral density waves, and tidal torques by circumbinary disks at the smallest separations (see eg Mayer 2013, Roedig et al. 2012; Farris et al 2015). The stochastic torque regime due to perturbations in an inhomogeneous interstellar medium (ISM), in particular, was pointed out only recently, and can delay or shorten orbital decay by orders of magnitude relative to the smooth disk case (Fiacconi et al. 2013; Mayer 2013). Finally, even in the simple case of a smooth CND there are indications that the orbital decay may stall, or at least slow down significantly, once the separations of the two holes drops below a tenth of a parsec (Mayer et al. 2008; Chapon et al. 2013).

As a result of the multiple regimes and physical parameters of galactic nuclei that can affect orbital decay, it is fair to say that black hole merger timescales are really uncertain overall, with values ranging from as small as 10 Myr (eg Khan et al. 2016) to as large as a few billion years (Khan et al. 2011;2012) depending on the specific regime and physical conditions considered. Indeed in this article we focus on this pivotal issue of timescales as it is highly relevant to define the science that will be possible with LISA.

2. Orbital decay in stellar-dominated hosts; a question of timescales

In the last few years a number of works, both numerical and semi-numerical, have provided evidence that there is no last-parsec problem in realistic galactic potentials, where non-axisymmetry and some degree of rotation are always present (Berczik et al. 2006; Vasiliev, Antonini & Merritt 2015; Khan et al. 2013b; Holley-Bockelmann & Khan 2015). Although convergence of the hardening rates is not yet demonstrated in collisional N-Body simulations (but see Holley-Bockelmann & Khan 2015 on highly rotating systems), alternative approaches such as the numerical convergence study of orbital diffusion in the loss cone using collisionless N-Body simulations (Gualandris et al. 2016) strongly support the notion that the loss cone should remain nearly full in realistic galaxy merger remnants. This in turn yields confidence in the results of collisional N-Body simulations of black hole binary hardening in galaxy mergers at the currently affordable resolutions (less than 10^7 particles, see eg Khan et al. 2012; Holley-Bockelmann & Khan 2015).

The latter simulations find that massive black hole binaries can merge on timescales shorter than an Hubble time. A particularly successful case is that of ultramassive black holes in stellar

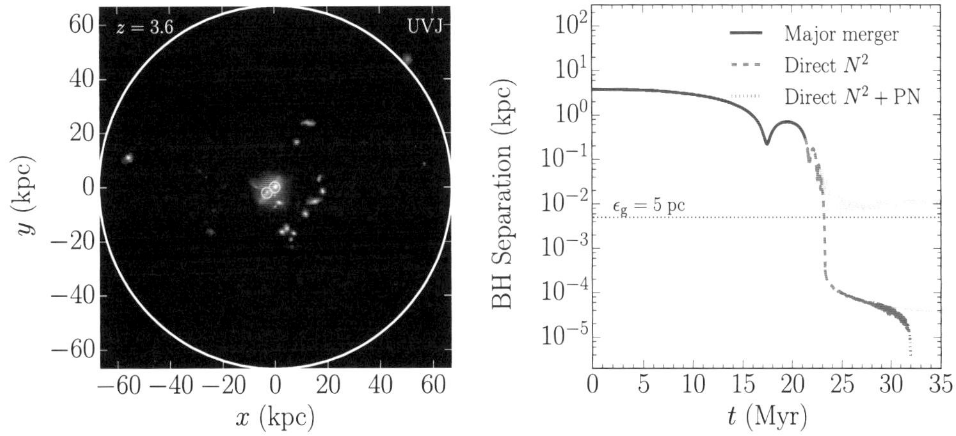

Figure 1. On the left we show a grey-scale luminosity map (in the UVJ filter) from the ARGO cosmological simulation showing the galaxy merger that was selected and re-simulated at much higher resolution (green circles). The white circle marks the virial radius of the group-sized halo, while the green circles mark the merging galaxies. From the spatial scale it is clear how the galaxies at study, which have stellar masses almost as large as those of our present-day Milky Way ($> 10^{10} M_\odot$) have sizes below 1 kpc, hence 4-5 times smaller than our Milky Way. This is expected based on the scaling arguments described in the text. On the right we show the complete orbital separation curve for the two massive black holes at the center of the re-simulated merger with pc scale resolution, ending with the in-spiral driven by GW emission (green dashed line ar 32 Myr). We plot also the results of simulations without the direct N-body integration, that captures the hardening phase, as well as with direct N-Body but without the post-newtonian corrections (Blanchet 2006), which are relevant in the very last phase. We refer to Khan et al. (2016) for details.

dominated hosts, for which dynamical friction is so effective to bring them all the way to the gravitational wave emission phase without the need of a long hardening stage (Khan & Holley-Bockelmann 2015). However, with this last exception, all published timescales are really long, ranging from slightly below 1 to a few Gyr, namely they encompass a significant fraction of the Hubble time. Note that the latter result is consistent with the simple analytical predictions of Gould & Rix (2000), which were assuming efficient loss cone refilling in a potential with the properties of present-day spheroids.

N-Body simulations mostly consider black holes at the high mass end, $M_{BH} > 10^7 M_\odot$, hence the coalescence timescale for LISA black holes, which are much lighter, would be even longer, assuming the same properties of the (stellar) background hold. If merging timescales of a few Gyr are the norm at any redshift, the consequence would be that black hole mergers would be exceedingly rare at $z > 2$ since the lookback time is a less than 3 Gyr then. This would be unfortunate since this is also the epoch for which the discovery potential of LISA is very high. The inefficient coalescence in stellar backgrounds has been and still is the motivation behind exploring the alternative regime in which orbital decay is driven by gasdynamical processes. However, as we will illustrate in section 3, coalescence does not become necessarily faster in gaseous backgrounds. Therefore it is fair to say we are facing a *timescale problem* rather than a *last parsec problem*.

We ought to make progress in order to obtain robust forecasts black hole coalescence rates in the LISA band. Uncertainties in the black hole seeding models would also generate important variations in the expected event rates (Sesana, Volonteri & Haardt 2007). Yet black hole seeding models tend to yield different black hole population properties only at at very high redshift, while afterwards the physics of black hole growth via gas accretion and mergers with other black holes become the dominant driver of the evolution of the black hole mass function ($z > 4$, see eg Volonteri & Natarajan 2009; Bonoli et al. 2014).

2.1. Coalescence on a fast track at high redshift

Sesana & Khan (2015) have pointed out that the hardening timescale is determined by the longest decay phase at sub-pc separations. This is the phase associated with the transition between dominance of 3-body encounters and dominance of energy loss by gravitational wave emission. The two processes have indeed rates proportional to a^2 and a^{-3}, respectively, where a is the semi-major axis (hereafter separation) of the binary. By defining a characteristic separation $a_{*/GW}$ at which the hardening rate by 3-body encounters and the hardening rate by gravitational wave emission are equal one obtains the lifetime of the binary at sub-pc scales as that corresponds to the orbital decay time in the longest phase of the decay. Following Sesana & Khan (2015). for the latter characteristic decay timescale one finds:

$$t_* = \frac{\sigma_*}{GH\rho_* a_{*/GW}} \tag{1}$$

where ρ_* and σ_* are, respectively, the characteristic stellar density and mean (1D) stellar velocity dispersion in the galactic nucleus, H is an adimensional parameter expressing how close the loss cone is to full ($F = 1$ for a full loss cone, $F \sim 0.7$ for triaxial systems employed in numerical simulations such as in Khan et al. 2012). The dependence on density and stellar velocity dispersion is linear. However this equation has no apparent connection with cosmology, although the parameters that appear in it need to be informed by cosmology in modern galaxy formation theory. Elucidating such connection is the main purpose of this section.

We start by recalling that in the galaxy formation scenario arising in a Universe dominated by Cold Dark Mater (CDM) cosmic structures form bottom-up via collapse of density fluctuations combined with hierarchical merging. The characteristic density of a dark matter halo is determined by the virial overdensity ρ_{vir}. Inside the region whose mean overdensity is the virial overdensity a halo is virialized, namely there is a well defined equlibrium structure for which the binding energy is set. In the simple spherical collapse model the virial overdensity can be calculated exactly once the cosmological parameters are known (eg Bryan & Norman 1998). Often for simplicity the redshift-independent overdensity defined to be 200 times the critical density of the Universe, ρ_{200}, is adopted. While such overdensity is only strictly valid in a Universe with matter density parameter $\Omega_m = 1$, quantitative statements concerning redshift evolution of structures are very weakly dependent on the presence or not of a cosmological constant (with corresponding density parameter Ω_Λ) until it begins to be dominant at $z < 1$. This is because the virial ovedensity varies only by factors of 2 as a function of redshift relative to ρ_{200} (Bryan & Norman 1998). In the remainder of this section we will thus drop Ω_Λ since equations are simplified considerably, and afterwards we will comment on the magnitude of the eventual corrections.

Inside a virialized halo baryons would collapse and form a luminous galaxy. Hence, as it is customary, one can assume that scaling relations between key galaxy structural parameters, such as mass, radius, circular velocity (or velocity dispersion), and thus also characteristic density, directly reflect those of the virialized host halo because this comprises most of the mass, and therefore most of the binding energy (White & Rees 1978; Mo,Mao & White 1998). Galaxy size will depend also on the amount of angular momentum, but this is also assumed to be proportional to the angular momentum of the halo and to be approximately conserved, introducing a simple scaling factor between galaxy size and halo virial radius. The virial overdensity ρ_{vir} is, by definition, a multiple of the mean density of the Universe at the corresponding epoch, thereby it has to increase as $(1+z)^3$ towards high redshift. With our assumption that $\rho_{vir} = \rho_{200}$ we are simply enforcing that the multiplication factor is constant over time. As a consequence, *if the characteristic density of galaxies follows the halo overdensity then it should also scale as* $(1+z)^3$. For instance at $z = 4$ galaxies should be approximately 5^3 times denser relative to their $z = 0$ counterparts. While this is essentially textbook structure formation one should appreciate the very important implications in the context of massive black hole binary coalescence. *Indeed based in eq. (1) this means that coalescence timescales can be shorter by two orders of magnitude at* $z = 4$ *relative to* $z = 0$. *The latter is a significant effect.*

This simple approach neglects two aspects. The first one is that, based on the same scaling relations between halos and galaxies, also the stellar velocity dispersion, which enters in equation (1), depends on redshift, as we will illustrate in a moment. The second one is that baryonic physics can decouple to some extent the dynamics of the baryons from that of the dark matter. Hence the scaling of halo and galaxy properties is not necessarily homologous (although cosmological hydro simulations show it is so to some degree, see eg Sokolowska et al. 2016). Indeed modern theory of galaxy formation is essentially based on the effort to try achieve the necessary decoupling suggested by observational constraints on galaxy properties via what we collectively indicate as "feedback mechanisms". Within the same framework, radiative cooling is also an obvious process that tends to decouple baryons from the collisionless dark matter fluid, with a tendency to increase density beyond the the inference of the scalings just described because important coolants such as recombination and atomic radiative transitions increase their rate with increasing density. These processes will shape the mass distribution in galaxies , hence they will determine the density profile as a function of radius. We will return to this second point in Section 3 but for now we can continue our reasoning as if galaxies were sufficiently described by a characteristic density, i.e. neglecting their actual mass profiles. Assuming an isothermal sphere for the velocity distribution we can move on easily with our calculations. This is an oversimplification given that (1) the NFW model, that is conventionally used to describe CDM halos, is close to the isothermal law only in the intermediate region, and that (2) both halos and galaxies are not sppherical so they will possess significant orbital anisotropy. Yet here we want to capture the essential notions and scalings rather than exact numbers. Hence continuing with the isothermal model one can write that the 1D velocity dispersion is $\sigma = V_{vir}/\sqrt{2}$, where we impose that the (constant) circular velocity V_c is equal to the virial velocity V_{vir} of the dark matter halo. The latter is automatically determined by the halo virial mass in dissipationless spherical collapse theory once the virial overdensity and the reference cosmological parameters are set. The virial radius is also automatically determined in the spherical collapse model as the radius containing a region with the virial overdensity, eg the radius at which the density equals ρ_{200}. The relation between halo virial mass M_{vir} and virial velocity V_{vir} is (see Mo, Mao & White 1998):

$$M_{vir} = \frac{V_{vir}^3}{GH(z)} \qquad (2)$$

The redshift dependence is contained in the Hubble parameter $H(z)$ which, in a generic cosmology with non-zero cosmological constant, is defined as:

$$H(z) = H_0[\Omega_{\Lambda,0} + (1 - \Omega_{\Lambda,0} - \Omega_0)(1+z)^2 + \Omega_0(1+z)^3]^{1/2} \qquad (3)$$

where Ω_0 and $\Omega_{\Lambda,0}$ are, respectively, the total density parameter and the cosmological constant density parameter at $z = 0$, and H_0 is the Hubble constant. Assuming $\Omega_{\Lambda,0} = 0$ and $\Omega_0 = 1$ in equation (3) and substituting the result into equation (2) one obtains the following scaling of the velocity dispersion with redshift for a fixed halo virial mass M_{vir}:

$$\sigma \sim V_{vir} \sim H(z)^{1/3} \sim (1+z)^{1/2} \tag{4}$$

Assuming the mass of the galaxy is simply proportional to halo mass, as strongly suggested by abundance matching (eg Behroozi et al. 2013), this simple scaling relation can also be used for the galaxy stellar velocity dispersion as long as the galactic potential does not modify greatly the relation between velocity and mass (which it cannot if the galaxy and halo are in equilibrium). At $z = 4$ the relation implies an increase in velocity dispersion of a factor of ~ 2.2 relative to $z = 0$ as opposed to an increase in density of more than a factor of a 100 (see above). Bringing the cosmological constant back in inside equation (3) would only change the result by about 6%. Hence, if we assume that the stellar velocity dispersion σ_* is of order of the velocity dispersion of the halo, σ, we can safely state that the main effect as redshift increases is that the increasing density should shorten the hardening timescale almost as fast as the density increases, namely as $(1+z)^3$.

The previous discussion is instrumental to understand the results of recent simulation work. Khan et al. (2016) have made the first attempt to use a a LCDM galaxy formation simulation as the initial condition to follow the evolution of a massive black hole binary into the gravitational wave emission regime. They selected the most massive pair of galaxies within the high resolution sub-volume of the cosmological zoom-in hydrodynamical simulation ARGO (Fiacconi & Feldmann 2015; Fiacconi, Feldmann & Mayer 2015), having stellar masses of a few times $10^{10} M_\odot$ in halos with $M_{vir} \sim 10^{12} M_\odot$. Then they increased the resolution further in two steps using particle splitting. Finally they converted into stars the very little residual gas after the major merger in order to continue the simulation as an N-body collisional calculation with post-newtonian corrections (Blanchet 2006). Uncertainties in baryonic processes,especially feedback, are of course present. Nevertheless the galaxy merger was chosen after verifying that it produces a galaxy whose key structural parameters match the available observational constraints, such as the relation between stellar mass and halo mass and the scaling relations between galaxy effective radius and its stellar velocity dispersion (Feldmann & Mayer 2015). In particular, the characteristic densities and stellar velocity dispersions, which enter eq (1), match reasonably those inferred for observed massive galaxies at high redshift (Szomoru et al. 2012; Bezanson et al. 2013). The selected merger starts at $z = 3.6$ and is over by $z = 3.2$ (Figure 1, left panel).

The timescale that we obtain for the coalescence after the galaxy merger is completed is surprisingly small, about 10 Myr,and corresponds to the duration of the hardening phase, as the previous phase in which the two galactic cores merge by dynamical friction inside the common halo lasts only about a Myr (Figure 1, right panel). Can we explain such a short timescale based on the scaling arguments just outlined? The answer is yes. Indeed we attempted the rescaling procedure suggested by the above equations in the non-cosmological models of Khan et al. (2012). Note that Khan et al (2012) used Dehnen models with with three different inner slopes (with exponents -1.5, -1 and 0.5, respectively). We found that the merger remnants of models with the steeper inner slope provide a better fit to the cosmological merger remnant of Khan et al. (2016), which reflects the effect of gas dissipation increasing the central mass density prior to the completion of the merger. Hence we first rescaled those remnants to the same mass of the Khan et al. (2016) cosmological merger remnant, and then rescaled the effective radius according to the expected redshift evolution so that the density was increased by a factor $(1+z)^3$ (the original models were constructed to represent at $z = 0$ galaxies). Using the characteristic density of the rescaled models in equation (1) we obtained an hardening timescale of about 30 Myr, as opposed to almost a Gyr in the mergers using the original models adopted in Khan et al. (2012), just slightly higher than the timescale we find in the cosmological simulation. This supports the notion that density scaling with redshift is the main reason behind our short timescales as a similar result can be obtained starting with simple idealized models once these

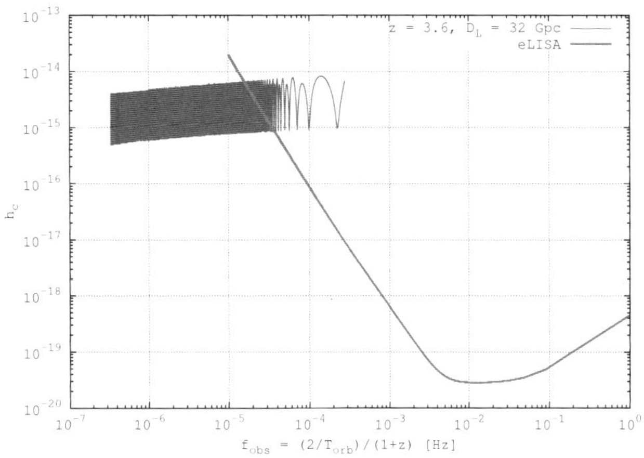

Figure 2. Wave-form for the massive black hole merger simulated in Khan et al. (2016), assuming a luminosity distance (32 Gpc) corresponding to $z = 3.6$ in the concordance LCDM cosmology. The LISA sensitivity curve is also shown in red, according to Klein et al. (2016).The merger should be detected by LISA in the last phase of the in-spiral.

are appropriately rescaled to account for the redshift-dependent effects (but note that we did not scale velocity dispersion).

The very central density and stellar velocity dispersion are the product of the balance between gas cooling and heating/stirring by feedback processes occurring during a nuclear starburst driven by the merger (see eg Capelo et al. 2015 for a recent large simulation survey of gas-rich galaxy mergers). The absence of black hole accretion and AGN feedback in the simulations of Khan et al. (2016) should not be a worry because we verified the local relation between black hole mass and central stellar velocity dispersion is reproduced. Furthermore, as already mentioned, central velocity dispersions are reasonable when compared to observed massive galaxies at $z > 2$ with similar luminosities. As there is mounting evidence that, at high z, black hole masses are shifted upwards relative to the relation, probably suggesting a lesser role of AGN feedback in regulating the concurrent growth of host galaxy and black hole masses (Trakhtebroth et al. 2015), the black hole masses chosen in these simulation are probably on the conservative side.

In lower mass galaxies, in which feedback has a stronger effect due to their shallower potential well, it is conceivable that the characteristic nuclear density might be reduced as a result of the starburst, an effect that has indeed been studied extensively for dwarf galaxies in cosmological hydrodynamical simulations. In this scale the evolution of the characteristic baryonic density with redshift might depart from the simple scaling arguments based on the dark halo evolution. In progenitors of Milky Way-sized galaxies, whose mass would be an order of magnitude lower than the current mass of the Milky Way at $z = 0$ at $z \sim 2 - 3$ (see eg Guedes et al. 2011;Agertz & Kravtsov 2015), the conditions in the host might be just intermediate. The study of black hole growth of a Milky Way-like galaxy carried out by Bonoli et al. (2016) using the ErisBH simulation does indeed reveal that nuclear gas inflows are rather weak even at $z \sim 2 - 4$, when the last significant mergers occur. This reflects the fact that feedback processes are very effective at counteracting radiative cooling in this case. Since such Milky-Way progenitors would likely host black holes in the range $10^5 - 10^6 M_\odot$, the typical target of LISA (Amaro-Seoane et al. 2013), massive black hole binary evolution in this type of galaxies should be studied thoroughly in the future.

3. The perilious journey of BH pairs in clumpy gaseous disks; from galactic to circumnuclear regions

The previous section focused on the sinking and coalescence of supermassive black holes ($M > 10^7 M_\odot$) in massive galaxies, a category of objects that will be only marginally in the detection window of LISA. In the early phase of their in-spiral they would, however, provide a very strong signal, with strong signal-to-noise ratio, at relatively low redshifts ($z < 4$) as in the case of the system studied by Khan et al. (2016) (see Fig.2).

Furthermore, the results presented in the last section are mostly relevant to gas-poor galaxies. The galaxy host merger remnant in Khan et al. (2016) indeed belongs to the category of galaxies that leave the star forming main sequence and become "red and dead" relatively early, at $z > 2$, such as the red nuggets identified in the CANDELS survey (Barro et al. 2013). While it is generally believed that more massive galaxies quench their star formation earlier, galaxy evolution is a nonlinear process which occurs along multiple tracks, hence even among massive galaxies star formation may continue or even peak at an epoch later than that of the merger remnant of Khan et al. (2016). Indeed at $z \sim 2-3$ we witness the existence of massive star forming galaxies that are very clumpy and disky (Tacconi et al. 2013; Forster-Schreiber et al. 2011, Wisnioski et al. 2015). This has often interpreted as evidence for massive, self-gravitating gas disks that fragment due to a Toomre-like instability (Bournaud et al. 2010; Ceverino et al. 2010). While the quantitative aspects of gravitational instability of massive gas disks at high-z are currently debated and may strongly depend on feedback processes (Tamburello et al. 2015; Mayer et al. 2016), it is important to ask how would massive black hole pairs evolve and sink in such clumpy disks . In particular, what would be the typical timescales for massive black hole coalescence compared to those in relatively smooth galactic disks at low redshift, or to those in gas-poor galaxies dominated by the stellar background physics?

In Tamburello et al. (2017) we considered the decay of two massive black holes that are already embedded in a gas-rich massive galactic disks subject to fragmentation into massive star forming clumps. We considered also the effect of black hole accretion and feedback, as well as star formation and feedback from supernovae explosions. We explored a large suite of simulations with galactic hosts having different masses and gas fractions, and choosing different eccentricities for the orbit of the secondary black hole (the mass ratio is fixed of the two black holes to 10:1, which should be statistically representative as the same mass ratio is most typical for halo/galaxy mergers). The main result is that the lighter black hole, with mass in the range $4 \times 10^7 - 10^8 M_\odot$ (1/5 of the mass of the primary) experiences repeated gravitational scattering by the most massive clumps as well as by strong spiral density waves often This often results in its ejection from the disk plane, slowing down its orbital decay by at least an order of magnitude as dynamical friction is suppressed in the low density envelope around disk midplane (Fig. 3). In a companion paper (Tamburello et al. 2016) we have shown how such clumps do reproduce the properties of observed star forming clumps in high-z massive galaxies once appropriate corrections for observational biases (limited sensitivity and resolution) are taken into account.

The suppression of orbital decay is exacerbated when AGN feedback is included as both dynamical friction and disk torques are weaker even when the secondary is still in the disk plane due to local heating of the gaseous background. Since it sinks slower, the secondary can be ejected when it is further away from the primary in this case. The secondary, following an ejection, can still wander at distances greater than 1 kpc more after 1 Gyr. decay has started. Taken at face value this would imply a possible complete abortion of the coalescence process. However, since Tamburello et al used galaxy models that lack the an extended bulge/spheroid component, dynamical friction was underestimated away from the galactic disk plane. By computing analytically the extra drag resulting from an Hernquist spheroid with

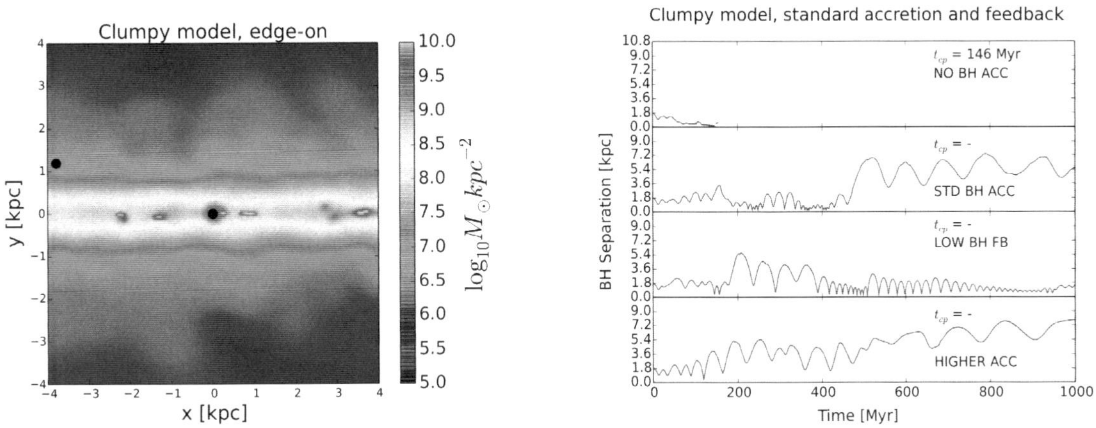

Figure 3. On the left we an edge-on color-coded density map of one of the clumpy hi-z galaxies from Tamburello et al. (2017), showing the ejection of the secondary black hole out of the disk plane (black dot on top left corner of panel). On the right we show a few examples of how the orbital separation of the secondary BH relative to the primary evolves with time. All runs shown have BH accretion and feedback except the one used in the top panel, which only includes BH accretion. Parameters related to the details of the accretion and feedback sub-grid models differ in the various runs (see labels in Figure and Tamburello et al. 2017). As it is seen there is a high degree of stochasticity in the orbital separation evolution, but in most cases the secondary keeps wandering at large distances even after a billion years from the beginning of the simulation (but see Tamburello et al. (2017) on the effect of a massive extended stellar bulge).

realistic structural parameters, it was found that, if no more ejections occur, , wandering secondaries should coalesce on timescales of order 1 billion year. This is still a very long timescale, comparable, if not longer, than the typical merging timescales of galaxies at $z > 2$.

The clumpiness of the ISM is an important factor even for the late stage of black hole pair decay, at 10 pc scales and below, near the phase in which a bound binary forms. This applies to gas-rich galactic hosts at any redshift as circumnuclear disks (CNDs) of gas and stars are ubiquitous at low z in merger remnants as well as in the central regions of Seyfert galaxies (Medling et al. 2014; Izumi et al. 2016) , which structurally typical late-type spiral galaxies. Fiacconi et al (2013) carried out the first study of black hole pairs in CNDs, documenting the dramatic effect of gravitational scattering by clumps and spiral density waves that we just illustrated. They considered black holes right in the LISA window ($10^5 - 10^7 M_\odot$). They also found that in some cases the decay can be accelarated due to the capture of a massive gas cloud ($M > 10^6 M_\odot$) by the secondary. They proposed the general notion that orbital decay is *stochastic* in a clumpy CND, with orbital decay timescales to 0.5 pc separations ranging from a few Myr to as much as 100 Myr. The shortest timescales agree with previous results for smooth CNDs (see Escala et al. 2005 , Dotti et al. 2006;2007, Mayer 2013). Statistically, though, a largest fraction of the simulations resulted in long timescales, in line with the same findings of Tamburello et al. (2017), especially for black holes on eccentric orbits.

These simulations were quite idealized in the physical treatment, lacking feedback mechanisms in particular, which should play a major role in CND, regions that are often subject to starbursts. Del Valle et al. (2015) and Souza Lima et al. (2016) recently carried out a similar studies but included a much richer inventory of physical processes. Del Valle et al. (2015) included star formation and a weak form of SN feedback, while Souza Lima et al. (2016) employed a popular effective sub-grid model of SN feedback, the blastwave feedback (Stinson et al. 2006), which is known to produce realistic stellar masses and disk sizes in galaxy formation simulations (Guedes et al 2011), and even added black hole accretion and AGN feedback in the same way as Tamburello et al. (2017). They confirmed the stochastic orbital decay picture and the prevalence of suppressed orbital decay with timescales in the range 50-100 Myr. The same long orbital decay timescales were found in a complimentary study modeling the entire massive black pair evolution before and after a major galaxy merger. This study also addresses the formation of the circumnuclear disk after the merger, but was restricted to only one initial condition for the galaxy merger due to the high computational burden introduced by deep the multi-scale nature of the calculation (Roskar et al. 2015). Souza Lima et al. (2016) is the only work, however, that includes also the effect of BH accretion and AGN feedback at CND-scales. Thanks to that they uncovered a new process, dubbed "wake evacuation effect", by which the secondary black hole carves a hole in the outer, lower density region of the CND, as it launches a hot pressurized bubble resulting from AGN feedback (Figure 5). The result is that dynamical friction and disk torques are stifled as the CND-black hole dynamical coupling is temporarily stifled by the presence of a large cavity. As a result, the secondary decays slower even before a strong interaction with a clump or spiral wave occurs, often resulting in an ejection when it is still far from the center. In the runs with AGN feedback the outcome is thus an even longer delay, of order a Gyr , which however should be regarded as an upper limit in absence of an extended massive stellar spheroid, for the same argument made in Tamburello et al. 2017 in the case of high-z galactic disks. When dynamical friction by an extended spheroid is added, for realistic parameters we obtained a decay timescale (to fraction of parsec separation, namely close to the resolution limit) of order a a few 10^8 yr. This is still almost two orders of magnitude longer than the decay timescale in a smooth CND and no feedback processes included (see eg Mayer 2013). These effects are only some of those that are possible in a complex multi-phase CND. For example, recently Park & Bogdanovic (2017) have shown that radiation pressure can also produce an extra drag pulling opposite to the dynamical friction wake, which again slows down the orbital decay.

In summary, the presence of a clumpy ISM as well as the concurrent effect of feedback processes complicate significantly the picture in gas-rich systems, suggesting that timescales in the range $10^8 - 10^9$ yr are likely for the decay to sub-pc separations from galactic scales, with the longer timescales more likely in the high-z clumpy galaxy phase. Despite uncertainties it is clear that, at high z, such timescales are much longer than those in gas-poor galaxies in which the decay is governed by the physics of the stellar dynamical background. At low z, instead, they turn out to be comparable or shorter than those in purely stellar backgrounds , as the simulations of Khan et al. (2012) found a typical timescale of order a Gyr for the hardening phase by 3-body encounters.

4. Summary and Outlook

The latest developments in modeling the orbital decay of massive black hole binaries that we just illustrated have turned upside-down the conventional notion that orbital decay of massive black hole binaries should be faster and more efficient in gas-rich galaxies relative to gas-poor galaxies. In particular at high reshift, while simulations finally show fast coalescence for massive black holes in gas-poor merger remnants of massive galaxies, with timescales at least an order of magnitude shorter than the merging time of galaxies themselves (10^7 yr as opposed to $10^8 - 10^9$ yr), the opposite seems to be the case for similarly massive gas-rich disks due to the perturbing action of the their inhomogeneous, clumpy interstellar medium. The existence of such clumpy medium is an inevitable consequence of the self-gravity of massive gas disks, although detailed properties such as the mass spectrum of cold clouds depend on the complicated interplay between gas physics and feedback processes. In these systems orbital decay timescales acquire a stochastic character, with values ranging from 10^8 yr to > 1 Gyr, namely longer than the galaxy merging timescales at $z > 1$. These estimates still do no account for the final phase of decay below 0.1 pc due to the resolution limitations of current multi-scale hydrodynamical simulations.

A word of caution must be spent in that there is still significant uncertainty in how to model feedback processes in galaxy formation and evolution, which in turn has a major impact on the nature of the ISM. The latest generation of strong feedback models, which are favoured by some constraints available on the rate of assembly of stellar masses in galaxies, do reduce significantly the clumpiness of the ISM even in the most massive high redshift disks (Mayer et al. 2016; Oklopcic et al. 2016). While massive star forming clumps are observed in massive disks at high-z, their actual nature is debated, especially whether they are bound coherent structures or the collection of much smaller, weakly bound structures (Behrendt et al. 2016; Tamburello et al. 2017). This of course will have an impact on how strong are the dynamical perturbations that disk substructure can induce on the orbits of massive black holes.

At low z it is established and inevitable that the ISM is clumpy at the scale of Giant Molecular Clouds (GMCs), namely at the scale of $10^6 - 10^7 M_\odot$ (the largest cloud masses are observed in galactic nuclei, eg in our Galactic Center, see Oka et al. 2001). These cloud masses appear to be large enough to cause significant perturbations to decaying massive black holes with masses best accessible to LISA, namely $< 10^7 M_\odot$, and their scattering action delays decay, in most cases leading to timescales $> 10^8$ yr to reach sub-pc scale separations (Souza Lima et al. 2016). More massive black holes would be unaffected as shown by simple analytical arguments (Fiacconi et al. 2013). New effects caused by AGN feedback, such as wake evacuation or anisotropic radiation pressure, can stifle the orbital decay further (Souza Lima et al. 2016; Park & Bogdanovic 2017).

The picture that is emerging is that the orbital decay of massive black holes is tightly linked to the properties of their galactic hosts, and to how they evolve in redshift. It has never been so evident as it is now that understanding the evolution of massive black holes requires control of the background galaxy formation theory, which currently is developed primarily with supercomputer simulations of galaxy formation in a cosmological context. This has two important implications, both crucial in preparation for LISA The first is that the potential of gravitational wave detection from massive black holes as a new probe of cosmic structure formation is much higher than we could ever suspect. The second is that to fully exploit such potential we need to devote significant resources to supercomputer simulations addressing all possible regimes of orbital decay, from large to small scales, and ranging across a wide variety of possible galaxy hosts, especially at $z > 1$ and for massive black holes below $10^8 M_\odot$, the regime where the discovery potential of LISA is highest.

In essence the latest simulations reveal a high complexity of massive black hole orbital decay, with multiple regimes and multiple physical effects involved. It is now clear that semi-analytical models describing evolving black hole populations in hierarchical galaxy formation, the tool that has been used so far to make predictions for LISA detections (eg Barausse 2012) need the input of such simulations to make reliable forecasts. Without that black hole merger rates could be off by orders of magnitude. The same applies to recent models coupling recipes for black hole orbital decay with large scale cosmological hydro simulation that cannot resolve better than kpc scales (Salcido et al. 2016).A joint effort of the different communities connected to the LISA science goals seems necessary to fill this gap and make future progress.

4.1. Acknowledgments

The author thanks his many collaborators, in particular Fazeel Khan, Pedro Capelo, Rok Roskar, Andreas Just, Peter Berczik, Silvia Bonoli, Romain Teyssier & Jillian Bellovary. A special thank goes to current and former PhD students, who have carried out a large part of the research described in this article, in particular Simone Callegari, Davide Fiacconi, Valentina Tamburello & Rafael Souza Lima. He acknowledges stimulating discussions over last years with Monica

Colpi, Piero Madau, Tamara Bogdanovic, Alberto Sesana, Scott Tremaine, Zoltan Haiman, Massimo Dotti, Andres Escala & Alessia Gualandris.

References

Amaro-Seoane, P., Aoudia, S., Babak, S., et al. 2013, *GW Notes*, **6**, 4
Agertz, O. & Kravtsov, A. V., 2015, *ApJ*, **804**, 18
Begelman, M. C., Blandford, R. D., & Rees, M. J. 1980, *Nature*, **287**, 307
Behrendt M., Burkert A. & Schartmann M., 2016, ApJ, **819**, L2
Barro, G., et al. 2013, *ApJ* **765**, 104
Barausse, E., 2012, *MNRAS*, **423**, 2533
Berczik, P., Merritt, D., Spurzem, R., & Bischof, H.-P. 2006, *ApJL*, **642**, L21
Behroozi P. S., Wechsler R. H., Conroy C., 2013, *ApJ*, **770**, 57
Bezanson, R., van Dokkum, P. G., Tal, T., et al. 2009, *ApJ*, 697, 1290 Bezanson, R., van Dokkum, P. G., van de Sande, J., et al. 2013, *ApJ*L, **779**, L21
Blanchet, L. 2006, *LRR*, **9**, 4
Bonoli, S., Mayer, L., & Callegari, S. 2014, *MNRAS*, **437**, 1576
Bonoli, S., Mayer, L., Kazantzidis, S., Madau, P., Bellovary, J., Governato, F., 2016, *MNRAS*, **459**, 2603
Bournaud F., Elmegreen B. G., Teyssier R., Block D. L., Puerari I., 2010, *MNRAS*, **409**, 1088
Bryan, G.L. & Norman, M.L., 1998, *ApJ*, **495**, 80
Capelo, P. R., Volonteri, M., Dotti, M., et al. 2015, *MNRAS*, **447**, 2123
Ceverino D., Dekel A., Bournaud F., 2010, *MNRAS*, **404**, 2151
Chapon, D., Mayer, L., & Teyssier, R. 2013, *MNRAS*, **429**, 3114
del Valle, L., Escala, A., Maureira-Fredes, C., et al. 2015, *ApJ*, **811**, 59
Dotti, M., Colpi, M. & Haardt, F., 2006, *MNRAS*, **367**, 103
Dotti, M., Colpi, M., Haardt, F., & Mayer, L. 2007, *MNRAS*, **379**, 956
Escala, A., Larson, R. B., Coppi, P. S. & Mardones, D., 2005, *ApJ*, **630**, 152
Farris, B., Duffel, P., MacFadyen, A,m & Haiman, Z., 2015, *MNRAS*, **447**, L80
Feldmann, R. & Mayer, L., 2015, *MNRAS*, **446**, 1939
Ferrarese, L., & Merritt, D. 2000, *ApJ*L, **539**, L9
Fiacconi, D., Feldmann, R., & Mayer, L. 2015, *MNRAS*, **446**, 1957
Fiacconi, D., Mayer, L., Roskar, R., & Colpi, M. 2013, *ApJ*L, **777**, L14
Gould, A., & Rix, H.W, 2000, *ApJ*L, **532**, L29
Graham, M. J., Djorgovski, S. G., Stern, D., et al. 2015, *Nature*, **518**, 74 Guedes, J., Callegari, S., Madau, P., & Mayer, L. 2011, *ApJ*, **742**, 76
Gualandris, A., Read, J. I., Dehnen, W., & Bortolas, E. 2016, *MNRAS*, **463**, 2301
Guedes, J., Callegari, S., Madau, P. & Mayer, L., 2011, *ApJ*, **742**, 76
Holley-Bockelmann K., Khan F. M., 2015, *ApJ*, **810**, 139
Izumi, T., Kawakatu, N., & Kohno, K. 2016, *ApJ*, **827**, 81
Khan F. M., Just A., Merritt D., 2011, *ApJ*, **732**, 89
Khan F. M., Preto M., Berczik P., Berentzen I., Just A., Spurzem R., 2012, *ApJ*, **749**, 147
Khan F. M., Holley-Bockelmann K., Berczik P., Just A., 2013a, *ApJ*, **773**, 100
Khan, F.M., Fiacconi, D., Mayer, L., Berczik, P., & Just, A., 2013b, *ApJ*, **828**, 73
Khan, F.M., Holley-Bockelmann, K., & Berczik, P., 2015, *ApJ*, **798**, 103
Klein, A., et al. 2016, *PhysRevD*, **93**, 2, id.0240004
Mayer, L. 2013, *Classical and Quantum Gravity*, **30**, 244008
Mayer, L., Kazantzidis, S., Madau, P., et al. 2007, *Science*, **316**, 1874
Mayer L., Tamburello V., Lupi A., Keller B., Wadsley J., Madau P., 2016, *ApJ*, **830**, L13

Medling, A. M., U., Vivian;, Guedes, J., Max, C. E., Mayer, L., Armus, L., Holden, B., Rokar, R., & Sanders, David, 2014, *ApJ*, **784**, 70
Merloni, A., Bongiorno, A., Bolzonella, M., et al. 2010, *ApJ*, **708**, 137
Milosavljevic, M., & Merritt, D. 2001, *ApJ*, **563**, 34
Mo, H.J, Mao, S. & White, S.D.M., 1998, *MNRAS*, **295**, 319
Oka, T., Hasegawa, T., Sato, F., Tsuboi, M., Miyazaki, A, & Sugimoto, M., 2001, *ApJ*, **562**, 348
Oklopcic A., Hopkins P. F., Feldmann R., Keres D., Faucher-Giguere C.-A., Murray N., 2016, preprint, (arXiv:1603.03778)
Park, K. & Bogdanovic, T. 2017, submitted to *ApJ*, (2017arXiv170100526P)
Roedig, C., Sesana, A., Dotti, M., Cuadra, J., Amaro-Seoane, P., Haardt, F. 2012, A&A, **545**, 127
Roskar, R., Fiacconi, D., Mayer, L., et al. 2015, *MNRAS*, **449**, 494
Sesana, A., & Khan, F. M. 2015, *MNRAS*, **454**, L66
Stewart, K. R., Bullock, J. S., Barton, E. J., & Wechsler, R. H. 2009, *ApJ*, **702**, 1005
Sesana, A., Volonteri, M., & Haardt, F., 2007, *MNRAS*, **377**, 1711
Salcido, J., Bower, R., et al. 2016, *MNRAS*, **463**, 870
Stinson, G., Seth, A., Katz, N., et al. 2006, *MNRAS*, **373**, 1074
Souza Lima, R., Mayer, L., Capelo, & Bellovary, J., 2016, *ApJ*, in press (arXiv:1610.01600)
Szomoru, D., Franx, M., & van Dokkum, P. G. 2012, *ApJ*, **749**, 121
Tacconi L. J., et al., 2013, *ApJ*, **768**, 74
Tamburello V., Mayer L., Shen S., Wadsley J., 2015, *MNRAS*, **453**, 2490
Tamburello, V. Rahmati, A., Mayer, L., Cava, A., Dessauges-Zavadsky, M., & Schaerer, D., 2016, *MNRAS*, submitted, (arXiv:1610.05304)
Tamburello, V., Capelo, P. R., Mayer, L., Bellovary, J. M., Wadsley, J. W. 2017, *MNRAS*, **464**, 2952
Trakhtenbrot, B., Urry, C. M., Civano, F., et al. 2015, *Sci*, **349**, 168
Vasiliev, E., Antonini, F., & Merritt, D. 2015, *ApJ*, 810, 49
Volonteri, M. & Natarajan, P., 2009, *MNRAS*, **400**, 1911
White, S.D.M. & Rees, M.J., 1978, *MNRAS* , **183**, 341
Wadsley, J. W., Stadel, J., & Quinn, T. 2004, *NewA*, **9**, 137
Wisnioski, E., Forster Schreiber, N. M., Wuyts, S., et al. 2015, *ApJ*, **799**, 209

IOP Conf. Series: Journal of Physics: Conf. Series **840** (2017) 012026 doi:10.1088/1742-6596/840/1/012026

"Enchilada" is back on the menu

Stanislav Babak

Max Planck Institute for Gravitational Physics (Albert Einstein Institute), Am Mühlenberg 1, Potsdam-Golm, 14476, Germany

E-mail: stas.babak@aei.mpg.de

Abstract. In this short proceedings we describe the new pipeline for generating simulated LISA data. The pipeline relies on the catalogues of sources which represent the realisation of the Universe in gravitational waves (GWs) for several astrophysical models. We have extended the set of GW models which could be used to simulate the GW signals. Finally we have adopted hdf5 as the main format for the storage the data and parameters.

1. Past Mock LISA Data Challenges

Mock LISA Data Challenge is an effort started in January 2006 by a group of people (task-force) [1, 2]. The main idea behind this effort was to develop the common framework for the LISA data analysis. This includes defining LISA's orbit, noise budget, and set of models for simulating GW signals. Such a common ground simplify the comparison between different data analysis techniques, which are applied to the *same* data. We have used several simulators to apply the LISA response function [3, 4, 5]. The simulated data were advertised to scientific community and were available for download to whoever wanted to take part in the challenge. Each challenge was not a competition, but rather a way to foster the development and improvement of the LISA data analysis methods. There were two data sets in each challenge: (i) open, where all details about the data (parameters of the noise and each GW source) was known, (ii) blind challenge where only partial information (ranges) about number of sources and their parameters was disclosed.

Challenges were issued roughly once per year and overall were increasing in complexity. A short summary of the challenges is given in the table 1 and we will give details of each challenge a bit later. The participation has varied from challenge to challenge but in total we had 70 participants from 25 institutions. Before describing the challenges we should recall the main GW sources of LISA and their basic features. We refer reader to [6] for a more detailed description of LISA sources and science return.

The strongest expected GW signal will come from the coalescing massive black hole (MBH) binaries. Those are BHs residing in the nuclei of (we believe) all galaxies, they gain mass through the gas accretion and through minor/major mergers with other MBHs. The mergers of MBH is the result of galactic collisions, two MBHs interact with the surrounding gas and stellar environment and form a close binary where the gravitational radiation is efficient to drive them to a merger. The notion of minor or major merger refers to the mass ratio of coalescing MBHs, where the major corresponds to two MBHs of comparable mass. LISA will be sensitive to the MBH binaries in a wide range of masses from $10^4 M_\odot$ to few $10^7 M_\odot$. The expected event rate is quite uncertain from astrophysical point of view and also depends on the final configuration of

LISA, but we could expect between few and few hundreds events per year. We refer to [7] and references therein for more details. The expected signal-to-noise ratio (SNR) for those events varies in a wide range between 10 (we use as the detection threshold) and ~ 1000. The GW signal could be conventionally split in three parts: (i) Inspiral, this is low-frequency part of the signal, where the two bodies spiral around each other and adiabatically approach each other due to loss of energy and angular momentum by GWs, this is the longest part of the signal and could be observed (typically) for several months. This part of signal could be described analytically [8]. (ii) Merger, this part is of relatively short duration, two MBHs approach each other and form a single highly deformed object, this is accompanied by a huge release of energy in GWs and will be seen by LISA throughout the Universe (iii) Ringdown, this is high frequency part of the signal, where the deformed remnant MBH releases its excitations through the gravitational radiation, which spectrum is the superposition of exponentially damped quasi-normal modes of a MBH. This part of the signal will dominated the SNR for the heavy systems with the total mass $\gtrsim 5 \times 10^6 M_\odot$. The BHs are expected to be spinning where the magnitude of the spins and their orientation depend on the environment and the past history (for example, persistent accretion disk significantly spins up BHs). The spins misaligned with the orbital angular momentum cause precession of the orbital plane which encoded in the GW signal.

The most interesting GW in the LISA band is the so-called extreme mass ratio inspiral (EMRI). We expect that the MBHs in quiescent galaxies is surrounded by a cusp of compact objects (neutron star, solar mass BHs) which have segregated to the nuclei by dynamical friction mechanism [9]. As a result of N-body interaction one of such compact object could be deflected to the orbit passing near a MBH and being captured, then it slowly spirals toward the central MBH until the final plunge. The parameters of the orbit and the event rate should provide us with the unique information about central parsec region around MBH, which cannot be obtained by other means. The expected event rate varies from few per year to almost a thousand and SNR lies between 20 (detection threshold) and up to few hundred. The signal consists of harmonics of three non-commensurate (in general) orbital frequencies slowly changing in time as a small compact object spirals toward the central MBH. The signl spends in LISA band $10^4 - 10^6$ cycles and lasts up to few years. Fitting the signal's phase allows ultra-precise measurement of the source' parameters and this also makes it difficult to model such a GW signal. We refer to [10] for detailed overview of EMRIs (and references therein).

The most numerous source of GWs is our own Galaxy Milky Way. Gravitational radiation form $\sim 10^7 - 10^8$ white dwarf binaries on the tight orbits spreads across the LISA band. Majority of these GW signals will not be detectable individually and form a stochastic astrophysical foreground signal below few mHz. At high frequencies the density of those signals is low enough to resolve and subtract each signal. We will refer to residual Galactic GW noise signal (after subtraction) as "reduced Galactic signal". Each GW signal coming from the white dwarf binary is almost monochromatic with slight drift in frequency due to GW emission and/or due to mass transfer in the interacting binaries. The signals are always present in the data and stay in band. From the electromagnetic observations we know a dozen of Galactic binary systems which fall into a LISA band and will be detectable in GWs, those binaries are called *verification* binaries. Those a guaranteed sources with known sky position and GW frequency, and could be used for monitoring the detector's performance.

The three GW signals from binary black holes detected by LIGO have opened the era of gravitational wave astronomy [11]. The biggest surprise was the high mass of detected BHs ($\sim 30 M_\odot$), and those binaries at their early stages of inspiral (5-10 years before the merger) should emit a GW signal detectable by LISA [12]. Those signals will be detectable around/above 10 mHz, and the population of binary BHs might create a detectable stochastic GW signal.

Besides the GW signal from coalescing binary systems described above we might be able to detect a stochastic GW signal from very energetic processes in early Universe [13]. Such a GW signal is expected to be isotropic and its detection is based on the cross-correlation of independent data streams constructed out of individual measurements from each link (laser signal connecting two spacecrafts). The stochastic signal from the astrophysical populations described above should have a specific power-law spectral index, which should allow to distinguish it from the stochastic signal of cosmological origin.

Finally, LISA might detect something exotic like GW burst signal coming from cusps or kinks formed on the cosmic strings (see [13] and references therein). those signals have a very specific shape in time and frequency domain which allows to search for them using matched filtering techniques.

Now we are in the position to describe the past MLDCs summarized in the table 1. The very first channel was the easiest, it contained several independent data sets (i) data set with few Galactic binaries (ii) data set with the verification binaries (iii) data set with few dozen of Galactic binaries (iv) data set with a single GW signal from non-spinning MBH binary with SNR$>$ 100. The second challenge was quite a big jump ahead, it contained again several independent data sets: (i) about 20 mln. monochromatic signals from Galactic binaries (ii) 4-6 (exact number was not disclosed) GW signals from non-spinning MBH binaries placed on top of the reduced Galactic signal (iii) 5 data sets each with a strong (SNR$>$ 50) GW signal from EMRIs. The next challenge was a repetition of the challenge 1 to allow new groups to join in and to allow further investigations and improvements in the data analysis algorithms. The MLDC3 had again many data sets each containing: (i) 60 mln. GW signals with frequency drift from Galactic binaries (ii) 4-6 GW signals from spinning (precessing) MBH binaries together with the reduces Galactic signal (iii) single data set with 5 weak (SNR$\approx$ 20) EMRIs (iv) data set with GW bursts from cusps on cosmic strings (v) data set with the isotropic stochastic GW signal. For the population of Galactic binaries we have used catalogue provided by G. Nelemans [14, 15].

Table 1. Summary of past mock LISA data challenges

Sources	MLDC1	MLDC2	MLDC1B	MLDC3
Gal. bin	Verification, few dozen	Galaxy, no freq. evolution	Verification, few dozen	Galaxy 6×10^7
MBH bin.	One, no spin	4-6, no spin, with red. Gal.	One, no spin	4-6 precess., with red. Gal
EMRIs	-	Isolated, strong	Isolated strong	Isolated, weak
Bursts	-	-	-	Cosmic strings
Stochastic	-	-	-	Isotropic, strong

2. Recipe for "Enchilada"

From the table 1 and from the previous description it is clear that the current data analysis methods have mainly concentrated on the detection and characterization of GW signals of the same type in the data sets where only few signals are present. Of course the analysis of Galactic binaries is an exception, there we meet the problem of the *signal confusion*, when multiple signals overlap in time and in frequency, but those signals are of the same type. Several data analysis methods were suggested to tackle problem of detecting Galactic binaries (see [16, 17] and references therein).

The "Enchilada" was suggested for the first time for the challenge 2 but it was premature, later it was a theme for the never-finished challenge 4. The "Enchilada" is a nick name (suggested by N. Cornish) for a challenge where the signals of the same and different kind are simultaneously present in the data. The Enchilada addresses the main problem of LISA' data analysis: confusion problem, where we need to detect and characterize (correctly) the maximum possible number of

IOP Conf. Series: Journal of Physics: Conf. Series **840** (2017) 012026 doi:10.1088/1742-6596/840/1/012026

GW signals. Besides the problem of detecting Galactic binaries, the only mix of sources was the GWs from MBH binaries plus reduced Galactic signal, but the reduced Galactic signal played role of the noise source.

The challenge 4 (which was never completed) was meant as a mild Enchilada problem, and here we want resurrect it but with improved "recipe". So what do we need to cook up the Enchilada data set.

- We need to simulate the Universe as it can be seen by LISA in GWs. We need astrophysical models which tell us how many signals we expect to have and distribution of parameters.
- We need reliable and fast to generate models of GW signals from anticipated sources.
- We need to simulate the instrument: mission configuration, noise level, instrumental and environmental artifacts.

We will walk through each of these items. The first item is about building catalogues of sources based on the current astrophysical models. The uncertainties in our knowledge/models will be reflected in several catalogues. For merging MBH binaries we used the models described in [7]. We also have catalogues for EMRIs [18] which was built on top of the MBH population [19, 20]. The novelty there (brought by LIGO) is that we have added heavy solar-origin BHs ($\sim 30M_{\odot}$). We have also improved catalogue of Galactic sources. The idea is to use the sources selected from these catalogues for the simulated data sets. The flexibility of a current data generating pipeline allows us to use a whole realization of a given model or just a selected number of sources.

The second bullet point is about simulating the GW signals. There are two key points which must be satisfied by a model: (i) it should be effectual representation of the GW signal (ii) it must be fast to generate. The former implies that model might not be necessarily very faithful, of course it is desirable. The later is required for the efficient exploration of the data analysis methods, and it wins over faithful but slow models. For coalescing MBH binaries we have extended the pool of models by including IMR models (models including all three parts of the signal inspiral-merger-ringdown), whereas previously we have used only inspiral part of the signal. We have added two models for precessing MBHs the EOB-based time domain model [21] and the phenomenological frequency-domain model [22]. The time domain model is convenient for the simulating the data, whereas the frequency domain model is more suitable for the data analysis purposes. This is an important extension, as the merger-ringdown is a very important part from the point of view of detection and parameter estimation. For EMRIs we use the model suggested in [23], however this model is notoriously unfaithful, so we have added also the model suggested in [24].

The third bullet point is actually very big. When we say GW signal model we usually mean two GW polarizations ($+, \times$), however LISA has a non-trivial response function (see, for example, [4]). What is used for data analysis are two (noise) independent TDI streams, called A, E, which are result of applying the response/transfer function to GW polarizations. The TDI stands for time delay interferometry, which is a linear combination (with time delays) of individual measurements which cancel laser frequency noise (the highest noise) [25]. We use LISACode [5] to apply the response and derive the A, E combinations for each GW signal. The LISACode has a flexibility to vary the LISA configuration (like arm-length, mission life time, size of collecting telescope, power of laser) as well as the basic noise budget. This is an important property which we want to have before the final LISA configuration is stable. We also have now LISA Pathfinder data [26] which exhibits fantastic performance. The measured data is however shows occasionally non-stationary features. We want to explore the effect of these noise artifacts on the data analysis algorithms which were developed with assumption of the Gaussian noise. For the Enchilada project we start with the Gaussian noise of a given color (derived from LISA Pathfinder measurements projected to LISA) and LISACode will be used to produce such a

f_ini	0.000427690264488	Hz
spin2x	-0.0891075135401	dimensionless
mass1	280533.8732	solar mass
mass2	39644.2998	solar mass
spin2z	0.576370782859	dimensionless
spin2y	-0.0366724501106	dimensionless
longitude	6.17673316892	rad
polarization	2.63132015174	rad
spin1x	0.181940330537	dimensionless
spin1y	0.132187387695	dimensionless
Lum_Distance	19299.8833863	Mpc
spin1z	0.762715833754	dimensionless
latitude	-0.0116491833204	rad
phi_ref	5.99308322852	rad
model	SEOBNRv3	approximant
cadence	3.0	sec
inclination	1.5333	rad

Figure 1. Example of the binary MBH parameter table using the small viewer supplied with the data.

noise data stream. Later we intend to extend this project by including the simulated realistic noise model.

3. Implementation

In this section we briefly discuss the implementation. The biggest difference with the previous MLDC pipeline is that we move to the new format, we intend to use hdf5 to store the parameters and the data. Those files will also contain some minimum metadata information (author, date, software version) which should be sufficient for reproducing the stored data. The hdf5 library has many APIs with different programming languages and we will provide a read/write interface with python and C/C++. There are standard hdf5 viewers which can easily parse the files and present the parameters as tables. We also provide a small viewer which allows to visualize the source parameters, an example for the a single MBH binary is given in the figure 1.

Previously we have used xml-format to store the metadata and links to the binaries with the simulate data [27], and indeed it is convenient format in some cases. We still can store the old xml files as a string in hdf5 files.

We have successfully produce several test data streams using new waveform models. We show example of two GW signals from merging MBH binaries in the figure 2. The red system (short one) is seen almost edge-on, and parameters of that system is given in the table of the figure 1, the clearly seen precession is caused by the misalignment of spins and the orbital angular momentum. For the long blue signal we have only very slight misalignment of spins and precession is not prominent.

In figure 3 we have put together GW signals from Galactic binaries, two MBH binaries (given in the figure 2) and one signal from EMRI. The left panel shows the simulated data in time domain and the right panel is power spectral density of the data. In particular we plot here X-TDI Michelson data combination. One can see in the time series that all signals are instantaneously below the noise level, even though they are detectable. The frequency domain plot shows the frequency range for each signal. Note that the merger part of MBHs signal and EMRIs could reach quite high frequency and require a small cadence (3-5 sec). This potentially could restrict the duration of the challenge data sets. This data set was obtained for the LISA

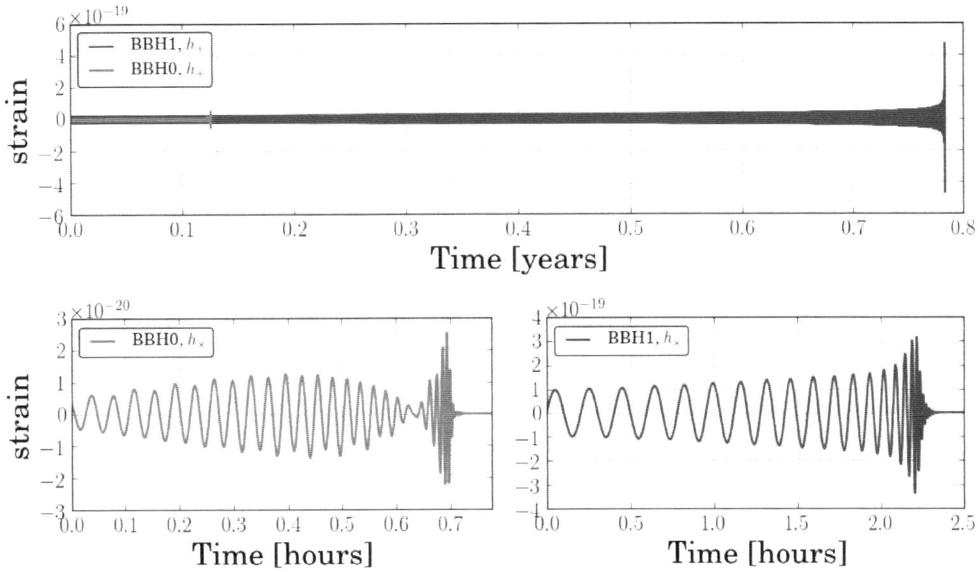

Figure 2. Two GW signals from MBH binaries. The top panel shows h_+ and lower bottom is a zoom in around the merger for othe polarization $h_\times$

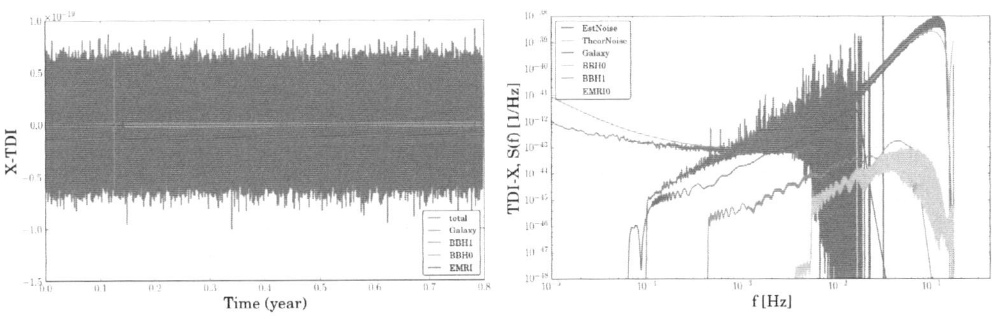

Figure 3. Example of the simulated data in time domain (left panel) and in frequency domain (power spectra density of the data in the right panel). The data contains simulated noise, 2 MBH binaries, 1 EMRI signal and a whole catalogue of the Galactic white dwarf binarie ($\sim 6 \times 10^7$ sources).

configuration with 1 mln. km armlength.

4. Discussion

We have pipeline for data set production, but we are not yet clear about the format of the new mock data challenges. There will be a core group which will analyze the data as it is required for the LISA project. With LISA on the clear roadmap, the funding agencies might be willing to support LISA data analysis related projects, and we want to make the challenges globally available. Another important question is about complexity of the data, we do not want to make a fake impression that the problem is solved by analysing the simplified data set, and we do not want to pass the wrong message that we cannot handle complexity of the LISA data. Probably we will make few initial trials "in-house" trying to analyze few data sets starting with a very simple one. In the mean time we might repeat single-source type challenge (but using the new

GW models) to dust off the tools used in the challenge 3, and get momentum for the "Enchilada" problem.

The data will be issued again in two formats: open, where all details about the data (noise, number of GW signals and their parameters) are known, and, closed data set which is blind data challenge, where only minimum information will be given to participants. We will make available all GW models (software) which were used for production of the challenge data sets, however some additional work (code optimization) might be required for the data analysis purposes.

Finally, we would like to say few words about the data analysis status. During the past MLDC several very promising methods were developed. Among the frequentist methods which aim at maximimization of the likelihood and getting the maximum likelihood estimators for parameters, we should mention the grid-based methods, Genetic Algorithm and particle swarm optimization. Bayesian methods were also used to analyse the data, those are various modifications of the Markov chain monte-carlo technique (simulated annealing, parallel tempering, etc.), and nested sampling (MultiNest). Note that there is a large community analysing the LIGO data, and some aspects of the LIGO data analysis are common with the LISA, so we can port the tools and expertise developed in LIGO to the MLDC. In addition to LIGO tools and methods, the problem of GW analysis of the Pulsar Timing Array (PTA) data is very similar to the Galactic GW foreground problem in LISA. One of the main source in the PTA band is a population of broad massive BH binaries in the local Universe. The GW signals which they emit are monochromatic over the observation time and they form a stochastic signal at low frequencies. Some GW signals (from nearby sources) could stand above of that GW background. At high frequencies the density of sources is dropping down and the GW signals should be detectable individually (when the sensitivity reaches the detectability level).

The new Mock LISA data challenge comes up this year. The most challenging problem which we will focus on is the signal confusion, also referred as Enchilada. We will used catalogues of sources based on the astrophysical models and we apply the new (extended) models for the GW signal simulation. Finally we adopt the new (hdf5) format for storing and distributing the data and associated metadata.

References

[1] Arnaud K A *et al.* (Mock LISA Data Challenge Task Force) 2006 *AIP Conf. Proc.* **873** 625–632 [,625(2006)] (*Preprint* `gr-qc/0609106`)
[2] Arnaud K A *et al.* 2006 *AIP Conf. Proc.* **873** 619–624 [,619(2006)] (*Preprint* `gr-qc/0609105`)
[3] Vallisneri M 2005 *Phys. Rev.* **D71** 022001 (*Preprint* `gr-qc/0407102`)
[4] Rubbo L J, Cornish N J and Poujade O 2004 *Phys.Rev.* **D69** 082003 (*Preprint* `gr-qc/0311069`)
[5] Petiteau A, Auger G, Halloin H, Jeannin O, Plagnol E, Pireaux S, Regimbau T and Vinet J Y 2008 *Phys. Rev.* **D77** 023002 (*Preprint* `0802.2023`)
[6] Amaro-Seoane P, Aoudia S, Babak S, Binetruy P, Berti E *et al.* 2013 *GW Notes* **6** 4–110 (*Preprint* `1201.3621`)
[7] Klein A *et al.* 2016 *Phys. Rev.* **D93** 024003 (*Preprint* `1511.05581`)
[8] Blanchet L 2014 *Living Reviews in Relativity* **17** URL `http://www.livingreviews.org/lrr-2014-2`
[9] Alexander T and Hopman C 2009 *Astrophys. J.* **697** 1861–1869 (*Preprint* `0808.3150`)
[10] Babak S, Gair J R and Cole R H 2015 *Fund. Theor. Phys.* **179** 783–812 (*Preprint* `1411.5253`)
[11] Abbott B P *et al.* (Virgo, LIGO Scientific) 2016 *Phys. Rev.* **X6** 041015 (*Preprint* `1606.04856`)
[12] Sesana A 2016 *Phys. Rev. Lett.* **116** 231102 (*Preprint* `1602.06951`)
[13] Babak S, Gair J R, Petiteau A and Sesana A 2011 *Class.Quant.Grav.* **28** 114001 (*Preprint* `1011.2062`)
[14] Nelemans G, Yungelson L R and Portegies Zwart S F 2001 *Astron. Astrophys.* **375** 890–898 (*Preprint* `astro-ph/0105221`)
[15] Nissanke S, Vallisneri M, Nelemans G and Prince T A 2012 *Astrophys. J.* **758** 131 (*Preprint* `1201.4613`)
[16] Babak S *et al.* (Mock LISA Data Challenge Task Force) 2008 *Class. Quant. Grav.* **25** 114037 (*Preprint* `0711.2667`)
[17] Babak S *et al.* (Mock LISA Data Challenge Task Force) 2010 *Class. Quant. Grav.* **27** 084009 (*Preprint* `0912.0548`)
[18] 2017 In preparation
[19] Sesana A, Barausse E, Dotti M and Rossi E M 2014 *Astrophys. J.* **794** 104 (*Preprint* `1402.7088`)

[20] Antonini F, Barausse E and Silk J 2015 *Astrophys. J.* **812** 72 (*Preprint* 1506.02050)
[21] Babak S, Taracchini A and Buonanno A 2017 *Phys. Rev.* **D95** 024010 (*Preprint* 1607.05661)
[22] Hannam M, Schmidt P, Bohé A, Haegel L, Husa S, Ohme F, Pratten G and Pürrer M 2014 *Phys. Rev. Lett.* **113** 151101 (*Preprint* 1308.3271)
[23] Barack L and Cutler C 2004 *Phys.Rev.* **D69** 082005 (*Preprint* gr-qc/0310125)
[24] Chua A J K and Gair J R 2015 *Class. Quant. Grav.* **32** 232002 (*Preprint* 1510.06245)
[25] Tinto M and Dhurandhar S V 2005 *Living Reviews in Relativity* **8** URL http://www.livingreviews.org/lrr-2005-4
[26] Armano M *et al.* 2016 *Phys. Rev. Lett.* **116** 231101
[27] Vallisneri M and Babak S 2008 *Computing in Science & Engineering* **10** 80–87 (*Preprint* http://aip.scitation.org/doi/pdf/10.1109/MCSE.2008.20) URL http://aip.scitation.org/doi/abs/10.1109/MCSE.2008.20

IOP Conf. Series: Journal of Physics: Conf. Series **840** (2017) 012027 doi:10.1088/1742-6596/840/1/012027

Supermassive black hole seeds: updates on the "quasi-star model"

Elena Maria Rossi

Leiden Observatory, Niels Bohrweg 2, NL-2333 CA Leiden,The Netherlands

E-mail: `emr@strw.leidenuniv.nl`

Abstract. LISA will allow us to see the infancy of the universe and gather unique constraints on an outstanding mystery: the formation of supermassive black holes. Here, I review a particular scenario, where black hole seeds are formed from direct collapse of gas at the centre of (proto)-galaxies. I will show that very massive ($> 10^5 M_\odot$)black hole seeds are possible but before confidently claim so the impact of physical ingredients, like internal rotation and super Eddington accretion, need to be fully understood.

1. Introduction

The Laser Interferometer Space Antenna (LISA) would be a unique probe for supermassive black hole (SMBH) science. In particular, it will be suited to directly detect massive black hole seeds — with masses between $10^4 - 10^6 M_\odot$ — at their formation epoch $z > 10$. This is one of my main motivations to study and try to assess the feasibility of the "direct collapse scenario", the leading model to form massive seeds at the centre of proto-galaxies. In addition, this scenario has the appealing feature of forming seeds where we actually observe SMBHs, contrary to the population III stars scenario, where mergers of the hosting mini-haloes need to eventually deliver a black hole to the bottom of the potential of an assembling galaxy with all the contingent complications nicely detailed in this recent review [1].

The challenge however remains: how actually do black holes form in first place? The issue is two-fold: (i) how to funnel this gas towards the centre of a galaxy from galactic scales avoiding the centrifugal barrier and fragmentation into stars, and (ii) once the gas is collected how actually the gas collapses into a black hole. In this contribution I will only address the second part of the puzzle, especially reviewing my recent results, published in [2] and [3].

Formation of massive seeds requires accretion rates $> 0.1 M_\odot$ s^{-1} (e.g. [4, 5, 6]). There are two main proposed mechanisms that may produce such high accretion rates and make massive seed formation possible. The first channel entails a major merger between two galaxies, that has been shown to effectively funnel $> 10^8 M_\odot$ in less than one Myr [7, 8], forming a structure which is gravitationally unstable and can directly collapse into a black hole that may retain all the mass [9, 10, 11]. This scenario however cannot explain the whole SMBH population and its entire range of observed black hole masses. Indeed major mergers are rare in the history of a galaxy and, at high redshifts considered here ($z > 10$), most of the galaxy mass is accreted in a secular process through cold gas streams that flow from the cosmic web into the galactic halo. These streams of dark and baryonic matter can contain substructures like dwarf galaxies, incorporating the concept of "minor mergers". Recent numerical cosmological simulations suggest that this is also the primary mode of SMBH growth [12]. In this scenario, the accumulated mass would first form a very massive star (MS) as nuclear reactions are ignited in their centre and, very much like in a regular massive star, its core would collapse after H exhaustion (lifetime ~ 1 Myr) into a black hole of maximum possible mass equal the total mass of the star: $0.1\ M_{\odot}\ \mathrm{yr}^{-1} \times 1$ Myr $\sim 10^5 M_{\odot}$.

2. Massive star formation as an intermediate step to seed formation

Let me now dwell more on the last scenario. The question I am interested in is: given the mass of a MS, what is the mass of the emerging black hole? The core of the MS has just a fraction of its mass. Therefore the previous question can be reformulated: after the core of the star collapses, what happens to the envelope? I believe it depends on the angular momentum re-distribution within the envelope, an important factor whose weight I came to appreciate in my last work on the subject [3]. I venture that the fate of a "slowly" rotating envelope is to follow the core to collapse directly into a black hole [3, 13]. The "fast" rotating ones will instead set as hydrostatic massive envelope around the black hole resulting from the collapse of the core: this structure is called a "quasi-star" and can foster the growth of the central black hole by super-Eddington accretion [14, 4, 15, 16, 17, 2]. In the following, I will first review the physical properties of a quasi-star (Section 3) and then physically explain the role of rotation in channelling a massive star towards either scenarios. It should then be clear what I mean by "slowly" ("fast") rotating envelope.

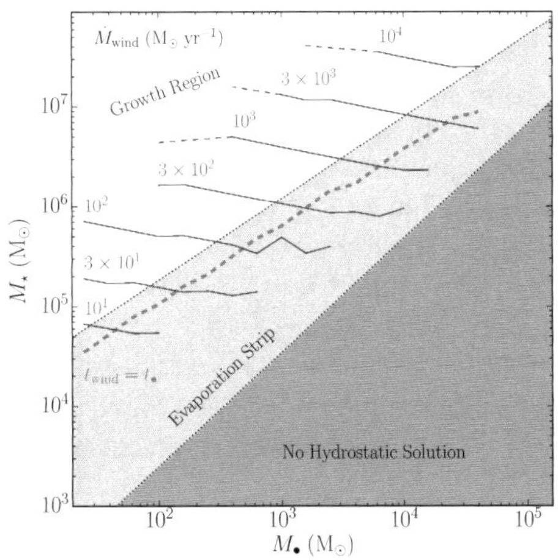

Figure 1. Quasi-star envelope mass vs. black hole mass plane. Three regions are highlighted: the "growth" region, the "evaporation strip", and the "no hydrostatic solution" region. The blue lines mark isocontours of the outflow rate $\dot{M}_{\mathrm{wind}}$ labelled by their value in $M_{\odot}\ \mathrm{yr}^{-1}$.

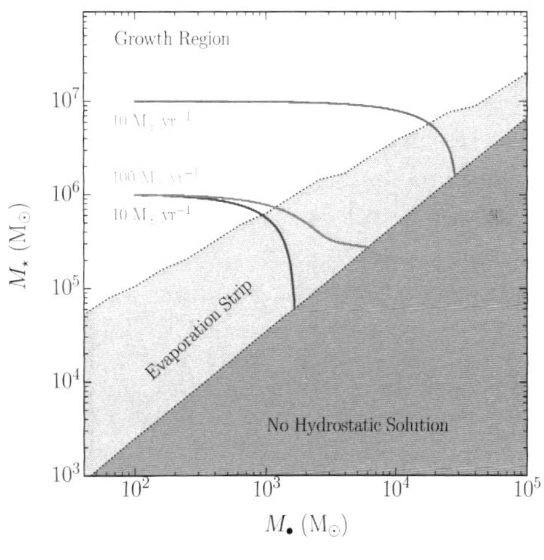

Figure 2. Evolutionary sequences of three quasi-stars in the envelope vs. black hole mass plane (time goes from left to right along the curves). The labels denote the assumed values for the (constant) inflow rates feeding the quasi-star envelopes.

3. The quasi-star scenario

As the core of a MS collapses into a black hole, the surrounding envelope follows the fall. However, if in the innermost region the gas has enough angular momentum to circularise outside the last stable circular orbit of the black hole and form an accretion disc, accretion energy may be injected back into the envelope stopping its collapse. The envelope would inflate and expand until equilibrium is found, which corresponds to a surface luminosity roughly equal to the Eddington luminosity for the *total mass of the envelope*. This is a highly super Eddington luminosity for the black hole, by a factor equal to the envelope to black hole mass ratio. This is possible because energy is transported through most of the envelope via convection, which is not subject to the Eddington limit [14]. The embryon black hole thus grows at a highly super-Eddington pace, until it has consumed enough envelope mass that the equilibrium for a quasi-star object is not longer possible. Here the feedback from the black hole is likely to blow away the remaining envelope, leaving a bear black hole, that will eventually accrete at a more subdued pace from the pre-galactic disc. Not all combinations of envelope mass M_* and black hole mass $M_{\rm BH}$ are allowed. On the lower right corner of Figure 1 lies the parameter space where no equilibrium solutions can be found: this occurs when the mass ratio between the envelope M_* and the $M_{\rm BH}$ is equal or lower than 20.

Interesting, [17] and [2] pointed out and study the possibility that the accretion power can even exceed the Eddington luminosity for the whole envelope, leading to mass loss from the quasi-star surface. A quasi-star is a radiation dominated object (as are all objects that emit around their Eddington limit, therefore energy losses (even slightly) above the Eddington limit unbound mass resulting in winds. We developed a self-consistent model for radiation dominated winds, valid in both the optically thin and thick regime [2]. Our main assumptions are steady state, spherical symmetry and an approximate description of the transition between the optically thick and the optically thin regimes. We find that, quite generally, the ultimate source of the wind kinetic energy is the advection energy, while the radiation energy (i.e. the radiation luminosity) remains constant. When applied to quasi-stars (with a self-consistent calculation of their envelope+wind structure), we find that very vigorous winds are blown from their surface. This can be appreciated in Figure 1, where we plot iso-contours of the mass loss rate: between 10 and 10 thousand solar masses per year can be lost by a quasi-star.

There are therefore two competing processes: the embryon black hole accretes from the inner edge of the envelope, while gas is blown from the surface. In Figures 1 the light grey strip diagonally running through the M_*-$M_{\rm BH}$ plane corresponds to models where the evaporation timescale (mass of the envelope divided by the calculated mass loss rate) is shorter than the accretion timescale (mass of the black hole divided by the accretion rate). For this reason it is called the *evaporation strip*: the envelope is consumed by the wind before the black hole can double its mass. The white region (the *growth* region) in that parameter space is instead where accretion is the faster process.

In Figure 2, we plot evolutionary tracks of quasi-stars fed by an external accretion flow: the envelope acquires mass from the pre-galactic disc and loses it because of its wind and the inner black hole accretion. The quasi-star moves through a series of equilibrium states until it hits the evaporation strip, where the envelope is rapidly consumed, leaving a bare black hole. The result is quite clear: massive seeds require quasi-star envelopes that are initially very massive $\sim 10^7\ M_\odot$ or that accrete at very high rates.

4. Rotation

A quasi-star to exist must be supplied by accretion energy. The internal rotation of the envelope sets the boundary condition for the accreting gas: are those always consistent with the existence of an accretion disc? In [3] we provide a first answer to this question. We use the so called "thermal wind equation" (e.g. [18, 19]) to model the internal rotation of a steady state and convective star, ignoring magnetic fields and turbulences. We then follow [20] by adopting a closure relation that allows one to reproduce data of the Sun's convective zone (e.g. [21]). Given this tentative framework, the result is clear: the rotation of the gas close to the black hole's Bondi radius is highly sub-Keplerian for a large range of parameters. We then check for each model in the parameter space shown in Figures 1 and 2 whether the gas angular momentum would be enough for the gas to circularise outside the black hole's innermost stable circular orbit. Surprisingly, we find that models in the "growth region" are *not* consistent with accretion disc formation at the centre of the envelope, while the opposite is true for models in the evaporation strip! The result is very striking, but of course our analytical modelling of the internal rotation can only be indicative. I think it is fair to say that our work proves that considering rotation (a so far neglected ingredient in MS models) in quasi-star modelling is important.

5. Discussion and Conclusions

We may now amuse ourselves and step on to more speculative grounds, by boldly taking at face value the results of the previous section and by foreseeing their full impact for SMBH formation. The above statement "models in the "growth region" are *not* consistent with accretion disc formation" may evoke a scenario in which, after the core of the MS has collapsed into a black hole, the envelope feels its huge gravitational pull and it does not have enough rotation to stop its quasi-radial fall. In the growth region envelopes are quite massive, therefore their direct collapse into a black hole may produce very massive $> 10^5$ $M_\odot$ black hole seeds. This can only happen in rare dark matter haloes with masses between $10^8 - 10^{10} M_\odot$ at $z > 10$ (see figure 5 in [3]). Quasi-stars can instead form in the evaporation strip, but this implies no substantial growth for the embryos black hole, that remains of the (relatively small) size of the MS core (i.e. $\sim 100 M_\odot$). For a given black hole mass, the envelopes in the evaporation strip are lighter than those in the growth region, and they can form in less massive haloes. There may therefore be two populations of seeds from the direct collapse scenario which live in different halo populations: very massive seeds $\geq 10^5 M_\odot$ and stellar size $\sim 100 M_\odot$ seeds. The observational impact of this dychotomy would require implementing this formation scenario in a cosmological evolution framework, which will be the topic of a follow up work.

References

[1] Volonteri M 2010 A&A Rev. **18** 279–315 (*Preprint* 1003.4404)
[2] Fiacconi D and Rossi E M 2016 MNRAS **455** 2–16 (*Preprint* 1509.07511)
[3] Fiacconi D and Rossi E M 2017 MNRAS **464** 2259–2269 (*Preprint* 1604.03936)
[4] Begelman M C 2010 MNRAS **402** 673–681 (*Preprint* 0910.4398)
[5] Hosokawa T, Yorke H W, Inayoshi K, Omukai K and Yoshida N 2013 ApJ **778** 178 (*Preprint* 1308.4457)
[6] Schleicher D R G, Palla F, Ferrara A, Galli D and Latif M 2013 A&A **558** A59 (*Preprint* 1305.5923)
[7] Mayer L, Kazantzidis S, Escala A and Callegari S 2010 Nature **466** 1082–1084 (*Preprint* 0912.4262)
[8] Mayer L, Fiacconi D, Bonoli S, Quinn T, Roškar R, Shen S and Wadsley J 2015 ApJ **810** 51 (*Preprint* 1411.5683)
[9] Shibata M and Shapiro S L 2002 ApJ **572** L39–L43 (*Preprint* astro-ph/0205091)
[10] Saijo M and Hawke I 2009 Phys. Rev. D **80** 064001 (*Preprint* 0908.3002)
[11] Montero P J, Janka H T and Müller E 2012 ApJ **749** 37 (*Preprint* 1108.3090)
[12] Di Matteo T, Khandai N, DeGraf C, Feng Y, Croft R A C, Lopez J and Springel V 2012 ApJ **745** L29 (*Preprint* 1107.1253)
[13] Ferrara A, Salvadori S, Yue B and Schleicher D 2014 MNRAS **443** 2410–2425 (*Preprint* 1406.6685)
[14] Begelman M C, Rossi E M and Armitage P J 2008 MNRAS **387** 1649–1659 (*Preprint* 0711.4078)
[15] Ball W H, Tout C A, Żytkow A N and Eldridge J J 2011 MNRAS **414** 2751–2762 (*Preprint* 1102.5098)
[16] Ball W H, Tout C A and Żytkow A N 2012 MNRAS **421** 2713–2721 (*Preprint* 1201.5560)
[17] Dotan C, Rossi E M and Shaviv N J 2011 MNRAS **417** 3035–3046 (*Preprint* 1107.3562)
[18] Kitchatinov L L and Ruediger G 1995 A&A **299** 446
[19] Thompson M J, Christensen-Dalsgaard J, Miesch M S and Toomre J 2003 ARA&A **41** 599–643
[20] Balbus S A, Bonart J, Latter H N and Weiss N O 2009 MNRAS **400** 176–182 (*Preprint* 0907.5075)
[21] Balbus S A and Latter H N 2010 MNRAS **407** 2565–2574 (*Preprint* 1005.5100)

IOP Conf. Series: Journal of Physics: Conf. Series **840** (2017) 012028 doi:10.1088/1742-6596/840/1/012028

MICROSCOPE: five months after launch

Joel Bergé[1], Pierre Touboul[1], Manuel Rodrigues[1], Françoise Liorzou[1]

[1] ONERA, the French Aerospace Lab, 29 avenue de la Division Leclerc, 92320 Châtillon, France

E-mail: `joel.berge@onera.fr`

Abstract. MICROSCOPE is a French Space Agency mission that aims to test the Weak Equivalence Principle in space down to an accuracy of 10^{-15}, two orders of magnitude better than the current constraints. The MICROSCOPE satellite was launched on April 25, 2016. We describe the MICROSCOPE mission, its measurement principle and instrument, and report on the status of its in-orbit commissioning as of September 2016.

1. Introduction

From the deviation of Mercury's perihelion to gravitational redshift to gravitational waves to the existence of black holes, General Relativity (GR) continues to pass experimental tests one century after its inception. The recent direct detection of gravitational waves by LIGO [1, 2] not only superbly matches expectations from GR, but it also directly proves the existence of back holes. So many successes would tend to consider GR as the correct theory for gravitation. However, fundamental physics has faced major challenges for decades: how to unify the (as successful) quantum standard model of particle physics with GR, and how to explain the acceleration of the cosmological expansion?

To answer those questions, we may need to look deeply in the foundations of GR themselves. A particularly important test is to look for a violation of the Weak Equivalence Principle (WEP), a cornerstone of GR. MICROSCOPE (Micro-Satellite à traînée Compensée pour l'Observation du Principe d'Equivalence [3, 4, 5]) is a drag-free microsatellite which aims to test the WEP down to the 10^{-15} level, and was launched on April 25, 2016. In this paper, we first briefly introduce the WEP in Sect. 2; we then give a short overview of MICROSCOPE in Sect. 3 before stating on its status (as of September 2016) in Sect. 4. We then briefly report on the ground segment status in Sect. 5 before concluding.

2. Weak Equivalence Principle

2.1. Weak Equivalence Principle and General Relativity

The Weak Equivalence Principle (aka universality of free fall) states that two bodies in the same gravitational field experience the same acceleration, independently of their mass and composition. Together with the local position invariance (laws of physics do not depend on the position) and the local Lorentz invariance (laws of physics do not depend on the speed of the observer in an inertial frame), this principle constitutes the Einstein Equivalence Principle (EEP), which is at the basis of Einstein's General Relativity. Any violation of the WEP (and of the EEP thereof), would indicate that GR is not the ultimate theory of gravity.

2.2. Tests of the Weak Equivalence Principle

The WEP has been tested throughout the 20th century with an increasing precision (see [6] for a review). Deviations from WEP are usually described by the Eötvös parameter

$$\eta = 2\frac{(m_g/m_i)_A - (m_g/m_i)_B}{(m_g/m_i)_A + (m_g/m_i)_B} \tag{1}$$

where "A" and "B" are two bodies experiencing the same gravitational field, m_g is the gravitational mass and m_i the inertial mass. If the WEP holds, then $m_g = m_i$ for all bodies, and $\eta = 0$. Figure 1 shows how the upper limit on η has decreased during the 20th century, as more accurate experiments have been put forth. The first measurement was made by Eötvös with a torsion pendulum, and allowed him to constrain the WEP at a level of 10^{-8}. Most recently, the monitoring of the Earth and Moon system with the Lunar Laser Range (LLR – e.g. [7, 8]), and the measurement by the Eöt-Wash group with a torsion pendulum, reached the best limits $\eta \leqslant 10^{-13}$ [9]. On-ground measurement are however reaching their limits in terms of signal-to-noise ratio, making a better measurement more difficult. As a consequence, efforts are underway to perform WEP tests with atomic interferometers (e.g. [10]). Another solution to increase the precision on the WEP is to test it in space: this is the goal of the MICROSCOPE mission. Theories currently developed to explain the accelerated expansion of the Universe, or to unify GR and quantum physics, predict that WEP is violated at a level $10^{-18} \leqslant \eta \leqslant 10^{-13}$ [11, 12]. MICROSCOPE is expected to measure η at the level 10^{-15}: it will thus allow us to probe a significant part of the parameter space under consideration by those new theories, and to start discriminating against theories.

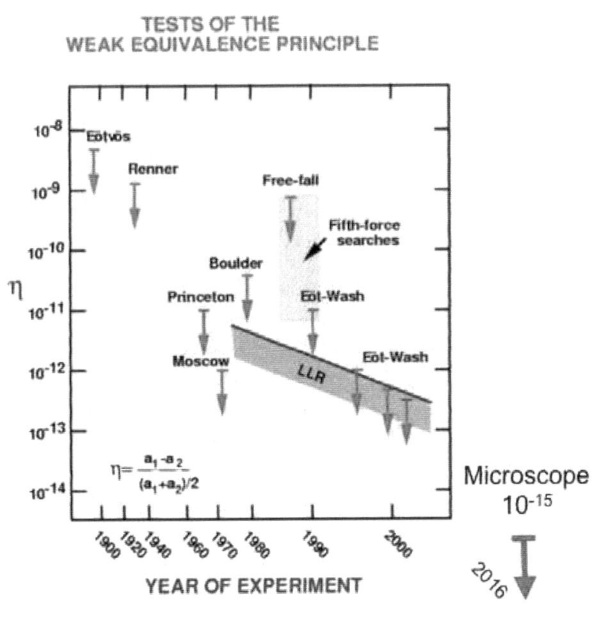

Figure 1. Tests of WEP throughout the 20th century. The arrow on the lower right corner shows the expectation for MICROSCOPE. Figure adapted from [6].

3. MICROSCOPE

MICROSCOPE, launched from Kourou on April 25, 2016, tests the WEP by comparing the acceleration experienced by two free-falling test masses in the Earth's gravity field. To this aim, it embarks two ultrasensitive electrostatic differential accelerometers, each of them consisting of two coaxial cylindrical test masses whose motion is electrostatically constrained.

In one (reference – 'REF') accelerometer, the test masses are made of the same material to demonstrate the experiment's accuracy; they are made of different materials in the second ('EP') accelerometer, which is used to test the WEP. The difference of electric potentials applied to keep the masses in equilibrium is a measure of the difference in the proof masses motion; hence, a non-zero difference of applied potentials is a measure of a WEP violation.

Figure 2 shows the measurement principle for the EP accelerometer. At leading order, its test masses experience the same gravitational field (red arrows) because their centers of mass are very close to each other (of the order of micrometers). In reality, the fact that the centers of mass are not exactly the same lead to a nuisance signal due to the gradient of the Earth's gravity field (see Eq. 2 below), that can easily be corrected for. If the WEP is violated, then the difference in accelerations along the EP test axis (horizontal black arrows, along which the test is performed) will be modulated by the instrument's motion around the Earth. We then expect to detect a sine wave corresponding to the modulation of the difference of the voltages applied in the two test-mass electrostatic configuration to keep them centered. Depending on the spacecraft's spin about the axis normal to the orbital plane (either null for an inertial session as depicted by the figure, or non-null for a spined session), the WEP violation signal will have a typical, expected frequency.

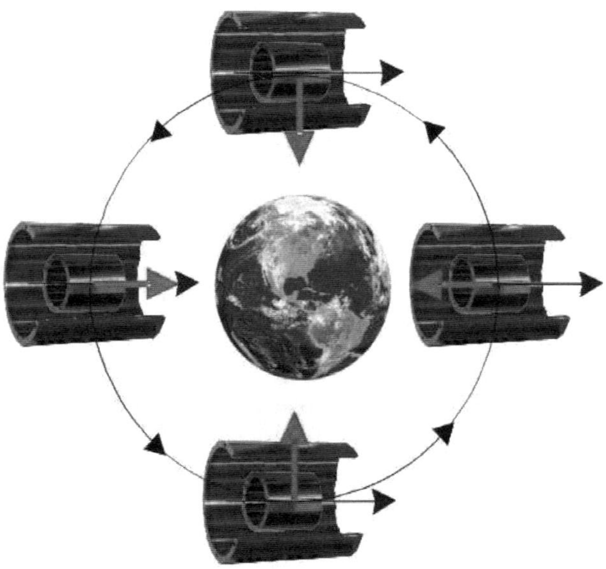

Figure 2. MICROSCOPE's measurement principle.

The difference of acceleration deduced by the measured difference in applied voltages can be written as:

$$\overrightarrow{\Gamma}_{\text{meas,d}} = [\mathcal{M}_c]\left(\eta\overrightarrow{g} + ([\mathcal{T}] - [\mathcal{I}n])\overrightarrow{\Delta} - 2[\Omega]\dot{\overrightarrow{\Delta}} - \ddot{\overrightarrow{\Delta}}\right) + \overrightarrow{K_{0,d}} + [\mathcal{M}_d]\overrightarrow{\Gamma}_{\text{App,c}} + \overrightarrow{\Gamma}_{\text{measquad,d}} + \overrightarrow{\Gamma}_{\text{n,d}} + [\mathcal{C}_d]\dot{\overrightarrow{\Omega}} \quad (2)$$

where η is the EP violation signal we are looking for; $\overrightarrow{g}$ is the Earth gravity field at the spacecraft's center-of-mass; $\overrightarrow{\Delta} = \overrightarrow{O_iO_j}$ is the distance between the center of the differential accelerometer's test masses; $\overrightarrow{K}_{0,d}$ is the accelerometer's bias; $[\mathcal{M}_c]$ is the common mode sensitivity matrix; $[\mathcal{T}]$ is the Earth gravity gradient tensor; $[\mathcal{I}n]$ is the inertial acceleration tensor, $[\Omega]\dot{\overrightarrow{\Delta}}$ is the Coriolis acceleration, with $[\Omega]$ the spacecraft's angular velocity matrix;

$[\mathcal{M}_d]$ is the differential mode sensitivity matrix; $\vec{\Gamma}_{\mathrm{App,c}}$ is the common mode acceleration and includes non-gravitational external accelerations; $\vec{\Gamma}_{\mathrm{measquad,d}}$ is the quadratic residuals; $\vec{\Gamma}_{\mathrm{n,d}}$ is the instrument's noise; and $[\mathcal{C}_d]$ is the differential angular to linear coupling matrix. All nuisance parameters are either corrected for through careful modeling (e.g. gravity gradient tensor) or calibrated in flight (e.g. bias [13, 14]), or minimized by design of the instrument and of the satellite (e.g. inertial tensor, instrument's noise).

MICROSCOPE's instrument (T-SAGE – Twin Space Accelerometer for Gravitation Experiment) and its performance have been described elsewhere (e.g. [3, 4]). As aforementioned, the instrument's mechanical core consists of two differential accelerometers (Sensor Units – SU), whose test masses are co-axial cylinders kept in equilibrium with electrostatic actuation. The test masses' materials were chosen carefully so as to maximize the scientific return of the experiment and to optimize their industrial machining: the EP test masses are made of alloys of Platinum-Rhodium (PtRh10 – 90% Pt, 10% Rh) and Titanium-Aluminium-Vanadium (TA6V – 90% Ti, 6% Al, 4% V), while the REF test masses are made of the same PtRh10 alloy. For each SU, the test masses are controlled electrostatically, through electrodes, without any mechanical contact; only a thin 7 μm-diameter gold wire, used to fix the masses' electrical potential to the electronics reference voltage, provides a mechanical contact (and associated, accounted for, damping noise) between the test masses and their cage. The test masses' control is performed by an electronic servo-loop. Two Front End Electronics Unit (FEEU) boxes (one per SU) include the capacitive sensing of masses, the reference voltage sources and the analog electronics to generate the electrical voltages applied to the electrodes; an Interface Control Unit (ICU) includes the digital electronics associated with the servo-loop digital control laws, as well as the interfaces to the satellite's data bus. Additionally, the same electronics' output is used by the drag-free system of the satellite.

Performance analyses predict a noise Amplitude Spectral Density of 10^{-12} m/s^2/Hz$^{1/2}$ in the frequency band $10^{-3} - 0.03$ Hz, compatible with a test of the WEP at a 10^{-15} accuracy (see [3] for a detailed uncertainty analysis).

The spacecraft is derived from the CNES's Myriad series of microsatellites. With a mass of 325 kg and dimensions of 1380x1040x1580 mm^3, it has been designed to be as symmetric as possible, with the T-SAGE instrument sitting at its center-of-mass so that the self-gravity is minimized. No moving mechanical parts can contaminate the Equivalence Principle measurement. Its propulsion system, based on cold gas, is derived from the GAIA mission and supplied by ESA. The MICROSCOPE mission uses a Sun-synchronous, very low eccentricity (1.5×10^{-3}), 710 km orbit.

4. MICROSCOPE's first few months in orbit: commissioning phase status

From early May 2016 to August 2016, the satellite was partially in eclipse. The periodic sun-shade transitions prevented a fine thermal control of the payload, and therefore a robust WEP test. This time was hence dedicated to the satellite's and instrument's commissioning phase. All the modes of the satellite operation were successfully verified. This section presents the status of the satellite and of the mission as of September 2016.

4.1. First in-orbit instrument health checks

The T-SAGE payload was first switched on on May 2nd 2016, one week after launch, after CNES checked orbital parameters and the main satellite's operational functions. The first operation consisted in a health verification in order to assess the integrity of the instrument, while it was still in a pre-launch configuration (i.e. with the test masses blocked in their electrostatic cage). The main functionalities verification were obtained through dedicated housekeeping data as well as through the science data themselves contained in the telemetry:

IOP Conf. Series: Journal of Physics: Conf. Series **840** (2017) 012028 doi:10.1088/1742-6596/840/1/012028

- at SU level: we checked that the gold wires survived the launch and did not break. The four of them were successfully detected through a specific telecommand. We also checked that all four masses were still blocked.
- at FEEU level: the "as expected" test masses position measurement validated the electronics, the SU/FEEU harness and the SU core integrity.
- at ICU level: we checked that the digital control loop through the open loop response, the data production, the command control, and the power supply were all nominal.

These operations allowed us to confirm the integrity of the instrument, and therefore to go ahead with the mission.

4.2. Test masses release and first control

The early operations' most critical step was the release of the test masses. Its failure would have directly translated into the failure of the mission. Until then, test masses were safely kept motionless by a dedicated mechanical blocking system to prevent any harm during launch. The cylindrical masses were maintained on their top and bottom faces by finger-shaped stops controlled by pressured force actuators. Their release was performed by delivering a specific voltage from the power bus of the satellite, allowing the perforation of the lower stops' valve membrane, therefore allowing the lower stops to get retracted. The instrument was then in its robust mode (also called "acquisition" mode) that delivers sufficient forces to displace the masses. Then, as soon as the blocking fingers were released, the test masses were automatically servo-controlled ("acquired") at the centre of the electrode set. Fig. 3 shows this first release and control of the internal test mass of the EP sensor. Within about 10 seconds and a few oscillations, the mass was brought from its blocked position (75 μm on the sensitive X-axis –blue curve) to the center of the electrostatic cage, where it became controlled electrostically. The four masses were successfully released and electrostatically levitated.

Since then, MICROSCOPE has been providing acceleration and position data for the four test masses' six degrees of freedom. Those data are used for the science output of the mission. The instrument is working and ready to test the Equivalence principle.

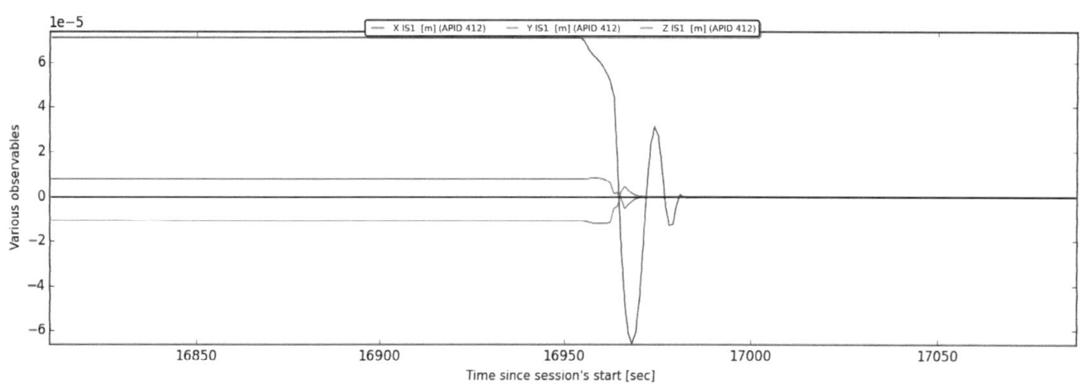

Figure 3. Release of the internal test mass of the EP sensor. The figure shows the position of the mass along its three linear axes (blue: X, green: Y, red: Z), from being blocked to being electrostatically controlled.

4.3. First switch on of the drag-free system

We checked that the simultaneous and continuous drag-free operation on six degrees of freedom is operating correctly. As the drag-free is controlled using a combination of the data provided by the instrument itself and of the satellite's star-tracker, it is a good indication that the instrument works as expected.

Fig. 4 provides 24 hours of data record (from midnight to midnight) during this technical session when the drag free system has been first started. The upper panel shows the acceleration measured along the X-axis, and the lower panel shows the accelerations measured along the Y-axis (blue) and Z-axis (green), as a function of time since midnight. The drag-free control is turned on around t=25000 seconds. Before that, the X-axis acceleration oscillates at the orbital frequency: this oscillation is due to the projection of the atmospheric drag on the X-axis (which lies in the orbital plane). The same behavior can be noticed for the Z-axis (which lies in the orbital plane too, and is therefore also subject to the drag). The Y-axis acceleration does not show any sinusoidal variation (lower panel, blue curve): because the Y-axis is normal to the orbital plane, the acceleration along it is insensitive to the drag. However, we can notice regular dips in the acceleration; they correspond to the satellite passing through the shade (these data have been taking during the eclipse season), when the solar radiation pressure (SRP) disappears. This shows that we can easily measure the SRP. After the drag-free is turned on, those regular patterns disappear, proving that the drag-compensation operates efficiently.

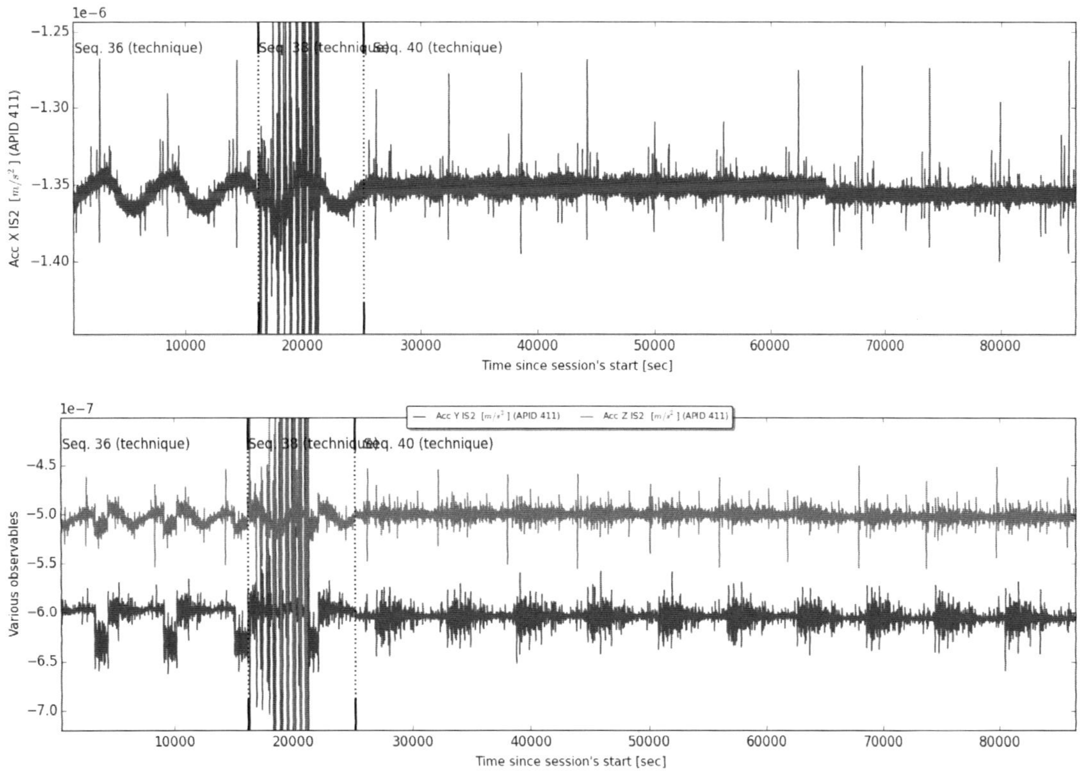

Figure 4. Measured linear accelerations along the X-axis (upper panel) and the Y- and Z-axes (lower panel), around the time the drag compensation was first turned on.

To conclude this section, we can note the periodic pattern of acceleration "bursts" (Fig. 4). Those spikes are due to crackles of the satellite multi-layer insulator, that reacts to the temperature changes as the satellite goes through the shade area once per orbit, then relaxes. As those spikes may contaminate the WEP test, we developed techniques to correct for them; we discuss them in the next section.

5. Ground Segment, data processing and data analysis status

5.1. Ground segment presentation and status

MICROSCOPE's ground segment operations are centered around three entities: the Command Control Center (CCC) and the Center of Expertise for Compensation of Drag (CECT), hosted by CNES-Toulouse, are in charge of the satellite's operations and drag-free and attitude control respectively, and the Science Mission Center (CMSM), hosted by ONERA-Palaiseau, is in charge of the monitoring and operations of MICROSCOPE's instrument, and of the data processing and analysis. In particular, the CMSM is responsible for (1) ensuring all operational functions to maximize the instrument's operation, (2) day-to-day instrument management and monitoring, (3) weekly mission performance check, (4) proposing modifications of the mission scenario to the Science Working Group, and (5) releasing and archiving the data.

Compatibility tests between CNES and ONERA have been performed in 2015, showing that the ground segment operation were ready. We could check this fact in real life operations since the MICROSCOPE launch; no problem has been found. Data processing and analysis softwares at CMSM are now routinely used to process and monitor the data on a daily basis. We are therefore confident that we can efficiently and immediately analyze the science data.

5.2. Mission scenario

From April 2016 to August 2016, the satellite and all subsystems (in orbit or on ground) have been tested. This acceptance phase was followed by a science acceptance phase until November 2016. During this last phase, calibration and WEP test sessions have been performed with their nominal duration, respectively 5 to 10 orbits and 120 orbits. This phase aims at performing a first guess on performance and at optimizing the procedures, the session's parameters and the ground segment data processes.

For instance, the matching of scale factors for the X axes was performed with less than 2×10^{-4} accuracy. To calibrate this parameter, the output of the drag-free accelerometer is biased by a sine signal at a very well-known frequency. Then the drag-free control system applies a thrust to compensate this stimulus. The accelerometer out of the drag-free loop fills then the sine thrust: the comparison with the initial stimulus at the known frequency gives the ration between the scale factors of the two accelerometers. Thus any common acceleration on the satellite is rejected by this 2×10^{-4} factor once the scale factors are corrected. It is worth noting that the difference of scale factors between the inner and outer tests masses was less than 1.5%, better than the 5% expected. Roughly the same result was obtained on the two SU.

The following step of the mission scenario is quite simple. From November 2016 to May 2017, the next Eclipse period of the year, science sessions should be cumulated in order to reach the best performance thanks to noise reduction with time integration.

5.3. Exorcising the curse of missing data

As mentioned in Sect. 3, a WEP violation signal can be detected as a sine wave corresponding to the modulation of the difference in accelerations experienced by the instrument's test-masses as the satellite orbits the Earth. The main task of the data analysis is therefore to extract this sine wave from the instrumental noise and to estimate its amplitude. Of foremost importance, we expect a small amount of data to be lost (e.g. because of short mechanical crackles –see Sect. 4); as a consequence, the noise will leak from the high frequencies (where its Power Spectral Density –PSD– increases) to the lower frequencies where the WEP violation signal is measured, therefore hampering the signal detection and estimation.

We investigated two ways to deal with missing data when estimating deterministic parameters: (1) accounting for them at the parameter estimation level and (2) fill in missing data before using ordinary parameter estimation techniques.

In Baghi et al. [15], we proposed the KARMA (Kalman-AR Model Analysis) method, a general linear regression method able to deal with incomplete data affected by unknown colored noise. It is based on an autoregressive (AR) fit of the noise that is used to whiten the data through a Kalman filtering process. Hence, the algorithm constructs a good approximation of the best linear unbiased estimator, conditionally to the AR model. This technique accounts for missing data without filling gaps, contrary to those that we present next. However, it does not allow us to visually correct the PSD for the leakage, making the assessment of the noise PSD difficult.

The Modified-Expectation-Condition-Maximization (M-ECM [16]) allows us to use KARMA results to fill in the missing data, and therefore reconstruct the full PSD. In this method, the missing data are efficiently estimated by their conditional expectation as in universal Kriging, based on the circulant approximation of the complete data covariance. After initialization with an autoregessive fit of the noise, a few iterations of estimation/reconstruction steps are performed until convergence of the regression and PSD estimates, in a way similar to the expectation-conditional-maximization algorithm. The estimation can be performed for an arbitrary PSD provided that it is sufficiently smooth.

Another possibility to fill in gaps is to use the inpainting algorithm [17, 18], which reconstructs missing data with a basis functions dictionary using a sparsity prior. We showed that inpainting allows us to recover the noise PSD and to use an ordinary least square method to look for a WEP violation signal.

Conclusion

MICROSCOPE aims to test the Weak Equivalence Principle in space down to an accuracy of 10^{-15}. This is two orders of magnitude better than the current constraints, and will allow us to watch for physics beyond General Relativity. Beside this science goal, MICROSCOPE is fulfilling a technology objective by showing that the technology is ready for extremely fine satellite attitude control and precise drag-free system. This will be of interest for future ambitious missions like eLISA. As of November 2016, the commissioning phase of MICROSCOPE has shown us that MICROSCOPE is on good tracks to complete its objectives.

Acknowledgments

The authors are grateful to the MICROSCOPE mission staff of CNES in Toulouse, the Observatoire de la Côte d'Azur in Grasse and ONERA in Châtillon, for all exchanges about the mission, the satellite, the payload and the science mission center. This work makes use of technical data from the CNES-ESA-ONERA-CNRS-OCA Microscope mission, and has received financial support from ONERA and CNES. JB acknowledges travel financial support from the LISA Symposium organizers, as well as from the UnivEarthS Labex program at Sorbonne Paris Cité (ANR-10-LABX-0023 and ANR-11- IDEX-0005-02)

References

[1] Abbott, B. P., Abbott, R., Abbott, T. D., et al. 2016, Physical Review Letters, 116, 061102
[2] Abbott, B. P., Abbott, R., Abbott, T. D., et al. 2016, Physical Review Letters, 116, 241103
[3] Touboul P. 2009, Space Sci. Rev., 148, 455
[4] Touboul P., Métris G., Lebat V., Robert A. 2012, Classical and Quantum Gravity, 29, 184010
[5] Bergé, J., Touboul, P., & Rodrigues, M., Journal of Physics Conference Series, 610, 012009
[6] Will, C. M. 2006, Living Reviews in Relativity, 9, 3
[7] Murphy, T. W., Jr., Adelberger, E. G., Battat, J. B. R., et al. 2012, Classical and Quantum Gravity, 29, 184005

[8] Williams, J. G., Turyshev, S. G., & Boggs, D. H. 2012, Classical and Quantum Gravity, 29, 184004
[9] Wagner, T. A., Schlamminger, S., Gundlach, J. H., & Adelberger, E. G. 2012, Classical and Quantum Gravity, 29, 184002
[10] Bonnin, A., Zahzam, N., Bidel, Y., & Bresson, A. 2013, Phys. Rev. A, 88, 043615
[11] Damour, T. 2012, Classical and Quantum Gravity, 29, 184001
[12] Overduin, J., Everitt, F., Worden, P., & Mester, J. 2012, Classical and Quantum Gravity, 29, 184012
[13] Hardy, É., Levy, A., Métris, G., Rodrigues, M., & Touboul, P. 2013, Space Sci. Rev., 180, 177
[14] Hardy, É., Levy, A., Rodrigues, M., Touboul, P., & Métris, G. 2013, Advances in Space Research, 52, 1634
[15] Baghi, Q., Métris, G., Bergé, J., et al. 2015, Physical Review D, 91, 062003
[16] Baghi, Q., Métris, G., Bergé, J., et al. 2016, Physical Review D, 93, 122007
[17] Bergé, J., Pires, S., Baghi, Q., Touboul, P., & Métris, G. 2015, Physical Review D, 92, 112006
[18] Pires, S., Bergé, J., Baghi, Q., Touboul, P., & Métris, G. 2016, Physical Review D, 94, 123015

IOP Conf. Series: Journal of Physics: Conf. Series **840** (2017) 012029 doi:10.1088/1742-6596/840/1/012029

Late time cosmology with LISA: Probing the cosmic expansion with massive black hole binary mergers as standard sirens

Nicola Tamanini

Institut de Physique Théorique, CEA-Saclay, CNRS UMR 3681, Université Paris-Saclay, F-91191 Gif-sur-Yvette, France

E-mail: `nicola.tamanini@cea.fr`

Abstract. This paper summarises the potential of the LISA mission to constrain the expansion history of the universe using massive black hole binary mergers as gravitational wave standard sirens. After briefly reviewing the concept of standard siren, the analysis and methodologies of Ref. [1] are briefly outlined to show how LISA can be used as a cosmological probe, while a selection of results taken from Refs. [1, 2] is presented in order to estimate the power of LISA in constraining cosmological parameters.

1. Introduction

The LISA mission [3] aims at measuring gravitational waves (GWs) in the frequency band around the mHz. Among the richness of astrophysical sources expected to produce a detectable GW signal at those frequencies, there are massive black hole binaries (MBHBs) from 10^4 to 10^7 solar masses. The inspiral, merger and ringdown of several MBHBs will be observed by LISA with an high signal to noise ratio (SNR), allowing for an accurate estimation of the parameters of the binary. The information gathered from these GW sources will lead to advances not only in astrophysics, but also in cosmology. In fact MBHBs can be used as reliable distance indicators since the luminosity distance to the source is one of the parameters entering the measured waveform. The joint observation of a GW signal and an electromagnetic (EM) counterpart, from which the redshift of the source can be extracted, will thus allow to test the cosmic expansion through the distance-redshift relation, in a similar fashion to type-Ia supernovae (SNIa) analyses. In what follows, after briefly reviewing how a compact binary system can be used to constrain the cosmic evolution, the analysis and techniques developed in [1] will be outlined and a selection of results from [1, 2] regarding forecasts for LISA, will be presented.

2. The concept of standard siren

At the lowest (Newtonian) order, the waveform produced by a binary astrophysical system and observed by an interferometric detector on Earth or in space, can be parametrized as (see e.g. [4])

$$h_\times = \frac{4}{d_L(z)} \left(\frac{G\mathcal{M}_c(z)}{c^2}\right)^{5/3} \left(\frac{\pi f}{c}\right)^{2/3} \cos\iota \sin[\Phi(f)]\,, \tag{1}$$

with a similar expression holding for the h_+ polarization. Here $\mathcal{M}_c(z)$ is the chirp mass of the binary, $d_L(z)$ is the luminosity distance, f is the GW frequency at the observer, ι is an angle characterizing the orientation of the binary orbital plane and $\Phi(f)$ is the phase of the GW. As pointed out for the first time by Schutz [5], the parameter estimation of the observed waveform (1) will directly yield the value of the luminosity distance d_L of the source, implying that binaries emitting GWs can be used as cosmological distance indicators. For this reason GW sources of this kind are called *standard sirens* [6], in analogy with SNIa which are called standard candles. If besides the value of d_L, obtained from the analysis of the detected GW signal, also the redshift of the GW source (or the one of its hosting galaxy) is measured, then one obtains a data point in the distance-redshift diagram, where the *distance-redshift* relation, defined as

$$d_L(z) = \frac{c}{H_0}\frac{1+z}{\sqrt{\Omega_k}}\sinh\left[\sqrt{\Omega_k}\int_0^z \frac{H_0}{H(z')}dz'\right] , \tag{2}$$

can be fitted against the observational data. Here Ω_k is the relative energy density of spacetime curvature and a FRW universe has been assumed. $H(z)$ is the Hubble rate whose redshift evolution depends on the specific cosmological model to be tested.

Thanks to the well-known theoretical understanding of binary inspirals, standard sirens have the advantage over SNIa to be free of unwanted systematic errors and calibration procedures, meaning that they provide a direct measure of the luminosity distance. Unfortunately determining the redshift of the GW source might be complicated. The easiest way to measure it is through the observation of an EM counterpart, which however is not guaranteed to be detected or even to be produced. Creation of an EM counterpart depends on the type of binary considered and on the environment in which the merger occurs. The current astrophysical models contain many uncertainties due to the lack of enough experimental data. Moreover a successful detection of an EM counterpart depends also on the sky angular resolution obtained by the GW detector and by the specifics of EM telescopes. Ideally what is needed is a network of GW detectors providing an accurate sky localization to several wide field telescopes able to observe possible EM transients. In general the situation with GWs is the opposite of the one experienced with EM waves: with standard sirens it is easy to measure d_L (directly from the waveform) but it is difficult to measure z (need an EM counterpart), with standard candles it is easy to measure redshifts (comparing EM spectra) but it is difficult to measure the luminosity distance (need objects of known absolute luminosity).

3. Using MBHB mergers as standard sirens with LISA

Any GW signal detected by LISA and produced by a binary astrophysical system, could in principle be used as standard siren. The expected list includes MBHBs from 10^4 to $10^7 M_\odot$, stellar BH binaries from 10 to 100 $M_\odot$ and extreme mass ratio inspirals (EMRIs). Nevertheless only MBHBs are expected to produce a detectable EM counterpart since they are supposed to merge in a gas rich environment and within the LISA frequency band. They are moreover characterized by a high LISA SNR and abundant at very high redshifts, up to $z \sim 15$, making them ideal standard sirens to probe the cosmic expansion at distances SNIa cannot reach. The following analysis will thus be focused on MBHBs, and it will be aimed at characterizing the number of GW detections with LISA and the fraction of sources for which a EM counterpart will likely be observed by future telescope facilities. Once these results are obtained, then constraints on cosmological parameters will be forecast. The procedure employed in [1] to simulate MBHB merger rates, compute the GW signal detected by LISA and model and observe EM counterparts, will be briefly summarised here. The reader interested in the details can find them in [1].

First of all, one needs to model realistically the expected MBHB sources. As a starting point the results of semi-analytical simulations of the evolution of the BH masses and spins during the hierarchical galaxy formation and evolution, are used in analogy with the analysis of [7].

This allows one to predict the rate and redshift distribution of MBHB merger events. Several variants of the semi-analytical model are produced by considering competing scenarios for the initial conditions of the massive BH population at high redshift – namely, a "light-seed" scenario in which the first massive BHs form from the remnants of population III (popIII) stars, and a "heavy-seed" one where massive BHs form from the collapse of protogalactic disks – and for the delays with which massive BHs merge after their host galaxies coalesce (Q3d and Q3nod). The simulations produce synthetic catalogues of MBHB merger events, including all information about the MBHBs and their host galaxies. The parameters of the MBHB systems of each catalogue are then inserted as input into a code that simulates the GW signal induced in LISA by the binaries' inspirals. Considering all phases of the observed signal (inspiral, merger and ringdown), the code computes the SNR of each merger event and the Fisher matrix of the corresponding waveform parameters, which includes in particular the 1σ error on the luminosity distance Δd_L, and on the sky location $\Delta\Omega$. Among the MBHB merger events, those that have SNR > 8 and $\Delta\Omega < 10\,\mathrm{deg}^2$ are selected. This guarantees that the events are measured with a directional uncertainty sufficiently small to allow EM telescopes to detect a counterpart, if it is present.

	SNR > 8	$\Delta\Omega <$ $10\,\mathrm{deg}^2$	EM count.	%
	360	35.4	28.2	7.83
N2A2M5L6	41.1	28.9	26.4	64.2
	610	214.	40.5	6.64
	683	75.2	45.8	6.71
N2A5M5L6	41.1	35.3	31.2	75.9
	611	385.	50.0	8.18

Table 1. *Average values (5 years) of counterpart detections for both N2A2M5L6 and N2A5M5L6 and all three MBHB formation models: popIII, Q3d and Q3nod (respectively from top to bottom in each cell). From left to right the table shows: the LISA detections, the LISA detections with sky location error below 10 deg^2, the optical counterparts (observed either with LSST or with SKA+ELT) and their percent fraction with respect to the total number of detection (first column). Results taken from Ref. [1].*

The following step is to select, among these events, those that are likely to provide a detectable optical counterpart. Several investigations (e.g. [8, 9, 10]) suggest that optical/radio emission is likely to happen at MBHB mergers in a transient fashion. In light of these results the counterpart generated by each MBHBs merger is modelled taking into account all the information on the binary and the hosting galaxy. Generically a quasar-like luminosity flare in the optical and magnetic field induced flare and jet in the radio, are produced at mergers. In order to be able to detect these associated EM emissions, ideally optical/radio telescopes should be pointed in the direction of the event prior to merger. Thus as soon as the event has been localised in the sky with the required precision, telescopes are alerted and pointed in that direction, looking for a distinctive flare occurring at merger. In the analysis presented here the specific of future realistic telescopes, such as LSST, SKA and ELT [11], are assumed to observe the counterpart and determine the redshift of the GW source; see [1] for details.

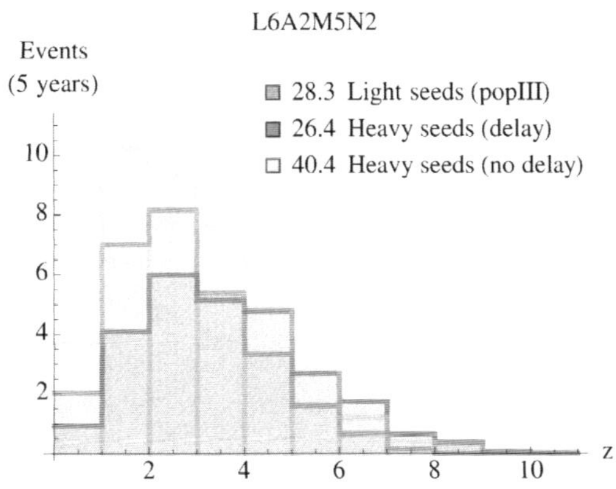

Figure 1. *Redshift distribution of MBHB standard sirens for N2A2M5L6. From Ref. [2].*

The LISA design considered in this analysis is labelled with N2A2M5L6 and defined by: armlength of 2 million km, number of active laser links (arms) fixed to six (three), mission duration of five years and low frequency noise level assumed to be the one expected to be tested by LISA pathfinder (N2 in [1, 7]). This choice is made in agreement with the first results obtained by the LISA pathfinder mission [12] and considering the actual LISA configuration that will be proposed at the L3 ESA call for mission, which has similar specifics: the armlength is set to 2.5 million km and the (nominal) mission duration is four years. Note that, according to [1, 2], the constraints on cosmological parameters forecast with N2A2M5L6 are basically equivalent to the ones obtained with N2A5M5L6 (same design with 5 million km). In what follows, unless otherwise specified, all results will refer to N2A2M5L6.

4. Forecast cosmological constraints

The number of MBHB mergers detected by LISA (with an SNR > 8), the ones with a sufficiently accurate sky localization ($\Delta\Omega < 10\,\mathrm{deg}^2$) and the ones for which an EM counterpart is observed, are exposed in Table 1. In the last column the fractional percentage between the number of standard sirens and the number of MBHB detected by LISA, is reported. The expected number of MBHB standard sirens is roughly 30 for N2A2M5L6 over the whole five year mission (Table 1 shows also numbers for N2A5M5L6 for comparison). These numbers are expected to scale linearly if a different mission duration is considered. The standard sirens distribution in redshift for N2A2M5L6 is given in Fig. 1. One can immediately notice that the bulk of data will appear in the range $1 < z < 4$, with a peak at $z \simeq 2$ and tails extending up to $z \simeq 8$. This implies that LISA will be able to directly probe the expansion of the universe at redshifts not explored by SNIa standard candles which are limited at $z \lesssim 2$.

	$\Delta\Omega_M$	$\Delta\Omega_\Lambda$	Δh
ΛCDM	0.0333	0.0333	0.0141
	0.0378	0.0378	0.0204
	0.0207	0.0207	0.0087
k-ΛCDM	0.0566	0.194	0.0370
	0.0954	0.234	0.0581
	0.0287	0.105	0.0179
	Δw_0	Δw_a	
DDE	0.173	0.935	
	0.247	1.17	
	0.108	0.582	

Table 2. *1σ errors on cosmological parameters for N2A2M5L6 and all three MBHB models: popIII, Q3d and Q3nod (respectively from top to bottom in each cell).*

At this point one can fit these standard sirens data with any cosmological model of interest in order to derive constraints on free parameters. Here three standard cosmological models will be first considered: **ΛCDM**: the standard concordance model with cold dark matter and a cosmological constant (free parameters Ω_M, h); ***k*-ΛCDM**: the standard concordance model plus curvature (free parameters $\Omega_M, h, \Omega_\Lambda$); **DDE**: a dynamical DE model where the DE equation

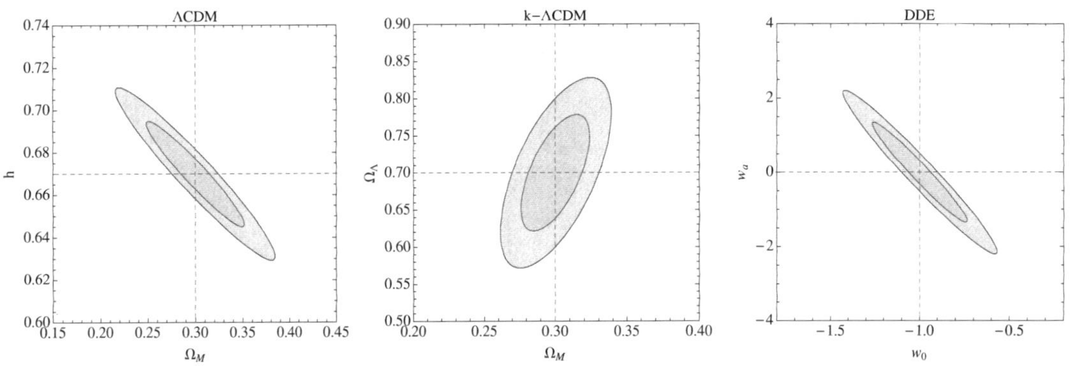

Figure 2. *1 (red) and 2σ (blue) contour plots in the parameter spaces of ΛCDM, k-ΛCDM (with h fixed to its fiducial value) and DDE, respectively from left to right. Here only results for N2A2M5L6 with the popIII MBHB model are shown.*

of state is $w(z) = w_0 + w_a z/(z+1)$ and all ΛCDM parameters are fixed (free parameters w_0, w_a). Fiducial values for these parameters are: $\Omega_M = 0.3$, $\Omega_\Lambda = 0.7$, $h = 0.67$, $w_0 = -1$ and $w_a = 0$. Each of these models gives rise to a different Hubble rate function $H(z)$ determined by the free parameters of the model. Once this function is inserted into (2), then the theoretical distance-redshift relation can be fitted against the standard sirens data and constraints on the cosmological parameters can be obtained statistically. The forecast constraints for N2A2M5L6 derived in [1], using Fisher matrix techniques, are reported in Table 2. In Fig. 2 one and two sigma parameter regions constrained by LISA (N2A2M5L6) for the three cosmological models are shown (for the middle panel, k-ΛCDM model, h has been fixed to its fiducial value). One can note the interesting constraint recovered on H_0 for ΛCDM, which is roughly at the 2% level. This might help in solving the current tension between local and CMB measurements of H_0, or at least in providing an independent check.

One can now compare the LISA forecasts with the results obtained by present cosmological probes such as SNIa analyses [13] and CMB experiments [14, 15]. For ΛCDM the constraint on H_0 obtained by Planck [14] is only a factor of about two better than the ones forecast for LISA, while the constraint on Ω_M is better by a factor of about three. On the other hand the forecast error on Ω_M once H_0 is fixed to its fiducial value, is $\Delta\Omega_M \simeq 0.009$, which is a factor of two better than the one obtained by present SNIa [13] (SNIa cannot measure H_0). Finally the constraints for DDE are comparable to the ones obtained with all present probes (CMB+SNIa+BAO) combined [13]. These numbers show the potential of the LISA mission: the constraints obtained by a single spaceborne GW probe and only with one population of sources (MBHBs), are roughly at the same level of the present constraints obtained by all combined cosmological probes. However one must keep in mind that future conventional EM probes, such as Euclid [16], will be able to obtain much better constraints than the present ones and thus to be more competitive than LISA. Nevertheless the results that will be obtained with LISA, and with GW experiments in general, will constitute a new way, independent from standard EM observations, to test the expansion of the universe.

The same analysis can be performed for other cosmological models. Following [2] the results for two simple alternative models will be presented. They constitute one-parameter extensions of ΛCDM and are defined as: **EDE**: an early DE model where the relative energy density of DE remains non negligible at early times [17, 18] (free extra parameter $\Omega_{\rm ede}$); **IDE**: a model where DE interacts with DM by an energy exchange proportional to the energy density of either DM (IDE1: free extra parameters ϵ_1) or DE (IDE2: free extra parameters ϵ_2), see [19] for a recent review. Both these models are assumed to effectively describe the cosmic expansion up to

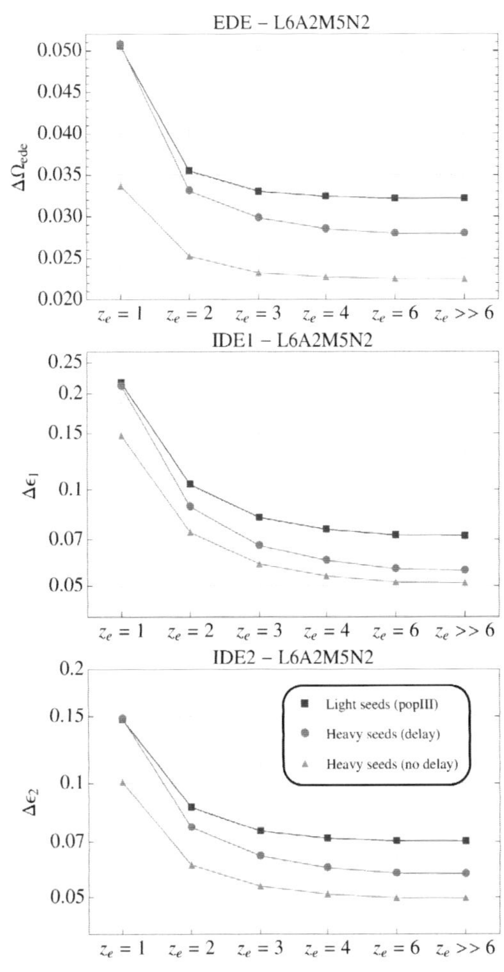

Figure 3. *1σ constraints from Ref. [2] on alternative cosmological models.*

some reference redshift z_e after which the standard ΛCDM dynamics is recovered (details can be found in [2]). This is done in order to test the ability of LISA to probe models where the deviations from ΛCDM happen only at late cosmological times. There are in fact for example some observational indications that an interaction in the dark sector might be present at late times, while being negligible at early times [20]. Assuming an exact prior on the ΛCDM parameters, the 1σ errors forecast for LISA on the extra alternative parameters are shown in Fig. 3 for different values of z_e. For each model the accuracy on the extra parameter reaches few 10^{-2}, worsening as z_e decreases and remaining constant for $z_e \gtrsim 6$. If the deviations from ΛCDM are relevant up to redshifts larger than about 10, then these constraints are worse, or at most comparable to, present CMB constraints [2]. However if these deviations are effectively important only after $z_e \simeq 10$, then CMB experiments cannot generally impose the same bounds they fix for z_e much bigger than 10; see [15]. This shows the main strength of LISA as a cosmological probe: it will test the cosmic expansion in the range $1 < z < 8$ where any deviation from ΛCDM will be directly constrained, irrespectively of it being negligible or not at earlier times.

5. Conclusions and future prospects

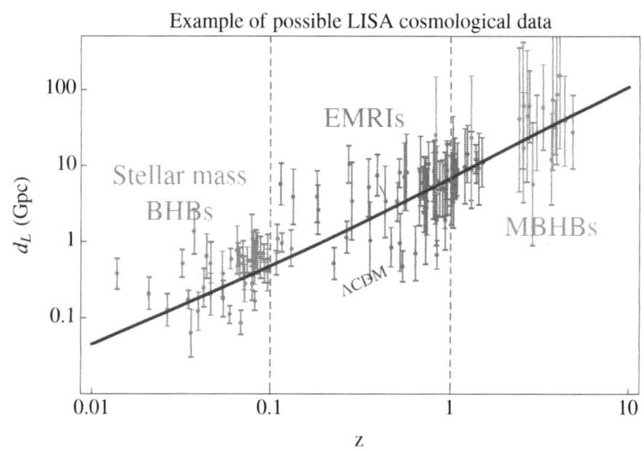

Figure 4. *Cosmology at all redshifts with LISA.*

GW standard sirens will in general offer a new independent way, complementary to common EM observations, to probe the cosmic expansion. Even if rare, GW standard sirens will complement, and increase confidence in, other cosmic distance indicators, in particular SNIa standard candles. MBHB mergers will constitute excellent standard sirens for LISA, allowing for systematic and calibration free distance indicators up to very high redshifts ($z \sim 8$). The biggest challenge remains the identification of an EM counterpart needed to measure the redshift of the GW source. Using a simple model of EM emission at merger and assuming the specifics of future telescopes, in [1] it was shown that LISA (N2A2M5L6) will be able to detect around 30 MBHB standard sirens over an observational period of five years. With these numbers the constraints forecast for standard and alternative cosmological parameters are roughly at the level of the present cosmological probes combined, but will likely be overcome by future observations. Nevertheless LISA will directly explore the expansion of the universe in a redshift range not accessible by other cosmic rulers, such as SNIa. For this reason the LISA mission can be used to obtain new and independent observational data to constrain standard and especially alternative models at very high redshift. On the other hand, MBHBs are not the only GW sources that LISA can use as standard sirens: stellar mass BHBs and EMRIs can also be used as distance indicators by LISA. Although no counterpart is expected for these sources, other (statistical) methods can be applied to overcome the lack of redshift information [21]. Interestingly stellar mass BHBs are usually detected in the range $z \lesssim 0.1$, while EMRIs appear at redshifts $0.1 \lesssim z \lesssim 1$. Considering the fact that MBHB standard sirens provide data in the $1 \lesssim z \lesssim 10$ range, this implies that LISA will become a unique cosmic probe able to test all redshift ranges up to $z \sim 10$, as qualitatively depicted in Fig. 4.

References

[1] Tamanini N, Caprini C, Barausse E, Sesana A, Klein A and Petiteau A 2016 *JCAP* **1604** 002 (*Preprint* 1601.07112)
[2] Caprini C and Tamanini N 2016 *JCAP* **1610** 006 (*Preprint* 1607.08755)
[3] Laser Interferometric Space Antenna (LISA): www.elisascience.org
[4] Maggiore M 2007 *Gravitational Waves. Vol. 1: Theory and Experiments* Oxford Master Series in Physics (Oxford University Press) ISBN 9780198570745, 9780198520740 URL http://www.oup.com/uk/catalogue/?ci=9780198570745
[5] Schutz B F 1986 *Nature* **323** 310–311
[6] Holz D E and Hughes S A 2005 *Astrophys. J.* **629** 15–22 (*Preprint* astro-ph/0504616)
[7] Klein A *et al.* 2016 *Phys. Rev.* **D93** 024003 (*Preprint* 1511.05581)
[8] Armitage P J and Natarajan P 2002 *Astrophys. J.* **567** L9–L12 (*Preprint* astro-ph/0201318)
[9] Palenzuela C, Lehner L and Liebling S L 2010 *Science* **329** 927 (*Preprint* 1005.1067)
[10] Giacomazzo B, Baker J G, Miller M C, Reynolds C S and van Meter J R 2012 *Astrophys. J.* **752** L15 (*Preprint* 1203.6108)
[11] Websites: www.lsst.org; www.skatelescope.org; www.eso.org/sci/facilities/eelt/.
[12] Armano M *et al.* 2016 *Phys. Rev. Lett.* **116** 231101
[13] Betoule M *et al.* (SDSS) 2014 *Astron. Astrophys.* **568** A22 (*Preprint* 1401.4064)
[14] Ade P A R *et al.* (Planck) 2016 *Astron. Astrophys.* **594** A13 (*Preprint* 1502.01589)
[15] Ade P A R *et al.* (Planck) 2016 *Astron. Astrophys.* **594** A14 (*Preprint* 1502.01590)
[16] Laureijs R *et al.* (EUCLID) 2011 (*Preprint* 1110.3193)
[17] Doran M and Robbers G 2006 *JCAP* **0606** 026 (*Preprint* astro-ph/0601544)
[18] Pettorino V, Amendola L and Wetterich C 2013 *Phys. Rev.* **D87** 083009 (*Preprint* 1301.5279)
[19] Wang B, Abdalla E, Atrio-Barandela F and Pavon D 2016 *Rept. Prog. Phys.* **79** 096901 (*Preprint* 1603.08299)
[20] Salvatelli V, Said N, Bruni M, Melchiorri A and Wands D 2014 *Phys. Rev. Lett.* **113** 181301 (*Preprint* 1406.7297)
[21] Kyutoku K and Setō N 2016 (*Preprint* 1609.07142)

IOP Conf. Series: Journal of Physics: Conf. Series **840** (2017) 012030 doi:10.1088/1742-6596/840/1/012030

Primordial Gravitational Waves with LISA

Angelo Ricciardone
Faculty of Science and Technology, University of Stavanger, 4036, Stavanger, Norway
E-mail: angelo.ricciardone@uis.no

Abstract. Primordial Gravitational Waves are the next target of modern cosmology. They represent a window on the early Universe and the only probe of the physics and microphysics of the inflationary period. When the production of GWs happens in scenarios richer than the standard single-field slow-roll, the GW signal becomes potentially detectable also on scales smaller than the Cosmic Microwave Background. LISA will be extremely complementary to CMB experiments to extract information about primordial inflationary models and in particular to probe phases of the inflationary period for which we have very poor knowledges.

1. Introduction

100 years ago Einstein wrote the theory of General Relativity and Gravitational Waves (GWs) were a precise prediction of his theory. In the Einstein theory GWs are wave solutions of the linearized weak-field equations; more precisely they were defined as transverse waves of spatial strain that travel at speed of light, generated by time variations of the mass quadrupole moment of a source. Many efforts during the last 100 years have been done to detect GWs, both through their direct or indirect effects. They are generated from cosmological and astrophysical sources and they represent a powerful tool to shed light on the physics responsible for their generation. In fact GWs of cosmological origin would allow for a clear understanding of the physics of the early Universe and they would give information on the quantum nature of gravity and fields; on the other hand GWs from astrophysical sources are important to test better General Relativity (GR) in the strong field limit and to study compact objects and their properties.

Concerning primordial GWs we can distinguish two main types of backgrounds: those expected from inflation, and those expected from mechanisms active between the end of inflation and the onset of Big Bang NucleoSynthesis (BBN). The first category includes the standard background of GWs due to tensor vacuum fluctuations h_{ij} of the Friedman-Lemaitre-Robertson-Walker (FLRW) metric during the inflationary period as well as the ones generated by other mechanisms still active during inflation, like GWs generated by the presence of extra-field(s), or by the effects of different symmetry pattern during inflation. In the second category we can mention GW backgrounds expected from very different sources: GWs produced during (p)reheating [1, 2, 3, 4, 5, 6, 7, 8], or by phase transitions [9, 10, 11, 12, 13, 14] and cosmic defects [15, 16, 17, 18], or even GWs coming from the merging of primordial black holes (PBHs) [19, 20, 21, 22, 23, 24]. In this contribution we will describe the GW backgrounds expected from different inflationary models, and we briefly mention about GW backgrounds expected from post-inflationary phenomena and GW production in alternative scenarios of inflation.

In the standard model of inflation, where the accelerated expansion is driven by a slow-roll scalar

IOP Conf. Series: Journal of Physics: Conf. Series **840** (2017) 012030 doi:10.1088/1742-6596/840/1/012030

field GWs are expected to be generated by the amplification of the tensor vacuum modes (metric fluctuations) and their spectrum amplitude, almost scale invariant, is directly connected with the energy scale of inflation. This GW backgound leaves a precise imprint on the Cosmic Microwave Background (CMB): a curl-like pattern in the polarization (B-modes). Up to now we do not have a detection of this primordial signal, but only an upper bound, from which we can infer that the spectral energy density at present time is $h_0^2\Omega_{\rm GW}(f) \approx 5 \cdot 10^{-16}(H/H_{\rm max})^2$, where H is the Hubble parameter during inflation and $H_{\rm max} \simeq 8.4 \cdot 10^{13}$ GeV is the current upper bound on the energy scale of inflation [25]. This GW background represents an irreducible contribution expected from any inflationary scenario and its detection would help to differentiate inflationary models from each other. Its tiny amplitude makes the detection extremely challenging both for CMB experiments and for the current and planned direct detection experiments like LIGO or LISA, since the amplitude of the spectrum is well below their sensitivities.

A way out to overcome this tiny amplitude is to consider inflationary scenarios richer than the standard single-field, for instance models where there are secondary fields with arbitrary spin [26] coupled to the inflaton either directly or only through a gravitational coupling [27, 28, 29, 30] or models where new symmetry patterns are allowed during inflation [31, 32, 33, 34].
Depending on the model, i.e. on the type of secondary field(s), the GW signal generated during inflation could be potentially interesting for the detectability by LISA. We can distinguish several cases, depending on the nature of the secondary field(s): (i) In the case in which the inflaton is a pseudo-scalar (axion) coupled with a gauge field, the latter are excited upon the rolling of the axion during inflation. Towards the end of inflation, the tensor spectrum is enhanced for the corresponding modes leaving the Hubble radius during the last e-folds of inflation. This results in a blue-tilted chiral spectrum of GWs, with an over production of one of the tensor polarizations versus the other; (ii) A spectator field during inflation, with a small speed of sound $c_s \ll 1$, and interacting only gravitationally with the inflaton field, sources second order tensor fluctuations and gives rise a blue tensor spectrum. (iii) Considering non-Abelian $SU(2)$ vector fields coupled to the inflaton through the interaction $\chi F\tilde{F}$ [35, 36] can also bring to an enhancement of tensor modes and parity violating signal. The time dependent background assumed during the inflationary period has suggested to assume that the inflaton is only time dependent during inflation. However considering the possibility to have (also) a space-dependent field during inflation can have potentially interesting features, in particular in the tensor sector. When, besides the time-reparameterization invariance ($t \to t + \xi_0$), also the space-reparameterization invariance ($x_i \to x_i + \xi_i$) is broken during inflation, tensors can acquire a non-trivial mass changing the dispersion relation [32]. This can enhance the spectrum of GWs on small scales becoming a potential interesting signal for interferometers, like LISA, and opening the possibility to test the symmetry patterns valid during the inflationary period at small scales.

A production of GWs is expected also during the post-inflationary period. An interesting case is when, in scenario like mild-waterfall hybrid inflation, large peaks appear in the matter power spectrum that later collapse forming primordial black holes, at horizon reentry during the radiation era. These PBHs can merge generating a stochastic background of GWs potentially detectable by the current aLIGO detector and most probably by LISA. Considering the period after the end of inflation, the so-called reheating phase, when the inflaton starts to oscillate and it couples with some "preheat" fields, there is the possibility, when the oscillations become large and coherent to generate a non-perturbative excitation of these fields and a consequent production of GWs. Typically the mechanisms through this happens are: i) parametric resonance [37] or ii) tachyonic preheating [38]. This depends strongly on the given scenario (inflationary potential/set-up), and on the specific form and strength of the interaction between the inflaton and the preheat field. It also depends on the spin of the preheat field, whether it is a fermion or a boson. The non-perturbative excitation of the preheat field only occurs within specific bands of momenta, so this translates into the development of rapidly evolving large inhomogeneities

of the energy density of the field. Hence, the large time-dependent gradients of the field provide an anisotropic-stress over the background, which naturally acts as a significant classical source of GW. The resulting spectrum of GW typically has a well defined peak, with an infrared slope decaying as $\propto k^3$, and an exponentially suppressed large-k tail. The position and height of the peak depend on the parameter space of each scenario. The amplitude of these backgrounds can be very large, but they naturally peaked at frequencies higher than the ones probed by interferometers like LISA. In the category of production of GWs during the post inflationary period can be mentioned also the possibility to generate GWs during a *stiff* phase after inflation, when the inflaton eventually behaves like a fluid with an equation of state $w > 1/3$ [39]. Potentially, this could enhance the inflationary signal so much that it becomes detectable in the range of frequencies probed by LISA, while being at the same time compatible with the CMB constraints. However this scenario deserves further analysis.
Interestingly, gravitational wave generation can happen also in the context of string cosmology, in particular in the pre-big bang scenario. In [40] has been shown that a generation of stochastic background of GWs with amplitude and frequency in the range probed by interferometers like aLIGO/Virgo and/or eLISA is possible, for a wide range of the parameter space. And in particular, this is consistent with the constraints coming from CMB about temperature anisotropies.
The mentioned scenarios are only a partial list of models where GWs produced in the early Universe are potentially detectable by an experiment like LISA, see i.e. [41] for a recent review. This represents a way to shed light on the early Universe and in particular to test the latest stage of inflation, or the presence of other field besides the inflaton during inflation, or to understand possible symmetry breaking pattern during inflation, or to study the post inflationary phase for which we have poor information or finally to confirm or disprove pre-big bang models or to impose new significant constraints on the related cosmological parameters.

2. Detection of GW

After the recent detection by Advanced LIGO of (almost) three GW signals generated by the merging of two massive BHs [42, 43], the detection of primordial GWs has become a feasible target of modern cosmology. Due to the weakness of the gravitational interaction, GWs cannot interact with the surrounding medium, and hence immediately after production they decouple and travel freely through the Universe until today. While CMB photons decouple at energy around 0.3 eV, primordial GWs are expected to decouple at very high energy scale (around the Planck scale). Then this would open a window on the very early Universe. Actually there are two main ways to try to detect GWs: through their indirect effect on the CMB and LSS or through their direct effect (of displacement) on the arms of interferometers. Let's briefly discuss about these ways, and report the most recent limits about the GW amplitude from different experiments:

Indirect detection The influence of GWs when photons decouple from ordinary matter forming the Last Scattering Surface (LSS) leaves a precise imprint: in particular the CMB radiation becomes linearly polarized showing a curl-like (B-mode) pattern. The amplitude of this signal is extremely small since only few CMB photons get polarized, then its detection is rather hard. The main contribution from primordial GWs is expected to be in the BB power spectrum mostly at very large scales ($\ell < 150$), but astrophysical foregrounds, galactic dust and gravitational lensing contaminate this signal making its detection even harder. Current data actually provide only an upper bound on the amplitude of the tensor power spectrum defined through the tensor-to-scalar ratio $r \equiv \mathcal{P}_T/\mathcal{P}_S$, where $\mathcal{P}_T$ and $\mathcal{P}_S$ are the amplitudes of the tensor and scalar power spectra respectively, evaluated at a reference scale. The joint analysis by BICEP2, Keck Array data, Planck polarization and WMAP9 23 GHZ and 33 GHZ maps gives, at a pivot reference scale $k_* = 0.05\ \mathrm{Mpc}^{-1}$, a bound on $r_{0.05} < 0.09$ at 95% C.L. [44], assuming a scale-invariant

GW power spectrum. Assuming the consistency relation between the tensor-to-scalar ratio and the tensor spectral tilt $r = -8n_T$, valid for single-field slow-roll models of inflation, the limit becomes tighter: $r_{0.05} < 0.07$ at 95% C.L. There are several current or forthcoming experiments devoted to B-mode detection: among the space-borne we can mention PRISM, COrE or PIXIE, while among the ground-based we can mention BICEP2, SPTPol, ACTPol, Polarbear, Spider. See the recent [45] for the next generation of ground-based CMB projects.

Direct detection Direct detection GW experiments are based on laser interferometry and they essentially account for the effect of distorsion that a passing GW can have on the length of the arms of the interferometer. They are mainly devoted to the analysis of GWs produced by compact binary systems or mergers of massive black holes, neutron stars or extreme-mass-ratio inspirals. However, as seen in a series of papers of the LISA collaboration, LISA will have the capabilities to detect or at least to constrain, not only astrophysical sources, but also cosmological sources, like GWs produced during first order phase transition, from cosmic defects [13] and probe the present expansion of the Universe [46]. Actual gravitational wave detectors already in operation on the Earth are LIGO, Virgo and GEO600. They are designed to be sensible to relative changes in the length of the arms of the order of $\Delta L/L \simeq 10^{-21}$. LIGO and Virgo have arm lengths of 4 and 3 Km respectively while GEO600 of about 600m. They were built to be sensitive in the range of frequency between few Hz and 10 kHz and they are mainly devoted to the detection of GWs generated by astrophysical objects. Their sensitivity however is limited by seismic noise and by practical reason as the length of the arm, so this pushed the community to propose the first gravitational wave observatory in space. LISA represents the main candidate for this mission [47]. The LISA concept, still based on laser interferometry, will be composed by a constellation of three spacecrafts, arranged in an equilateral triangle with million-kilometre arms (5 million km or 2 million km) flying along an Earth-like heliocentric orbit with a duration of the mission between 2 and 5 years. It will have the main sensitivity in the frequency domain between mHz and Hz and it will be extremely complementary to Earth based interferometers [48].
Other forthcoming or planned gravitational wave detectors are KAGRA [49], LIGO-India [50] and Einstein Telescope [51] that will be also mainly devoted to study astrophysical sources of GWs. In order to have GW interferometers sensible to a primordial GW signal generated in a standard inflationary scenario, then we have to refer to more futuristic experiments like BBO and DECIGO [52].
Nowadays, from a non-detection of primordial signal from the aformentioned detectors we have current and forecasted upper bounds on the GW spectral energy density: the LIGO/Virgo collaboration provided an upper limit $\Omega_{GW} < 5.6 \times 10^{-6}$ at 95% C.L. for $f \in (41.4, 169.25)$. The upgraded aLIGO detector is expected to lower this bound to $\Omega_{GW} < 10^{-9}$ [53], while LISA (best configuration) is expected to reach $\Omega_{GW} \lesssim 10^{-14}$ [54].

Effect on LSS Primordial GWs can leave their imprint also on Large Scale Structures (LSS), in particular a primordial long-wavelength tensor mode may leave its imprint on the primordial scalar power spectrum and consequently on the matter power spectrum [55, 56]. The typical effects are the generation of a local power quadrupole in the matter power spectrum, that gives rise to an apparent local departure from statistical isotropy, and a non-Gaussian four-point function. These effects, incorporated in the $\langle\gamma\zeta\zeta\rangle$ (tensor-scalar-scalar) correlation function, have observational consequences, even if the tensor perturbations cannot be detected directly. In particular the quadrupole anisotropy is generated by a primordial GW with wavelength larger that the size of the galaxy survey while the non-Gaussian effects may arise from GWs of wavelength comparable to the survey size. These observables are useful quantities to discriminate among the plethora of inflationary models and the next galaxy surveys will have the possibility to probe this *indirect* GW effect []. From this brief description of the three main strategies to detect GWs it becomes clear the importance of the complementary among the different experiments in

IOP Conf. Series: Journal of Physics: Conf. Series **840** (2017) 012030 doi:10.1088/1742-6596/840/1/012030

	k $[\mathrm{Mpc}^{-1}]$	$N_{\mathrm{estim.}}$
CMB / LSS	$10^{-4} - 10^{-1}$	$56 - 63$
$y-$ & $\mu-$distortions	$10^{-1} - 10^{4}$	$45 - 56$
$P_\zeta \to$ PBH $\to$ GW @ LISA	$10^{5} - 10^{7}$	$38 - 41$
$P_\zeta \to$ PBH $\to$ GW @ AdvLIGO	$10^{7} - 10^{8}$	$35 - 37$
$P_{\delta g} \to$ GW @ LISA	$10^{11} - 10^{14}$	$22 - 28$
$P_{\delta g} \to$ GW @ AdvLIGO	$10^{16} - 10^{17}$	$15 - 17$

Table 1. List of observational windows on inflation in correspondence of the wavenumber of the primordial modes and estimated number of efolds at which those modes exited the horizon. Taken from [57].

order to probe different (energy) scales and automatically to test several stages of the primordial Universe, as clearly explained in Table 1.

3. Inflationary consistency relation

The majority of the models that produce a GW signal during the inflationary period are characterized by a primordial power spectrum with a power law shape, that can be directly connected to the present spectral GW energy density. This possibility to parametrize the GW energy-density spectrum by a power law opens the possibility to constrain cosmological parameters which are strictly connected with the inflationary period. In particular, we can constrain the GW spectral index n_T and tensor-to-scalar ratio r. Then assuming a power law spectrum like $\Omega_{GW}(f) = \Omega_{GW}^{CMB}\left(\frac{f}{f_{CMB}}\right)^{n_T}$, we can re-express Ω_{GW}^{CMB} in terms of the tensor-to-scalar ratio r and the amplitude of the primordial scalar power spectrum at CMB scales for which we have precise limits, given by Planck [25]. In this way we can combine constraints on r and n_T from the CMB scale with constraints to $\Omega_{GW}(f)$ and n_T from direct detection experiments.
Up to now a constraint on n_T have been obtained combining BICEP2/Keck Array and Planck (BKP), Planck 2013, WMAP low ℓ polarization, HST data, BAO measurements from SDSS and the upper limit on the energy density of stochastic GW background from LIGO: $n_T = 0.06^{+0.63}_{-0.89}$ at 95% C.L. [58], in correspondence of a best fit for the tensor-to-scalar ratio of $r_{0.01} = 0.02$.
Recently, it has been shown how CMB experiments alone are not able to put strong constraints on the spectral tilt ($n_T \lesssim 5$ at 95% C.L. for $r_{0.01} = 0.02$ [59]), even in the case of a detection of B-modes, since they focus on a narrow range of frequency around 10^{-16}Hz; so, it becomes clear the importance of the combination of several experiments that cover different range of scales. In [59, 60] it has been pointed out how the combination of GW experiments on a large range of frequencies, as ground and space-based interferometers, indirect measurements and CMB, BAO, BBN, puts tighter constraints on the tensor spectral index.
As forecasted in [61], LISA will have the ability to constrain better the spectral tilt n_T, considering r consistent with the current CMB upper bound. The results are shown in Fig. 1, where are shown different LISA sensitivities; assuming the best LISA configuration with six links, five million km arm length and 5 year mission, the constraint for the spectral index turns out to be $n_T \lesssim 0.2$, see also [62].This ability of LISA in constraining the tensor spectral tilt, can be seen as a test of the so-called consistency relation $r = -8n_T$ for primordial inflationary models. In fact, we know that, all single-field slow-roll models of inflation predict such a relation. Any evidence of violation of such a relation would be an indication of deviation from single-field slow-roll models.

4. Gravitational Waves from Inflation

4.1. The standard inflationary GW background

Gravitational Waves are with no doubt the smoking gun of an inflationary period. After the several predictions verified by the Planck satellite about a period of accelerated expansion [63],

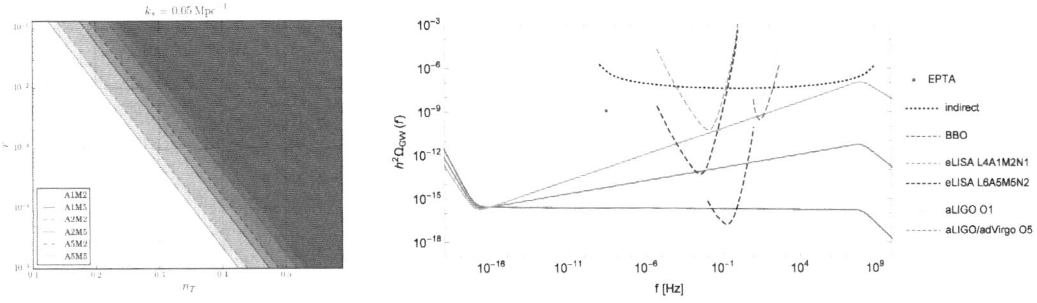

Figure 1. Left: limits on the tensor spectral tilt n_T and the tensor-to-scalar ratio r for six LISA configurations. Taken from [61]. Right: GW energy spectrum for different value of the tensor spectral tilt vs sensitivities of different experiments (orange $n_T = 0.36$, red $n_T = 0.18$, brown $n_T = -r/8$). Taken from [41]; see also [59].

GWs represent the key test to finally confirm inflation. They are ubiquitous in all the inflationary models and are represented by the tensor perturbation h_{ij} of the FLRW metric $g_{ij} = a^2(\delta_{ij}+h_{ij})$. They carry two physical degrees of freedom represented by the two polarizations and this is mathematically given by the trasverse-traceless conditions on the tensor perturbation ($\partial_i h^i{}_j = 0$, $h^i_i = 0$). Their dynamics is described by a wave equation in an expanding background. These vacuum metric fluctuations are expected to be amplified during the inflationary period, when small-scale fluctuations are stretched to superhorizon scales. Being massless (or very light) quantum fields in an expanding background, they fluctuate with a magnitude proportional to their stored energy density, which through Einstein equations, is related to the (square) of the Hubble parameter during inflation. This fixes also the amplitude of the GW power spectrum and shows the importance of a detection of primordial GW signal. In fact in the standard inflation scenario, where the accelerated expansion is driven by a slowly rolling scalar field, the inflaton, the amplitude of the primordial GW signal, expressed by the tensor-to-scalar ratio r, defines two important relations: $V^{1/4} \simeq 10^{16}\mathrm{GeV}\left(\frac{r}{0.1}\right)^{1/4}$, which clarifies the relation between the energy scale of inflation and the amplitude of the GW spectrum, and $\Delta\phi/M_{Pl} \gtrsim 1.06\left(\frac{r}{0.1}\right)^{1/2}$ which connects the amplitude of the tensor spectrum with the excursion of the inflaton during inflation. These two relations are a clear implication of the importance of the detection of inflationary GWs since they would represent a power way to discriminate among inflationary models, to fix the energy scale regime of the inflationary period and related issues about quantum gravity effects.

In single-field scenarios such a signal is expected to have a small amplitude and to peak at large scales where the level of foreground contamination is quite large, that makes it detection quite challenging. Nowadays we do not have a detection of such a primordial signal but only current limits as seen before. Considering this bound and the almost scale-invariance of the spectrum, predicted by the simplest models of inflation involving a slow rolling scalar field, constrains this GW signal in a range of frequencies and amplitude out of current direct GW detectors and also from the LISA capabilities.

4.2. Beyond the standard inflationary GW background

As soon as the inflationary scenario becomes richer than standard single-field slow-roll inflation the GW predictions can change. In particular when there are additional fields during inflation or when the pattern of symmetry is modified with respect to the usual time-dependent background,

IOP Conf. Series: Journal of Physics: Conf. Series **840** (2017) 012030 doi:10.1088/1742-6596/840/1/012030

the GW signal can become large and blue tilted, becoming a possible target for detection at scales of interferometers like LISA and/or aLIGO.
Whenever secondary field are present during inflation, coupled directly or only gravitationally to the inflaton, the wave equation for the tensor gets modified by a source term, which generates an extra "sourced" contribution which adds to the irreducible background signal given by vacuum metric fluctuations. In particular the equation governing the GW evolution becomes $\ddot{h}_{ij}(\mathbf{k},t)+3H\,\dot{h}_{ij}(\mathbf{k},t)+k^2\,h_{ij}(\mathbf{k},t)=\frac{2}{M_{Pl}^2}\Pi_{ij}^{TT}(\mathbf{k},t)$, where a dot denotes derivative with respect to t, H is the Hubble rate, M_{Pl} is the reduced Planck mass, k is the physical momentum, and Π_{ij}^{TT} is the source of GWs, corresponding to the transverse-traceless part of the anisotropic stress Π_{ij}. The latter is directly connected to the nature of the additional sources, i.e. if they are scalars or vectors. In the following models we will see its specific form. In many cases this sourced contribution can be larger that the vacuum contribution, generating a GW signal large at small scales and still compatible with CMB bounds.
Particle production during Inflation Among the different mechanisms of particle production during inflation and subsequent enhancement of the GW signal potentially detectable by LISA interferometer, still compatible with the constraints from the CMB, the models involving $U(1)$ gauge field, coupled with a pseudo-scalar inflaton field, look the more promising. These models solve also a key problem of inflation that is the protection of the flatness of the inflaton potential from radiative corrections. In fact the protection is naturally present since the pseudo-scalar axion respects an approximate shift symmetry.
In the simple model [64] the inflaton ϕ is a pseudo-scalar field and it is coupled with a gauge field A_μ with the relevant Lagrangian operator and the source tensor given by $\mathcal{L} \supset -\frac{\phi}{4f}F_{\mu\nu}\tilde{F}^{\mu\nu} \Rightarrow \Pi_{ij}^{TT} \propto (E_iE_j + B_iB_j)^{TT}$ where f is the axion decay constant, $F_{\mu\nu} \equiv \partial_\mu A_\nu - \partial_\nu A_\mu$ is the field strength of the gauge field A_μ and $\tilde{F}_{\mu\nu}$ is its dual; $\mathbf{E} = -\mathbf{A}'/a^2$, $\mathbf{B} = \nabla \times \mathbf{A}/a^2$ and the source term arises not from the $F\tilde{F}$ but from the term FF also present in this kind of models. The dynamics of the scenario is the following: the (slowly-rolling) inflaton excites, through its coupling, one helicity of the vector field, that in turn sources the GW of one definite chirality in an interaction like $\delta A + \delta A \rightarrow \delta g$. However also the inflaton perturbations are sourced through the inverse decay of the gauge field $\delta A + \delta A \rightarrow \delta\phi$. This suggests a high level of scalar non-Gaussianity that, being strongly constrained by the non-observation in the CMB, puts a limit on the parameter ξ [63]. In fact the production is exponentially proportional to the parameter $\xi \equiv \frac{\dot{\phi}}{2fH}$, that sets the exponential amplification of the gauge mode ($A_+ \propto e^{\pi\xi}$) and is constrained by observations from CMB and LSS to be smaller than 2.5 at 95% C.L. in order not to generate too much non-Gaussianity. Such small value suppresses the amplitude of the GW signal, however such a quantity ξ is depending on the time derivative of the inflaton that towards the end of inflation speeds up. Therefore, in this model we naturally expect a growth of ξ, and a consequent exponential increase of GW production, to at level that may be probed at LISA and at terrestrial interferometer scales [26, 65, 64, 66], like shown in Fig. 2. A detailed analysis of this model in connection with a GW signal potentially detectable at LISA scales has been carried out in [61]. In particular they used a local approach, which allows to study the phenomenology of the model within a well-defined observational range, and a global approach which cover a full observational range. Both approaches highlight the ability of the LISA detector to extract relevant information about the parameters of the model giving in this way the possibility to test stage of inflation that are far from the scales probed by CMB. In particular the analysis shows the complementary between CMB observables, coming from power spectrum and bispectrum analysis and parameters related to the model and connected to the GW spectrum. As anticipated, this scenario shows peculiar features that are extremely useful as observational test: in particular a GW signal that is completely chiral [67], which means that one of the polarizations of the gravitational field shows a larger spectrum amplitude, and a

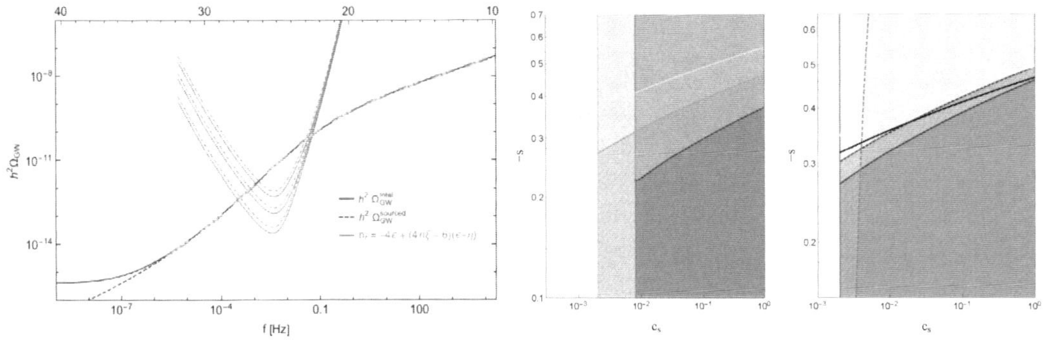

Figure 2. Spectrum of GWs in particle production model (left panel) and parameter space that LISA can probe with different configurations for the spectator field scenario (right panel). Taken from [61].

highly tensor non-Gaussian signal with a precise relation between the bispectrum and the power spectrum: $k^6 \langle h^3 \rangle'_{\rm equil} \simeq 23 P_{\rm GW}^{3/2}$. If detected, these properties would be a smoking gun for the cosmological origin of this GW signal and its nature.

GWs from spectator field during inflation Also the presence of other fields, not directly coupled to the inflaton, during inflation can generate interesting features in the GW spectrum. In particular considering a *spectator* field with a small speed of sound during inflation, where a spectator is usually referred to a field that does not influence the background inflaton dynamics, we can generate a blue-tilted primordial power spectrum for tensor modes, allowing to a possible detection of this signal at LISA scales. The presence of a spectator field, however, is expected to source also scalar fluctuations, as it happens in particle production by gauge field as seen before, and these are well constrained by current CMB measurements [25]. This mechanism is able to generate a small value of the tensor-to-scalar ratio on CMB scales, as shown in a particular realization [68], while can instead produce more tensor power on smaller scales thanks to the presence of a *source* term in the equation of motion of the tensor, given by the spectator.
Considering the model introduced in [68] and analyzed also in [61] with $\mathcal{L} \supset P(X, \sigma) \Rightarrow \Pi_{ij}^{TT} \propto P_X (\partial_i \delta\sigma \partial_j \delta\sigma)^{TT}$, where σ is the spectator field, $X = \frac{1}{2}\partial_\mu \sigma \partial^\mu \sigma$ and P is a generic function of X and σ, an interesting case appears when the spectator sound speed $c_s \equiv P_X / \left(P_X + P_{XX}\dot{\sigma}_0^2\right)$ (where σ_0 is the background value) and its time variation $s \equiv \dot{c}_s / H c_s \neq 0$ are very small, in particular when $c_s \ll 1$ and $|s| < 1$. Both in the scalar and tensor sectors the power spectra will be the sum of the usual vacuum contribution and of the sourced contribution generated by the presence of the spectator field perturbation. Particular regimes for the parameter c_s and s allow to obtain a sourced GW contribution with a sufficiently large amplitude, in principle detectable by LISA, and at the same time keeping a small amplitude at CMB scales. The parameter space of the model is characterized by the Hubble rate during inflation, the speed of sound of the spectator c_s and its time variation s, since: $P_S \equiv P_S(\epsilon, H, c_s)$, $P_{GW} \equiv P_{GW}(H, c_s)$ and the spectral tilts $n_T^{(\sigma)} = -4\epsilon - 3s$, $n_S^{(\sigma)} - 1 = -4\epsilon - 7s$. So taking into account the actual limits on ϵ from CMB ($\epsilon < 0.0068$ at 95% C.L. [25]) and on the amplitude of the scalar perturbation ($A_{0.05} = 2.21 \cdot 10^{-9}$ at 68% C.L. [69]), it is possible to scan the parameter space $s - c_s$ probed by the six-different LISA configurations for different values of the Hubble rate during inflation. Fig. 2 reports some of the results found in [61], showing the ability of different LISA configurations to scan the parameter space of the model. The region below each curve represents the parameter space left open by a non-detection of a GW signals associated to the spectator scenario. Bounds for this scenario come also from other current observations like Big

Bang Nucleosynthesis, or from the aLIGO O1 which both provide upper bound on the spectral energy density of GW. These bounds are also reported in the right panel of Fig. 2 and is possible to compare how much the LISA detector will eventually improve these current bounds.
As a consequence of the increased power in the tensor sector due to the spectator field, also the scalar sector, and in particular the sourced contribution is expected to increase at small scales. However on such small scales we have to be consistent with bounds coming from PBHs. This bound would reduce drastically the parameter space accessible to LISA, however the modellization of PBH physics has many uncertaintes which allow to still consider the future constraints by LISA as the more robust. Whenever this modellization of the behaviour of PBH will become more solid, then LISA will still be a validation test for the modellization. An interesting perspective on how much PBH constraints limit the ability of observation of GWs by LISA is addressed in [57].
GWs in the Effective Field Theory of space-reparameterization There are also interesting case of large GW signal at small scales without invoking the presence of an extra field. In particular, this can be realized in models where inflationary fields acquire vacuum expectation values along space-like directions [31, 70]. While time translation invariance ($t \to t+\xi_0$) is surely broken by the cosmological expansion during inflation, it is interesting to explore the possibility that space-reparameterization ($x_i \to x_i + \xi_i$) are also broken during the inflationary era. One writes the most general set of operators that satisfy the symmetry requirements, and connects the coefficients of such operators with observable quantities. The effective field theory of inflation (EFTI) provides a powerful tool for obtaining model independent predictions of large classes of inflationary scenarios. It requires only information about the symmetries broken during the inflationary era, and on the number and nature of fields that drive inflation. The advantage of EFTI is that one has not to commit on specific realizations in order to deduce observable consequences. If space-reparameterizations are broken, the graviton can acquire a mass and a non-trivial sound speed during inflation [33]. At the quadratic level in perturbations, in an EFT approach, the action for graviton fluctuations h_{ij} around a conformally flat FLRW background can be expressed as in [32, 33, 34]: $\mathcal{L}_h = \frac{M_{Pl}^2}{8}\left[\dot{h}_{ij}^2 - \frac{c_T^2(t)}{a^2}\left(\partial_l h_{ij}\right)^2 - m_h^2(t)\, h_{ij}^2\right]$ Using this action, the tensor power spectrum and its related spectral tilt are found to be: $\mathcal{P}_T = \frac{2\,H^2}{\pi^2\,M_{Pl}^2\,c_T^3}\left(\frac{k}{k_*}\right)^{n_T}$ and the spectral tilt $n_T = -2\epsilon + \frac{2}{3}\frac{m_h^2}{H^2}\left(1+\frac{4}{3}\,\epsilon\right)$. Hence, if the quantity m_h/H is sufficiently large, one can get a blue tensor spectrum with no need to violate any energy condition in the early Universe (null energy condition). This is a feature common in scenarios where space reparameterization is broken during inflation. In [61] also this scenario has been analyzed finding interesting results. In Fig. 3 we report the spectral GW energy density computed starting from the power spectrum. This shows how the possibility to have a non trivial mass and sound speed for the graviton during inflation can enhance the tensor signal on small scales becoming a possible target for the LISA mission. In Fig. 3 we report the perspective of LISA in constraining the parameter space ($c_T - m_h/H$) for the best configuration of LISA: A5M5, with 5 million arm length, 5 year mission (left panel) . This analysis highligths the ability of LISA to probe different range of masses for the graviton, giving the possibility to test pattern of symmetries valid during the inflationary period.

5. Beyond the power spectrum: parity violation, non-Gaussianity and extra-polarization states

As seen in the previous sections a peculiar signature of some primordial GW signals is the chirality, then it is natural to ask if LISA will be able to detect parity violating GW signal. In particular we have seen that the breaking of parity invariance is a peculiar feature of models including a coupling between an axion-like field and gauge field(s). This has as a consequence

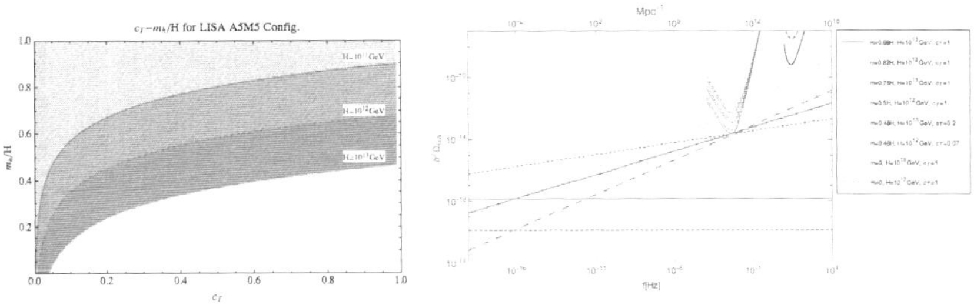

Figure 3. Parameter space that LISA can probe with the "best" (A5M5) configuration (left panel) and spectrum of GWs in the EFT framework (right panel). Taken from [61].

that the spectra of the left and right-handed tensor modes generated by the amplification of the gauge field are different. This is a distinctive signature that could distinguish GWs of cosmological origin from astrophysical ones and also the GWs produced by these mechanisms from the vacuum ones. Parity violation signal can arise also in sectors of string theory (*i.e.* type IIB, I, heterotic), in loop quantum gravity, leading to the question of whether the gravitational interaction is parity invariant in the strong field. Prospects of measuring parity violation from the early Universe using gravitational terrestrial wave detectors have been done in [65, 71].
At the same level of scalar also tensor non-Gaussianity represents a powerful tool in order to help in constraining inflationary physics, see e.g. [72]. From CMB data we already have strict bounds on different configurations of the scalar bispectrum, and also (weaker) constraints on the tensor three-point function [63]. Being in many models the GW signal produced by non-linear sources it is natural to ask if a detector like LISA will be able to measure the GW three-point function. As we have seen in the axion-gauge model a large tensor bispectrum can be generated and the amplitude can provide upper bounds on the model parameter ξ, complementary to those coming from scalar CMB bispectrum [73, 29]. Moreover in that specific scenario a precise relation between the power spectrum and the bispectrum is predicted which represents a smoking gun for such kind of GW signals.
Finally, as last open question, there is the possibility to consider additional GW polarizations. In fact, in alternative theories of gravity, GWs could have up to four additional (apart the "plus" and "cross" polarization models within standard General Relativity) polarization modes. Then it is natural to ask if LISA will have the capabilities to detect extra GW polarization states.

6. Conclusions

Primordial GWs are one of the next targets of modern cosmology since they represent a unique opportunity to shed light on the physics of the early Universe and in particular to probe the microphysics of inflation. An irreducible GW background is an ubiquitous prediction of all the inflationary models and this represents a smoking gun of the primordial accelerated expansion. The primary probe of primordial GW is the polarization of the CMB and in particular the curl-free polarization pattern (B-modes). When the inflationary scenario is enriched by secondary fields, besides the inflaton, or by assuming new symmetry pattern, GWs become a target also on scales smaller than CMB, and in particular for interferometers like LISA. While the GW by vacuum fluctuations is not visible by the next generation interferometers, many well-motivated inflationary scenarios produce a signal that can be visible or that can be extremely useful to reduce the parameter space of such scenarios. It becomes clear that a detection of GW signal on the small scales accessible to LISA, will offer a window on scales of inflation on which we currently have little knowledge and will become of fundamental importance in order to provide constraints on tensor perturbations complementary to the CMB.

References

[1] S. Y. Khlebnikov and I. I. Tkachev, "Relic gravitational waves produced after preheating," *Phys. Rev.* **D56** (1997) 653–660, `arXiv:hep-ph/9701423 [hep-ph]`.

[2] R. Easther and E. A. Lim, "Stochastic gravitational wave production after inflation," *JCAP* **0604** (2006) 010, `arXiv:astro-ph/0601617 [astro-ph]`.

[3] J. Garcia-Bellido and D. G. Figueroa, "A stochastic background of gravitational waves from hybrid preheating," *Phys. Rev. Lett.* **98** (2007) 061302, `arXiv:astro-ph/0701014 [astro-ph]`.

[4] J. F. Dufaux, A. Bergman, G. N. Felder, L. Kofman, and J.-P. Uzan, "Theory and Numerics of Gravitational Waves from Preheating after Inflation," *Phys. Rev.* **D76** (2007) 123517, `arXiv:0707.0875 [astro-ph]`.

[5] D. G. Figueroa and T. Meriniemi, "Stochastic Background of Gravitational Waves from Fermions – Theory and Applications," *JHEP* **10** (2013) 101, `arXiv:1306.6911 [astro-ph.CO]`.

[6] D. G. Figueroa, "A gravitational wave background from the decay of the standard model Higgs after inflation," *JHEP* **1411** (2014) 145, `arXiv:1402.1345 [astro-ph.CO]`.

[7] D. G. Figueroa, J. Garcia-Bellido, and F. Torrenti, "Gravitational wave production from the decay of the standard model Higgs field after inflation," *Phys. Rev.* **D93** no. 10, (2016) 103521, `arXiv:1602.03085 [astro-ph.CO]`.

[8] S. Antusch, F. Cefala, and S. Orani, "Gravitational waves from oscillons after inflation," `arXiv:1607.01314 [astro-ph.CO]`.

[9] A. Kosowsky, M. S. Turner, and R. Watkins, "Gravitational radiation from colliding vacuum bubbles," *Phys. Rev.* **D45** (1992) 4514–4535.

[10] M. Kamionkowski, A. Kosowsky, and M. S. Turner, "Gravitational radiation from first order phase transitions," *Phys.Rev.* **D49** (1994) 2837–2851, `arXiv:astro-ph/9310044 [astro-ph]`.

[11] C. Caprini, R. Durrer, and G. Servant, "Gravitational wave generation from bubble collisions in first-order phase transitions: An analytic approach," *Phys.Rev.* **D77** (2008) 124015, `arXiv:0711.2593 [astro-ph]`.

[12] C. Caprini, R. Durrer, T. Konstandin, and G. Servant, "General Properties of the Gravitational Wave Spectrum from Phase Transitions," *Phys. Rev.* **D79** (2009) 083519, `arXiv:0901.1661 [astro-ph.CO]`.

[13] C. Caprini *et al.*, "Science with the space-based interferometer eLISA. II: Gravitational waves from cosmological phase transitions," *JCAP* **1604** no. 04, (2016) 001, `arXiv:1512.06239 [astro-ph.CO]`.

[14] D. J. Weir, "Revisiting the envelope approximation: gravitational waves from bubble collisions," *Phys. Rev.* **D93** no. 12, (2016) 124037, `arXiv:1604.08429 [astro-ph.CO]`.

[15] T. Vachaspati and A. Vilenkin, "Gravitational Radiation from Cosmic Strings," *Phys.Rev.* **D31** (1985) 3052.

[16] D. G. Figueroa, M. Hindmarsh, and J. Urrestilla, "Exact Scale-Invariant Background of Gravitational Waves from Cosmic Defects," *Phys. Rev. Lett.* **110** no. 10, (2013) 101302, `arXiv:1212.5458 [astro-ph.CO]`.

[17] P. Binetruy, A. Bohe, C. Caprini, and J.-F. Dufaux, "Cosmological Backgrounds of Gravitational Waves and eLISA/NGO: Phase Transitions, Cosmic Strings and Other Sources," *JCAP* **1206** (2012) 027, `arXiv:1201.0983 [gr-qc]`.

[18] S. A. Sanidas, R. A. Battye, and B. W. Stappers, "Constraints on cosmic string tension imposed by the limit on the stochastic gravitational wave background from the European Pulsar Timing Array," *Phys. Rev.* **D85** (2012) 122003, `arXiv:1201.2419 [astro-ph.CO]`.

[19] J. Garcia-Bellido, A. D. Linde, and D. Wands, "Density perturbations and black hole formation in hybrid inflation," *Phys. Rev.* **D54** (1996) 6040–6058, `arXiv:astro-ph/9605094 [astro-ph]`.

[20] S. Clesse and J. García-Bellido, "Detecting the gravitational wave background from primordial black hole dark matter," `arXiv:1610.08479 [astro-ph.CO]`.

[21] S. Clesse and J. García-Bellido, "The clustering of massive Primordial Black Holes as Dark Matter: measuring their mass distribution with Advanced LIGO," *Phys. Dark Univ.* **10** (2016) 002, `arXiv:1603.05234 [astro-ph.CO]`.

[22] S. Bird, I. Cholis, J. B. Muñoz, Y. Ali-Haïmoud, M. Kamionkowski, E. D. Kovetz, A. Raccanelli, and A. G. Riess, "Did LIGO detect dark matter?," *Phys. Rev. Lett.* **116** no. 20, (2016) 201301, `arXiv:1603.00464 [astro-ph.CO]`.

[23] I. Cholis, E. D. Kovetz, Y. Ali-Haïmoud, S. Bird, M. Kamionkowski, J. B. Muñoz, and A. Raccanelli, "Orbital eccentricities in primordial black hole binaries," *Phys. Rev.* **D94** no. 8, (2016) 084013, `arXiv:1606.07437 [astro-ph.HE]`.

[24] R. Dong, W. H. Kinney, and D. Stojkovic, "Gravitational wave production by Hawking radiation from rotating primordial black holes," *JCAP* **1610** no. 10, (2016) 034, `arXiv:1511.05642 [astro-ph.CO]`.

[25] **Planck** Collaboration, P. A. R. Ade *et al.*, "Planck 2015 results. XX. Constraints on inflation," *Astron. Astrophys.* **594** (2016) A20, `arXiv:1502.02114 [astro-ph.CO]`.

[26] J. L. Cook and L. Sorbo, "Particle production during inflation and gravitational waves detectable by ground-based interferometers," *Phys. Rev.* **D85** (2012) 023534, arXiv:1109.0022 [astro-ph.CO]. [Erratum: Phys. Rev.D86,069901(2012)].
[27] M. Biagetti, E. Dimastrogiovanni, M. Fasiello, and M. Peloso, "Gravitational Waves and Scalar Perturbations from Spectator Fields," *JCAP* **1504** (2015) 011, arXiv:1411.3029 [astro-ph.CO].
[28] R. Namba, M. Peloso, M. Shiraishi, L. Sorbo, and C. Unal, "Scale-dependent gravitational waves from a rolling axion," *JCAP* **1601** no. 01, (2016) 041, arXiv:1509.07521 [astro-ph.CO].
[29] M. Shiraishi, A. Ricciardone, and S. Saga, "Parity violation in the CMB bispectrum by a rolling pseudoscalar," *JCAP* **1311** (2013) 051, arXiv:1308.6769 [astro-ph.CO].
[30] I. Ben-Dayan, "Gravitational Waves in Bouncing Cosmologies from Gauge Field Production," *JCAP* **1609** no. 09, (2016) 017, arXiv:1604.07899 [astro-ph.CO].
[31] S. Endlich, A. Nicolis, and J. Wang, "Solid Inflation," *JCAP* **1310** (2013) 011, arXiv:1210.0569 [hep-th].
[32] D. Cannone, G. Tasinato, and D. Wands, "Generalised tensor fluctuations and inflation," *JCAP* **1501** no. 01, (2015) 029, arXiv:1409.6568 [astro-ph.CO].
[33] N. Bartolo, D. Cannone, A. Ricciardone, and G. Tasinato, "Distinctive signatures of space-time diffeomorphism breaking in EFT of inflation," *JCAP* **1603** no. 03, (2016) 044, arXiv:1511.07414 [astro-ph.CO].
[34] A. Ricciardone and G. Tasinato, "Primordial gravitational waves in supersolid inflation," arXiv:1611.04516 [astro-ph.CO].
[35] P. Adshead and M. Wyman, "Chromo-Natural Inflation: Natural inflation on a steep potential with classical non-Abelian gauge fields," *Phys. Rev. Lett.* **108** (2012) 261302, arXiv:1202.2366 [hep-th].
[36] **CLEO** Collaboration, I. Obata and J. Soda, "Chiral primordial gravitational waves from dilaton induced delayed chromonatural inflation," *Phys. Rev.* **D93** no. 12, (2016) 123502, arXiv:1602.06024 [hep-th].
[37] L. Kofman, A. D. Linde, and A. A. Starobinsky, "Reheating after inflation," *Phys. Rev. Lett.* **73** (1994) 3195–3198, arXiv:hep-th/9405187 [hep-th].
[38] G. N. Felder, J. Garcia-Bellido, P. B. Greene, L. Kofman, A. D. Linde, and I. Tkachev, "Dynamics of symmetry breaking and tachyonic preheating," *Phys. Rev. Lett.* **87** (2001) 011601, arXiv:hep-ph/0012142 [hep-ph].
[39] B. Spokoiny, "Deflationary universe scenario," *Phys. Lett.* **B315** (1993) 40–45, arXiv:gr-qc/9306008 [gr-qc].
[40] M. Gasperini, "Observable gravitational waves in pre-big bang cosmology: an update," arXiv:1606.07889 [gr-qc].
[41] C. Guzzetti, M., N. Bartolo, M. Liguori, and S. Matarrese, "Gravitational waves from inflation," *Riv. Nuovo Cim.* **39** no. 9, (2016) 399–495, arXiv:1605.01615 [astro-ph.CO].
[42] **Virgo, LIGO Scientific** Collaboration, B. P. Abbott *et al.*, "Observation of Gravitational Waves from a Binary Black Hole Merger," *Phys. Rev. Lett.* **116** no. 6, (2016) 061102, arXiv:1602.03837 [gr-qc].
[43] **Virgo, LIGO Scientific** Collaboration, B. P. Abbott *et al.*, "GW151226: Observation of Gravitational Waves from a 22-Solar-Mass Binary Black Hole Coalescence," *Phys. Rev. Lett.* **116** no. 24, (2016) 241103, arXiv:1606.04855 [gr-qc].
[44] **BICEP2, Keck Array** Collaboration, P. A. R. Ade *et al.*, "Improved Constraints on Cosmology and Foregrounds from BICEP2 and Keck Array Cosmic Microwave Background Data with Inclusion of 95 GHz Band," *Phys. Rev. Lett.* **116** (2016) 031302, arXiv:1510.09217 [astro-ph.CO].
[45] **CMB-S4** Collaboration, K. N. Abazajian *et al.*, "CMB-S4 Science Book, First Edition," arXiv:1610.02743 [astro-ph.CO].
[46] N. Tamanini, C. Caprini, E. Barausse, A. Sesana, A. Klein, and A. Petiteau, "Science with the space-based interferometer eLISA. III: Probing the expansion of the Universe using gravitational wave standard sirens," *JCAP* **1604** no. 04, (2016) 002, arXiv:1601.07112 [astro-ph.CO].
[47] P. Amaro-Seoane *et al.*, "eLISA/NGO: Astrophysics and cosmology in the gravitational-wave millihertz regime," *GW Notes* **6** (2013) 4–110, arXiv:1201.3621 [astro-ph.CO].
[48] A. Sesana, "Prospects for Multiband Gravitational-Wave Astronomy after GW150914," *Phys. Rev. Lett.* **116** no. 23, (2016) 231102, arXiv:1602.06951 [gr-qc].
[49] **KAGRA** Collaboration, K. Somiya, "Detector configuration of KAGRA: The Japanese cryogenic gravitational-wave detector," *Class. Quant. Grav.* **29** (2012) 124007, arXiv:1111.7185 [gr-qc].
[50] http://gw-indigo.org/tiki-index.php?page=LIGO-India/.
[51] B. Sathyaprakash *et al.*, "Scientific Objectives of Einstein Telescope," *Class.Quant.Grav.* **29** (2012) 124013.
[52] S. Kawamura *et al.*, "The Japanese space gravitational wave antenna: DECIGO," *Class. Quant. Grav.* **28** (2011) 094011.
[53] **Virgo, LIGO Scientific** Collaboration, B. P. Abbott *et al.*, "GW150914: Implications for the stochastic gravitational wave background from binary black holes," *Phys. Rev. Lett.* **116** no. 13, (2016) 131102,

arXiv:1602.03847 [gr-qc].

[54] A. Petiteau, "in preparation,".

[55] E. Dimastrogiovanni, M. Fasiello, D. Jeong, and M. Kamionkowski, "Inflationary tensor fossils in large-scale structure," *JCAP* **1412** (2014) 050, arXiv:1407.8204 [astro-ph.CO].

[56] P. Creminelli, J. Gleyzes, J. Norena, and F. Vernizzi, "Resilience of the standard predictions for primordial tensor modes," *Phys. Rev. Lett.* **113** no. 23, (2014) 231301, arXiv:1407.8439 [astro-ph.CO].

[57] J. Garcia-Bellido, M. Peloso, and C. Unal, "Gravitational waves at interferometer scales and primordial black holes in axion inflation," arXiv:1610.03763 [astro-ph.CO].

[58] P. D. Meerburg, R. Hložek, B. Hadzhiyska, and J. Meyers, "Multiwavelength constraints on the inflationary consistency relation," *Phys. Rev.* **D91** no. 10, (2015) 103505, arXiv:1502.00302 [astro-ph.CO].

[59] P. D. Lasky *et al.*, "Gravitational-wave cosmology across 29 decades in frequency," *Phys. Rev.* **X6** no. 1, (2016) 011035, arXiv:1511.05994 [astro-ph.CO].

[60] G. Cabass, L. Pagano, L. Salvati, M. Gerbino, E. Giusarma, and A. Melchiorri, "Updated Constraints and Forecasts on Primordial Tensor Modes," *Phys. Rev.* **D93** no. 6, (2016) 063508, arXiv:1511.05146 [astro-ph.CO].

[61] N. Bartolo *et al.*, "Science with the space-based interferometer LISA. IV: Probing inflation with gravitational waves," arXiv:1610.06481 [astro-ph.CO].

[62] Y.-T. Wang, Y. Cai, Z.-G. Liu, and Y.-S. Piao, "Probing the primordial universe with gravitational waves detectors," *JCAP* **1701** (2017) 010, arXiv:1612.05088 [astro-ph.CO].

[63] **Planck** Collaboration, P. A. R. Ade *et al.*, "Planck 2015 results. XVII. Constraints on primordial non-Gaussianity," arXiv:1502.01592 [astro-ph.CO].

[64] N. Barnaby, E. Pajer, and M. Peloso, "Gauge Field Production in Axion Inflation: Consequences for Monodromy, non-Gaussianity in the CMB, and Gravitational Waves at Interferometers," *Phys. Rev.* **D85** (2012) 023525, arXiv:1110.3327 [astro-ph.CO].

[65] S. G. Crowder, R. Namba, V. Mandic, S. Mukohyama, and M. Peloso, "Measurement of Parity Violation in the Early Universe using Gravitational-wave Detectors," *Phys. Lett.* **B726** (2013) 66–71, arXiv:1212.4165 [astro-ph.CO].

[66] V. Domcke, M. Pieroni, and P. Binétruy, "Primordial gravitational waves for universality classes of pseudoscalar inflation," *JCAP* **1606** (2016) 031, arXiv:1603.01287 [astro-ph.CO].

[67] L. Sorbo, "Parity violation in the Cosmic Microwave Background from a pseudoscalar inflaton," *JCAP* **1106** (2011) 003, arXiv:1101.1525 [astro-ph.CO].

[68] T. Fujita, J. Yokoyama, and S. Yokoyama, "Can a spectator scalar field enhance inflationary tensor mode?," *PTEP* **2015** (2015) 043E01, arXiv:1411.3658 [astro-ph.CO].

[69] **Planck** Collaboration, P. A. R. Ade *et al.*, "Planck 2013 results. XVI. Cosmological parameters," *Astron. Astrophys.* **571** (2014) A16, arXiv:1303.5076 [astro-ph.CO].

[70] N. Bartolo, S. Matarrese, M. Peloso, and A. Ricciardone, "Anisotropy in solid inflation," *JCAP* **1308** (2013) 022, arXiv:1306.4160 [astro-ph.CO].

[71] N. Seto and A. Taruya, "Measuring a Parity Violation Signature in the Early Universe via Ground-based Laser Interferometers," *Phys. Rev. Lett.* **99** (2007) 121101, arXiv:0707.0535 [astro-ph].

[72] N. Bartolo, E. Komatsu, S. Matarrese, and A. Riotto, "Non-Gaussianity from inflation: Theory and observations," *Phys. Rept.* **402** (2004) 103–266, arXiv:astro-ph/0406398 [astro-ph].

[73] J. L. Cook and L. Sorbo, "An inflationary model with small scalar and large tensor nongaussianities," *JCAP* **1311** (2013) 047, arXiv:1307.7077 [astro-ph.CO].

IOP Conf. Series: Journal of Physics: Conf. Series **840** (2017) 012031 doi:10.1088/1742-6596/840/1/012031

Acoustic waves and the detectability of first-order phase transitions by eLISA

David J. Weir
Department of Physics and Helsinki Institute of Physics, FI-00014 University of Helsinki, Finland
E-mail: david.weir@helsinki.fi

Abstract. In various extensions of the Standard Model it is possible that the electroweak phase transition was first order. This would have been a violent process, involving the formation of bubbles and associated shock waves. Not only would the collision of these bubbles and shock waves be a detectable source of gravitational waves, but persistent acoustic waves could enhance the signal and improve prospects of detection by eLISA. I summarise the results of a recent campaign to model such a phase transition based on large-scale hydrodynamical simulations, and its implications for the eLISA mission.

1. Introduction

The direct observation of gravitational waves from an astrophysical source [1] has also increased interest in direct observation of cosmological, stochastic sources of gravitational radiation. These include inflation and the processes which take place at the end of inflation, dynamics of networks of cosmic strings, and – perhaps most promisingly for future space-based gravitational wave detectors – first-order thermal phase transitions. Such a phase transition at the electroweak scale is perhaps the most promising cosmological source of directly observable gravitational waves.

A first-order phase transition proceeds by the nucleation and collision of bubbles of the new phase [2]. These bubbles may experience friction due to the bubble dynamics taking place in a bath of lighter particles. These interactions can slow the bubble walls considerably, so that they move subsonically in the medium of the plasma. Interactions between the plasma and the bubble can also, in principle at least, facilitate baryogenesis (see Ref. [3] for a recent review).

In the Standard Model, the electroweak phase transition has long been known to be a crossover [4, 5, 6], but it is quite easy to find extended models in which a first-order phase transition is possible. This therefore motivates thorough study of the processes and observable consequences of a first-order phase transition. One such consequence is the production of gravitational waves.

In the following sections I will summarise the various sources considered in the eLISA Cosmology Working Group report [7], as well as the ongoing efforts to simulate and model the underlying processes.

We assume that the phase transition took quickly on the scale of the expansion history of the universe. In the sections that follow, we will consider the energy density in gravitational waves $h^2\Omega_{\mathrm{GW}}$. This depends on a number of parameters, which we summarise here. We assume that the gravitational waves are produced when the universe is at a temperature T_*, and that this can be assumed to be the same as the nucleation temperature. The system initially has g_* relativistic degrees of freedom. The nucleation rate β is then usually quoted relative to the Hubble rate H_* when the universe has temperature T_*. The terminal wall velocity of the bubbles is v_w. Energy densities are usually expressed in terms of three ratios. First, the ratio of vacuum energy to radiation energy is $\alpha = \rho_\mathrm{vac}/\rho_\mathrm{rad}$. This quantity is typically combined with the efficiencies of conversion into fluid and scalar gradient energy, defined as $\kappa_v = \rho_v/\rho_\mathrm{vac}$ and $\kappa_\phi = \rho_\phi/\rho_\mathrm{vac}$ respectively. Analogously, one can also define the efficiency of conversion into turbulent flow, $\kappa_\mathrm{turb} \approx 0.05\kappa_v$.

With the above quantities, usually calculable from phenomenological studies of the phase transition in a given model [8], we can make predictions about the expected gravitational radiation that will result, and that is our focus here.

In this short summary, the dynamics of the thermal phase transition will be separated into three stages: the initial collision of the bubbles, the acoustic waves set up as the bubbles collide, and then the onset of turbulence. We take each of these phenomena in turn and come up with a fiducial form of the gravitational wave power spectrum informed by the latest simulation results, before turning our attention briefly to how these relate to predictions for specific models.

2. The envelope approximation and beyond

Most predictions of gravitational wave power spectra from first-order phase transitions make use of a set of approximations which, taken together, are termed the 'envelope approximation' [9, 10, 11]. This assumes that the shell of stress-energy around the bubbles (whether gradient energy of the bubble wall or fluid kinetic energy in the plasma) is very thin on the scale of the bubble radii, and that the collision happens instantaneously at any point on the bubble surface. It is also assumed that the stress energy disappears and does not propagate into the interior of the merged bubble – in other words, that the 'envelope' of the bubble is all that matters.

These assumptions seem to work well for bubbles where there is no plasma – for example, a near-vacuum phase transition – but they are not adequate to describe the full physics of the shock waves that develop around the bubbles. Therefore we take the envelope approximation to account only for the bubble walls colliding. Their contribution to the gravitational wave power spectrum is approximately

$$h^2\Omega_\phi(f) = 1.67\times10^{-5}\left(\frac{H_*}{\beta}\right)^2\left(\frac{\kappa_\phi\alpha}{1+\alpha}\right)^2\left(\frac{100}{g_*}\right)^{\frac{1}{3}}\frac{0.11v_\mathrm{w}^3}{0.42+v_\mathrm{w}^2}S_\mathrm{env}(f) \qquad (1)$$

with $S_\mathrm{env}(f)$ determined from semi-numerical evaluations of the envelope approximation,

$$S_\mathrm{env}(f) = \frac{(p+q)(f/f_\mathrm{env})^q}{1+q(f/f_\mathrm{env})^{p+q}} \qquad (2)$$

with a rising power law $q\approx3$ and falling power law $p\approx1$ (Ref. [7] takes $q=2.8$ and $p=1$). There is some detail in the peak structure which may make these indices slightly smaller. The frequency f_env is set by the size of the bubbles at the time of collision, which is itself a function of the nucleation rate β and wall velocity v_w.

Figure 1 shows that the envelope approximation works well for large numbers of colliding scalar field bubble walls, for $v_\mathrm{w}<1$.

One could replace κ_ϕ with κ_v to get an estimate of the gravitational waves from colliding shocks; indeed, this was for a long time the standard approach. However, it was demonstrated explicitly in Ref. [12] that the qualitative shape of the envelope approximation gravitational wave power spectrum is incorrect, because the acoustic wave source rapidly dominates.

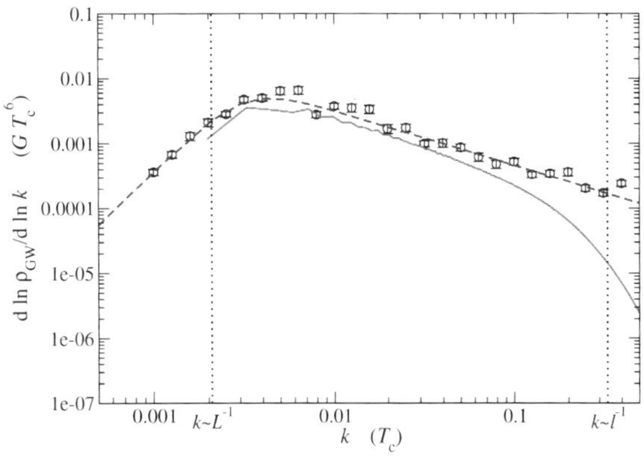

Figure 1. Plot comparing the gravitational wave energy density (which we denote $\rho_{\rm GW}$ here) from colliding scalar field walls ($v_{\rm w} = 0.44$) at the time of the phase transition as measured by two different simulation techniques. The envelope approximation result (square symbols, fit to dashed blue curve) is compared to a lattice simulation of a field-fluid system where the metric perturbations are sourced by the scalar field (solid red line). Where the two simulations are comparable, the envelope approximation adequately describes the collision of the scalar field bubble walls, although this is not usually the dominant source in a realistic scenario. For more information and the parameters used, see Ref. [12].

3. Acoustic waves

We cannot, therefore, treat the plasma shocks that surround each bubble by assuming that all the fluid kinetic energy lies in an infinitesimally thin region. Furthermore, we cannot assume that the kinetic energy dissipates immediately after the bubbles have collided. On the contrary, the collision is just the start [13, 14, 15]. After the bubbles have collided, the shock waves are no longer tied to the bubble wall and so slow down (or speed up) to move with the speed of sound in the medium.

These acoustic waves continue to interact and overlap, until expansion, shear viscosity or turbulence substantially change the dynamics. Generically it is the expansion of the universe that attenuates this source, on much longer timescales than the phase transition. The source appears stationary from the creation of the acoustic waves until this time. We concentrate on this stationary source of gravitational waves here, ignoring the initial transient effects.

We have carried out large-scale simulations of a model consisting of a relativistic scalar field, modelling the Higgs, coupled to a relativistic ideal fluid, modelling the plasma. Our most recent simulations of this system used up to simulation volumes up to 4200^3 points and took around a million core-hours per run [16].

Based on fits to our simulation results (see Fig. 2), we give the gravitational wave power as approximately

$$h^2\Omega_{\rm sw}(f) = 2.65 \times 10^{-6} \left(\frac{H_*}{\beta}\right) \left(\frac{\kappa_v \alpha}{1+\alpha}\right)^2 \left(\frac{100}{g_*}\right)^{\frac{1}{3}} v_{\rm w} S_{\rm sw}(f). \tag{3}$$

The form of $S_{\rm sw}(f)$ is the same as that for Eq. 2,

$$S_{\rm sw}(f) = \frac{(p+q)(f/f_{\rm sw})^q}{1 + q(f/f_{\rm sw})^{p+q}} \tag{4}$$

but this time we take $q = 3$ and $p = 4$. The peak frequency $f_{\rm sw}$ here is likely to be quite close

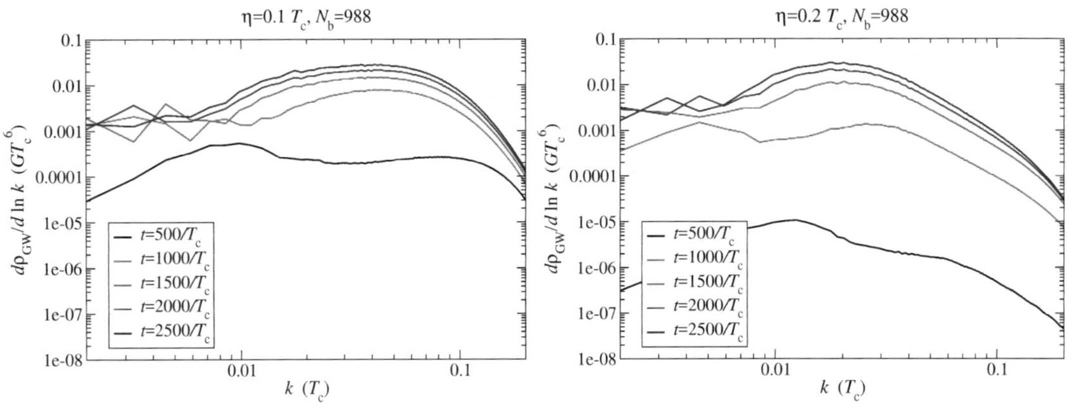

Figure 2. Gravitational wave power spectra for two example cases. At left is a simulation with wall velocity $v_\mathrm{w} = 0.83$, a detonation; at right is a simulation with $v_\mathrm{w} = 0.44$, a deflagration. Results like these were used to produce the fitting ansatz Eq. 3. For full details of the simulation parameters and interpretation of the results, see Ref. [14].

to f_env, but could differ by an order of magnitude, depending on the thickness of the fluid shell associated with the bubble wall.

The resulting gravitational wave power spectrum has a steeper high-frequency power law than the envelope approximation described above, although the overall amplitude is much higher, for a generic thermal phase transition. Work remains to be done on establishing the exact power law index but we expect to see $p \in [3, 4]$.

4. Turbulence

A thermal phase transition dumps energy into the plasma on long length scales. However, the kinetic and magnetic Reynolds numbers of the plasma are likely to be very large and so turbulence will develop in the plasma. This in turn amplifies magnetic fields and produces MHD turbulence [17]. Both turbulent processes redistribute the energy supplied to the plasma by the phase transition onto shorter length scales until a stationary spectrum is reached. The resulting gravitational wave power spectrum is [17, 18]

$$h^2\Omega_\mathrm{turb}(f) = 3.35 \times 10^{-4} \left(\frac{H_*}{\beta}\right) \left(\frac{\kappa_\mathrm{turb}\alpha}{1+\alpha}\right)^{\frac{3}{2}} \left(\frac{100}{g_*}\right)^{\frac{1}{3}} v_\mathrm{w} S_\mathrm{turb}(f) \tag{5}$$

where the shape function $S_\mathrm{turb}(f)$ is rather different from the broken power law form for S_sw and S_env, with a dependence on the Hubble rate h_* due to the duration of the source,

$$S_\mathrm{turb}(f) = \frac{(f/f_\mathrm{turb})^3}{[1 + (f/f_\mathrm{turb})]^{\frac{11}{3}}(1 + 8\pi f/h_*)}. \tag{6}$$

The characteristic frequency f_turb, like f_sw and f_env, depends on the average bubble radius at the time of collision, amongst other considerations.

In addition to our use of an ideal fluid without viscosity, the length scales captured by and duration of our simulations are currently insufficient to capture the onset of turbulent behaviour (see Fig. 3), so we are reliant upon these analytical results and treat them as a totally separate source from the acoustic waves above.

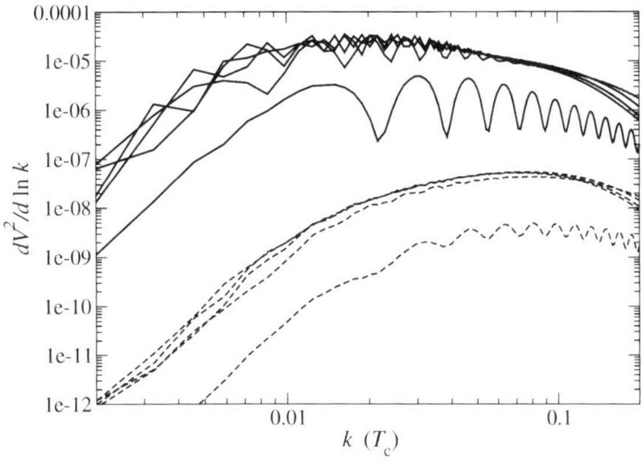

Figure 3. Comparison between rotational and longitudinal parts of the fluid velocity power spectrum computed in the simulations discussed in the main text. The viscosity, length scales and the duration of the simulation are insufficient to model the development of turbulence and so we rely on well-established analytic results from the literature [17]. The simulation parameters are the same as used to produce the plot in Fig. 2, right, and the curves are separated by the same time intervals.

5. Putting it all together

We combine the three sources described above, taking into account their relative magnitude for the scenario of interest,

$$h^2\Omega_{\rm GW}(f) = h^2\Omega_\phi(f) + h^2\Omega_{\rm sw}(f) + h^2\Omega_{\rm turb}(f). \tag{7}$$

As an example of this process, in Fig. 4 the gravitational wave power spectrum for an example of a strong, relativistic thermal phase transition is shown, along with the sensitivity curves for several eLISA mission profiles.

Note that this simple addition of the three sources neglects some subtleties. The initial collision of the fluid shock profiles is not modelled separately, and is assumed to be less important than either the colliding scalar field walls or the acoustic source. In addition, the onset of turbulence is ignored, which will in reality modulate the acoustic source. Finally, in simulations there is little separation between length scales and so considerable extrapolation is required to the large separation between wall thickness and bubble radius that would actually exist in a realistic electroweak phase transition.

6. Conclusions

We have summarised recent work on simulating and modelling the processes which take place in a thermal phase transition, as well as recent efforts to turn the results of those simulations into meaningful predictions for LISA.

The key result is that a wide variety of models and scenarios are detectable by LISA, perhaps wider than previously anticipated. The acoustic source being more significant than the mere collision of fluid shells (to which the envelope approximation was previously most commonly applied) would suggest. The result is an enhancement of order $60\beta/H_*$ in $h^2\Omega_{\rm GW}$.

Much remains to be done: the onset of turbulence has not been modelled; the power laws arising from acoustic waves need to be better understood; and the qualitative results are mostly from numerical simulations. Nevertheless there is cause for optimism in the ability of LISA to detect a very wide variety of first-order phase transitions at or around the electroweak scale. It should prove to be an important tool for high energy physics in the coming decades.

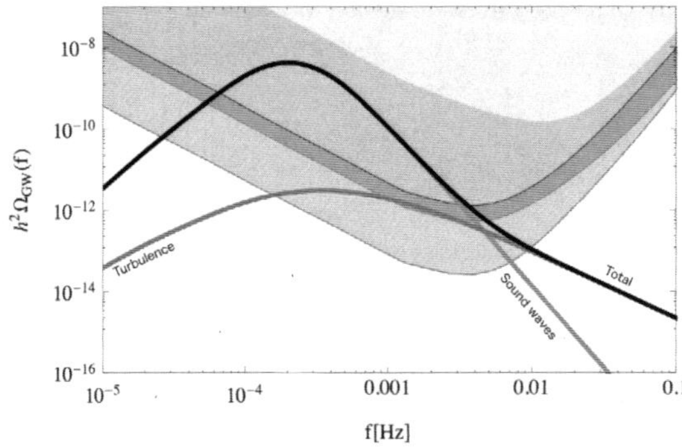

Figure 4. Gravitational wave power spectrum with transition temperature $T_* = 100\,\mathrm{GeV}$, $\alpha_{T_*} = 0.5$, $v_\mathrm{w} = 0.5$ and $\beta/H_* = 10$. For this parameter choice, the acoustic source is substantially responsible for the potential detectability of such a transition. Further details can be found in Ref. [7].

Acknowledgments

I am grateful to my collaborators Mark Hindmarsh, Stephan Huber and Kari Rummukainen for their collaboration on much of the simulation work summarised above. I am also grateful to my collaborators within the eLISA Cosmology Working Group and its coordinators, Chiara Caprini and Germano Nardini. I acknowledge PRACE for awarding access to resource HAZEL HEN based in Germany at the High Performance Computing Center Stuttgart (HLRS). My work was supported by the People Programme (Marie Skłodowska-Curie actions) of the European Union Seventh Framework Programme (FP7/2007-2013) under grant agreement number PIEF-GA-2013-629425. This research was supported by the Munich Institute for Astro- and Particle Physics (MIAPP) of the DFG cluster of excellence "Origin and Structure of the Universe".

References

[1] Abbott B P *et al.* 2016 *Phys. Rev. Lett.* **116** 061102 (*Preprint* 1602.03837)
[2] Steinhardt P J 1982 *Phys. Rev* D **25** 2074
[3] Morrissey D E and Ramsey-Musolf M J 2012 *New J. Phys.* **14** 125003 (*Preprint* 1206.2942)
[4] Kajantie K, Laine M, Rummukainen K and Shoposhnikov M E 1996 *Phys. Rev. Lett.* **77** 2887 (*Preprint* hep-ph/9605288)
[5] Gurtler M, Ilgenfritz E M and Schiller A 1997 *Phys. Rev.* D **56** 3888 (*Preprint* hep-lat/9704013)
[6] Csikor F, Fodor Z and Heitger J 1999 *Phys. Rev. Lett* **82** 21 (*Preprint* hep-ph/9809291)
[7] Caprini C *et al.* 2016 *J. Cosmol. Astropart. Phys.* JCAP04(2016)001 (*Preprint* 1512.06239)
[8] Espinosa J R, Konstandin T, No J M and Servant G 2010 *J. Cosmol. Astropart. Phys.* JCAP06(2010)028 (*Preprint* 1004.4187)
[9] Kosowsky A, Turner M S and Watkins R 1992 *Phys. Rev. Lett.* **69** 2026
[10] Kamionkowski M, Kosowsky A and Turner M S 1994 *Phys. Rev.* D **49** 2837 (*Preprint* astro-ph/9310044)
[11] Huber S J and Konstandin T 2008 *J. Cosmol. Astropart. Phys.* JCAP09(2008)022 (*Preprint* 0806.1828)
[12] Weir D J 2016 *Phys. Rev.* D **93** 124037 (*Preprint* 1604.08429)
[13] Hindmarsh M, Huber S J, Rummukainen K and Weir D J 2014 *Phys. Rev. Lett* **112** 041301 (*Preprint* 1304.2433)
[14] Hindmarsh M, Huber S J, Rummukainen K and Weir D J 2015 *Phys Rev.* D **92** 123009 (*Preprint* 1504.03291)
[15] Hindmarsh M 2016 *Preprint* 1608.04735
[16] Hindmarsh M, Huber S J, Rummukainen K and Weir D J (*In preparation*)
[17] Caprini C, Durrer R and Servant G 2009 *J. Cosmol. Astropart. Phys.* JCAP12(2009)024 (*Preprint* 0909.0622)
[18] Binetruy P, Bohe A, Caprini C and Dufaux J F 2012 *J. Cosmol. Astropart. Phys* JCAP06(2012)027 (*Preprint* 1201.0983)

Massive Primordial Black Holes as Dark Matter and their detection with Gravitational Waves

Juan García-Bellido

Instituto de Física Teórica UAM-CSIC, Universidad Autónoma de Madrid, Cantoblanco, 28049 Madrid, Spain

E-mail: juan.garciabellido@uam.es

Abstract. Massive Primordial Black Holes (MPBH) can be formed after inflation due to broad peaks in the primordial curvature power spectrum that collapse gravitationally during the radiation era, to form clusters of black holes that merge and increase in mass after recombination, generating today a broad mass-spectrum of black holes with masses ranging from 0.01 to $10^5\ M_\odot$. These MPBH could act as seeds for galaxies and quick-start structure formation, initiating reionization, forming galaxies at redshift $z > 10$ and clusters at $z > 1$. They may also be the seeds on which SMBH and IMBH form, by accreting gas onto them and forming the centers of galaxies and quasars at high redshift. They form at rest with zero spin and have negligible cross-section with ordinary matter. If there are enough of these MPBH, they could constitute the bulk of the Dark Matter today. Such PBH could be responsible for the observed fluctuations in the CIB and X-ray backgrounds. MPBH could be directly detected by the gravitational waves emitted when they merge to form more massive black holes, as recently reported by LIGO. Their continuous merging since recombination could have generated a stochastic background of gravitational waves that could eventually be detected by LISA and PTA. MPBH may actually be responsible for the unidentified point sources seen by Fermi, Magic and Chandra. Furthermore, the ejection of stars from shallow potential wells like those of Dwarf Spheroidals (DSph), via the gravitational slingshot effect, could be due to MPBH, thus alleviating the substructure and too-big-to-fail problems of standard collisionless CDM. Their mass distribution peaks at a few tens of $M_\odot$ today, and could therefore be detected also with long-duration microlensing events, as well as by the anomalous motion of stars in the field of GAIA. Their presence as CDM in the Universe could be seen in the time-dilation of strong-lensing images of quasars. The hierarchical large scale structure behaviour of MPBH does not differ from that of ordinary CDM.

1. Introduction

The discovery of several massive black hole mergers by the LIGO interferometer [1, 2, 3, 4, 5, 6] has opened the new era of Gravitational Wave Astronomy, exactly one century after Albert Einstein proposed his theory of gravitation. The masses of these black holes are somewhat larger than expected from remnants of supernovae explosions and stellar evolution [5, 7], so it is possible that LIGO has discovered a whole new population of massive black holes, formed in the early universe. Our aim is to understand how to generate such a population, and search for any signatures that will distinguish them from stellar black holes. It was first proposed in Ref. [8] that massive primordial black holes (MPBH) could arise from high peaks in the primordial matter power spectrum that collapse gravitationally during the radiation era. More recently, we found [9] that broad peaks in the power spectrum would form clusters of black holes, which would merge and accrete mass (gas and smaller black holes) since recombination, reaching tens to hundreds (a few of them can reach billions) of solar masses today. These MPBH could be the dominant of the Dark Matter in the Universe [8]; they would act as seeds for the first galaxies, reionizing the intergalactic medium and forming galaxies and quasars at high redshift [9]. Here we will explore the consequences of this new scenario for the understanding of dark matter and extend the very rich phenomenology that primordial black holes induce in the late Universe.

Since Zwicky proposed it to explain the stability of clusters of galaxies [10], the nature of dark matter has puzzled both physicists and astronomers alike. Today we have evidence of dark matter from the rotation curves of galaxies [11], from gravitational lensing on large scale structures [12] and from the temperature anisotropies in the CMB [13]. According to the usual ΛCDM paradigm, structures like galaxies and clusters of galaxies form by hierarchical accretion of ordinary gas on dark matter halos. All observations so far rely solely on the gravitational interaction of dark matter, but gives no hint about its other interactions. Many different hypothesis have been put forward on its nature, with ninety orders of magnitude in the range of masses of its possible components, from ultralight axions [14] to massive black holes [15]. Until recently, the most popular candidates have been some weakly interacting particles (WIMP) predicted in certain extensions of the standard model of particle physics, but no evidence of any such particle has been found so far at the Large Hadron Collider (LHC) at CERN, neither in underground facilities, with cryogenic detectors like CDMS or liquid Xenon detectors like LUX, nor by indirect searches of their annihilation with astronomical satellites like Fermi [15].

An alternative and very natural hypothesis, considered for a long time, is that of MAssive Compact galactic Halo Objects (MACHOs), such as planets, brown dwarves, neutron stars or black holes. Such compact objects were searched for decades ago, thanks to the microlensing of millions of stars [16, 17] in nearby galaxies like the Large Magellanic Cloud (LMC) or Andromeda [18, 19, 20, 21, 22, 23, 24, 25]. The search stopped ten years ago when the experiments failed to find MACHOs below 10 solar masses as the main component of the dark matter halo of our galaxy [21], since compact objects with larger masses were unexpected [22]. Stellar black holes are the end points of stellar evolution under gravitational collapse after a supernova explosion. Most of these black holes have stellar masses, but only a few are above 10 $M_\odot$. Their masses are usually determined from the X-ray emission of their accretion disks. The more massive BH are supposed to arise from the gravitational collapse of population III supermassive stars, with very low metallicity, otherwise most of their mass will be blown away and will not end up inside the black hole. Nevertheless, it is difficult to produce BH from stellar evolution with masses above 10 $M_\odot$, even in the progenitor star had a mass above 250 $M_\odot$ [7]. It is therefore rather surprising that the first direct detection of gravitational waves from the merger of two black holes involved not just one but two $\sim 30\ M_\odot$ black holes. So, how did they form in the first place? and second, how did they find each other and merge within the lifetime of the Universe?

With the recently born gravitational wave astronomy thanks to LIGO and the first detection of 30 $M_\odot$ black holes, we have opened again the window on the search for MPBH as the main component of dark matter in the Universe [9, 26, 27, 28, 29]. Furthermore, massive PBH could be the missing link between early star and galaxy formation soon after recombination and the supermassive black holes at the centers of quasars and active galactic nuclei observed at high and intermediate redshifts. The use of gravitational wave interferometers as astronomical tools will help characterize this new population of black holes and measure their mass and spin distribution, which will allow astronomers to identify their progenitors and formation rates.

In fact, LIGO has detected not just one massive BH binary merger, but *three* events, in the course of run O1, from September 2015 to January 2016, see Figs. 1 and 2. The masses of the inspiralling BH before merging are not all equal, they range from 8 to 36 $M_\odot$, and the final BH masses range from 23 to 62 $M_\odot$. If there is a population of BHs out there with large masses, which do not arise from stellar evolution, where did they come from?

A natural hypothesis is that these massive black holes are of primordial origin. That is, they were there before the stars started to shine. The first proposal of primordial black holes dates back to Hawking's work on black hole evaporation. He expected tiny black holes, of

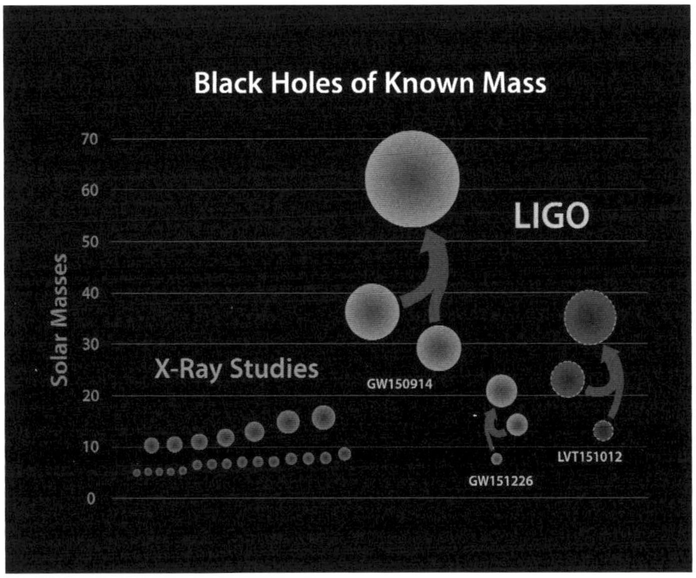

Figure 1. The Black Holes of Known Mass detected by LIGO. It is clear that they correspond to a new population of black holes unheard off before. While IMBH and SMBH were known to populate the centers of globular clusters and galaxies, respectively, this new class of black holes in binaries had not been detected before. Figure from [LIGO webpage].

masses of order 10^{15} g, to be evaporating today, emitting in their last stages a burst of energy that he thought at the time could be the observed gamma ray bursts [30, 31]. Now we know that GRB have a different origin and, moreover, these PBH are too light to account for the LIGO observations. Since then, a series of proposals have been made. For instance, if the QCD quark-hadron transition were first order, then the resulting zero-pressure fluid may have allowed primordial fluctuations to collapse to form black holes below [32] and around a solar mass, corresponding to the mass within the horizon at that time. This was further studied by Jedamzik [33, 34], to account for the measured microlensing events towards the LMC, found by the MACHO collaboration [21]. Now we know that the QCD transition is not first order, but actually is a crossover, so the equation of state is non-zero, and the small fluctuations generated during inflation cannot give rise to PBH at that stage.

Our scenario of massive PBH is a different scenario, based on the fundamental physics of inflation, which has important consequences for the whole of Astronomy and Cosmology. We would like to understand the nature of these massive black holes detected by LIGO and determine their origin, whether primordial or not. If the merging rate is as high as the LIGO collaboration seem to suggest, then we will have hundreds of events in the next few years, enough to convince ourselves that we are in front of a new class of BH. This new scenario opens up a multi-probe, multi-epoch and multi-wavelength approach to the nature of Dark Matter in the form of MPBH.

In this brief review we explore the very rich phenomenology that arises in this scenario, and describe a broad range of new signatures that could be measured in a wide variety of astrophysical and cosmological observations in order to test this hypothesis. In section **II**, I describe the various formation scenarios, with different ways of generating peaks in the matter power spectrum, studying the different models of PBH growth via gas accretion and BH merging. In section **III**, I describe the very rich phenomenology that such a scenario puts forward and connects many different aspects of Astrophysics and Cosmology. It is possible that in the next

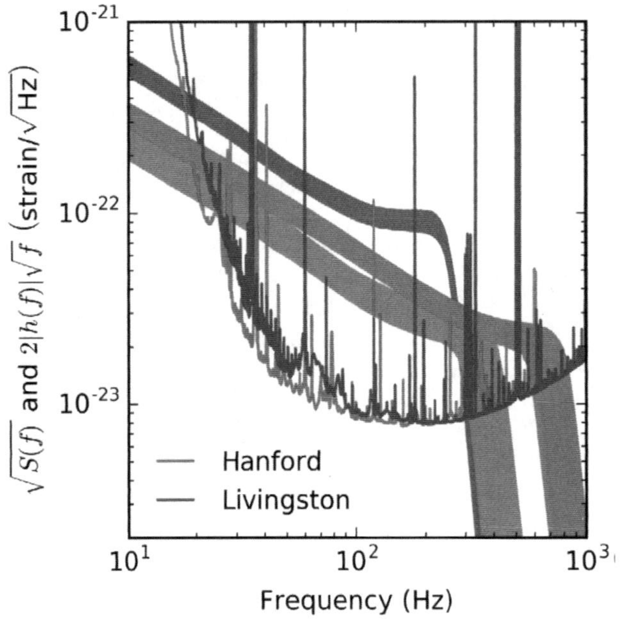

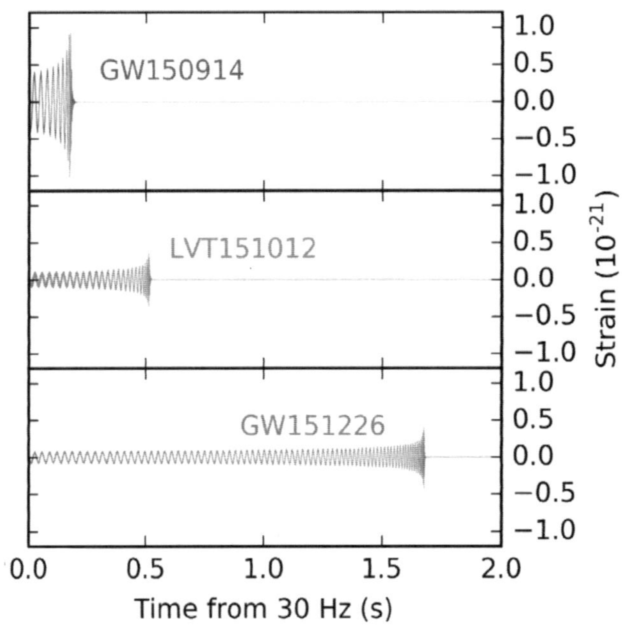

Figure 2. The strain sensitivity and amplitude of the three GW events detected by advanced LIGO as they inspiral towards merging. The inspiralling binary black hole (BBH) coalescence time-span after the 30 Hz mark of the three events. More massive binaries would not be seen by LIGO, since $f_{\mathrm{ISCO}} = 44$ Hz $(100\ M_{\odot}/M_{\mathrm{tot}}) > 30$ Hz implies $M_{\mathrm{tot}} < 147\ M_{\odot}$, but better seismic attenuation by Virgo and KAGRA may help explore the higher mass range. Figures from Ref. [6].

decade we may have a final picture of the nature of dark matter. In section **IV**, I explore one particular aspect of the whole list of phenomenological signatures, that of the anomalous motion of stars in the solar neighborhood, which may be detectable in the near future by the GAIA mission, and entertain on the phenomena of gravitational slingshot for ejecting stars from Globular Clusters and Dwarf Spheroidals (DSph). In section **V**, I describe evidence for microlensing in the time residuals of strong lensing of distant quasars. In section **VI**, I explore the gamma and X-ray emissions from MPBH in our galaxy and local group as seen by Fermi and Chandra respectively. In section **VII**, I study the different sources of GW from MPBH. In section **VIII**, I conclude with a summary and some discussion.

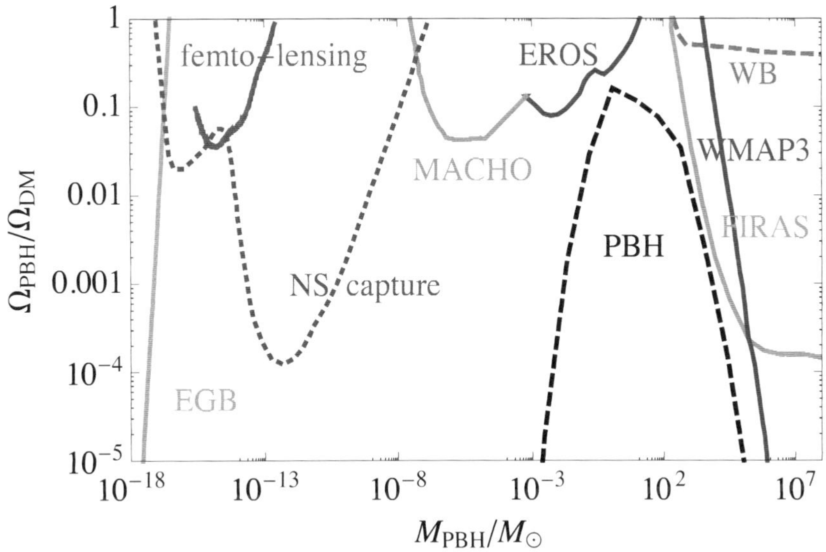

Figure 3. Limits on the abundance of PBH today, from extragalactic photon background (orange), femto-lensing (red), micro-lensing by MACHO (green) and EROS (blue), from wide binaries (light brown), and CMB distortions by FIRAS (cyan) and WMAP3 (purple). The constraints from star formation and capture by neutron stars in globular clusters are displayed for $\rho_{\rm DM}^{\rm Glob.Cl.} = 2 \times 10^3$ GeV/cm^3 (dark brown). The black dashed line corresponds to a particular realization of our scenario of PBH formation. Figure adapted from Ref. [9].

2. Massive PBH formation scenarios

In 1996, together with Linde and Wands [8], we realized that, if the primordial curvature fluctuation spectrum had a pronounced peak at some particular scale, it was possible to generate PBHs whenever those high curvature fluctuations reentered the horizon during the radiation era, since nothing could prevent their gravitational collapse. The mass of those PBH would be given very approximately by the total mass within the horizon at the time of reentry. Moreover, in certain realizations of inflation, it was possible that the number of PBH would be so large that these could become the bulk of the dark matter in the universe [8]. At the time, the constraints on BHs of a solar mass were not very strong. Thanks to MACHO and EROS collaborations, the window on solar mass BHs as the sole (100%) contributor to the Milky Way halo was ruled out, and very stringent bounds were put at the level of less than 10% of the halo in masses above

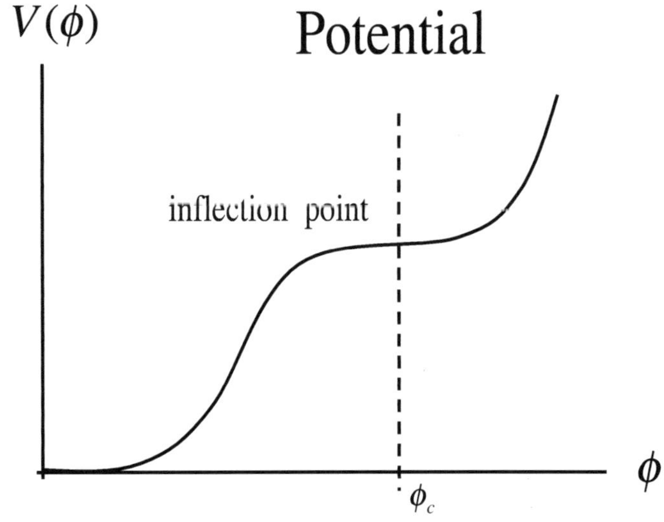

Figure 4. Generic inflationary potential giving rise to peaks in the power spectrum. In two-field models like hybrid inflation the inflection point is substituted by the symmetry breaking point. In both cases there are a few-folds of inflation after the critical point ϕ_c.

10 $M_\odot$. This seemed to close the window on massive PBH as the origin of dark matter, and thus encouraged the search for particle dark matter (PDM) as the main contributor to the DM of the Universe. Some groups still consider the possibility of low-mass PBH (below planetary size) as the main component of DM [35], although there are very stringent constraints on these, see e.g. the reviews Ref. [36, 29].

More recently, Clesse and García-Bellido proposed a scenario [9] where, rather than a single sharp peak in the matter power spectrum, one could generate a *broad spectrum* of PBH masses from a large peak in the primordial spectrum of curvature fluctuations from inflation. This broad mass spectrum could account for all of the dark matter, and nevertheless pass all the bounds from wide binaries and microlensing events of our halo, since any given range of masses is always below the 10% fraction of the halo mass, see Fig. 3.

2.1. Peaks in the primordial Power Spectrum

The width of the peak in the matter power spectrum that gives rise to the mass range of PBH depends on the particular inflationary model behind the origin of curvature fluctuations that gave rise to the formation of structure in the universe. Some models of inflation may have a phase transition 20 e-folds before the end of inflation, triggered by a symmetry breaking field, like in hybrid inflation [8], and this will give rise to a very pronounced peak in the spectrum, whose width will depend on the parameters of the model. Another possibility is to have the inflaton coupled to gauge fields, which grow exponentially and themselves backreact on the curvature, producing high peaks in the spectrum [37]. Alternatively, one can think of single-field models with an inflection point in the potential, see Fig. 4, which slows down the scalar field driving inflation and produces a broad peak in the spectrum [38].

While the amplitude of anisotropies in the cosmic microwave background cannot exceed

IOP Conf. Series: Journal of Physics: Conf. Series **840** (2017) 012032 doi:10.1088/1742-6596/840/1/012032

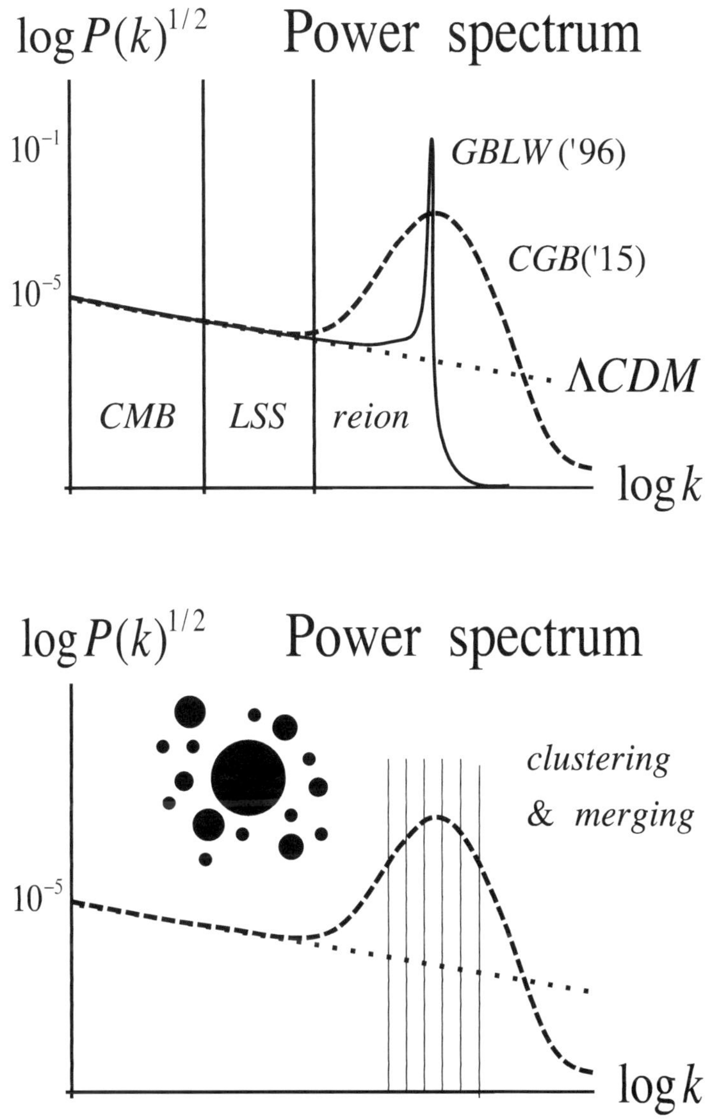

Figure 5. The primordial power spectrum of curvature fluctuations induced during inflation. Those fluctuations that enter during the radiation era collapse to form black holes within a range of masses, and spatially clustered around the more massive ones. The dotted line corresponds to the Gaussian primordial spectrum predicted by inflation and at the core of the ΛCDM paradigm, and consistent with the observed CMB anisotropies; the continuous line corresponds to the original model of GBLW (1996) [8], with a sharp peak in the spectrum, that gives rise to a monochromatic mass spectrum, and the dashed line corresponds to the recent proposal of CGB (2015) [9] for a broad peak in the spectrum, giving rise to a broad mass spectrum of PBH, which are strongly clustered.

a few parts in 10^{-5} on the scale of the horizon, the fluctuation spectrum on much smaller scales is largely unconstrained. The CMB only probes the 50-to-60 efolds range, depending on the actual temperature of reheating after inflation, and future 21cm surveys with intensity mapping will probe the reionization epoch of 50-to-40 efolds range, see Fig. 5. The peak in the primordial spectrum would begin to become important around 40-to-20 efolds before the end of inflation, on scales never directly probed by any astronomical survey, since on those scales, density fluctuation quickly become non-linear and gravitational collapse washes out any trace of the primordial spectrum. Nevertheless, if there is a large peak, as in Fig. 5, the gravitational collapse of those fluctuations upon reentry will form a broad mass distribution of PBH, that could act as seeds for massive stars and galaxies.

Observations will soon determine the mass and spin spectrum of the MPBH responsible for the measured Gravitational Waves from BBH coalescence. This could then be used as a test of the mechanisms for PBH production, opening a new window into the last 40 e-folds of inflation, and thus help us access the ultra high energy fundamental physics domain that is completely inaccessible by particle physics accelerators [38].

2.2. Accretion and merging

In this subsection we study the evolution through accretion and merging of the population of PBH. On the one hand, during the radiation era, they can accrete gas and radiation via Bondi-Hoyle accretion but this rate is negligible. Later, during the matter era, they can start to accrete faster, but always below the Eddington limit. Alternatively, if they acquire an accretion disk of gas, and the outflows are along the axis, like in quasars and microquasars [63], they can accrete much faster. On the other hand, since MPBH originate in clusters, they can find each other and merge more easily, by gravitational attraction, emitting gravitational waves in the process, and generating a stochastic background [39].

Therefore, their masses grow at a significant rate, both through accretion and merging, thus shifting their mass distribution towards larger values since recombination, and increasing the high-mass tails of the distribution. On top of that, since PBH form a fluid of zero pressure, their density contrast (i.e. inhomogeneities in the spatial distribution of PBH, arising from the standard metric fluctuations induced during inflation) grows like any matter component, linear with the scale factor during the matter era. Apart from changes in their mass distribution, we expect also changes in their spin distribution. PBH are born without spin since they arise from the gravitational collapse of an isotropic gas upon horizon crossing of order-one fluctuations in curvature. However, the spin will build up from subsequent mergers. As PBH masses start to build up through merging of smaller PBH, they can acquire a significant spin from the angular momentum of the inspiralling orbit. However, subsequent mergers occur from a random distribution of orientations of the orbits, and it is likely that the final spin distribution will be centered around the zero-spin configuration, with a dispersion that depends on the merging history and mass distribution. We are performing numerical relativity simulations in order to derive a prediction for the mass and spin distributions of present black holes.

3. Signatures of massive PBH

In this section we will discuss the possible signatures of primordial black holes as the dominant component of Dark Matter. I will itemize here some of those signatures that I believe could be used to distinguish PBH from stellar black holes. They range from early to late epochs of the evolution of the Universe, with different probes (electromagnetic and gravitational waves) and different frequencies (from nHz to GHz and from meV to tens of GeV). Some of these signatures are ready to be explored; others will take a few years, or even decades. There is a lot of work ahead to thoroughly test the scenario.

- **CMB distortions and anisotropies**. Massive PBH formation during the radiation era has important consequences on the CMB. Gas falling into PBH through Bondi-Hoyle accretion will reinject some fraction of its energy into the strongly-coupled plasma before recombination, which will then distort the blackbody spectrum of photons at decoupling [40]. There are stringent limits on the CMB y and z-distortions, which can be evaded if the PBH have small masses before recombination, which then grow during the matter era. The most stringent bounds come from a paper by Ricotti et al. [41], which has an error in Eq.(44): There is a factor $(1+z)^2$ missing, which changes the bounds by several orders of magnitude towards larger PBH masses, leaving a range of values around $M \sim 100\,M_\odot$ unconstrained, see also [42, 43]. This is the reason why we shifted these bounds (labeled FIRAS and WMAP) to the right in Fig. 3.
- **First stars in the Universe**. The random distribution of PBH will act as seeds for gas to fall and initiate star formation and the reionization of the universe. This will generate a UV and gamma background at high redshift ($z \sim 20$) that could be seen today redshifted into the infrared and soft X-ray, respectively. The recent measurement of strong cross-correlations between fluctuations in the CIB and the diffuse X-ray background [44] suggests that a population of PBH could have initiated star formation and reionization at high redshift and also be responsible for the sources generating the present X-ray background.
- **Reionization and 21cm Intensity mapping**. While present CMB experiments give us a measurement of the integrated optical depth from the last scattering surface to the present [13], future 21cm intensity mapping surveys like SKA will be able to measure the whole history of reionization, from redshift $z \sim 20$ to $z \sim 6$, thus constraining the local sources of reionization at high redshift [45]. The power spectrum of fluctuations on 21cm maps will put very stringent limits on the mass distribution of PBHs.
- **Early galaxy formation**. A population of massive primordial black holes would act as centers of accretion of ordinary gas, and give rise to galaxies at high redshift ($z \sim 10$), very massive black holes at the centers of quasars ($z \sim 6$) and massive clusters today ($z \sim 1$). See for instance the recent discovery of a galaxy cluster at $z = 2.5$ [46]. In general, massive PBH quick start structure formation, compared to the standard ΛCDM paradigm, since they seed structures much more efficiently than a rare 5σ-fluctuations in the Gaussian primordial spectrum [40].
- **Formation of SMBH and IMBH**. The successive PBH merging and gas accretion since recombination produces supermassive black holes (SMBH) at the centers of galaxies, and intermediate mass black holes (IMBH) at the centers of globular clusters. Their PBH origin is common to both types, and is the agent responsible for the velocity dispersion of the system (stars in the bulge and cluster respectively). Not surprisingly, there seems to be a simple power law relation between the mass of the black holes at the center of those systems and their velocity dispersion [47], see Fig. 8. In the MPBH scenario *all* DM halos have a black hole at its center, which grows by accretion of gas and the differences in masses are related to the time since first accretion. The fact that there is an underlying mechanism common to all massive black holes, from microquasars to SMBH, is seen here as a correlation which is universal.
- **Superluminal SNe**. A distribution of massive PBH acting as DM in the universe should be a source of femtolensing of distant powerful sources like supernovae explosions. Some very luminous and strongly time-delayed supernovae have been observed, with emissions in the infrared that last for months before decaying. They are called superluminous SNe and there are a handful of them [48, 49]. It could be that some of these are microlensed

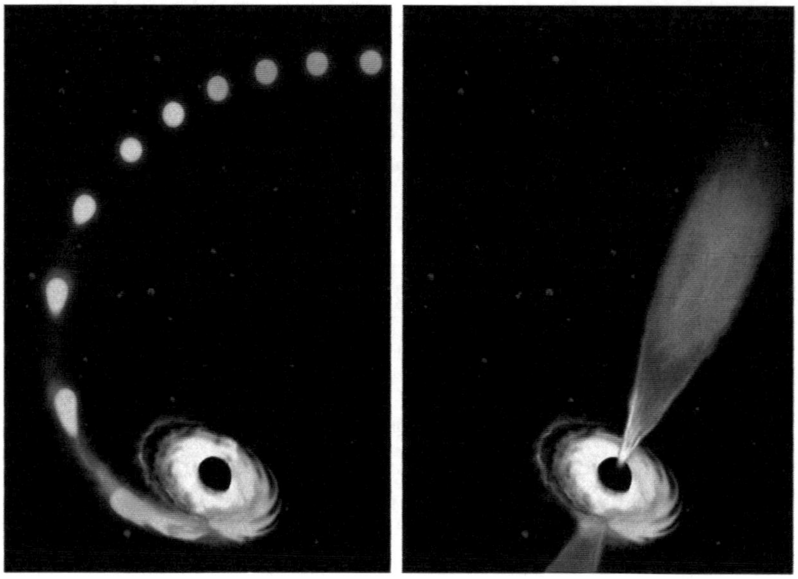

Figure 6. The tidal disruption of a star captured by a PBH and subsequent flash of X-rays. This may end up on a ULX-ray transient of an accretion disk like in a microquasar. *Credit: modified from an original image by Amadeo Bachar.*

SN of known types, see e.g. [50], whose light has been amplified by a femtolensing event on massive PBH in the line of sight.

- **Long-duration microlensing**. The duration of microlensing events towards the LMC is proportional to the mass of the lensing object [16, 17]. If the halo of the Milky Way dark matter is constituted mainly by PBH with a range of masses peaked at around 50 $M_\odot$, the typical full duration is of the order of a decade. The stars in the LMC have not been monitored long enough by microlensing experiments to exclude a halo made of PBH with those masses if PBH at the peak constitute less than 10% of the halo, see Fig. 3 and Refs. [21, 24]. Moreover, the light curves due to microlensing of clustered PBH is expected to have multiple caustics due to smaller PBHs orbiting the main deflector, and this may be difficult to distinguish from planets around stars in LMC, so they may have passed unnoticed. Also, the fact that these PBH are clustered decreases the optical depth towards a relatively compact structure like the LMC, by increasing the relative separation between clusters of PBH.
- **Residual microlensing in time-matched strong-lensed multiple images of quasars**. Quasars that suffer strong lensing by an intervening galaxy show two or more images with identical short-time variations delayed by a constant time-shift between the images [51], typically of order 100 days [52]. The residual long-range modulation in amplitude has a peak and a duration above a decade, which matches a possible microlensing by a cluster of massive PBH, see also [53].
- **Tidally disrupted stars and ULX-ray transients**. It has been observed in both the Milky Way and nearby galaxies like Andromeda the disruption of a star by a black hole. Some of the ultraluminous X-ray transients are believed to arise from these kind of tidal-disruption events, although there seems to be more events than expected if those black holes come from stellar origin, see Fig. 6 and Ref. [54].

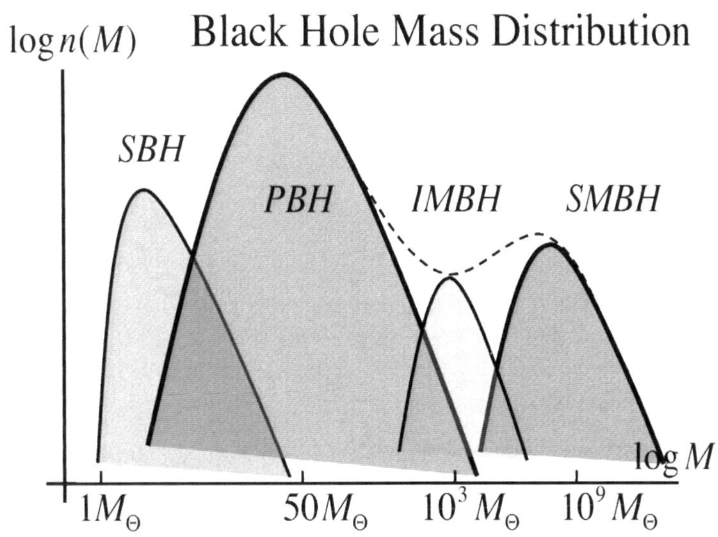

Figure 7. The mass distribution of black holes in our universe can be classified in three large groups, Stellar Black Holes, that arise from the gravitational collapse of stellar systems, the Primordial Black Holes that were formed in the early Universe, the Supermassive Black Holes at the centers of Quasars and the Intermediate Black Holes at the centers of Globular Clusters. The scenario described here considers the possibility that the last three kinds are actually closely related. There could be another population of BHs at the center of microquasars with tens of solar masses and accretion disks arising from stellar capture. Their distribution is still rather uncertain, see however [55].

- **Large Scale Structure N-body simulations**. The scenario of MPBH cannot be distinguished at cosmological superstructures scales from that of Particle DM, since the evolution of dark matter on large scales behaves in a way characterized by the primordial spectrum set during inflation and in accordance with the measured anisotropies in the CMB. The growth of structure occurs via the gravitational collapse of a zero-pressure fluid which, at the resolution scales of the simulations (particle mass $\sim 10^6 - 10^9\ M_\odot$), cannot distinguish PBH of 100 $M_\odot$ from particle DM [56].
- **Substructure and too-big-to-fail problems**. At small galactic scales, numerical simulations of structure formation based on CDM seem to give many more smaller structures than observed. We realized in [39] that a core of MPBH at the center of shallow potential wells like those of dwarf spheroidals could be enough to expel solar-mass stars from these structures, by inducing higher than escape velocities, through the gravitational slingshot effect, see below. This could be the reason for the high mass-to-light ratios of DSph, and why we don't observe the stellar light from those substructures. These substructures have recently been observed with dedicated high exposures in sensitive cameras like DECam [65]. Similarly with large objects that should have been seen and only recently have been detected via gravitational lensing. Thus MPBH could be responsible for the solution to the substructure and too-big-to-fail problems of Cold Dark Matter.
- **Missing baryon problem**. Perhaps PBH have eaten up most of the baryons since recombination. The last time they were measured in the CMB were in agreement with

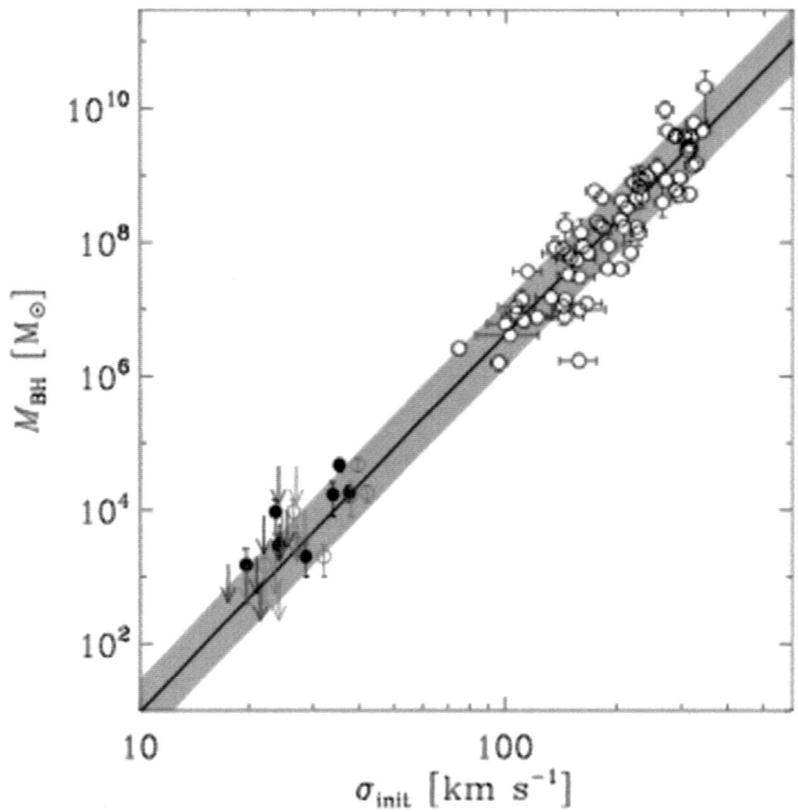

Figure 8. The power law relation between the mass of the central black hole and the velocity dispersion of the system extends all the way from Supermassive Black Holes at the centers of Quasars to the Intermediate Black Holes at the centers of Globular Clusters. It is expected that further searches will show that these two classes are not distinct, but actually correspond to a common origin. In fact, a third class of BHs at the centers of microquasars in our galaxy may also have a similar relation between the central BH mass and the dispersion velocity of the gas in the accretion disk around it. From Ref. [47].

BBN light element abundance. There seems to be less in the present universe in hot plasmas around clusters and in galaxies [57]. They may simply not shine or they could be locked inside the PBH that account for the CDM in the Universe.

- **Cluster collisions and cross-sections**. The bullet cluster is used to constrain models of PDM by setting stringent limits on the interaction cross-sections of DM particles, $\sigma/M < 1\ \mathrm{cm}^2/g$ [76], which can be evaded by PBH with Schwarzschild-Born cross-sections and $M \sim 50\ M_\odot$, by many orders of magnitude. That is, PBHs are essentially the best candidates for collisionless dark matter.
- **Wide binaries in the Milky Way**. These are very sensitive probes of massive objects passing by, like PBHs, since they can disrupt the orbits of far-apart components of the binary and break it up. Stringent bounds have been recently updated and allow for less than a few percent halo fraction in the form of 50 $M_\odot$ PBH, in agreement with Fig. 3 and Ref. [77].

- **Compact stellar clusters in dwarf galaxies**. Dynamical friction tend to move both massive black holes and stellar clusters to the center of dwarf galaxies, where they interact and heat up, puffing up the stellar clusters [78]. Long lived star clusters at the centers of dwarf galaxies therefore can be used to detect massive PBH, although the constraints weaken if there is an intermediate mass BH at the center of the dwarf galaxy, which provides stability and increases the velocity dispersion [79].
- **Lensed Fast Radio Bursts**. Strong lensing of extragalactic fast radio bursts (FRB) induces a repetition of the burst signal with time delays of a few milliseconds. A survey of tens of thousands of sources with future radio facilities like CHIME could be used to constrain or detect MPBH in the $20 - 100\ M_\odot$ [80].
- **X-ray binaries and microquasars**. We don't know the origin of X-ray binaries, where a BH accretes gas from a companion star, forming an accretion disk and emitting in X-rays. They come in two classes, low-mass and high-mass XRB. They could arise from stellar binaries that have evolved together and the companion survived the supernova explosion that lead to the BH, although typically those BHs acquire huge kicks and move away from the disk of the galaxy, most of the times loosing their stellar companion. In fact, very massive BH at the centers of accretion disks are rare outcomes from supernova explosions, since these are usually low mass BHs. It is possible to imagine a different origin of High-Mass XRB, as a companion star being captured by a passing-by isolated MPBH and initiating an accretion disk from gas outflows as the star orbits the black hole. This could be a scenario close to that of ULX-ray transients, where the impact parameter is not so small and the star survives the BH encounter loosing momentum and forming a bound system. A recent detection by Chandra of a new population of low-mass XRB [55] could hint in that direction. Moreover, microquasars and High-Mass XRB, that do not move at high speeds inside the galaxy, could have their origin in PBH that have captured a star and formed an accretion disk. In that case, there should be a similar relation between the central BH mass and the dispersion velocity of the gas around it, like in IMBH and SMBH, see Fig. 8.
- **Fermi 3FGL and Chandra point sources**. The fourth year Fermi-LAT point source catalog shows 30% of unknown gamma ray sources [81]. Most of them are probably blazars and quasars, but many could be nearby sources within our galaxy. Some could be massive primordial black holes in dense environments that reinject energy into the ISM. The Chandra satellite also finds large numbers of compact X-ray sources in our galaxy and other nearby galaxies like Andromeda. A fraction of those could very well be MPBH. It is necessary to design methods to distinguish between the signatures of these PBH from ordinary BH from stellar evolution. There are also the EGRET third catalog of unidentified sources, many of which are in the Galaxy. Identifying the spatial distribution of those sources may give a clue as to their origin.
- **GAIA anomalous astrometry**. If the halo of dark matter in our galaxy is not made up of a diffuse gas of particles but rather of compact objects like MPBH, one would expect close encounters of stars with those hard cores, such that individual stellar motions will be affected, inducing anomalous deviations with respect to their surroundings. A careful monitoring of positions and velocities of billions of stars like that planned by GAIA [82] will be able to discover those anomalous motions and predict the location and velocities of those MPBH in our galaxy that constitute the DM.
- **Dynamical friction towards the center**. Massive objects tend to concentrate via dynamical friction towards the center of potential wells. MPBH within a range of masses will thus tend to accumulate at the centers of all intermediate structures, from globular clusters, through dwarf spheroidals, to our own galaxy. In particular one expects MPBH to merge and form more massive IMBH, which will tend to orbit today around the SMBH

at center of all galaxies. There is a chance that we may discover some IMBH around the SMBH at the center of our galaxy, the Milky Way, see e.g. [83].

- **Emission of GW in PBH binaries (LIGO)**. In their dynamical evolution as a pressureless fluid, MPBH will naturally merge emitting gravitational waves like those recently detected by LIGO. Now we know that there has been at least three GW merging events detected by LIGO in three months. Their masses range from 60 $M_\odot$ (after merging) to 8 $M_\odot$ and therefore cannot be described by a monochromatic spectrum of PBH masses, rather by a wide mass distribution. It is true that three events are not enough to characterize the whole mass distribution, but more events have already been reported by LIGO in Run O2, so we expect within a decade to have enough events to make statistical claims [27].
- **Mass and spin distribution of PBH**. For the moment, we are probably only detecting with LIGO the tip of the MPBH distribution, where the majority of the PBH live, but soon we will have of the order of one BBH coalescence per day, with the expected advanced LIGO nominal sensitivity, which will allow us to measure their mass distribution with a few percent accuracy. Unfortunately, due to the seismic noise barrier of terrestrial interferometers, we may not be able to map well the mass distribution of PBH between 100 and 10^4 $M_\odot$, since Advanced LIGO is sensitive only to masses $M_{\rm PBH} < 150$ $M_\odot$, due to strain sensitivities above 30 Hz and a mass-dependent ISCO frequency, $f_{\rm ISCO} = 44$ Hz $(100 M_\odot/M)$, see Fig. 2. Larger masses could only be probed by future interferometers like Einstein Telescope. Moreover, with better BBH merging templates and GW detections, we will be able to measure not just their mass distribution, but also the spin distribution. Together this will help us distinguish the origin of these BBH mergers. Primordial BHs naturally form by gravitational collapse of a spherical overdensity the size of the horizon when a primordial fluctuation reentered during radiation epoch, and therefore it is expected that PBH were born without spin, in a cluster of many more PBH which merge successively in pairs. It is expected that the spin of the resulting black hole from each merging will have a non-negligible spin, although by the time several isotropic mergings have occurred, those spins could be averaged out. Future N-body simulations will be performed to study this issue.
- **Stochastic background of GW (LISA)**. The merging of PBH soon after recombination to form larger mass black holes is a powerful source of gravitational waves, with efficiency of order few permil, which redshifts faster than matter and thus can reach $\Omega_{\rm GW} h^2 \sim 10^{-6}$ today, from unresolved mergers, going back to at least $z \sim 200$, and frequencies in a broad range, from the nanoHertz, possibly detectable by pulsar timing arrays, through miliHertz, easily detectable by the proposed LISA satellite [84, 39], down to kiloHertz seen by terrestrial interferometers like LIGO, VIRGO, KAGRA, etc. There is also the stochastic background generated at PBH formation. The violent gravitational collapse to form PBH at horizon reentry produces also a small fraction of energy, of order 10^{-6} of the total, in the form of gravitational waves, or $\Omega_{\rm GW} h^2 \simeq 10^{-11}$ today.

4. Anomalous motion of stars

In this section we describe the effect of PBH on the nearby starts, as could be measured by the satellite GAIA, as well as the effect on stars of shallow potential wells of dwarf spheroidals, which could have acted by expelling stars from them via the gravitational slingshot effect.

4.1. GAIA and anomalous astrometry

We first compute the anomalous acceleration, velocity and displacement induced by a population of PBH in the solar neighborhood. The dark matter density near the Sun was

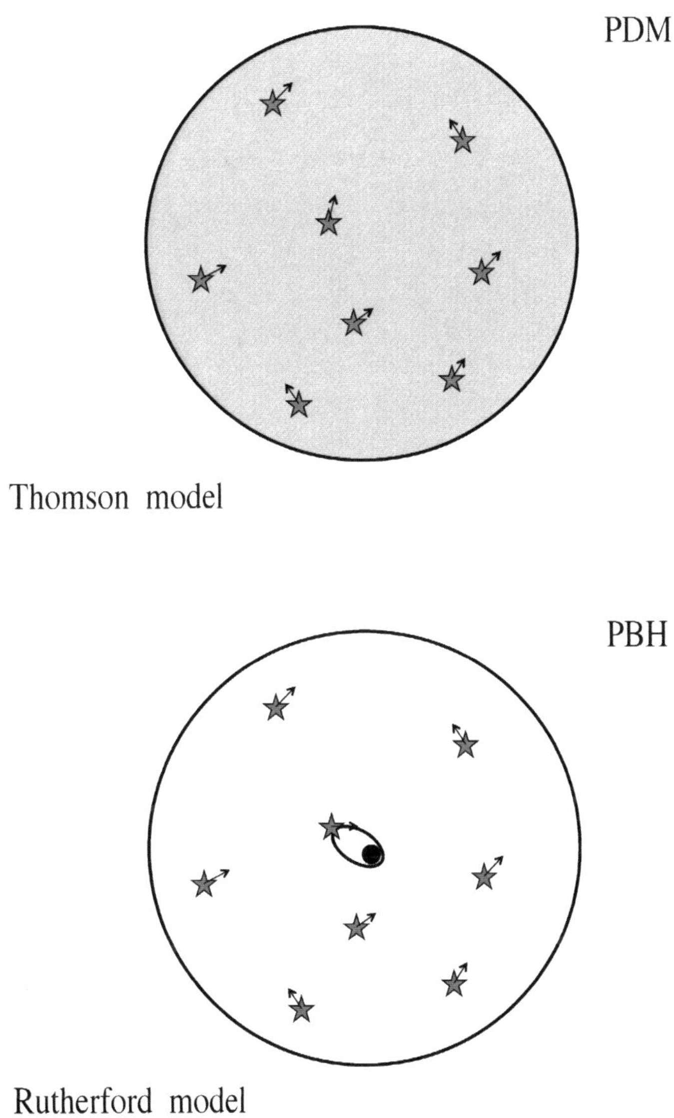

Figure 9. The standard paradigm of particle dark matter assumes a uniform distribution of DM particles around stars. This new PBH scenario substitutes a large volume for a compact central massive object. There is a natural analog between the PDM scenario and the Thomson model of the atom and the PBH scenario and the Rutherford model. While stars move in the PDM halo without any effort, in the new scenario a few stars may have close encounters with the PBH at the center and abruptly change their trajectories.

measured recently by Ref. [74]

$$\rho_\odot^{\rm DM} = 0.0088 \pm 0.0020\, M_\odot/{\rm pc}^3\,, \tag{1}$$

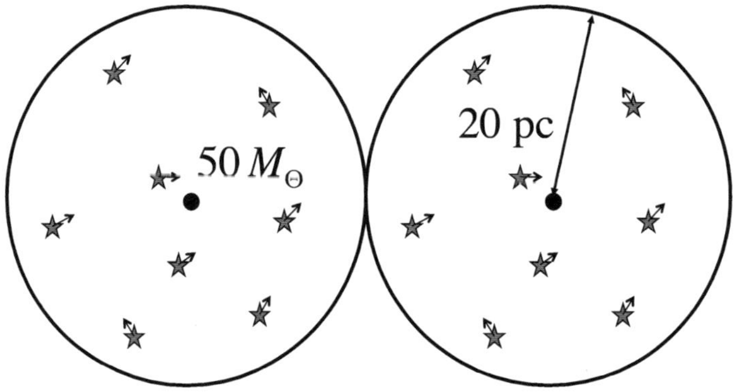

Figure 10. The average distance between PBH depends very much on their clustering. Assuming PBH form the bulk of the DM in the halo, we can compute their typical separation. For isolated PBH, of mass $M \sim 50\ M_\odot$, the typical distance is 40 pc. Such a volume comprises close to 100,000 stars separated approximately 1 pc from eachother. Only a few stars in $\sim 10^5$ will have close encounters (of less than a parsec) with the compact PBH forming the halo.

which gives a typical distance between black holes of

$$d_{\rm PBH} \equiv 2 \times \left(\frac{\rho_{\rm halo}}{M_{\rm PBH}}\right)^{-1/3} = 40\,{\rm pc}\left(\frac{M_{\rm PBH}}{50\ M_\odot}\right)^{1/3}. \tag{2}$$

This means that in a halo volume with a single PBH of 50 $M_\odot$ in the solar neighborhood, there are about 10^5 stars, separated from each other typically by 1 pc. Most of these stars will not feel the presence of the nearby PBH. The gravitational attraction on each of these stars is given by

$$a_i = \frac{GM_{\rm PBH}}{b^2} + \sum_j \frac{Gm_j}{r_{ij}^2}\,\hat{u}_j\,, \tag{3}$$

where the first term is due to the MPBH at a distance b from the star,[1] and the second term is due to the rest of the stars in the neighborhood. The presence of a dark compact object like a MPBH can be deduced from the *anomalous* acceleration induced by the MPBH,

$$\Delta a = 1 \times 10^{-15}\,m\,s^{-2}\left(\frac{M_{\rm PBH}}{50\ M_\odot}\right)\left(\frac{20\,{\rm pc}}{b}\right)^2, \tag{4}$$

which seems utterly negligible. However, for impact parameters $b < 0.1$ pc, which may happen for one in 10^5 stars in the field, one can reach significant accelerations. The anomalous relative velocity induced in those cases is

$$\frac{\Delta v}{v_0}(0.1\,{\rm pc}) \sim 3 \times 10^{-4}\,, \tag{5}$$

[1] For a fluid of dark matter particles permeating the solar neighborhood this term would be absent by Gauss law.

for typical halo velocities of $v_0 \sim 200$ km/s. These induce anomalous displacements which, at a distance of approximately $x_0 \sim 1$ kpc, could be seen as transverse displacements of order

$$\frac{\Delta x}{x_0}(0.1\,\mathrm{pc}) \sim 0.56 \text{ marcsec}\,, \tag{6}$$

which are in principle detectable with GAIA [82].

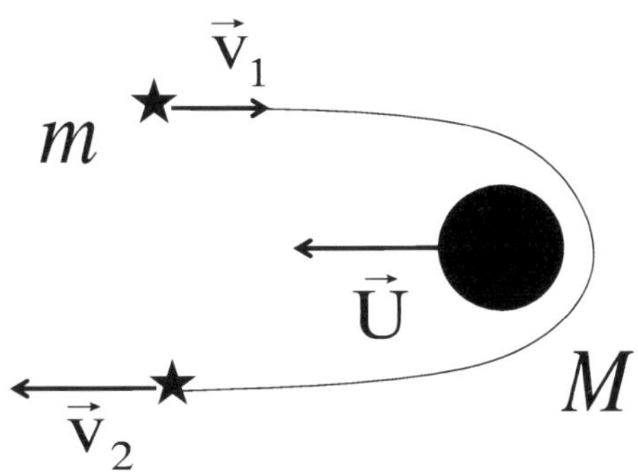

Figure 11. The gravitational slingshot effect converts a small peculiar velocity $\vec{v}_1$ of a star of mass m into a larger than escape velocity $\vec{v}_2$, thanks to the exchange of energy and momentum with the MPBH of mass M.

4.2. Ejecting stars from Globular Clusters and DSph with the gravitational slingshot effect

Apart from anomalous motion of stars in our solar neighborhood, one can also compute the effect of close encounters of stars with massive PBH in globular clusters and dwarf spheroidals with shallow potential wells, where the typical velocities of stars are tens of km/s, while those of MPBH (at the core of the potential well, due to dynamical friction) are close to a hundred km/s. Thanks to energy-momentum conservation, stars that pass close to these MPBH will be expelled from the cluster. They will suffer the well known gravitational slingshot effect: a star of mass m with incoming velocity $\vec{v}_1$ encounters a MPBH with mass M and velocity $\vec{U}$, and through gravitational recoil the outgoing star acquires a velocity $\vec{v}_2$, see Fig. 11, with magnitude given by

$$v_2 = \frac{2U + (1-q)\,v_1}{1+q}\,, \tag{7}$$

where $q = m/M$. If the MPBH has a large velocity (since it probes the core of the potential well) then it will imprint on the star a velocity twice as large. If the star and the black hole encounter each other at an angle θ, the recoil velocity will be slightly different,

$$v_2(\theta) = \sqrt{v_1^2 \sin^2\theta + \left(\frac{2U + (1-q)\,v_1\cos\theta}{1+q}\right)^2}\,, \tag{8}$$

but marginalizing over angles,

$$\begin{aligned}\bar{v}_2 &= \frac{1}{2}\int_0^\pi d\theta\,\sin\theta\;v_2(\theta) = \\ &= \frac{v_1}{8}\left[-\frac{2\beta(1-6\alpha+\alpha^2)}{1+\alpha}+\frac{(1+\alpha)(1+\alpha\beta^2)}{\sqrt{\alpha}}\times\right. \\ &\left.\left(\arctan\frac{\sqrt{\alpha}(2+\beta-\alpha\beta)}{2\alpha\beta+\alpha-1}+\arctan\frac{\sqrt{\alpha}(2-\beta+\alpha\beta)}{2\alpha\beta-\alpha+1}\right)\right] \\ &\simeq \frac{2U}{1+q}, \end{aligned} \tag{9}$$

which is almost identical to Eq. (7), for $\alpha = 1/q \gg 1$, and $\beta = U/v_1 \gg 1$.

With a few of these BH bypasses (slingshots), the star will acquire a velocity above the escape velocity of the globular cluster or the dwarf spheroidal, and be expelled from the shallow potential well.[2] This may be the reason why small substructures like dwarf galaxy satellites, with masses below 10^6 to 10^8 $M_\odot$, do not shine: they could have lost most of their stars by ejections, and thus present today large mass-to-light ratios, of order 300 to 1000, as recently measured [65]. This may explain both the substructure and the too-big-to-fail problems of standard CDM scenarios.

5. Microlensing within strong lensing of QSO

Well known quasars like QSO 0957+4561 have two or more images, whose short-time magnitude pattern matches, once we take into account a constant time delay due to the different paths that light has travelled from each source. However, the residual long-time pattern is not constant, but follows a characteristic amplification light curve of more than a decade [53], which suggests that one of the lines of sight has been microlensed by a black hole of a few tens of solar masses. Detailed studies of several of these strong-lensed quasars with multiple images may allow us to determine not only the rate of expansion [51], from the time delay between images [52], but also the dark matter distribution through the long-time residuals, if these happen to show microlensing amplification [53].

6. Gamma and X-ray emission

Ever since satellites have explored the sky in X-rays and gammas, they have discovered multiple bright sources, some point-like and others more extended, and even a diffuse background. Some of the sources have been identified with distant AGN and QSO sources, outside our galaxy. Others are neutron stars and X-ray binaries. There is, however, a growing list of unidentified sources in all point source catalogs, from EGRET to Chandra/XMM and Fermi LAT. Some of these sources are extragalactic, but many are unknown galactic sources, with hard spectra. There is the natural assumption that at the core of these sources lie a black hole, some more massive than others, but in any case there seems to be too many BH to be accounted for by the standard stellar evolution and subsequent supernova explosions. Extrapolated to the history of the galaxy, it would imply that there should have been a larger than observed cosmic ray flux and thus a larger UV and IR background.

Moreover, the recent Fermi LAT measurements of the diffuse gamma-ray background towards the galactic center (GC) has an uncertain origin. Some researchers speculated that it could have

[2] A similar mechanism has been invoked in Ref. [70] to explain the ejected stars from the galactic center by the presence of an IMBH [71].

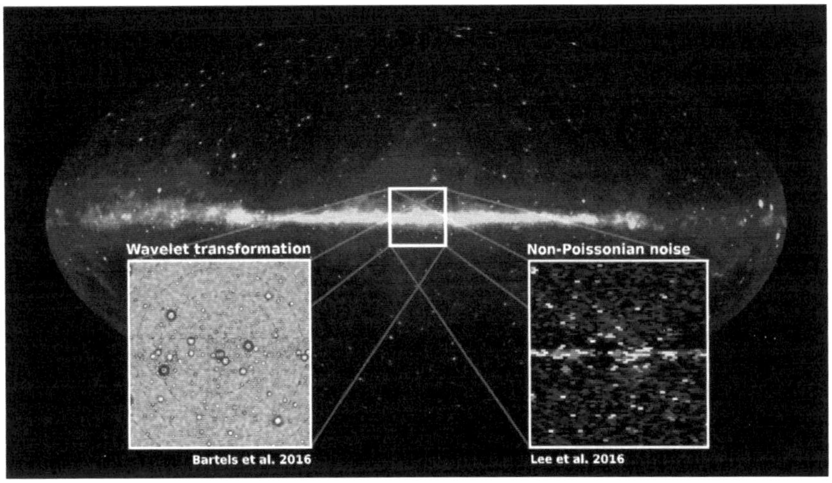

Figure 12. The Fermi 3FGL point source catalog presents a large number of unidentified sources, as well as a diffuse background that could be consistent with bright point sources due to gamma emission around PBH. Figure from CERN Courier [67].

arised from PDM annihilations, but this has been discarded as a reasonable explanation by two independent groups [68, 69], which have found that the diffuse GC background is better described by a population of unresolved point sources, adding extra support to compact massive objects like black holes as the more plausible origin of the GC background.

It is therefore worth studying these individual point sources, e.g. in the 3FGL catalog of Fermi-LAT, to probe their nature and determine whether a PBH could be their source of energy, rather than a pulsar or a massive star. Some of these sources are correlated with molecular clouds close to the galactic center and these shine in X-rays and many other wavelengths. Moreover, since the spatial distribution of PBH should follow that of the bulk of the dark matter in the halo, it is expected that some spatial correlation between the X-ray and gamma-ray sources of EGRET, Chandra and Fermi and a generic NFW or Einasto profile should exist. Furthermore, since the power spectrum of PBH dark matter has a broad peak, responsible for the clustering of the initial primordial black holes, one should expect to see higher spatial correlations at short distances of order a parsec or smaller.

7. Gravitational wave emission

The MPBH scenario has three different sources of gravitational waves at very different frequencies. The broad and high peak in the curvature fluctuation power spectrum is generated during inflation from the slow-roll motion of the inflaton field and possibly other fields coupled to it. Scalar fluctuations have a large enough amplitude that backreact on both scalar and tensor modes, creating a sizeable gravitational wave background with a peak at frequencies associated with the size of the horizon at reentry that will have redshifted today. Then there is the GW stochastic background from the formation of PBH themselves. The gravitational collapse that gives rise to the PBH at horizon reentry during the radiation era produces also a small fraction of energy, of order 10^{-6} of the total, in the form of gravitational waves, which have redshifted (from the time of formation at $z = z_f$) to $\Omega_{\rm GW} h^2 \simeq 10^{-11}$ today. Such a background is similar to that produced at violent first order phase transitions, and will have a peak at a typical frequency today

$$f^{\rm peak} \simeq 1.4 \times 10^{-8}\, Hz \left(\frac{30\, M_\odot}{M_{\rm PBH}^{\rm peak}}\right)^{1/2},$$

and an amplitude at the peak given by

$$h_c(f^{\rm peak}) \simeq 2.85 \times 10^{-16} \left(\frac{M_{\rm PBH}^{\rm peak}}{30\, M_\odot}\right)^{1/2},$$

for PBHs of mass peaked at $M_{\rm PBH}^{\rm peak} \simeq 30\, M_\odot \left(\frac{7\times 10^{11}}{1+z_f}\right)^2$, which falls in the range of sensitivity of future Pulsar Timing Arrays.

7.1. Stochastic background of GW

Finally there is the GW stochastic background arising from the inspiralling and merging of PBH during the matter era and detected as unresolved sources by LIGO, LISA and PTA. For a constant rate of merging since recombination at $R \sim 55$ events/yr/Gpc^3 [4, 6], in a ΛCDM universe, one finds [39]

$$h_c(f) \simeq 1 \times 10^{-24} \left(\frac{\mathcal{M}_c}{M_\odot}\right)^{5/6} \left(\frac{f}{Hz}\right)^{-2/3}, \tag{10}$$

where $\mathcal{M}_c^{5/3} = M_1 M_2 (M_1 + M_2)^{-1/3}$ is the chirp mass, and

$$h^2\Omega_{\rm GW}(f) \simeq 8 \times 10^{-13} \left(\frac{\mathcal{M}_c}{M_\odot}\right)^{5/3} \left(\frac{f}{Hz}\right)^{2/3}, \tag{11}$$

which is just below the Advanced LIGO sensitivity but could eventually be detected by LISA, at lower frequencies, see Fig. 13.

The spectrum of gravitational waves from MPBH thus covers a huge range of frequencies and amplitudes, see Fig. 14. Some may be accessible through terrestrial GW interferometers, in the kHZ domain, others through GW antennas in space, in the mHz, and yet another by Pulsar Timing Arrays, in the nHz range.

8. Discussion and Conclusions

I have presented here a new paradigm, based on massive PBH, rather than fundamental particles, as the main component of Dark Matter. MPBH could arise from peaks in the spectrum of matter fluctuations produced during inflation. With the future terrestrial and space GW interferometers we may be able to characterize their distribution and properties, providing clues about their origin and the early stages of the evolution of the Universe. Those MPBH have typically a broad mass distribution, from below a solar mass to several tens of solar masses. Large masses are difficult to detect with the present GW interferometer LIGO, due to the seismic wall at low frequencies. However, if in the near future a black hole is detected by LIGO in a Compact Binary Coalescense with a mass smaller than solar, it could never have been produced by stellar collapse, due to the fundamental Chandrasekhar limit, and therefore is a clear signal of primordial origin.

There is already a lot of evidence in favor of PBH as DM. The strong correlation between fluctuations in the infrared and the diffuse X-ray backgrounds suggests that a large population

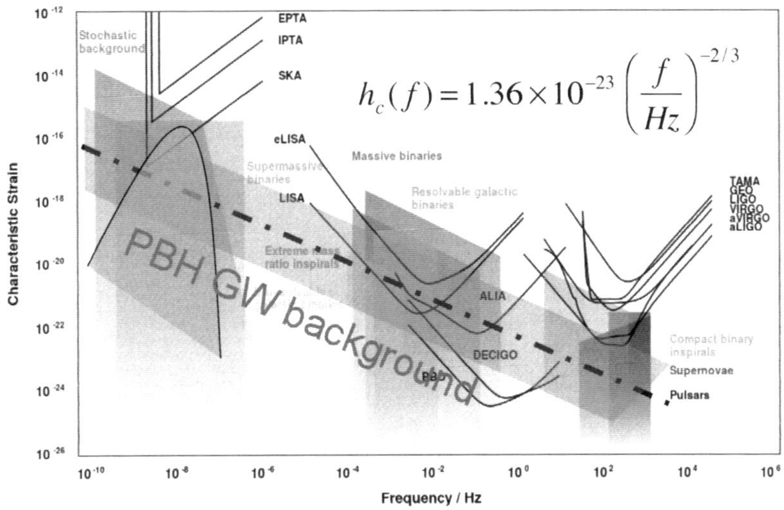

Figure 13. The stochastic background of gravitational waves from the merging of primordial black holes since recombination spans many orders of magnitude, from below the nanoHertz to above the kiloHertz. The range is covered by ground interferometers, satellites and pulsar timing arrays. Also shown in grey is the stochastic background of gravitational waves created from the violent gravitational collapse that formed the PBH upon horizon entry of large fluctuations during the radiation era. Adapted from Ref. [75].

of PBH could have initiated early structure formation and reionization, and at the same time be responsible for the present X-ray background. This speed-up in structure formation may explain the existence of fully formed galaxies and clusters at high redshift, much before expected from the standard ΛCDM paradigm. Moreover, PBH could also be responsible for the observed ultra luminous x-ray sources, the SMBH and IMBH at centers of galaxies and globular clusters, and alleviate the substructure and too-big-to-fail problems of the standard CDM paradigm.

Moreover, in this work I describe multiple signatures within a large variety of phenomena in Astrophysics and Cosmology where the presence of MPBH have an impact. Some are well known, others are new avenues of exploration. Here is a comprehensive list: CMB distortions and anisotropies; First stars in the Universe; Reionization and 21cm Intensity mapping; Early galaxy formation; SMBH and IMBH at the centers of galaxies and globular clusters; Superluminal SNe; Long-duration microlensing; Residual microlensing in time-matched strong-lensed multiple images of quasars; Tidally disrupted stars and Ultra Luminous X-ray transients; Substructure and too-big-to-fail problems in LSS N-body simulations; Missing baryon problem; Cluster collisions and cross-sections; Wide binaries in the Milky Way; Compact star clusters in dwarf galaxies; Lensed Fast Radio Bursts; X-ray binaries and microquasars; Fermi 3FGL and Chandra point source catalogs; X-ray and Gamma Diffuse backgrounds; GAIA anomalous astrometry; Dynamical friction and formation rate of SMBH; Emission of GW in MPBH binaries; Mass and spin distribution of PBH; Stochastic background of GW. All of them are not only tests of the new paradigm, but also opportunities to constrain the Early Universe scenario behind the MPBH origin.

The fact that quantum fluctuations of the inflaton field can backreact on space-time and form classical inhomogeneities, giving rise to CMB anisotropies and big structures like galaxies

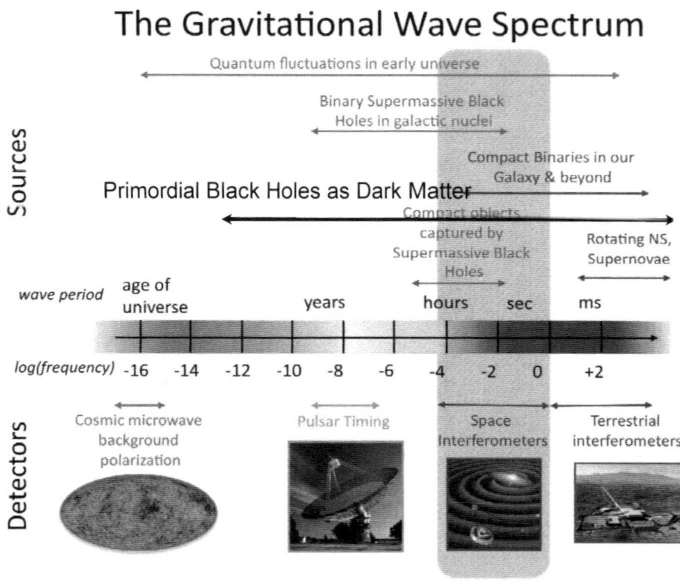

Figure 14. The spectrum of gravitational waves from the merging of primordial black holes spans many orders of magnitude, from below the nanoHertz to above the kiloHertz. The range is covered by ground interferometers, satellites and pulsar timing arrays. *Credit: modified from an original image by NASA Goddard Space Flight Center.*

and clusters, is a fascinating property of Quantum Field Theory in curved space, and one of the great successes of inflation. What is even more surprising is that a local feature in the inflaton dynamics can give rise to large amplitude fluctuations in the spatial curvature, which collapse to form black holes upon reentry during the radiation era, and that these PBH may constitute today the bulk of the matter in the universe. In this new scenario, Dark Matter is no longer a particle produced after inflation, whose interactions must be deduced from high energy particle physics experiments, but an object formed by the gravitational collapse of relativistic particles in the early universe when subject to high curvature gradients, themselves produced by quantum fluctuations of a field and stretched to cosmological scales by inflation.

In conclusion, Massive Primordial Black Holes are the perfect candidates for collisionless CDM, in excellent agreement with present observations. This new scenario opens up a very rich phenomenology that can be tested in the near future from CMB, LSS, X-rays and GW observations, in a multi-probe, multi-epoch and multi-wavelength approach to study the nature of Dark Matter. Furthermore, if MPBH are indeed responsible for DM, and its properties are measured, it will open a new window into the Early Universe, where the dynamics occurring in the last stages of inflation may be probed by astrophysical and cosmological observations.

Acknowledgments

I thank Sebastien Clesse, Andrei Linde, Marco Peloso, Ester Ruiz Morales, Caner Unal and David Wands, for a very fruitful collaboration that has led to the massive PBH scenario described here. These proceedings are a transcription of the various talks I have given during 2015 and

2016 in several workshops and conferences, with further insights and new research avenues. I also thank Lam Hui, Alan Heavens, Miguel Zumalacárregui, Ruth Durrer, Alvise Raccanelli, Lenny Susskind, Peter Michelson, Peter Graham, Javier Rico, Manel Martinez, Valenti Bosch-Ramon, Emma de Oña, German Vargas, Savvas Nesseris, José Pedro Mimoso, Julian Muñoz, Jim Annis, Tamara Davis, Alex Drlica-Wagner, Alexandre Refregier, Mikhail Shaposhnikov, Marti Raidal, Kfir Blum and many others for very insightful discussions on the physics of PBHs. This work is supported by the Research Project of the Spanish MINECO FPA2015-68048-C3-3-P and the Centro de Excelencia Severo Ochoa Program SEV-2012-0249.

References

[1] B. P. Abbott *et al.* [LIGO Scientific and Virgo Collaborations], "Observation of Gravitational Waves from a Binary Black Hole Merger," Phys. Rev. Lett. **116**, 061102 (2016) [arXiv:1602.03837 [gr-qc]].
[2] B. P. Abbott *et al.* [LIGO Scientific and Virgo Collaborations], "GW151226: Observation of Gravitational Waves from a 22-Solar-Mass Binary Black Hole Coalescence," Phys. Rev. Lett. **116**, 241103 (2016) [arXiv:1606.04855 [gr-qc]].
[3] B. P. Abbott *et al.* [LIGO Scientific and Virgo Collaborations], "Supplement: The Rate of Binary Black Hole Mergers Inferred from Advanced LIGO Observations Surrounding GW150914," arXiv:1606.03939 [astro-ph.HE].
[4] B. P. Abbott *et al.* [LIGO Scientific and Virgo Collaborations], "GW150914: Implications for the stochastic gravitational wave background from binary black holes," Phys. Rev. Lett. **116**, 131102 (2016) [arXiv:1602.03847 [gr-qc]].
[5] B. P. Abbott *et al.* [LIGO Scientific and Virgo Collaborations], "Astrophysical Implications of the Binary Black-Hole Merger GW150914," Astrophys. J. **818**, L22 (2016) [arXiv:1602.03846 [astro-ph.HE]].
[6] B. P. Abbott *et al.* [LIGO Scientific and the Virgo Collaborations], "Binary Black Hole Mergers in the first Advanced LIGO Observing Run," arXiv:1606.04856 [gr-qc].
[7] K. Belczynski, D. E. Holz, T. Bulik and R. O'Shaughnessy, "The first gravitational-wave source from the isolated evolution of two 40-100 Msun stars," Nature **534**, 512 (2016), arXiv:1602.04531 [astro-ph.HE].
[8] J. García-Bellido, A. D. Linde and D. Wands, "Density perturbations and black hole formation in hybrid inflation," Phys. Rev. D **54**, 6040 (1996), [astro-ph/9605094].
[9] S. Clesse and J. García-Bellido, "Massive Primordial Black Holes from Hybrid Inflation as Dark Matter and the seeds of Galaxies," Phys. Rev. D **92**, 023524 (2015), arXiv:1501.07565 [astro-ph.CO].
[10] F. Zwicky, "The redshift of extragalactic nebulae," Helv. Phys. Acta **6**, 110 (1933).
[11] V. C. Rubin and W. K. Ford, Jr., "Rotation of the Andromeda Nebula from a Spectroscopic Survey of Emission Regions," Astrophys. J. **159**, 379 (1970).
[12] R. Massey *et al.*, "Dark matter maps reveal cosmic scaffolding," Nature **445**, 286 (2007), [astro-ph/0701594].
[13] P. A. R. Ade *et al.* [Planck Collaboration], "Planck 2015 results. XIII. Cosmological parameters," A&A (2016), arXiv:1502.01589 [astro-ph.CO].
[14] L. Hui, J. P. Ostriker, S. Tremaine and E. Witten, "On the hypothesis that cosmological dark matter is composed of ultra-light bosons," arXiv:1610.08297 [astro-ph.CO].
[15] G. Bertone *et al.*, "Particle Dark Matter: Observations, Models and Searches," Cambridge Univ. Press, UK (2010) 738 p
[16] B. Paczynski, "Gravitational microlensing by the galactic halo," Astrophys. J. **304**, 1 (1986).
[17] B. Paczynski, "Gravitational microlensing of the Galactic bulge stars," Astrophys. J. **371**, L63 (1991).
[18] The MACHO Collaboration, C. Alcock et al., Nature **365** 621, (1993).
[19] The EROS Collaboration, E. Aubourg et al., Nature **365** 623, (1993).
[20] The OGLE Collaboration, A. Udalski, Acta Astron. **43**, 289 (1993).
[21] C. Alcock *et al.* [MACHO Collaboration], "The MACHO project: Microlensing results from 5.7 years of LMC observations," Astrophys. J. **542**, 281 (2000), [astro-ph/0001272].
[22] R. A. Allsman *et al.* [MACHO Collaboration], "MACHO project limits on black hole dark matter in the 1-30 solar mass range," Astrophys. J. **550**, L169 (2001), [astro-ph/0011506].
[23] S. Calchi Novati *et al.* [AGAPE Collaboration], "POINT-AGAPE pixel lensing survey of M31: Evidence for a MACHO contribution to galactic halos," Astron. Astrophys. **443**, 911 (2005), [astro-ph/0504188].
[24] P. Tisserand *et al.* [EROS-2 Collaboration], "Limits on the Macho Content of the Galactic Halo from the EROS-2 Survey of the Magellanic Clouds," Astron. Astrophys. **469**, 387 (2007) [astro-ph/0607207].
[25] L. Wyrzykowski *et al.*, "Black Hole, Neutron Star and White Dwarf Candidates from Microlensing with OGLE-III," Mon. Not. Roy. Astron. Soc. **458**, 3012 (2016) [arXiv:1509.04899 [astro-ph.SR]].

[26] S. Bird, I. Cholis, J. B. Muõz, Y. Ali-Haïmoud, M. Kamionkowski, E. D. Kovetz, A. Raccanelli and A. G. Riess, "Did LIGO detect dark matter?," Phys. Rev. Lett. **116**, 201301 (2016), [arXiv:1603.00464 [astro-ph.CO]].
[27] S. Clesse and J. García-Bellido, "The clustering of massive Primordial Black Holes as Dark Matter: measuring their mass distribution with Advanced LIGO," Phys. Dark Univ. **10**, 002 (2016), [arXiv:1603.05234 [astro-ph.CO]].
[28] M. Sasaki, T. Suyama, T. Tanaka and S. Yokoyama, Phys. Rev. Lett. **117**, no. 6, 061101 (2016), [arXiv:1603.08338 [astro-ph.CO]].
[29] B. Carr, F. Kuhnel and M. Sandstad, "Primordial Black Holes as Dark Matter," arXiv:1607.06077 [astro-ph.CO].
[30] B. J. Carr and S. W. Hawking, "Black holes in the early Universe," Mon. Not. Roy. Astron. Soc. **168**, 399 (1974).
[31] B. J. Carr, "The Primordial black hole mass spectrum," Astrophys. J. **201**, 1 (1975).
[32] A. Dolgov and J. Silk, "Baryon isocurvature fluctuations at small scales and baryonic dark matter," Phys. Rev. D **47**, 4244 (1993).
[33] K. Jedamzik, "Primordial black hole formation during the QCD epoch," Phys. Rev. D **55**, 5871 (1997), [astro-ph/9605152].
[34] K. Jedamzik and J. C. Niemeyer, "Primordial black hole formation during first order phase transitions," Phys. Rev. D **59**, 124014 (1999), [astro-ph/9901293].
[35] P. H. Frampton, M. Kawasaki, F. Takahashi and T. T. Yanagida, "Primordial Black Holes as All Dark Matter," JCAP **1004**, 023 (2010), arXiv:1001.2308 [hep-ph].
[36] M. Y. Khlopov, "Primordial Black Holes," Res. Astron. Astrophys. **10**, 495 (2010), arXiv:0801.0116 [astro-ph].
[37] J. Garcia-Bellido, M. Peloso and C. Unal, "Gravitational waves at interferometer scales and primordial black holes in axion inflation," JCAP **1612**, no. 12, 031 (2016), [arXiv:1610.03763 [astro-ph.CO]].
[38] J. Garcia-Bellido and E. Ruiz Morales, "Primordial black holes from single field models of inflation," arXiv:1702.03901 [astro-ph.CO].
[39] S. Clesse and J. García-Bellido, "Detecting the gravitational wave background from primordial black hole dark matter," arXiv:1610.08479 [astro-ph.CO].
[40] K. J. Mack, J. P. Ostriker and M. Ricotti, "Growth of structure seeded by primordial black holes," Astrophys. J. **665**, 1277 (2007), [astro-ph/0608642].
[41] M. Ricotti, J. P. Ostriker and K. J. Mack, "Effect of Primordial Black Holes on the Cosmic Microwave Background and Cosmological Parameter Estimates," Astrophys. J. **680**, 829 (2008), [arXiv:0709.0524 [astro-ph]].
[42] Y. Ali-Haïmoud and M. Kamionkowski, "Cosmic microwave background limits on accreting primordial black holes," arXiv:1612.05644 [astro-ph.CO].
[43] D. Aloni, K. Blum and R. Flauger, "CMB constraints on primordial black hole dark matter," arXiv:1612.06811 [astro-ph.CO].
[44] A. Kashlinsky, "LIGO gravitational wave detection, primordial black holes and the near-IR cosmic infrared background anisotropies," Astrophys. J. **823**, L25 (2016), arXiv:1605.04023 [astro-ph.CO].
[45] C. A. Blake, F. B. Abdalla, S. L. Bridle and S. Rawlings, "Cosmology with the Square Kilometre Array," New Astron. Rev. **48**, 1063 (2004), [astro-ph/0409278].
[46] T. Wang *et al.*, "Discovery of a galaxy cluster with a violently starbursting core at z=2.505," Astrophys. J. **828**, 56 (2016), [arXiv:1604.07404 [astro-ph.GA]].
[47] J. M. D. Kruijssen and N. Lützgendorf, "Did massive black holes in globular clusters initially satisfy galactic scaling relations?," Mon. Not. Roy. Astron. Soc. **434**, 41 (2013), [arXiv:1306.0561 [astro-ph.CO]].
[48] R. M. Quimby *et al.*, "Hydrogen-poor superluminous stellar explosions," Nature **474**, 487 (2011), [arXiv:0910.0059 [astro-ph.CO]].
[49] J. Cooke *et al.*, "Super-luminous supernovae at redshifts of 2.05 and 3.90," Nature **491**, 228 (2012), [arXiv:1211.2003 [astro-ph.CO]].
[50] M. Smith *et al.* [DES Collaboration], "DES14X3taz: A Type I Superluminous Supernova Showing a Luminous, Rapidly Cooling Initial Pre-Peak Bump," Astrophys. J. **818**, L8 (2016), [arXiv:1512.06043 [astro-ph.SR]].
[51] S. Refsdal, "On the possibility of determining Hubble's parameter and the masses of galaxies from the gravitational lens effect," Mon. Not. Roy. Astron. Soc. **128**, 307 (1964).
[52] H. Dahle, M. D. Gladders, K. Sharon, M. B. Bayliss and J. R. Rigby, "Time Delay Measurements for the Cluster-lensed Sextuple Quasar SDSS J2222+2745," Astrophys. J. **813**, 67 (2015), [arXiv:1505.06187 [astro-ph.CO]].
[53] M. R. S. Hawkins, "The case for primordial black holes as dark matter," Mon. Not. Roy. Astron. Soc. **415**,

2744 (2011), arXiv:1106.3875 [astro-ph.CO].
[54] M. Servillat, A. Coleiro and S. Chaty, "Herschel views on ultra-luminous X-ray sources," PoS INTEGRAL **2012**, 101 (2013), [arXiv:1302.4452 [astro-ph.HE]].
[55] B. E. Tetarenko *et al.*, "The first low-mass black hole X-ray binary identified in quiescence outside of a globular cluster," Astrophys. J. **825**, 10 (2016), [arXiv:1605.00270 [astro-ph.HE]].
[56] V. Springel, C. S. Frenk and S. D. M. White, "The large-scale structure of the Universe," Nature **440**, 1137 (2006) [astro-ph/0604561].
[57] A. H. Maller and J. S. Bullock, "Multi-phase galaxy formation: High velocity clouds and the missing baryon problem," Mon. Not. Roy. Astron. Soc. **355**, 694 (2004), [astro-ph/0406632].
[58] P. S. Cowperthwaite *et al.*, "A DECam Search for an Optical Counterpart to the LIGO Gravitational Wave Event GW151226," arXiv:1606.04538 [astro-ph.HE].
[59] M. Soares-Santos *et al.* [DES Collaboration], "A Dark Energy Camera Search for an Optical Counterpart to the First Advanced LIGO Gravitational Wave Event GW150914," Astrophys. J. **823**, L33 (2016), arXiv:1602.04198 [astro-ph.CO].
[60] A. Raccanelli, E. D. Kovetz, S. Bird, I. Cholis and J. B. Munoz, Phys. Rev. D **94**, no. 2, 023516 (2016), [arXiv:1605.01405 [astro-ph.CO]].
[61] I. Cholis, E. D. Kovetz, Y. Ali-Hamoud, S. Bird, M. Kamionkowski, J. B. Muñoz and A. Raccanelli, Phys. Rev. D **94**, no. 8, 084013 (2016), [arXiv:1606.07437 [astro-ph.HE]].
[62] D. Blais, C. Kiefer and D. Polarski, "Can primordial black holes be a significant part of dark matter?," Phys. Lett. B **535**, 11 (2002), [astro-ph/0203520].
[63] I. F. Mirabel and L. F. Rodriguez, "Microquasars in our Galaxy," Nature **392**, 673 (1998),
[64] I. F. Mirabel, R. Mignani, I. Rodrigues, J. A. Combi, L. F. Rodriguez and F. Guglielmetti, "The Runaway black hole GRO J1655-40," Astron. Astrophys. **395**, 595 (2002), [astro-ph/0211445].
[65] K. Bechtol *et al.* [DES Collaboration], "Eight New Milky Way Companions Discovered in First-Year Dark Energy Survey Data," Astrophys. J. **807**, 50 (2015), [arXiv:1503.02584 [astro-ph.GA]].
[66] Q. Z. Liu, J. van Paradijs and E. P. J. v. d. Heuvel, "A catalogue of low-mass X-ray binaries in the Galaxy, LMC, and SMC. 4th Edition," Astron. Astrophys. **469**, 807 (2007), arXiv:0707.0544 [astro-ph].
[67] Marc Türler, CERN Courier, 18 Marzo 2016.
[68] R. Bartels, S. Krishnamurthy and C. Weniger, "Strong support for the millisecond pulsar origin of the Galactic center GeV excess," Phys. Rev. Lett. **116**, 051102 (2016), arXiv:1506.05104 [astro-ph.HE].
[69] S. K. Lee, M. Lisanti, B. R. Safdi, T. R. Slatyer and W. Xue, "Evidence for Unresolved γ-Ray Point Sources in the Inner Galaxy," Phys. Rev. Lett. **116**, 051103 (2016), arXiv:1506.05124 [astro-ph.HE].
[70] Y. Levin, "Ejection of high-velocity stars from the Galactic Center by an inspiraling intermediate-mass black hole," Astrophys. J. **653**, 1203 (2006), [astro-ph/0508193].
[71] B. Hansen and M. Milosavljevic, "The need for a second black hole at the Galactic center," Astrophys. J. **593**, L77 (2003), [astro-ph/0306074].
[72] A. F. Heckler and E. W. Kolb, "Searching for stellar mass black holes in the solar neighborhood," Astrophys. J. **472**, L85 (1996), [astro-ph/9605199].
[73] J. R. Chisholm, S. Dodelson and E. W. Kolb, "Stellar-mass black holes in the solar neighborhood," Astrophys. J. **596**, 437 (2003), [astro-ph/0205138].
[74] P. R. Kafle, S. Sharma, G. F. Lewis and J. Bland-Hawthorn, "On the Shoulders of Giants: Properties of the Stellar Halo and the Milky Way Mass Distribution," Astrophys. J. **794**, 59 (2014), [arXiv:1408.1787 [astro-ph.GA]].
[75] `http://rhcole.com/apps/GWplotter/`
[76] A. Robertson, R. Massey and V. Eke, "What does the Bullet Cluster tell us about Self-Interacting Dark Matter?," arXiv:1605.04307 [astro-ph.CO].
[77] D. P. Quinn, M. I. Wilkinson, M. J. Irwin, J. Marshall, A. Koch and V. Belokurov, "On the Reported Death of the MACHO Era," Mon. Not. Roy. Astron. Soc. **396**, 11 (2009), [arXiv:0903.1644 [astro-ph.GA]].
[78] T. D. Brandt, "Constraints on MACHO Dark Matter from Compact Stellar Systems in Ultra-Faint Dwarf Galaxies," Astrophys. J. **824**, L31 (2016), [arXiv:1605.03665 [astro-ph.GA]].
[79] T. S. Li *et al.* [DES Collaboration], "Farthest Neighbor: The Distant Milky Way Satellite Eridanus II," [arXiv:1611.05052 [astro-ph.GA]].
[80] J. B. Muñoz, E. D. Kovetz, L. Dai and M. Kamionkowski, "Lensing of Fast Radio Bursts as a Probe of Compact Dark Matter," arXiv:1605.00008 [astro-ph.CO].
[81] D. Schoonenberg, J. Gaskins, G. Bertone and J. Diemand, "Dark matter subhalos and unidentified sources in the Fermi 3FGL source catalog," JCAP **1605**, no. 05, 028 (2016), [arXiv:1601.06781 [astro-ph.HE]].
[82] M. I. Wilkinson *et al.*, "Spectroscopic survey of the galaxy with GAIA. The Expected science yield from the radial velocity spectrometer," Mon. Not. Roy. Astron. Soc. **359**, 1306 (2005), [astro-ph/0506083].
[83] T. Oka, R. Mizuno, K. Miura, S. Takekawa, "Signature of an intermediate mass black hole in the central

molecular zone of our galaxy," Astrophys. J. Lett. **816**, L7 (2016).
[84] N. Bartolo *et al.*, "Science with the space-based interferometer LISA. IV: Probing inflation with gravitational waves," JCAP **1612**, no. 12, 026 (2016), [arXiv:1610.06481 [astro-ph.CO]].

IOP Conf. Series: Journal of Physics: Conf. Series **840** (2017) 012033 doi:10.1088/1742-6596/840/1/012033

Primordial GWs from universality classes of pseudo-scalar inflation

M. Pieroni

Laboratoire AstroParticule et Cosmologie (APC), Université Paris Diderot
Paris Centre for Cosmological Physics (PCCP), F75205 Paris Cedex 13

E-mail: mauropieroni@gmail.com, mauro.pieroni@apc.in2p3.fr

Abstract. In this contribution we discuss the possibility of generating an observable gravitational wave (GW) background by coupling a pseudo-scalar inflaton to some Abelian gauge fields. This analysis is performed by dividing inflationary models into universality classes. We find that of the most promising scenario is a Starobinsky-like model, which may lead to the generation of observational signatures both in upcoming CMB detectors as well as for direct GW detectors. The signal which can be produced in these models would both be observable in ground-based detectors, such as advanced LIGO, and in space-based detectors, such as LISA. The complementarity between the CMB and direct GW detection may be used to extract informations on the microphysics of inflation. Interestingly the mechanism discussed in this contribution may also be relevant for the generation of Primordial Black Holes (PBHs).

1. Introduction

Inflation is nowadays accepted as a cornerstone of modern cosmology and in particular its simplest realization in terms of single scalar field slow-rolling in its potential appears to match with cosmological observations at CMB scales. During inflation, quantum fluctuations of the inflaton field and of the metric are stretched on macroscopic scales by the exponential expansion. This mechanism naturally provides an explanation to the anisotropies in the CMB and it is also expected to generate a background of primordial GWs. Because of the weakness of their interactions, GWs are practically freely traveling through the Universe and thus they can provide precise information on times as early as inflation. As a consequence the detection of this primordial GW background at direct GW detectors would deepen our understanding of the physics of the early Universe. Unfortunately, the spectrum of GWs produced during slow-roll inflation lies below the range of current and upcoming direct GW detectors.

While this minimal setup is both simple and effective, it is not completely satisfactory from a theoretical point of view. For this reason, several generalized landscape (such as multifields models or models with non-standard kinetic terms) have been proposed in order to define a realistic model of inflation that is also consistent with our knowledge of the fundamental interactions. In this contribution we discuss a realization of inflation where the inflaton ϕ (which is considered to be a pseudo-scalar) is coupled to some Abelian gauge fields through a generic higher-dimensional coupling [1] $\sim \phi\, F_{\mu\nu}\, \tilde{F}^{\mu\nu}$ (where $F_{\mu\nu}$ is the usual field strength tensor and $\tilde{F}^{\mu\nu}$ is the dual field strength tensor). Remarkably, this term generates an instability, which leads to an exponential production of the gauge fields [2–4]. As a consequence, the gauge fields produced with this mechanism induce a back-reaction both on the background dynamics [5, 6] and on the

perturbations [7, 8] (both the scalar and tensor power spectra are exponentially enhanced at small scales).
This generalized framework leads to a wide set of observational consequences such as: the production of a GW background which can be observed at GW interferometers such as advanced LIGO/VIRGO and LISA [1, 6, 11, 12], the presence of a non-Gaussian component in the scalar power spectrum [6–9, 11, 12], the generation of a distribution of PBHs [8, 11, 12] and the generation of μ-distortions [10–12].
In this contribution, we discuss the shape of the scalar and tensor spectra that are generated in this generalized framework. In particular we show that because of the exponential enhancement of the tensor spectrum at small scales, GW interferometers may be used to probe different inflationary models. The analysis presented in this contribution is carried out by grouping models into universality classes [12, 13]. As this characterization is based only on the asymptotic properties of the potential, it leads to a more general description of the problem. The results presented in this proceeding are based on work with Valerie Domcke and Pierre Binétruy [11].

2. Pseudo-scalar inflation in presence of gauge fields.

We consider the action [2–8] for a pseudo-scalar inflaton ϕ coupled to a certain number $\mathcal{N}$ of Abelian gauge fields A^a_μ associated to some $U(1)$ gauge symmetries:

$$\mathcal{S} = \int \mathrm{d}^4x\sqrt{|g|}\left[m_p^2\frac{R}{2} - \frac{1}{2}\partial_\mu\phi\partial^\mu\phi - V(\phi) - \frac{1}{4}F^a_{\mu\nu}F^{\mu\nu}_a - \frac{\alpha^a}{4\Lambda}\phi F^a_{\mu\nu}\tilde{F}^{\mu\nu}_a\right] , \qquad (2.1)$$

where $m_p \simeq 2.4 \cdot 10^{18}$ GeV denotes the reduced Planck mass, $V(\phi)$ is the scalar potential, $F^a_{\mu\nu}$ is the usual field-strength tensor, $\tilde{F}^{\mu\nu}_a$ is the dual field-strength tensor, Λ is a mass scale (that suppresses the higher-dimensional scalar-vector coupling) and α^a are the dimensionless coupling constants of the Abelian gauge fields. In the following we consider $\alpha^a = \alpha$ for all $a = \{1, 2, ..\mathcal{N}\}$. The equations of motion which describe the classical evolution of ϕ and $\vec{A}^a$ read:

$$\ddot{\phi} + 3H\dot{\phi} + \frac{\partial V}{\partial\phi} = \frac{\alpha}{\Lambda}\langle\vec{E}^a \cdot \vec{B}^a\rangle , \qquad \frac{d^2}{d\tau^2}\vec{A}^a - \nabla^2\vec{A}^a - \frac{\alpha}{\Lambda}\frac{\mathrm{d}\phi}{\mathrm{d}\tau}\nabla \times \vec{A}^a = 0 , \qquad (2.2)$$

where dots are used to denote derivatives with respect to cosmic time t, τ is the conformal time defined as $\mathrm{d}\tau = \mathrm{d}t/a(t)$, the brackets $\langle\cdot\rangle$ denote a spatial mean and the vectors $\vec{E}^a$ and $\vec{B}^a$ are the "electric" and "magnetic" fields defined as:

$$\vec{E}^a \equiv -\frac{1}{a^2}\frac{\mathrm{d}\vec{A}^a}{\mathrm{d}\tau} = -\frac{1}{a}\frac{\mathrm{d}\vec{A}^a}{\mathrm{d}t} , \qquad \vec{B}^a \equiv \frac{1}{a^2}\vec{\nabla} \times \vec{A}^a . \qquad (2.3)$$

As we show in the following, assuming $\dot{\phi}$ to be slowly varying (that indeed is satisfied during slow-roll inflation), the equation of motion for $\vec{A}^a$ can be solved analytically. Once this solution is found, we can plug it into the equation of motion for ϕ and study the back-reaction. By performing a spatial Fourier transform, the equation of motion for the gauge fields reads:

$$\frac{\mathrm{d}^2\,\tilde{A}^a_\pm(\tau,\vec{k})}{\mathrm{d}\tau^2} + \left[k^2 \pm 2k\frac{\xi}{\tau}\right]\tilde{A}^a_\pm(\tau,\vec{k}) = 0 , \qquad (2.4)$$

where, given the two helicity modes $\hat{e}_\pm$, we have defined $\tilde{A}^a_\pm(\tau,\vec{k})$ as $\tilde{\vec{A}}^a_\pm = \hat{e}_\pm\tilde{A}^a_\pm \exp(i\vec{k}\cdot\vec{x})$ and we have introduced the dimensionless parameter ξ defined by $\xi \equiv \alpha|\dot{\phi}|/(2\Lambda H)$. For slowly varying ξ this equation describes a tachyonic instability in the A_+ mode that induces an exponential growth for modes at low k. In particular, the mode $\tilde{A}_+$ can be expressed as [5, 6, 9]:

$$\tilde{A}^a_+ \simeq \frac{1}{\sqrt{2k}}\left(\frac{k}{2\xi aH}\right)^{1/4} e^{\pi\xi - 2\sqrt{2\xi k/(aH)}} . \qquad (2.5)$$

At this point it is interesting to notice that $\xi \sim \dot{\phi}/H \sim \sqrt{\epsilon_H}$ (where ϵ_H is the first slow-roll parameter $\epsilon_H \equiv -\dot{H}/H^2 \simeq \epsilon_V \equiv (V_{,\phi}/V)^2/(2m_p^2)$). As we have $\epsilon_H \ll 1$ at CMB scales and $\epsilon_H = 1$ at the end of inflation, we may have a negligible gauge field production at CMB scales (consistently with CMB observations) and a strong production towards the end of inflation.
The back-reaction on the equation of motion for ϕ is given by $\langle \vec{E}^a \cdot \vec{B}^a \rangle = \mathcal{N} \cdot 2.4 \cdot 10^{-4} H^4/\xi^4 e^{2\pi\xi}$. This term plays the role of an additional friction term that does not effect the evolution at CMB scales (where $\xi \propto \sqrt{\epsilon_H} \ll 1$) but it strongly affects the last part of the evolution. It is important to stress that, by modifying the background dynamics, this mechanism affects the standard inflationary predictions [11,12]. In particular the predicted value of scalar spectral index n_s gets increased and the predicted value of the tensor-to-scalar ration r gets decreased with respect to the case with $\alpha/\Lambda = 0$.
As already mentioned in Sec. 1, the strong production of gauge fields is not only affecting the background evolution but it is also modifying the scalar and tensor power spectra. In particular, the presence of the gauge fields induces a source term in the equation of motion for perturbations leading to an additional contribution to the spectra.
Let us start by considering the expression for the scalar power spectrum [8,11]:

$$\Delta_s^2(k) = \Delta_s^2(k)_{\text{vac}} + \Delta_s^2(k)_{\text{gauge}} = \left(\frac{H^2}{2\pi|\dot{\phi}|}\right)^2 + \left(\frac{\alpha\langle \vec{E}^a \cdot \vec{B}^a \rangle/\sqrt{\mathcal{N}}}{3\Lambda b H \dot{\phi}}\right)^2 . \tag{2.6}$$

where we have defined $b \equiv 1 - 2\pi\xi\alpha\langle \vec{E}^a \cdot \vec{B}^a \rangle/(3\Lambda H\dot{\phi})$. As $\langle \vec{E}^a \cdot \vec{B}^a \rangle \simeq \mathcal{N} \cdot \langle \vec{E} \cdot \vec{B} \rangle$ ($\langle \vec{E} \cdot \vec{B} \rangle$ is the value of $\langle \vec{E}^a \cdot \vec{B}^a \rangle$ for $\mathcal{N} = 1$), it is possible to show that for large values of ξ (*i.e.* at small scales) we get:

$$\Delta_s^2(k) \simeq \frac{1}{\mathcal{N}(2\pi\xi)^2} . \tag{2.7}$$

Interestingly, with a proper choice of the parameters of the model, we recover the usual scale-invariant power spectrum of inflation at large scales (*i.e.* scales probed through CMB observations) and a nearly flat spectrum at small scales. The strong amplification of the scalar spectrum at small scales may lead to the generation of a distribution of massive PBHs [8].
The normalized density of GWs at present time can be expressed as [6,11]:

$$\Omega_{GW} \equiv \frac{\Omega_{R,0}}{24}\Delta_t^2 \simeq \frac{1}{12}\Omega_{R,0}\left(\frac{H}{\pi m_p}\right)^2\left(1 + 4.3 \cdot 10^{-7}\mathcal{N}\frac{H^2}{m_p^2\xi^6}e^{4\pi\xi}\right) , \tag{2.8}$$

where $\Omega_{R,0} = 8.6 \cdot 10^{-5}$ denotes the radiation energy density today. Before concluding this section it is crucial to stress that the contribution of one of the two modes of the GWs is suppressed by a 10^{-3} factor with respect to the other. As a consequence the new contribution to the GW spectrum is strongly chiral. This is a peculiar characteristic of this model and it offers a method to distinguish this signal from other GW backgrounds.

3. Universality classes and GWs.

While in the existing literature [2–8] these models have always been studied by assuming a chaotic potential [14] $V(\phi) = m^2\phi^2/2$ for the inflaton, in this contribution (following the analysis of [11]) we extend the analysis to a broader class of models. For this purpose it is useful to notice that a large set of single-field slow-roll models can be expressed in terms of the parametrization [12,13]:

$$\epsilon_H \simeq \epsilon_V \simeq \frac{\beta_p}{N^p} + \mathcal{O}(1/N^{p+1}) , \tag{3.1}$$

where $N \equiv -\int H dt$ denotes the number of e-foldings from the end of inflation, $\beta_p > 0$ and $p > 1$ are constants. It is important to stress that using this parametrization we are not completely

specifying the potential $V(\phi)$ but only its asymptotic expression. As a single asymptotic behavior can be reached by several models, with this parametrization we can produce an analysis that does not holds for single models but rather for sets of theories that share a single asymptotic behavior *i.e.* for universality classes of models.
By noticing that the parameter ξ that controls the instability is proportional to the square root of the first slow-roll parameter ϵ_H, we can thus conclude that it is particularly convenient to use the parametrization of Eq. (3.1) to produce a general analysis of the model introduced in Sec. 2. Indeed (as explained in [11]) using the parametrization of Eq. (3.1) we have direct control on all the parameters that affect the shape of the scalar and tensor spectra *i.e.* $\mathcal{N}$, α/Λ, β_p, p and V_0 (normalization of the potential). This allows for the definition of a systematic method to scan the parameter space and to identify the different classes of models.

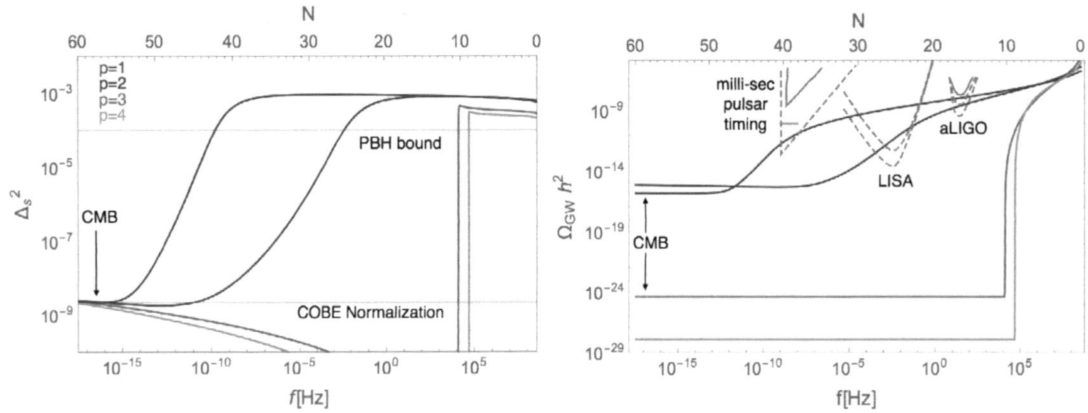

Figure 3.1: Scalar (left) and tensor (right) power spectra for different universality classes of models with $\mathcal{N} = 1$. The upper horizontal line in the left plot is the estimated PBH bound, the lower horizontal line indicates the COBE normalization. In the plot of the tensor power spectrum we present the sensitivity curves of some present and future direct GW detectors.

The models shown in Fig. 3.1 are: chaotic inflation [14] ($p = 1$), Starobinsky-like[1] inflation [17] ($p = 2$) and two implementations of hilltop inflation [18] ($p = 3, 4$). For all the universality classes shown in this plots we have fixed $\mathcal{N} = 1$. Consistently with the N-dependence expressed by the parametrization of Eq. (3.1), by increasing the value of p we get a smaller value of ξ at CMB scales and a larger value at the end of inflation (implying a steeper increase). This explains the steeper increase in the spectra for models with large values of p (see Fig. 3.1). In order to be consistent with the constraints set by CMB observations [19] the potential V_0 is fixed in order to respect the COBE normalization that fixes the amplitude of the scalar spectrum at CMB scales. At a given value for p (which specifies the different universality classes of inflation) we are thus left with only two parameters *i.e.* α/Λ and β_p. The parameter α/Λ controls the strength of the interactions between the inflaton and the gauge fields. As a consequence by increasing α/Λ, the gauge field production gets faster and the enhancement of the spectrum is shifted at earlier times (*i.e.* larger frequencies). Conversely, by varying the parameter β_p, we change the vacuum amplitude of the tensor spectrum. In particular, by the increasing the value of β_p we "vertically " shift the GW spectra. The parameters α/Λ and β_p for the models shown in Fig. 3.1 are chosen in order to maximize the GW spectrum while obeying the CMB constraints [19]. For example the late rise in both the scalar and tensor spectra for the hilltop models ($p = 3, 4$), is due to the

[1] Starobinsky-like models of inflation with pseudo-scalar fields can be effectively implemented in the context of supergravity by employing a shift-symmetry in the Kähler potential [15] (see for example [16]).

constraints on n_s scalar spectral index[2].
By observing the plot of the scalar spectra (where the amplitude of the spectrum at CMB scales is fixed) we can see that a larger value for p implies a smaller amplitude at small scales (towards the end of inflation). Such a behavior is consistent with the analytical prediction given in Eq. (2.7). Notice that all the models shown in this plot are in tension (by a $\mathcal{O}(1)$ factor) with the estimate of the PBH bound [8]. This discrepancy can be eased by taking into account the theoretical uncertainties in the derivation of the PBH bound. Moreover, by considering models with $\mathcal{N} > 1$ it is possible to produce a GW background that can be observed at upcoming direct GW detectors, while respecting this bound [11, 12].
For tensor spectra we can see that larger values of p imply a suppression of the amplitude at CMB scales but also a steeper increase of the spectrum towards higher frequencies. It is interesting to notice that the tensor spectra are approaching a nearly universal value in the strong gauge field regime. As discussed in [11], this behavior can be explained on solid grounds using the condition $\epsilon_H \simeq 1$ at the end of inflation. In particular this condition can be used to impose an absolute upper bound on the GW spectrum Ω_{GW}:

$$\Omega_{GW} h^2 \lesssim 2.4 \cdot 10^{-5} \mathcal{N}^{-1} , \tag{3.2}$$

Given the weakness of the assumptions to derive this bound ($\epsilon_H \simeq 1$ at the end of inflation), in practice it can be seen as model independent.

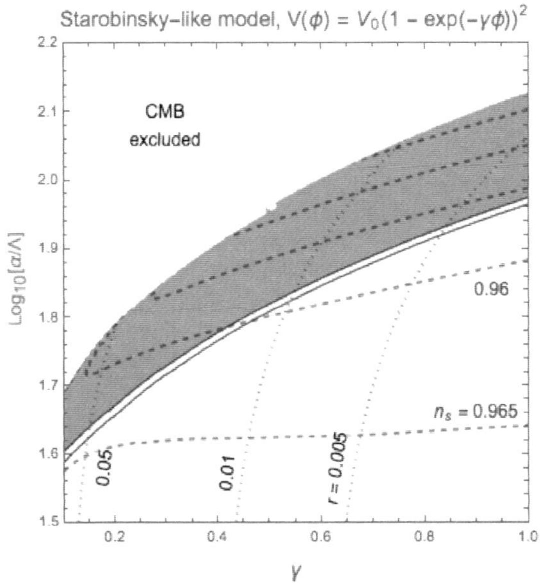

Figure 3.2: Plot of the $(\gamma, \alpha/\Lambda)$ parameter space for Starobinsky-like models where we have set $\beta_p = 1/(2\gamma^2)$ (to lighten the notation we have set $m_p = 1$ and thus Λ is dimensionless). The yellow and orange regions respectively correspond to the regions of the parameter space that can be probed with the A5M5 and A1M2 LISA configurations. Dashed blue lines correspond to fixed values for n_s (from 0.965 to 0.945) and dotted gray lines correspond to fixed values of r (0.05, 0.01, 0.005). The white region in the top left corner is excluded because of the non-observation of non-gaussianities in the CMB. A similar plot can be produced with the projected sensitivity of the different runs of Advanced LIGO.

As from the point of view of potential observations Starobinsky-like models ($p = 2$) appears to be the most promising among the universality classes shown in Fig. 3.1, we proceed by restricting our analysis to this particular class. In particular, in Fig. 3.2 we show a scan plot for the parameters of these models that correspond to:

$$V(\phi) \simeq V_0 \left[1 - \exp(\gamma\phi)\right]^2 , \tag{3.3}$$

[2] For the models described by the parametrization of Eq. (3.1) we have $n_s \simeq 1 - p/N$ and thus larger values of p implies smaller values for n_s. As already mentioned in Sec. 2, the presence of gauge fields effectively reduces the value of n_s and thus CMB measurements [19] set a rather severe constraint on the values of α/Λ that are allowed for these models.

where $\beta_p = 1/(2\gamma^2)$. Interestingly this plot shows a complementarity between CMB measurements and GW interferometers. Different measurements are starting to probe the viable parameter space of these models from different directions: while searches for non-Gaussianities in the CMB and direct GW detection probe the region of large α/Λ, searches for GWs in the CMB constraints the small γ region.

4. Conclusions.

The general higher-dimensional coupling between a pseudo-scalar inflaton and some Abelian gauge fields induces a tachyonic instability that leads to an exponential production of gauge fields during inflation. The strong production of these gauge fields back-reacts on the system and introduces an additional friction term that affects the last part of inflation. In general this affects the standard inflationary predictions for a given model by decreasing the predicted value of n_s and increasing r. The exponential production of gauge fields also leads to a modification of the equation of motion for scalar and tensor perturbations. In particular it gives rise to a strong enhancement of the scalar and tensor power spectra at high frequencies (small scales). This mechanism may lead to the generation of a distribution of massive PBHs which contribute to the fraction of Dark Matter present in the Universe [20] and to the production of chiral primordial GWs that can be observed at direct GWs detectors. The complementarity between different measurements provide a powerful method to probe these models. Moreover, as direct GWs detectors may access a very wide range of frequencies, they can potentially distinguish between different universality classes. Such a measurement would provide precious information on the microphysics of inflation.

Acknowledgements

We acknowledge the financial support of the UnivEarthS Labex program at Sorbonne Paris Cité (ANR-10-LABX-0023 and ANR-11-IDEX-0005-02) and the Paris Centre for Cosmological Physics.

References

[1] J. L. Cook and L. Sorbo, Phys. Rev. D **85**, 023534 (2012). Erratum: [Phys. Rev. D **86**, 069901 (2012)]
[2] M. S. Turner and L. M. Widrow, Phys. Rev. D **37**, 2743 (1988).
[3] W. D. Garretson, G. B. Field and S. M. Carroll, Phys. Rev. D **46**, 5346 (1992).
[4] M. M. Anber and L. Sorbo, JCAP **0610**, 018 (2006).
[5] M. M. Anber and L. Sorbo, Phys. Rev. D **81**, 043534 (2010).
[6] N. Barnaby, E. Pajer and M. Peloso, Phys. Rev. D **85**, 023525 (2012), N. Barnaby, R. Namba and M. Peloso, JCAP **1104**, 009 (2011).
[7] M. M. Anber and L. Sorbo, Phys. Rev. D **85**, 123537 (2012).
[8] A. Linde, S. Mooij and E. Pajer, Phys. Rev. D **87**, no. 10, 103506 (2013).
[9] N. Barnaby and M. Peloso, Phys. Rev. Lett. **106**, 181301 (2011).
[10] P. D. Meerburg and E. Pajer, JCAP **1302**, 017 (2013).
[11] V. Domcke, M. Pieroni and P. Binétruy, JCAP **1606**, 031 (2016).
[12] M. Pieroni, arXiv:1611.03732 [gr-qc].
[13] V. Mukhanov, Eur. Phys. J. C **73**, 2486 (2013), D. Roest, JCAP **1401**, 007 (2014), P. Binétruy, *et. al* JCAP **1504**, no. 04, 033 (2015).
[14] A. D. Linde, Phys. Lett. **129B**, 177 (1983).
[15] M. Kawasaki, M. Yamaguchi and T. Yanagida, Phys. Rev. Lett. **85**, 3572 (2000).
[16] G. Dall'Agata and F. Zwirner, JHEP **1412**, 172 (2014), R. Kallosh and A. Linde, JCAP **1011**, 011 (2010), R. Kallosh, A. Linde and T. Rube, Phys. Rev. D **83**, 043507 (2011)
[17] A. A. Starobinsky, Phys. Lett. **91B**, 99 (1980).
[18] L. Boubekeur and D. H. Lyth, JCAP **0507**, 010 (2005).
[19] P. A. R. Ade *et al.* [Planck Collaboration], arXiv:1502.02114.
[20] S. Clesse and J. García-Bellido, Phys. Dark Univ. **10**, 002 (2016), S. Clesse and J. García-Bellido, arXiv:1610.08479.

LISA Pathfinder: Optical Metrology System monitoring during operations

Heather E Audley[1], for the LISA Pathfinder collaboration

[1] Albert-Einstein-Institut, Max-Planck-Institut für Gravitationsphysik and the Leibniz Universität Hannover, 30167 Hannover, Germany

E-mail: heather.audley@aei.mpg.de

Abstract. The LISA Pathfinder (LPF) mission has demonstrated excellent performance. In addition to having surpassed the main mission goals, data has been collected from the various subsystems throughout the duration of the mission. This data is a valuable resource, both for a more complete understanding of the LPF satellite and the differential acceleration measurements, as well as for the design of the future Laser Interferometer Space Antenna (LISA) mission. Initial analysis of the Optical Metrology System (OMS) data was performed as part of daily system monitoring, and more in-depth analyses are ongoing. This contribution presents an overview of these activities along with an introduction to the OMS.

1. Introduction

The LISA Pathfinder satellite is a technology demonstrator that has proven the feasbility of some key technologies required for the future Laser Interferometer Space Antenna (LISA) mission [1]. These technologies include the use of picometre precision interferometry and microNewton thrusters to achieve control of a free-falling test mass (TM), such that it is influenced only by the local geodesics of space-time. The differential acceleration of the two LPF test masses is the scientific product that defines how well free-fall has been achieved.

The differential acceleration is calculated from the differential longitudinal displacement of the TMs. This, along with the angular fluctuations of the TMs around the sensitive axis, is derived from interferometric measurements made by the Optical Metrology System (OMS) [2].

2. The Optical Metrology System

The subsystems that make up the OMS are shown in Figure 1. A brief overview of the measurement chain is given in this section. The monitoring of the system health, and other relevant activities are presented in subsequent sections.

Light from the Reference Laser Unit (RLU) is coupled via a fibre into the Laser Modulation Unit (LMU), where it is split into two beams. Each beam passes through an Acousto-Optic Modulator (AOM) and the beams are shifted in frequency such that the difference between them is 1 kHz, the heterodyne frequency. The beams are transmitted onto the optical bench where they form four interferometers, as shown in Figure 2:

- The X1 interferometer determines the position of TM1 relative to the optical bench.
- The X12 interferometer determines the position of TM1 relative to TM2.

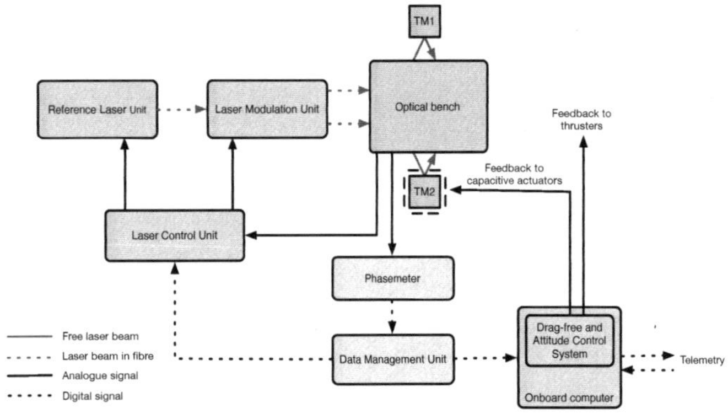

Figure 1. A basic block diagram showing subsystems of the optical metrology system in the blue, green, and yellow boxes. The gray boxes show how the signals from the optical metrology system are used as inputs for the Drag-Free and Attitude Control System (DFACS) to maintain the suspended condition of test mass two, and the drag-free condition of test mass one. Image taken from [2].

- The frequency (Freq.) interferometer has a deliberate pathlength mismatch between the two interfering beams, allowing measurement of laser frequency noise. The processed output from the frequency interometer is the input to the nested frequency control loop. This loop suppresses laser frequency noise by acting on the laser crystal piezo actuator and the crystal temperature, and is described in more detail in the corresponding proceeding [3].
- The two beams of the reference interferometer (Ref.) traverse paths of equal length, and are confined to the optical bench. The processed signal from this interferometer is subtracted from the other interferometer outputs to remove common-mode noise. The processed phase output from the reference interferometer also acts as the input for the Optical Pathlength Difference (OPD) control loop. The signal is fed back to piezo actuators in the Laser Modulation Unit (LMU) to adjust the relative length of the measurement and reference beams to counteract fluctuations in the pathlengths due to environmental disturbances that occur before they reach the stable optical bench. This control loop is described in more detail in [4].

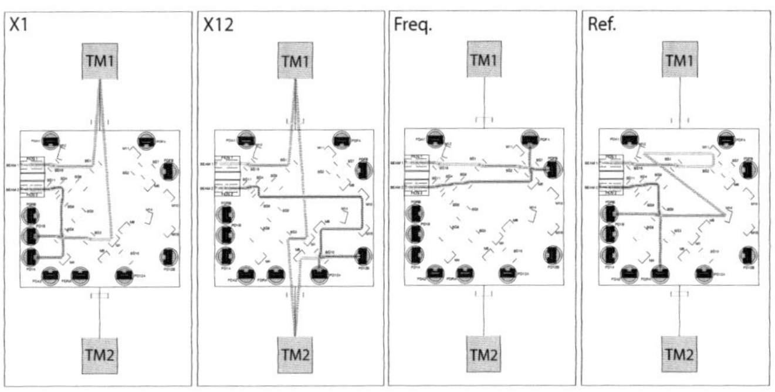

Figure 2. Optical diagrams showing the beampaths of the four interferometers on the LISA Pathfinder optical bench. The measurement beam is shown in red, and the reference beam in blue. Picture courtesy of Airbus Space and Defense and the Institute for Gravitational Research, Glasgow.

Each interferometer signal is detected by two quadrant photodiodes, one nominal and one redundant. The output signals from the individual quadrants are the inputs to the phasemeter. In the phasemeter, each input is converted to a voltage via a trans-impedance amplifier, then low-pass filtered, digitised, and scaled. A single-bin discrete Fourier transform is performed at the heterodyne frequency, giving the real and imaginary components of the complex amplitude. The total power per quadrant is also calculated. These three products are passed from the OMS to the Data Management Unit (DMU), where they are processed, nominally using the mean of the nominal and redundant photodiode channels, to give the main OMS outputs in the form of telemetry. These outputs include the longitudinal fluctuations of TM1, and of TM1 relative to TM2, the angular fluctuations of TM1 and TM2, and the contrast of each photodiode.

In order to measure the power on the optical bench, a beamsplitter splits off a fraction of each beam directly after the fibre injectors. This is then read out by a single element photodiode, one each for the measurement and reference beams. This signal also provides the input for the laser power control loop. This loop was designed as a nested loop, with two 'fast' analogue loops that take the signal from each of the photodiodes to the corresponding AOM to change the amount of light that passes through, and a 'slow' loop, which uses an average of the two signals, calculated in the DMU, to feedback to the RLU and directly control the DC output of the laser. During ground-based test campaigns, and in-flight, the decision was made to use only the 'fast' loops, as the 'slow' control loop did not function as required [2].

3. Laser Monitoring during operations

In flight, housekeeping telemetry from the OMS subsystems is downloaded from the satellite. This telemetry is monitored regularly in order to:

- check system health and functionality
- monitor system performance, and any changes throughout the mission
- identify noise sources, particularly those that might impact the differential acceleration measurement
- provide information that may be useful when considering the LISA design

Responsibility for the monitoring activities is allocated according to the country that provided each unit, along with support from industry. As such, the monitoring for the laser subsystem falls to the Max Planck Institute for Gravitational Physics (AEI), Hannover.

Housekeeping telemetry for the laser subsystems includes the laser power and temperature inside the RLU, currents and voltages of the RLU and the laser electronics, the laser power measured on the optical bench, as well as error flags. The telemetry is downloaded monthly, and a report produced which is available to the full collaboration. This report includes:

- the mean, maximum, and minimum values
- a time series plot
- the trend of the data over the month
- a power spectral density along with a reference, if appropriate
- alerts if the parameter crosses into predefined warning or critical regions

Alerts and anomalies are investigated and reported to Mission Operations Control. A technical note is produced each month to summarise the performance and show the longer term behaviour.

4. Conclusions

A huge amount of data relating to the laser subsystems are available, and analyses are ongoing to extract the maximum amount of useful information. These include and are not limited to:

- laser stability during each of the acceleration measurement investigations
- the laser power stability with temperature
- differences in noise levels between different operational periods
- investigations into the spikes seen in the power control loop signals

Acknowledgments
This work has been made possible by the LISA Pathfinder mission, which is part of the space-science program of the European Space Agency. We gratefully acknowledge support by the European Space Agency (ESA) (22331/09/NL/HB, 16238/10/NL/HB), by the Deutsches Zentrum fur Luft- und Raumfahrt (DLR) with funding of the Bundesministerium fur Wirtschaft und Energie with a decision of the Deutschen Bundestag (DLR project reference numbers FKZ OQ 0501 and FKZ 50 OQ 1601), and thank the German Research Foundation for funding the Cluster of Excellence QUEST (Centre for Quantum Engineering and Space-Time Research).

References

[1] Armano M et al. 2016 *Phys. Rev. Lett.* **23** 231101
[2] Audley H 2014 *Thesis* Preparing for LISA Pathfinder Operations: characterisation of the Optical Metrology System
[3] Paczkowski S, and the LPF collaboration 2017 *Journal of Physics: Conference Series* to be published in this issue.
[4] Born M, and the LPF collaboration 2017 *Journal of Physics: Conference Series* to be published in this issue.

A two-stage torsion pendulum for ground testing free fall conditions on two degrees of freedom

M Bassan[1], A Cavalleri[2], M De Laurentis[3,4], F De Marchi[5], R De Rosa[3,4], L Di Fiore[3], R Dolesi[2], N Finetti[6,7], F Garufi[3,4], A Grado[3,8], M Hueller[2], L Milano[3,4], G Pucacco[1], R Stanga[6,9], D Vetrugno[2], M Visco[1,10], S Vitale[2] and JW Weber[2]

1 University and INFN of Roma Tor Vergata , 2 University of Trento and INFN-TIFPA, 3 INFN Napoli, 4 University of Napoli "Federico II", 5 DIMA Università di Roma Sapienza, 6 INFN Firenze-Urbino, 7 University of l'Aquila, 8 INAF-OACN, 9 University of Florence, 10 INAF-ISS

luciano.difiore@na.infn.it

Abstract. Ground testing with torsion pendulums played a key role in the development and characterization of the Gravitational Reference Sensor (GRS) of LISA-Pathfinder (LPF). We report on a torsion pendulum facility with 2 soft degrees of freedom (DOF), realized by off-axis cascading two torsion fibers. This instrument, developed for testing on two DOFs the LPF GRS, allows simultaneous measurement of force and torque acting on the suspended test mass (TM), approaching free-fall condition on two DOFs down to a few mHz. We will report on the results of some measurement campaigns devoted in particular to the characterization of force to torque and torque to force actuation cross-talks (CT).

1. Introduction

PETER (PEndolo Translazionale E Rotazionale, namely Translational and Rotational Pendulum) is a two DOFs torsion pendulum facility [1]. A first fiber hangs a cross shaped stage; at one end of the cross a second fiber supporting the TM is suspended. Three counter weights are supported by rigid rods attached to the other arms of the cross. The system shows two soft DOFs, namely the rotation around the upper fiber (φ_a) and the rotation around the lower fiber (φ_b). These coordinates are simply related to two TM DOFs, i.e. displacement and rotation. With respect to these, the TM can be considered, above the resonance frequencies (1.3 and 2 mHz), in free fall. The suspended TM is enclosed in an engineering model of the LPF GRS which provides capacitive readout of the 6 DOFs of the TM as well as electrostatic actuation. Optical Read-Out (ORO) sensors, based on optical levers [2], provide independent measurement of TM and upper stage displacement. The GRS is mounted on motorized stages that allow to move it with respect to the TM in 5 DOF (2 translations and 3 angles). The GRS electronics, provided by ETHZ, is very similar to the flight model.

2. Application to ground testing of LPF and LISA: actuation CT measurement

As an example of how PETER can provide useful ground testing for LPF, and future LISA mission, we describe here the measurement of actuation CT. We recall that the LPF test mass is in free fall along the x axis, while under feedback control on the remaining DOFs: an undue CT of these feedback

IOP Conf. Series: Journal of Physics: Conf. Series **840** (2017) 012035 doi:10.1088/1742-6596/840/1/012035

signals into the x channel would spoil the quality of geodesic motion. We show here actuation CTs measured by acting in force (F) on the suspended TM and measuring the CT torque (τ) and vice-versa. We report, as an example, the results of a $CT_{F\rightarrow\tau,}$ and a $CT_{\tau\rightarrow F}$ measurement runs performed in May and June 2015 respectively. In both cases, we moved the GRS along a regular patterns in x and y, around the TM, for a total of 25 positions. For each position the CT measurement is averaged for 2 h. The total measurement time is 50 h. In figures 1 and 2, we show the data in 3D plots compared to the GRS analytical model and contours plots of second order surfaces fitted to the data for the two measurement runs. For further details on the measurement technique see reference [2].

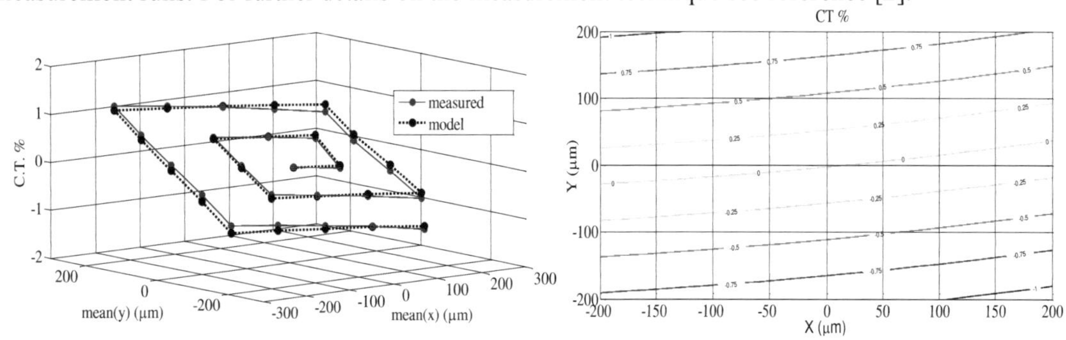

(a) (b)

Figure 1. measured $CT_{F\rightarrow\tau,}$ (solid line) compared to the analytical model (dotted line). (b) measured $CT_{F\rightarrow\tau,}$ contour plot (0.25 % level spacing). The CT is quite small and in good agreement with the model.

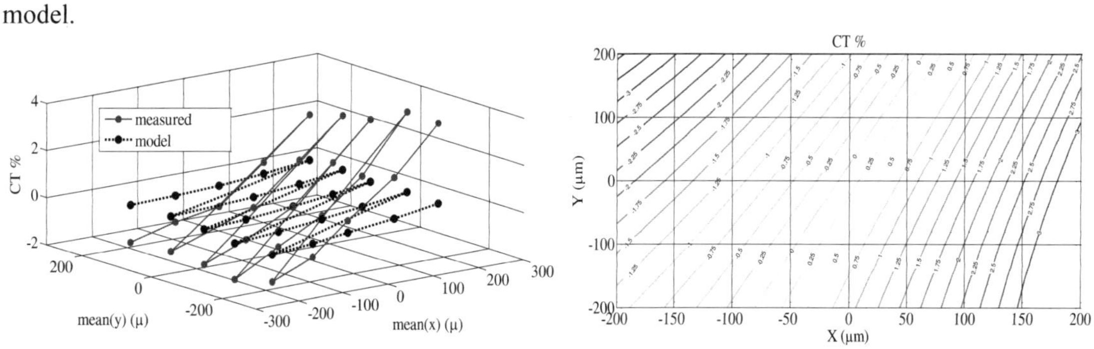

(a) (b)

Figure 2. (a) measured $CT_{\tau\rightarrow F}$ (solid line) compared to the analytical model (dotted line). (b) measured $CT_{\tau\rightarrow F}$ contour plot (0.25 % level spacing).. In this case, the CT is larger than expected, but, in the central part of the GRS measurement range, it drops to small values (<0.3 % within $\pm$ 20 μm from GRS center).

3. Conclusion and perspectives

We have performed several repeatable measurement of actuation CT, the first available for the LPF GRS before flight. The CTs measured with PETER are, in the central part of the GRS, quite small and within specs; they should not represent a problem for GRS operation. We think that the simulation of free fall on ground of multiple DOFs, first demonstrated with PETER, will be very useful for the testing of flight hardware of LISA and other space missions requiring drag-free control.

References

[1] Bassan M et al. 2013 *Phys. Rev.* D **87** 122006.
[2] De Rosa R et al. 2011 Astrop. Phys.**34** 394.
[3] Bassan M et al. 2016 *Phys. Rev. Lett.* **116** 051104.

IOP Conf. Series: Journal of Physics: Conf. Series **840** (2017) 012036 doi:10.1088/1742-6596/840/1/012036

LISA Pathfinder: OPD loop characterisation

Michael Born[1] **on behalf of the LPF collaboration**

[1] Albert-Einstein-Institut, Max-Planck-Institut für Gravitationsphysik und Universität Hannover, 30167 Hannover, Germany

E-mail: michael.born@aei.mpg.de

Abstract. The optical metrology system (OMS) of the LISA Pathfinder mission is measuring the distance between two free-floating test masses with unprecedented precision. One of the four OMS heterodyne interferometers reads out the phase difference between the reference and the measurement laser beam. This phase from the reference interferometer is common to all other longitudinal interferometer read outs and therefore subtracted. In addition, the phase is fed back via the digital optical pathlength difference (OPD) control loop to keep it close to zero. Here, we analyse the loop parameters and compare them to on-ground measurement results.

1. Introduction

LISA Pathfinders (LPF) main measurement [1] is the acceleration between two test masses (TM). For this purpose, the relative TM positions are determined by the optical metrology system (OMS) using heterodyne interferometry [2]. Here, the phase difference of the so-called reference and the measurement laser beam is measured at the heterodyne frequency of f_{het} = 1 kHz. Both beams are generated from a single laser source (Nd:YAG, λ=1064 nm) at the laser modulator unit (LMU). They are transmitted via fibres onto the optical bench which houses 4 ultra-stable bonded interferometers. The optical properties of fibres are sensitive to thermal and mechanical stress. This results in a variation of the optical pathlength difference (OPD) between the reference and the measurement beam. The varying phase (Ψ_r) between the beams is measured in the reference interferometer. As this phase is common to the other 3 interferometers, it is always subtracted from their longitudinal read out.

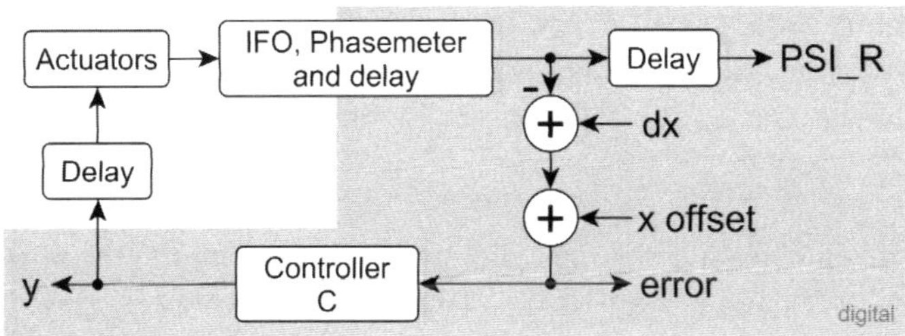

Figure 1. Schematic of the OPD control loop. The reference interferometer (IFO) measures the optical pathlength difference PSI_R(Ψ_r). Arrows (dx, ...) pointing towards the loop show injection and offset points. Arrows (y, ...) pointing away from the loop symbolise parameters that can be accessed.

In addition, there is a digital control loop using the phase $-\Psi_r$ as the error point (see Figure 1). The response of the digital controller is fed back to 2 piezo actuators in the LMU. Here, a push-pull configuration allows to double the actuation range and provides redundancy in case of a piezo failure. The reference interferometer then measures the changed OPD and the loop is closed.

2. OPD noise performance

Spectra of the longitudinal signal Ψ_r are shown in Figure 2. The data from the on-station thermal tests (OSTT) in 2011 shows the best performance that was measured on ground [3]. As the LPF spacecraft at the Lagrange point L1 provides a thermally and mechanically stable environment for the OMS, the OPD noise can be expected to be lower. First data looks promising but needs to be further analysed.

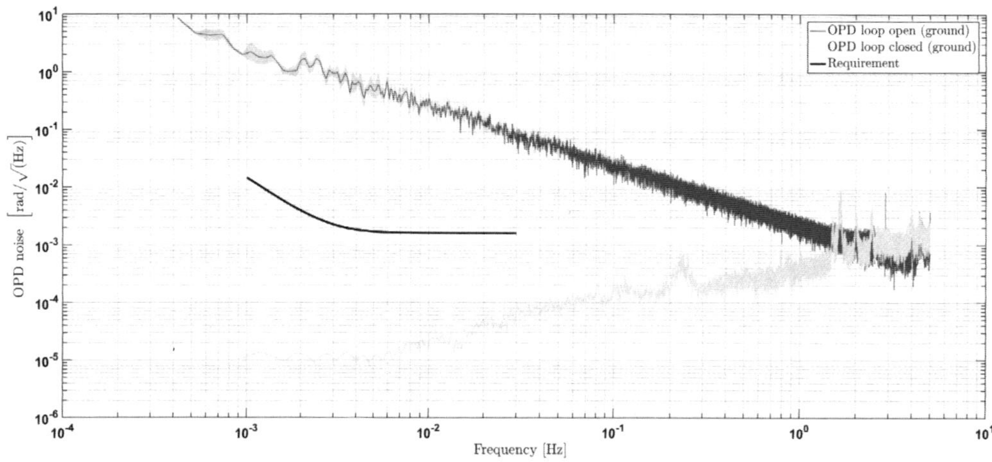

Figure 2. OPD noise spectra measured during OSTT campaign on-ground.

3. OPD system identification

The OPD system identification is done by measuring transfer functions (TF) of sinusoidal guidance injections (Figure 3). The 8 sine waves (f=[0.011 .. 1.123] Hz) are injected into the closed loop at dx in Figure 1. During the injections PSI_R is available at a sample rate of 10 Hz. The other parameters have only a 1 Hz rate and are so-called housekeeping data (HK). Therefore, only the 6 injections below the Nyquist frequency of 0.5 Hz can be analysed. The other 2 signals require the use of IDL data which has the full 100 Hz rate that is used by the digital system. The 100 Hz data requires special commanding and is only available for a limited time because of memory constraints. One of the challenges of the ongoing analysis is the timing correlation between the different clocks of the 10 Hz and the HK parameters.
Furthermore, the long-term stability of the piezo actuator gain will be investigated.

4. Conclusion

The OPD noise data looks very promising. With the digital OPD control loop working as expected the noise is further reduced. As the main mission goal is already achieved [1], further investigations of the very good OMS performance are being done. For the OPD analysis, a stable timing correlation between the involved parameters needs to be achieved for precise phase values in the transfer function measurements. After the end of the extended mission, final results and an analysis of the long term performance of the OPD loop will be published.

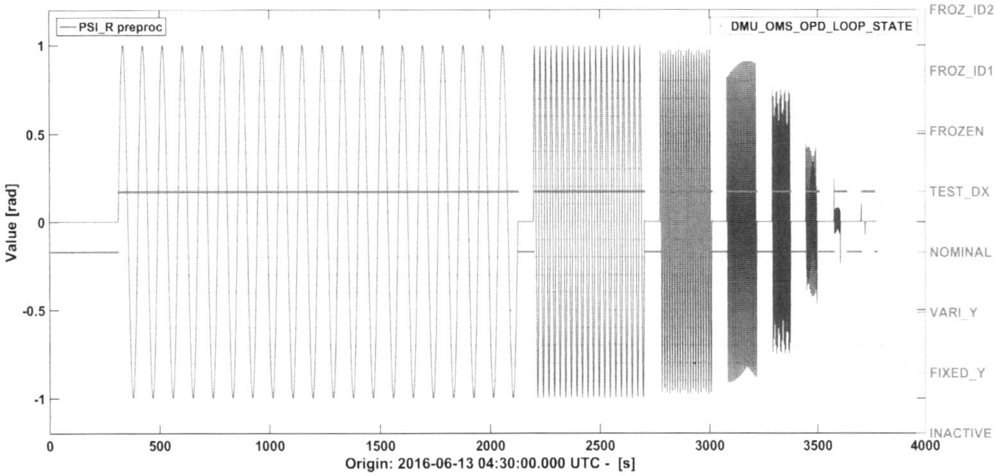

Figure 3. The 8 guidance injections into the OPD loop are seen in PSI_R. For each injection the control loop is switched to the 'TEST_DX' state.

5. Acknowledgements

The Albert-Einstein-Institut acknowledges the support of the German Space Agency, DLR. The work is supported by the Federal Ministry for Economic Affairs and Energy based on a resolution of the German Bundestag (FKZ 50OQ0501 and FKZ 50OQ1601).

References

[1] Armano M *et al.* 2016 *Phys. Rev. Lett.* **116**(23) 231101
[2] Audley H 2017 *Journal of Physics: Conference Series to be published in this issue*
[3] Audley H 2014 Preparing for LISA pathfinder operations : characterisation of the optical metrology system

IOP Conf. Series: Journal of Physics: Conf. Series **840** (2017) 012037 doi:10.1088/1742-6596/840/1/012037

GCR flux 9-day variations with LISA Pathfinder

C Grimani[1,2] for the LISA Pathfinder Collaboration

and S Benella[1,2], M Fabi[1], N Finetti[2,3], D Telloni[2,4]

[1] DiSPeA, University of Urbino, 61029 Urbino, Italy
[2] National Institute for Nuclear Physics (INFN), Section in Florence, 50019 Sesto Fiorentino, Italy
[3] Department of Physical and Chemical Sciences, University of L'Aquila, 67100 L'Aquila, Italy
[4] National Institute for Astrophysics, Astrophysical Observatory of Torino, 10025 Pino Torinese, Italy

E-mail: catia.grimani@uniurb.it

Abstract. Galactic cosmic-ray (GCR) energy spectra in the heliosphere vary on the basis of the level of solar activity, the status of solar polarity and interplanetary transient magnetic structures of solar origin. A high counting rate particle detector (PD) aboard LISA Pathfinder (LPF) allows for the measurement of galactic cosmic-ray and solar energetic particle (SEP) integral fluxes at energies > 70 MeV n^{-1} up to 6500 counts s^{-1}. Data are gathered with a sampling time of 15 s. A study of GCR flux depressions associated with the third harmonic of the Sun rotation period ($\sim$ 9 days) is presented here.

1. Introduction

Solar activity level and Global Solar Magnetic Field (GSMF) polarity modulate galactic cosmic-ray (GCR) fluxes in the heliosphere (see for instance [1, 2, 3, 4, 5]). A direct correlation between high solar activity and solar energetic particle (SEP) event occurrence is also observed [6, 7].

The symmetric model in the *force field approximation* by Gleeson and Axford [8] allows for the estimate of the energy spectra of cosmic rays at a distance r from the Sun, at a time t during periods of positive polarity, when time-independent interstellar intensities are assumed. An energy loss related to the charge of cosmic-ray particles and a *solar modulation parameter*, ϕ, above about 100 MeV are considered. The role played by the GSMF polarity on the drift of positive (negative) particles in the heliosphere during periods of negative (positive) polarity must be also taken properly into account. Protons and nuclei reaching Earth during negative polarity periods come mainly from the ecliptic regions along the Heliospheric Current Sheet (HCS). An opposite situation holds during positive magnetic field polarity periods when positive particles come from the polar region of the Sun. Particles propagating along the HCS lose more energy than those coming from the poles.

Nominal quasiperiodicities of 27, 13.5 and 9 days related to the Sun rotation period and its higher harmonics are also observed in the cosmic-ray flux, the interplanetary magnetic field (IMF), the solar wind plasma and the geomagnetic activity parameters [9]. GCR observations gathered in space above a few tens of MeV by Helios 1, Helios 2 and IMP-8 experiments indicated that the effects of corotating interaction regions, generated by high-speed solar wind streams

IOP Conf. Series: Journal of Physics: Conf. Series **840** (2017) 012037 doi:10.1088/1742-6596/840/1/012037

overtaking preceding slower solar wind, are at the origin of the 9-day GCR modulations [10]. A high counting rate particle detector (PD) aboard LISA Pathfinder (LPF) presents a typical response of more than 100 counts s^{-1} in 15 s sampling time, allowing for limiting the statistical uncertainty on one-hour binned GCR proton and helium nucleus flux data at percent level above 70 MeV n^{-1}. The aim of this work is to report on the characteristics of $\sim$ 9-day GCR short-term variations, compatible with the third harmonic of the solar rotation period, during the first three months of the LPF mission.

2. The Particle Detector aboard LPF

LPF [11, 12, 13] is the ESA (European Space Agency) key technology demonstrator mission for LISA [14], the first interferometer devoted to gravitational wave detection in space. The LPF spacecraft was successfully launched from the Kourou base in French Guiana on December 3rd, 2015 aboard a Vega rocket. The satellite reached its six-months, final orbit around the Earth-Sun Lagrangian point L1 at about 1.5 million km from Earth on January 2016. The LPF orbit is inclined at about 45 degrees on the ecliptic plane. Orbit minor and major axes are approximately of 0.5 million km and 0.8 million km, respectively. LPF carries two 2-kg cubic platinum-gold free-floating test masses that play the role of mirrors of the interferometer. The charging process due to GCRs and SEPs results in spurious force noise on both test masses [15]. The effects of the charging process were properly estimated before the mission launch with Monte Carlo simulations [16, 17] carried out on the basis of GCR and SEP occurrence projections for the period orbits about the L1 point [18]. The minimum energy of protons and ions (98% in composition of both GCRs and SEPs [19]) penetrating and charging the LPF test masses is 100 MeV n^{-1}. The conservative choice of monitoring GCR above 70 MeV n^{-1} on LPF was made in order not to underestimate the overall particle flux charging the test-masses. The LPF PD is mounted behind the spacecraft solar panels and oriented along the Sun-Earth direction. This device consists of two, 300 μm thick, silicon wafers of 1.4 x 1.05 cm^2 area, separated by 2 cm and placed in a telescopic arrangement [20, 21]. For particle energies $>$ 100 MeV n^{-1} the instrument geometrical factor is found to be energy independent and equal to 9 cm^2 sr for particle isotropic incidence on each silicon wafer. In coincidence mode the detector geometrical factor is about one tenth of this value. A shielding, copper box surrounds the silicon wafers. The box thickness is 6.4 mm. This shielding material stops particles with energies smaller than 70 MeV n^{-1}. The PD allows for the counting of particles crossing each silicon layer and for the measurement of ionization energy losses of particles passing through both silicon wafers (coincidence mode). The PD data are stored in the form of histograms over periods of 600 s and then sent to the on-board computer. The maximum allowed PD counting rate is 6500 counts s^{-1}, corresponding to an event integrated proton fluence of 10^8 protons cm^{-2} at energies $>$ 100 MeV. In coincidence mode, up to 5000 energy deposits per second can be stored in each histogram. The expected occurrence of SEP events with fluence $> 10^8$ protons cm^{-2} at energies $>$ 30 MeV, for instance, is less than 1 per year as it follows from the Nymmik model [22]. None of SEP events, characterized by proton differential fluxes above 100 MeV n^{-1} overcoming that of galactic origin, were observed since the LPF launch and up to the time of writing. This scenario makes the whole period of the LPF mission the most favourable for GCR short-term variation investigation. Data from the PD are available since mid-February 2016.

3. GCR and interplanetary solar wind plasma parameter short-term variations

The parameterization of the GCR energy spectra adopted here follows from the work by Papini, Grimani & Stephens [19]:

$$F(E) = A\ (E + b)^{-\alpha}\ E^{\beta} \quad \text{particles/(m}^2\ \text{sr s GeV/n).} \tag{1}$$

IOP Conf. Series: Journal of Physics: Conf. Series **840** (2017) 012037 doi:10.1088/1742-6596/840/1/012037

The values of the parameters A, b, α and β for proton and helium nucleus energy spectra at the beginning of the LPF mission (corresponding to a solar modulation parameter ϕ=550 MV c^{-1} [23]) appear in Table 1.

Table 1. Parameterization of proton and helium energy spectra during the first part of the LPF mission.

	A	b	α	β
p	18000	1.19	3.66	0.87
He	850	0.96	3.23	0.48

The fractional variations of GCR fluxes of duration compatible with a 9-day periodicity corresponding to the third harmonic of the Sun rotation period, observed with the LPF PD during the Bartels rotations 2490-2492 (February 6th, 2016 - April 27th, 2016) are shown in Figs. 1 - 6. In Figs. 1, 3 and 5 GCR data are compared to the magnetic field and solar wind plasma parameters gathered simultaneously in L1 by the ACE experiment [24]. A striking majority of the observed GCR flux depressions show longer recovery than decrease periods and appear correlated with the interplanetary magnetic field intensity above 10 nT (fourth and bottom panels in Figs. 1-3-5) and solar wind plasma speed above 400 km s^{-1} (second and bottom panels in figures 1-3-5). HCS crossing is also indicated in the middle panels of figures 1-3-5. HCS are modulated by incoming high-speed solar wind streams. Changes of IMF polarity associated with plasma density enhancements are observed when LPF crosses the HCS. In Figs. 2-4-6 the LPF GCR flux fractional variations are compared to analogous polar and near-polar neutron monitor observations. The stations considered here present vertical geomagnetic cutoffs < 0.8 GV c^{-1} and effective energies > 5.5 GeV. It is worthwhile to recall that neutron monitors allow for a direct measurement of the GCR flux at energies larger than the effective energy, when fractional variations of neutron monitor countings are identical to those of the GCR flux. In general, the LPF GCR flux depressions show a trend similar to those observed on polar and near-polar neutron monitors. Although the LPF spacecraft remained below the ecliptic for the period of the study reported here, no evident better agreement was found with South neutron monitor (i. e. Mc Murdo and Terre Adelie stations) observations with respect to those of northern stations (i. e. Thule and Oulu stations). Average depression characteristics are reported in Table 2. The interval of time during which the flux of GCRs remains depressed between decrease and recovery phases is called here *plateau*. The depression intensity was observed to be of 6%, on average. Earth neutron monitor observations, being representative of GeV GCR fluxes, present maximum depression intensities of 2%. In general, the dynamics of

Table 2. Characteristics of GCR short-term depressions observed with the PD aboard LPF.

	Duration (days)	%
Decrease	2.6±1.0	
Plateau	1.3±0.8	
Recovery	3.7±1.1	
Total	7.6±1.6	
Intensity		6.0±1.7

IOP Conf. Series: Journal of Physics: Conf. Series **840** (2017) 012037 doi:10.1088/1742-6596/840/1/012037

short-term variations with intensities smaller than 3% on LPF are not observed on Earth.

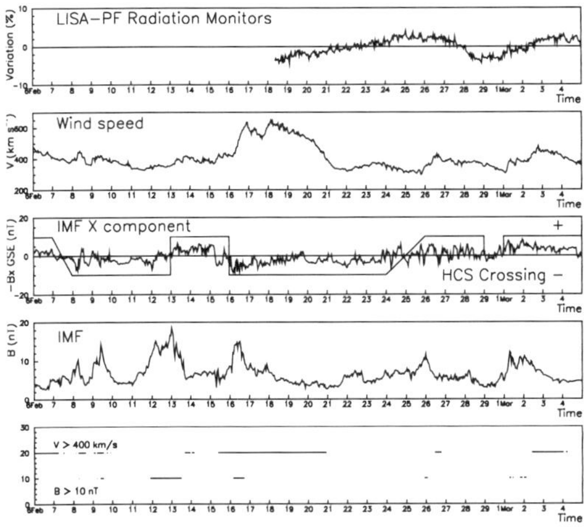

Figure 1. Comparison of LPF PD counting rate fractional variations with simultaneous ACE observations in L1 of solar magnetic field and wind plasma parameters during the Bartels rotation 2490 (February 6th, 2016 - March 4th, 2016).

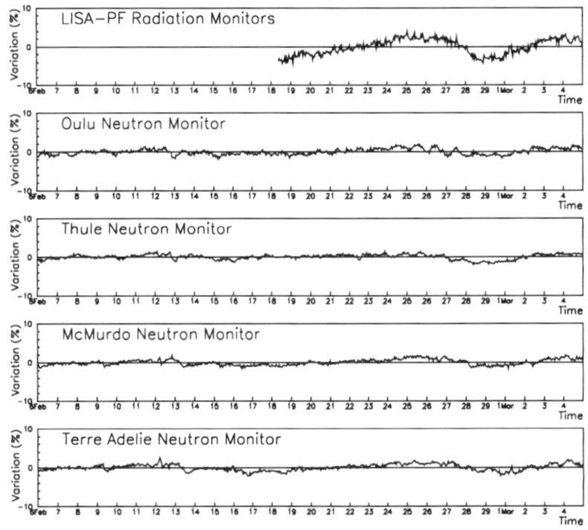

Figure 2. Comparison of LPF PD with neutron monitor data during the Bartels rotation 2490.

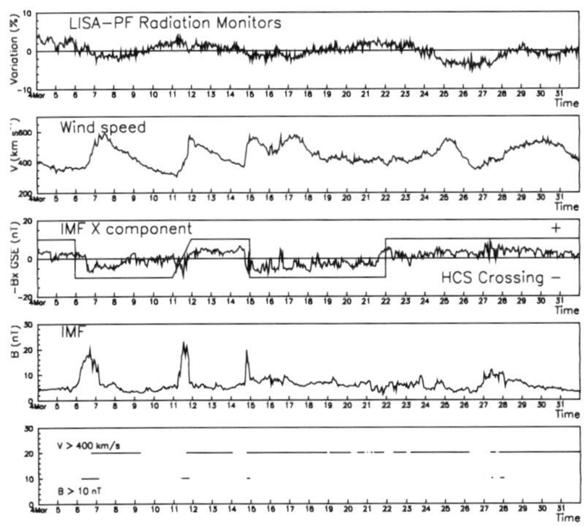

Figure 3. Same as Fig. 1, but during the Bartels rotation 2491 (March 4th, 2016 - March 31st, 2016).

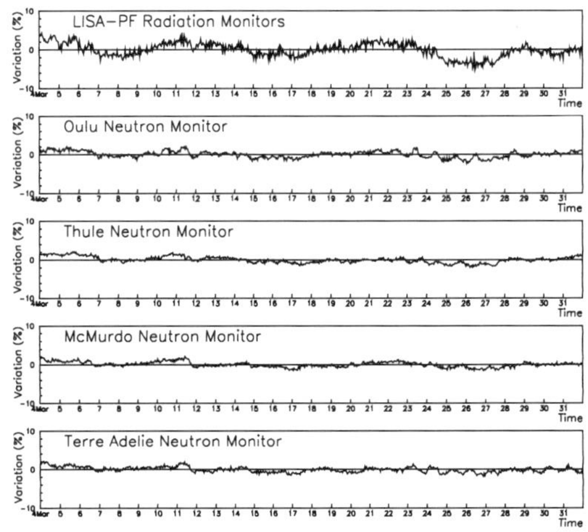

Figure 4. Same as Fig. 2, but during the Bartels rotation 2491.

4. Conclusions

The high counting rate PD placed aboard LPF allows for the measurement of GCR proton and helium flux oscillations compatible with the third harmonic of the Sun rotation period above 70

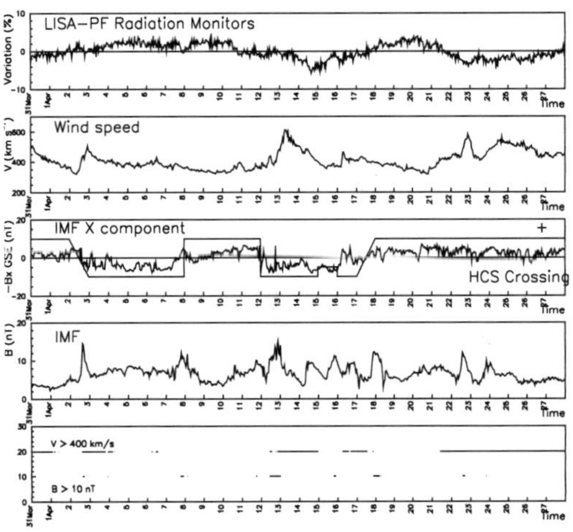

Figure 5. Same as Fig. 1, but during the Bartels rotation 2492 (March 31st, 2016 - April 27th, 2016).

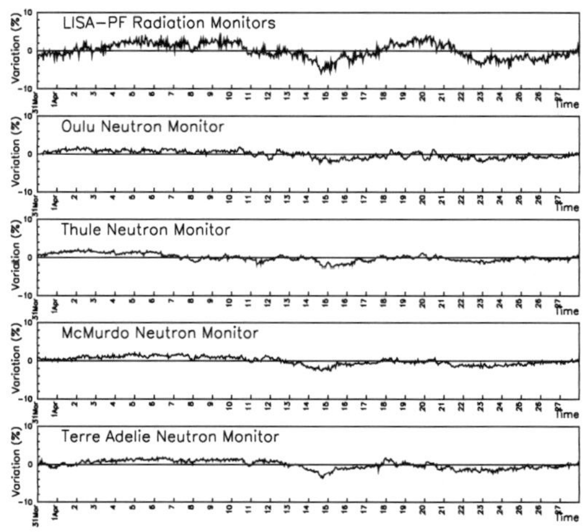

Figure 6. Same as Fig 2, but during the Bartels rotation 2492.

MeV n^{-1}. One-hour binned data present errors at percent level. GCR depressions under study appear to be correlated with an IMF intensity > 10 nT and/or a solar wind speed > 400 km s^{-1}. Their average intensity and duration are of (7.6±1.6) days and (6.0±1.7)%, respectively. Low-vertical geomagnetic cutoff neutron monitor observations show that these same depressions appear of just 2% above 5.5 GeV.

References

[1] Beer J 2000 *Cosmic Rays and Earth* J W Bieber, E Eroshenko, P Evenson, E O Flückiger and R Kallenbach eds. *Sp. Sc. Rev.* (Elsevier) **93**
[2] Shikaze Y *et al* 2007 *Astropart. Phys.* **28** 154-167
[3] Grimani C *et al* 2007 *Proc. 30th Int. Cosmic Ray Conf. (Merida)* **1** 485-488
[4] Grimani C 2004 *A&A* **418** 649-653
[5] Grimani C 2007 *A&A* **474** 339-343
[6] Nymmik R A *et al* 1999 *Proc. 26th Int. Cosmic Ray Conf. (Salt Lake City)* **6** 268-271
[7] Nymmik RA *et al* 1999 *Proc. 26th Int. Cosmic Ray Conf. (Salt Lake City)* **6** 280-283
[8] Gleeson L J and Axford W I 1968 *Ap. J.* **154** 1011-1026
[9] Sabbah I and Kudela K 2011 *J. Geophys. Res.* **116** A04103
[10] Richardson I G 2004 *Sp. Sc. Rev.* **111** 267-376
[11] Antonucci F *et al* 2011 *Class. Quantum Grav.* **28** 094001
[12] Antonucci F *et al* 2012 *Class. Quantum Grav.* **29** 124014
[13] M Armano *et al* 2016 *Phys. Rev. D* **116** 231101
[14] https://wolke7.aei.mpg.de/index.php/s/i6qxc8zNnOgGzKQ
[15] Shaul D *et al* 2006 *AIP Conf. Proc.* **873** 172-178
[16] Araújo H M *et al* 2005 *Astropart. Phys.* **22** 451-469
[17] Grimani C *et al* 2005 *Class. Quantum Grav.* **22** S327-S332
[18] Grimani C *et al* 2015 *Class. Quantum Grav.* **32** 035001
[19] Papini P, Grimani C and Stephens A S 1996 *Nuovo Cimento* **19** 367-387
[20] Cañizares P *et al* 2011 *Class. Quantum Grav.* **28** 094004
[21] Mateos I *et al* 2012 *J. Phys.: Conf. Ser.* **363** 012050
[22] Storini M *et al* 2008 *OPOCE publisher for COST 724 action* 63-69
[23] http://cosmicrays.oulu.fi/phi/Phi_mon.txt
[24] https://cdaweb.sci.gsfc.nasa.gov/index.html/

IOP Conf. Series: Journal of Physics: Conf. Series **840** (2017) 012038 doi:10.1088/1742-6596/840/1/012038

LISA Pathfinder closed-loop analysis: a model breakdown of the in-loop observables

LISA Pathfinder collaboration

E-mail: henri.inchauspe@apc.in2p3.fr

Abstract. This paper describes a methodology to analyze, in the frequency domain, the steady-state control performances of the LISA Pathfinder mission. In particular, it provides a technical framework to give a comprehensive understanding of the spectra of all the degrees of freedom by breaking them down into their various physical origins, hence bringing out the major contributions of the control residuals. A reconstruction of the measured in-loop output, extracted from a model of the closed-loop system, is shown as an instance to illustrate the potential of such a model breakdown of the data.

1. Introduction

The LISA Pathfinder mission [3] has been operating since March 2016 and has demonstrated the feasibility of sub-femto-g free-fall of test masses necessary to build a LISA-like gravitational waves observatory in space [2]. While the publication [2] is mostly concerned by acceleration noise performances along the most sensitive x-axis, LISA Pathfinder has shown impressive performances and stability for the other measurement channels. Those extra degrees of freedom, though not the main channel of interest, are of great importance for the satellite's closed-loop control framework, the so-called *Drag-Free and Attitude Control System (DFACS)* [6]. This core subsystem has been designed by *Airbus Defence & Space* to achieve the specific requirements of LISA Pathfinder. To reach such an acceleration sensitivity along x-axis implies a very quiet and stable environment, and in particular, a very stable test-mass position and orientation, along and around all the geometrical axes, in order to avoid coupling to the sensitive x-axis through cross-talks and stiffnesses, for instance.

The left-hand diagram of the figure 1 shows the scheme of the experiment. Two cubic test masses of around 2kg are almost free-floating inside housings whose inside walls have electrodes facing the test masses. These electrodes, forming a set of capacitances with the test mass surface, are solicited for sensing and control of the mass inside its housing. While the test mass positions can be corrected through electrostatic actuation along x_2-axis and θ_2, η, ϕ angles (*suspension control*), the other degrees of freedom x_1, y, z and θ_1 are controlled exclusively through spacecraft micro-propulsion (*drag-free control*). The only exception to that is made for the *attitude control*, for which test masses are actuated along y, z and θ_1, in order to induce a torque command on the spacecraft from drag-free control. The role of the *DFACS* is to maintain the test masses parallel to the electrode surfaces, and centered in their housing, except for test mass 2 along the x-axis which is constrained to maintain null relative displacement w.r.t the test mass 1 (x_{12} coordinate is in-loop instead of x_2).

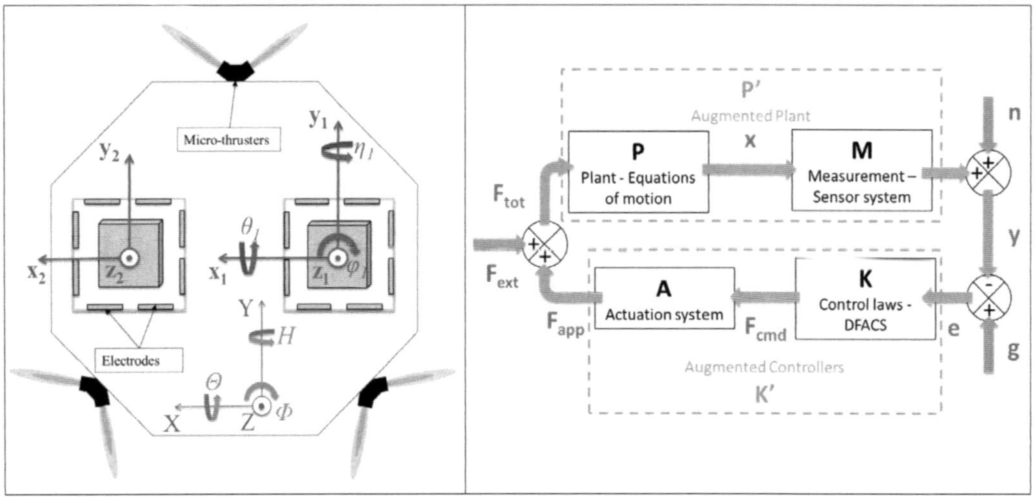

Figure 1: *Left-hand figure*: Simplified sketch of the experiment scheme, representing essentially the system of the three bodies involved and the coordinate systems used to describe the motion of the bodies one w.r.t the others. *Right-hand figure*: Diagram of the closed loop system. The closed loop system is excited in 3 different channels: the guidance signal g, the sensing noise n, and the external forces F_{ext}.

2. Study of LISA Pathfinder closed loop system

2.1. The sensitivity closed-loop transfer functions

The LISA Pathfinder closed loop system, approximated as a linear system in this document, can be fully represented by a set of four (non-independent) closed loop transfer function, often named as "The Gang of Four" in the literature [4], denoting the influence of system inputs coming from system environment (noise, disturbances) over the in-loop controlled variables (measurements, commanded forces, system states...). These transfer functions are all functions of every open-loop transfer functions involved in the closed loop system (Plant, Measurement and Actuation blocks, Controllers etc. see block diagram in figure 1):

- The *sensitivity function* $S(f)$, or *S-gain*, denotes the sensitivity of the in-loop sensors outputs $\vec{o}$ to the sensing noise $\vec{n}$ of all the sensors. $S(f)$ is a 15x15 matrix of transfer functions in LISA Pathfinder's case.

- The *load disturbance sensitivity function* $L(f)$, or the *L-gain*, calls for the sensitivity of the state variable $\vec{x}$ (displacements for the dynamical states) to the direct force and torque noises $\vec{F}_{ext}$ applied on the test masses (along every degree of freedom). $L(f)$ is a 15x18 matrix of transfer functions.

- The *complementary sensitivity function* $T(f)$, or the *T-gain*, stands for the sensitivity of the state variable $\vec{x}$ to the sensing noise $\vec{n}$ of all the sensors.. $T(f)$ is a 15x15 matrix of transfer functions.

The fourth and last one is not be used in the present analysis. The relationships between these closed loop functions and the sub-blocks in figure 1 are given by equations 1:

$$S(f) = \frac{1}{1+P'K'} \qquad L(f) = \frac{P'}{1+P'K'}$$
$$T(f) = \frac{P'K'}{1+P'K'} \tag{1}$$

where it can be seen that the functions are not independent (for instance, S and T are constrained by the identity $S + T = 1$).

From this definition, it follows that measuring this set of transfer functions allows to fully characterize the closed loop behavior of the linear system. Conversely, having a sufficiently reliable model of these characteristic functions allows one to predict the spectral behavior of the (physical) in-loop variables, provided that one has some knowledge of the external disturbing signals exciting the loop. Both those aspects will be considered in the next section.

2.2. In-loop sensing outputs and states breakdowns

The LISA Pathfinder collaboration has implemented a full linear state space modeling of the closed loop system [8], from which the closed loop functions discussed in section 2.1 can be extracted. Hence, an interesting comparison can be performed, using the state space model and a measurement of the external signals to reconstruct an estimate of the in-loop variables. By focusing on the sensing outputs $\vec{Y}$ (equation 2) and state variables $\vec{X}$ (equation 3) , one can express those quantities w.r.t. the external signals in the diagram of the figure 1:

$$\tilde{Y}^q = \sum_{p=x,y,z,\theta,\eta,\phi} L^{qp}_{gain}\tilde{F}^p_{ext} + \sum_{p=x,y,z,\theta,\eta,\phi} S^{qp}_{gain}\tilde{n}^p + \sum_{p=x,y,z,\theta,\eta,\phi} T^{qp}_{gain}\tilde{g}^p \tag{2}$$

$$\tilde{X}^q = \sum_{p=x,y,z,\theta,\eta,\phi} L^{qp}_{gain}\tilde{F}^p_{ext} - \sum_{p=x,y,z,\theta,\eta,\phi} T^{qp}_{gain}\tilde{n}^p + \sum_{p=x,y,z,\theta,\eta,\phi} T^{qp}_{gain}\tilde{g}^p \tag{3}$$

$$\tilde{F}^q_{cmd} = \sum_{p=x,y,z,\theta,\eta,\phi} L^{qp}_{gain}\tilde{F}^p_{ext} - \sum_{p=x,y,z,\theta,\eta,\phi} T^{qp}_{gain}\tilde{n}^p + \sum_{p=x,y,z,\theta,\eta,\phi} T^{qp}_{gain}\tilde{g}^p \tag{4}$$

One notes that the sensing noise does not influence $\vec{Y}$ and $\vec{X}$ in the same manner. Indeed, while the double integrator behavior of the plant makes it respond efficiently at low frequency, the control gain is also important at low frequency, resulting in a very low *S-gain* at low frequency. Hence, one does not see any sensing noise in the low frequency part of $\vec{Y}$ spectrum. However, for the very same reason, the *T-gain* is maximal at very low frequency, thus suggesting a perfect correlation between the state $\vec{X}$ and the noise $\vec{n}$ in that domain. This can be viewed even more clearly in considering that the high control gain in that frequency domain tends to "shake" the test mass in order to cancel the sensing noise, the latter being undistinguishable from real motion from *DFACS*'s point of view.

From equations 2, 3 and 4, one can extract two sorts of information. First, one can identify where the sensing outputs and the commanded forces, both delivered by the telemetry, can reflect the out-of-loop signals, i.e. the sensing and force noises, thus constituting a direct measurement of those. As an example, at sufficiently high frequency the plant gain is so large that any commanded forces applied on the masses will result in no displacement at all, meaning that one directly observes the sensing noise from the sensing output. That property is directly inferable from the value of the S-function at high frequency, which tends towards unity. For the force noise though, one can equally well subtract the calibrated commanded forces to retrieve the external noisy force, thus avoiding control in-loop features.

3. Output decomposition from model: instance of the angle ϕ_1

Once the out-of-loop signals have been measured or estimated, one can reconstruct assessed versions of the outputs. One can also decomposed those quantities w.r.t. the contributions from each individual noise sources. This technic is a very useful tool for diagnose purposes, as will be seen below. In confronting these estimates to the actually observed quantities, one verifies that the behavior of each controlled degree of freedom is well understood. The figure 2 exhibits such a comparison between modeled and measured displacement spectrum, in the case of the ϕ_1 degree of freedom (i.e. Cardan rotation angle around the z_1-axis, cf. figure 1), measured by the interferometer through differential wavefront sensing [7]. In the same plot is drawn the estimated contributions to the spectrum from the various noise sources.

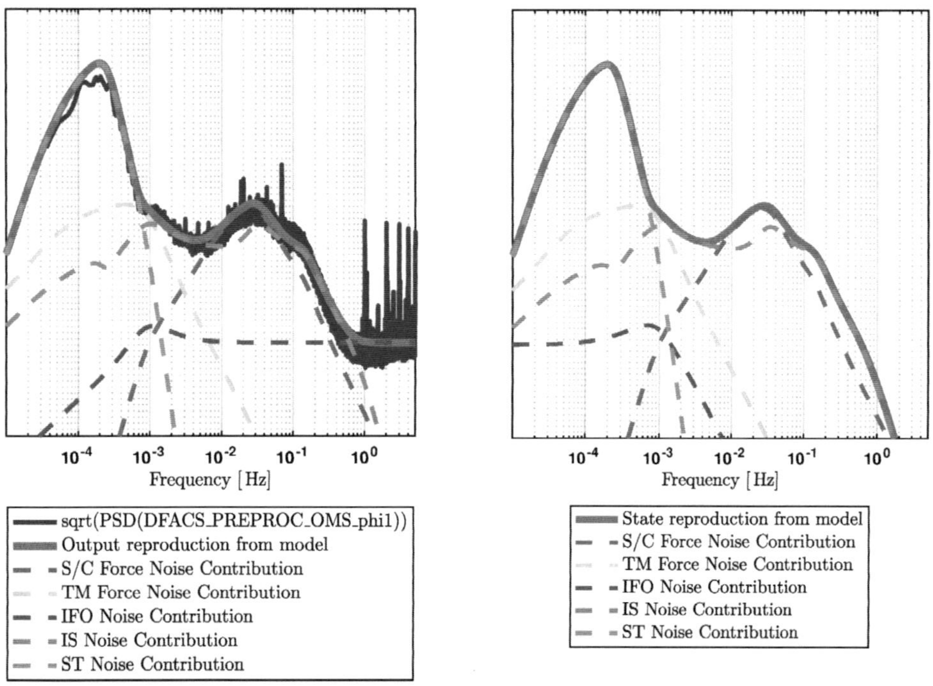

Figure 2: Decomposition in the frequency domain of the measurement outputs (on the left) and the modeled true dynamical states (on the right) into the various contributions from the sensing noise and the external disturbances. The blue curve comes directly from LPF data corresponding to the long noise run (i.e. free of any calibration signal) in nominal science mode of April 2016 (**dataset from 2016-04-04 00:00 UTC to 2016-04-14 00:00 UTC**[1]). The red curve is built from modeled closed loop transfer functions (SSM model) with noise level listed in the table above. Dashed curves give the individual contributions of each of the noise channels. These plots help to identify what are the physical phenomena explaining the output spectrum in the different frequency band.

The case of ϕ angular displacement of the test masses is of particular interest. Indeed, in analyzing what are the major contributing noise sources to ϕ_1 observed displacement in figure 2, one notes that various sources come equally into play, depending on the frequency bandwidth of focus. At very low frequency, a large bump is clearly imputed to the star tracker noise, which indeed, though very indirectly, affects the in-loop test masses angular

[1] The y-scales are voluntarily hidden to keep data exclusivity for an upcoming publication [5]

variables. The attitude control strategy of LISA Pathfinder is such that the spacecraft reacts to star tracker error signals, say around Φ, by first pushing the test masses away one from the other along y_1 and y_2. Then the Drag-Free control immediately sees the induced differential motion between test masses, and commands the spacecraft to rotate in order to compensate that small linear motion. The spacecraft rotation entails in its turn a misalignment between the spacecraft and the test masses reference frame, and finally the suspension control, i.e. the capacitive actuation system, commands torques on the test masses to reorient them inside their housing. Hence, the link between the star tracker signal and the test masses rotation is far from being direct, though obviously large in the figure 2. The level of that contribution is both due to an increase of the star tracker noise and a significantly high authority of the attitude control gain typically below 10^{-3}Hz. In the intermediate frequency band, i.e. between 10^{-2}Hz and 10^{-1}Hz, the Drag-Free control gain decreases significantly and reaches its minimum, thus allowing spacecraft to test masses relative displacements, both linear and angular. In the latter case, it will force the suspension control to correct these noisy rotations. However, because the control gain is also relatively weak in that frequency band, angular displacements are allowed to pass through the loop. At very high frequency finally, the commanded forces become inefficient because of the test masse inertia and one observes the interferometer sensing noise only which provides, as already discussed in section 2.2, a way to measure it.

4. Conclusion

A strategy to analyze *DFACS* performances and interpret control residuals has been developed and is presented in this paper. It allows for a full understanding of the LISA Pathfinder general in-loop dynamics, which has a significant impact on the noise performances along the optical sensitive axis, and more generally, on the future LISA observatory [1]. According to figure 2, the state space model developed by the collaboration is shown to be remarkably reliable in explaining the observed system in-loop behavior, hence giving the state space model a very interesting power of diagnosis. The reliability of the model seems to argue in favor of the assumptions of linearity and stationarity of the system. This capacity of the model to reliably represent LISA Pathfinder's closed loop system is also very interesting in the LISA context, especially for the upcoming task of transferring the knowledge acquired from LISA Pathfinder experiment to LISA simulations and design.

Acknowledgements

E.P. and H.I would like to acknowledge the financial support of the UnivEarthS Labex program at Sorbonne Paris Cité (ANR-10-LABX-0023 and ANR-11-IDEX-0005-02) and the French space agency CNES for its financial support during these studies.

References

[1] Pau Amaro-Seoane et al. "Laser Interferometer Space Antenna". In: *arXiv:1702.00786 [astro-ph]* (Feb. 2017). arXiv: 1702.00786. URL: `http://arxiv.org/abs/1702.00786` (visited on 02/16/2017).

[2] M. Armano et al. "Sub-Femto-g Free Fall for Space-Based Gravitational Wave Observatories: LISA Pathfinder Results". In: *Physical Review Letters* 116.23 (June 2016), p. 231101. DOI: `10.1103/PhysRevLett.116.231101`. URL: `http://link.aps.org/doi/10.1103/PhysRevLett.116.231101` (visited on 10/24/2016).

[3] M. Armano et al. "The LISA Pathfinder Mission". en. In: *Journal of Physics: Conference Series* 610.1 (May 2015), p. 012005. ISSN: 1742-6596. DOI: `10.1088/1742-6596/610/1/012005`. URL: `http://iopscience.iop.org/1742-6596/610/1/012005` (visited on 06/19/2015).

[4] Karl Johan Åström and Richard M. Murray. *Feedback systems: an introduction for scientists and engineers.* anglais. Princeton, Royaume-Uni de Grande-Bretagne et d'Irlande du Nord, 2008. ISBN: 978-0-691-13576-2.

[5] LISA Pathfinder collaboration. "LISA Pathfinder drag-free performances: frequency domain analysis of the in-loop dynamics". In: *N.D.* (Upcoming publication).

[6] Walter Fichter et al. "LISA Pathfinder drag-free control and system implications". en. In: *Classical and Quantum Gravity* 22.10 (Apr. 2005), S139. ISSN: 0264-9381. DOI: 10.1088/0264-9381/22/10/002. URL: http://iopscience.iop.org/article/10.1088/0264-9381/22/10/002/meta (visited on 02/16/2017).

[7] G. Heinzel et al. "The LTP interferometer and phasemeter". en. In: *Classical and Quantum Gravity* 21.5 (Feb. 2004), S581. ISSN: 0264-9381. DOI: 10.1088/0264-9381/21/5/029. URL: http://iopscience.iop.org/article/10.1088/0264-9381/21/5/029/meta (visited on 02/16/2017).

[8] Mario Weyrich. "Modelling and Analyses of the LISA Pathfinder Technology Experiment". PhD thesis. University of Stuttgart, 2008.

IOP Conf. Series: Journal of Physics: Conf. Series **840** (2017) 012039 doi:10.1088/1742-6596/840/1/012039

Radiation pressure calibration and test mass reflectivities for LISA Pathfinder

Natalia Korsakova* and Brigitte Kaune on behalf of LPF collaboration**

* SUPA, University of Glasgow, Glasgow, G12 8QQ, United Kingdom

** Albert-Einstein-Institut, Max-Planck-Institut für Gravitationsphysik und Universität Hannover, Callinstrasse 38, 30167 Hannover, Germany

E-mail: natalia.korsakova@glasgow.ac.uk

Abstract. This paper describes a series of experiments which were carried out during the main operations of LISA Pathfinder. These experiments were performed by modulating the power of the measurement and reference beams. In one series of experiments the beams were sequentially switched on and off. In the other series of experiments the powers of the beams were modulated within 0.1% and 1% of the constant power. These experiments use recordings of the total power measured on the photodiodes to infer the properties of the Optical Metrology System (OMS), such as reflectivities of the test masses and change of the photodiode efficiencies with time. In the first case the powers are back propagated from the different photodiodes to the same place on the optical bench to express the unknown quantities in the measurement with the complimentary photodiode measurements. They are combined in the way that the only unknown left is the test mass reflectivities. The second experiment compared two estimates of the force applied to the test masses due to the radiation pressure that appears because of the beam modulations. One estimate of the force is inferred from the measurements of the powers on the photodiodes and propagation of this measurement to the test masses. The other estimation of the force is done by calculating it from the change in the main scientific output of the instrument – differential displacement of the two test masses.

1. Introduction

LISA Pathfinder (LPF) is a technology demonstration mission for the future gravitational wave observatory in space – Laser Interferometer Space Antenna (LISA). The Optical Metrology System (OMS) on board LPF measures relative position of two free falling test masses and the position of one of the test masses relative to the optical bench. The measurements are performed with interferometers that are bonded on the ultra-stable bench between the test masses. The OMS of LPF has four interferometers bonded on the optical bench:

- *X12* – interferometer that measures relative displacement of the two test masses;
- *X1* – interferometer that measures the displacement of the first test mass relative to the optical bench;
- *Frequency* – interferometer with the deliberate pathlength mismatch used for the frequency stabilisation of the laser;
- *Reference* – interferometer, which measurement is subtracted from the other interferometer measurements to get rid of the common mode noise.

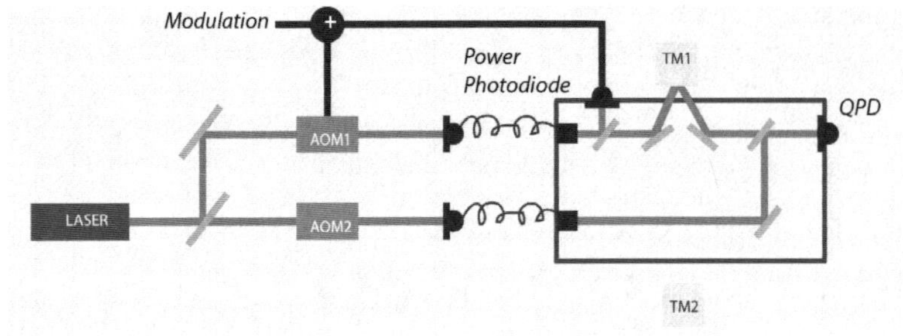

Figure 1. This is an outline of the experiment that presents how the modulation is done within the OMS (Optical Metrology System) which comprises of laser assembly, Acousto-Optical Modulators (AOMs), Test Masses (TMs), modulation bench and optical bench.

There are two quadrant photodiodes for each interferometer. In addition there are two single element photodiodes that measure the power transmitted to the optical bench. These measurements are used for the stabilisation of the power that goes to the optical bench. Other components on the optical bench are beam splitters and mirrors, the reflectivities and transmissivities for which we know within errors. The outline of the optical bench with the labels for its components is shown on the Figure 2.

2. Experiment setup

Two types of measurements were performed:

(i) Sequence of the beam on/beam of measurements. We measure the sum of the power on the quadrants of the Quadrant Photodiodes (QPDs) and use the ratio of the measurements from two interferometers to estimate the test mass reflectivities.

(ii) Modulation of the measurement beam, modulation of the reference beam and counter modulation of both beams. The modulations were done with 0.1% and 1% of the original power at frequencies around 2 mHz.

The schematic of the experiment setup is presented on the Figure 1. The power is modulated by adding signal to the error point of the power stabilisation loop.

3. Test mass reflectivities

To calculate test mass reflectivities we treat test mass as one of the components in the light path that goes from Fibre Injector Optical Subassemblies (FIOS) to the respective QPDs. To put it simply we need to divide the measured power by the injected power. However, we do not know the efficiencies of the photodiodes and values of the power injected on the optical bench well enough. Therefore we need to use a combination of the measurements to get rid of the unknown values. To eliminate QPD efficiencies we use the measurements from the same QPDs but for the experiments when either measurement of reference beams were on. To measure the powers we use the QPDs that do not have a test mass in the path of their beam such as the measurements from the *Reference* or from the *Frequency* interferometer. We propagate the power back through component from the QPDs to the place where it is injected on the optical bench. The powers propagated from either *X1* or *X12* interferometers we express through the powers propagated from either *Reference* or *Frequency*. We use *Reference* interferometer for the calculations, for the other one equations can be derived in the similar way and will give us redundant measurements. Moreover we will use only A photodiodes, for the B photodiodes calculations are done similarly and they also provide redundant measurements.

Let us derive the equation for the first test mass reflectivity only, for the second test mass it is done in a similar way.

First, let us define the convention for the naming that is going to be used throughout this section:

- E will stand for the efficiencies with the respective photodiode name in the subscript;
- Σ will stand for the total power on the photodiode with the respective photodiode name in the subscript;
- subscripts "m" or "r" for the total power measurements will define whether measurement or reference beam is on, respectively.

Let us propagate the power for the measurement $P_{\rm m}$ (beam 1) and reference $P_{\rm r}$ (beam 2) beams to the *Reference interferometer* photodiode PDRA:

$$P_{\rm m} \cdot L_{{\rm m}RA} = E_{RA} \cdot \Sigma_{{\rm m}RA} \qquad (1)$$
$$P_{\rm r} \cdot L_{{\rm r}RA} = E_{RA} \cdot \Sigma_{{\rm r}RA}$$

where the optical paths are

$$L_{{\rm m}RA} = BS16t \cdot BS1t \cdot BS2r \cdot M11 \cdot M12 \cdot M14 \cdot BS5r \qquad (2)$$
$$L_{{\rm r}RA} = BS11t \cdot BS4t \cdot BS9r \cdot BS5t.$$

Let us propagate the power for the measurement $P_{\rm m}$ (beam 1) and reference $P_{\rm r}$ (beam 2) beams to the *X1 interferometer* photodiode:

$$P_{\rm m} \cdot L_{{\rm m}1A} \cdot TM1r \cdot WIN1t^2 = E_{1A} \cdot \Sigma_{{\rm m}1A} \qquad (3)$$
$$P_{\rm r} \cdot L_{{\rm r}1A} = E_{1A} \cdot \Sigma_{{\rm r}1A}$$

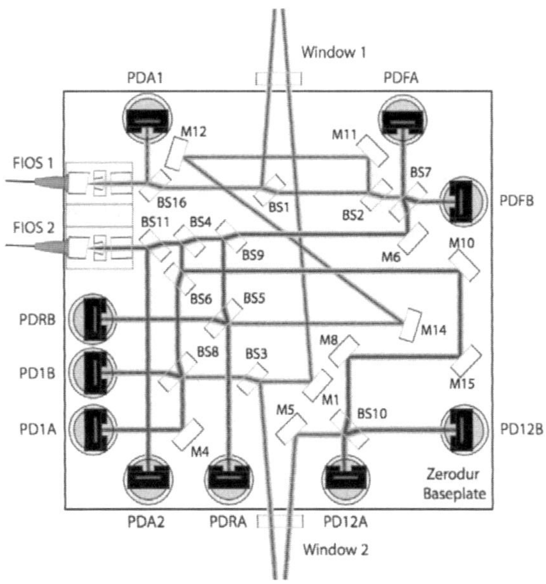

Figure 2. Optical bench schematic with the names of the components and photodiodes. Image courtesy of Institute for Gravitational Research, Glasgow and Albert Einstein Institute, Hannover

where optical paths are

$$L_{\mathrm{m}1A} = BS16t \cdot BS1r \cdot M1 \cdot BS3t \cdot BS8r \cdot M4 \tag{4}$$
$$L_{\mathrm{r}1A} = BS11t \cdot BS4r \cdot BS6t \cdot BS8t \cdot M4$$

Then we take ratios of $P_{\mathrm{m}}/P_{\mathrm{r}}$ for both equations 1 and 3

$$\frac{P_{\mathrm{m}}}{P_{\mathrm{r}}} \cdot \frac{L_{\mathrm{m}RA}}{L_{\mathrm{r}RA}} = \frac{\Sigma_{\mathrm{m}RA}}{\Sigma_{\mathrm{r}RA}} \qquad \frac{P_{\mathrm{m}}}{P_{\mathrm{r}}} \cdot \frac{L_{\mathrm{m}1A}}{L_{\mathrm{r}1A}} \cdot TM1r \cdot WIN1t^2 = \frac{\Sigma_{\mathrm{m}1A}}{\Sigma_{\mathrm{r}1A}} \tag{5}$$

and equate them. This provides the way to express test mass one reflectivity in terms of powers measured with the QPDs of the *X1* and *Reference* interferometers and the optical components

$$TM1r \cdot WIN1t^2 = \frac{L_{\mathrm{r}1A}}{L_{\mathrm{m}1A}} \cdot \frac{\Sigma_{\mathrm{m}1A}}{\Sigma_{\mathrm{r}1A}} \cdot \frac{\Sigma_{\mathrm{r}RA}}{\Sigma_{\mathrm{m}RA}} \cdot \frac{L_{\mathrm{m}RA}}{L_{\mathrm{r}RA}}. \tag{6}$$

It should be noted that we cannot independently measure the properties of the optical window. Therefore we measure a combination of the test mass reflectivity multiplied by the square of the optical window transmissivity.

4. Photodiode calibration

Next set of experiments are performed by modulating the power of the measurement and reference beams with 0.1 and 1% around the constant power. It introduces additional force from the radiation pressure at the frequencies of the modulation. There are two independent ways to estimate this force. First, from the differential displacement measured by the interferometer. Differential acceleration can be inferred from the differential displacement in the following way

$$\Delta g(t) = \Delta\ddot{x}(t) - \Delta g_c + \omega_{12}^2 x_1(t) + \omega_2^2 \Delta x(t) - \delta_{\mathrm{ifo}} \ddot{x}_1(t), \tag{7}$$

where $x_1(t)$ is the measurement of the position of the first test mass to the optical bench. $\Delta x(t)$ is the relative position of the two test masses. ω_2^2 and ω_{12}^2 are the relative stiffness and stiffness for the second test mass which couple spacecraft motion into $\Delta g(t)$. $\Delta g_c(t)$ is the commanded force on the test mass two and δ_{ifo} is the cross coupling term from the interferometer readout [2].

Second way to calculate the force acting on the test masses from the radiation pressure, is by estimating the radiation pressure that laser beam applies on the test masses. This can be inferred from by propagating the power measured with the photodiodes to the test masses in the following way

$$a_{\mathrm{est}} = \frac{1 + TM1r}{Mc} P_{\mathrm{TM1}} + \frac{1 + TM2r}{Mc} P_{\mathrm{TM2}}. \tag{8}$$

Here c is the speed of light, M is the mass of the test masses, $TM1r$ is the reflectivity of the test mass one and $TM2r$ is the reflectivity of the test mass two. Variables P_{TM1} is the power propagated to the test mass one and P_{TM2} power propagated to the test mass two. For example, powers from the QPDs propagated to the test masses are

$$P(TM1) = \Sigma_1/(M1 \cdot BS3t \cdot 0.5(BS8t + BS8r \cdot M4)), \tag{9}$$

$$P(TM2) = \Sigma_{12}/(M5 \cdot 0.5(BS10r + BS10t)). \tag{10}$$

The radiation pressure measurement is a completely independent check of the differential acceleration measurement. At the same time the residual differential acceleration measurement can be used to calibrate the measurement of the power at the photodiodes which involve unknown efficiency of the photodiode. Repeating this experiment several time during the mission allows to measure the decay of the photodiodes.

5. Acknowledgments

The Albert-Einstein-Institut acknowledges the support of the German Space Agency, DLR. The work is supported by the Federal Ministry for Economic Affairs and Energy based on a resolution of the German Bundestag (FKZ 50OQ0501 and FKZ 50OQ1601 NK would like to acknowledge support of the Newton International Fellowship from the Royal Society and the Scottish University Physics Alliance (SUPA).

References

[1] Robertson D.I. *et al*, Construction and testing of the optical bench for LISA Pathfinder, CQG 2013
[2] Armano, M. *et al*, Sub-Femto-g Free Fall for Space-Based Gravitational Wave Observatories: LISA Pathfinder Results, PRL 2016

IOP Conf. Series: Journal of Physics: Conf. Series **840** (2017) 012040 doi:10.1088/1742-6596/840/1/012040

GRS electronics for a space-borne gravitational wave observatory

D Mance[1], P Zweifel[1], L Ferraioli[1], J ten Pierick[2], N Meshksar[1] and D Giardini[1] for the LISA Pathfinder Collaboration

[1] Department of Earth Sciences, Institute of Geophysics, ETH Zurich, Zurich, Sonneggstrasse 5, Switzerland

[2] Department of Physics, University of Zurich, Winterthurerstrasse 190, Zurich, Switzerland

E-mail: davor.mance@erdw.ethz.ch

Abstract. The Gravitational Reference Sensor (GRS) electronics is a crucial element of the future space-borne gravitational wave observatory. Together with the optical metrology system, it provides position measurements of the sensor's reference body, a Test Mass (TM), for all axes. This is needed for precise spacecraft control. In addition, the GRS electronics can actuate the TM using electrostatic forces, which is used to keep the TM centered in its enclosure or to follow a certain guidance. The GRS electronics has been successfully tested during the LISA Pathfinder mission, launched in December 2015. The electronics has been designed in Switzerland by RUAG and HES-SO under supervision of ETH Zurich and University of Zurich. The paper describes the working principle and the adopted technical solutions for the LISA Pathfinder GRS electronics and for the LISA GRS electronics prototype. Both confirm the readiness of the technology for LISA.

1. GRS Front End Electronics

The Gravitational Reference Sensor (GRS), the associated Front End Electronics (FEE) and the laser metrology system are the core of the future space-borne gravitational wave observatory, the Laser Interferometer Space Antenna (LISA) [1]. To verify the technology readiness for a space-borne gravitational waves observatory, ESA has launched the LISA Pathfinder (LPF) spacecraft in December 2015, which completed its main mission in 2016 and is currently in extended mission phase until May 2017 [2]. The LPF GRS FEE, also called Inertial Sensor FEE (IS FEE), has been designed in Switzerland by RUAG Zurich and HES-SO Sion under supervision of ETH Zurich and University of Zurich.

The FEE, provides position measurements of the GRS's reference body, a Test Mass (TM), for all axes [3]. This is needed for precise drag-free spacecraft control on those axes not controlled by the optical metrology. In addition, the FEE can actuate the TM using electrostatic forces, which is used to keep the TM centered in its enclosure or to follow a certain guidance.

Figure 1 shows simplified block diagram of the FEE and the TM in two Degrees of Freedom (DoF) x and φ, with the remaining DoF omitted for clarity. The TM is electrostatically suspended between surrounding sensing and actuation electrodes and biased by a stable 100 kHz AC signal via separate injection electrodes, i.e. by capacitive coupling. The TM position sensing is based on measurement of capacitances existing between electrodes and TM along all DoF. Its motion causes an imbalance in capacitance, and thus an imbalance in AC currents flowing in primary windings of both transformer bridges (between front/rear electrodes A+/- and B+/-). The preamplifier, following the bridge, detects the differential current in the transformer secondary winding and converts it to an AC voltage, which is then band-pass filtered (BPF), converted into DC voltage by a synchronous demodulator and then finally digitized by an Analog to Digital Converter (ADC).

To allow for a high sensitivity of capacitive measurement, the transformer with a high quality factor Q is operated at a resonance, matching the injection bias frequency, for which capacitors (Cp) are added in parallel with inductance of the primary windings. The sensing noise has its minimum at resonance and needs to be precisely tuned. This tuning capacitance includes the capacitance of the cables, which requires individual tuning of every sensing channel.

In order to reduce the stray acceleration on TM or to add some guidance, force signals are provided by numerically synthesized AC waveforms at audio frequencies (60 – 270 Hz), not to interfere with the sensing circuit operated at 100 kHz. To apply strong forces on TM during its release or spacecraft maneuvers, large AC voltages are needed (120 V) and operated in the Wide-Range (WR) actuation mode. The force signal for each electrode is converted into voltage by a Digital to Analog Converter (DAC) and amplified by a corresponding amplifier. Actuation signal is then filtered and applied on the same electrodes used for sensing, such that actuation and sensing circuits operate simultaneously with minimum crosstalk.

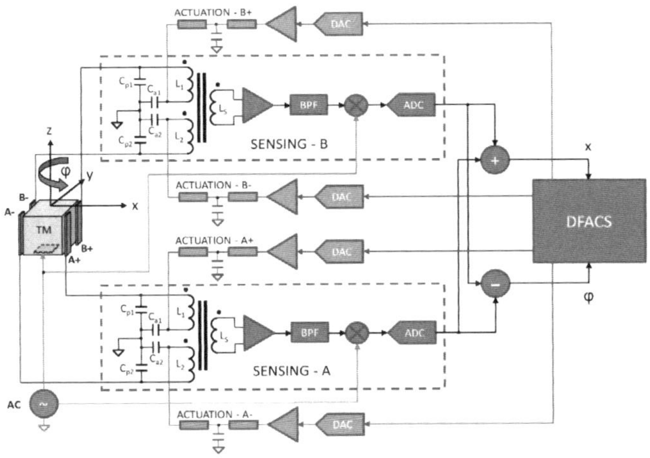

Figure 1. A block diagram of the TM sensing and actuation electronics associated with the x and φ DoF. Two pairs of electrodes (A+/A-, B+/B-) allow simultaneous measurement of TM translation and rotation, which is achieved by measurement of the gaps between TM and electrodes at TM corners. The capacitance is proportional to the respective TM - electrode gap. Actuation circuit (one for each electrode) electrostatically applies forces and torques on TM using same electrodes.

As shown on figure 1, two capacitive measurements are combined to calculate TM translation (x) and rotation (φ). Similarly, other 8 electrodes (not shown for clarity) are used to derive remaining TM motion in y – θ and z – φ DoF. TM attitude and control forces are calculated by the Drag-Free Attitude Control System (DFACS) software located in the LPF spacecraft On-Board Computer (OBC).

2. LISA Pathfinder GRS FEE sensing performance

The sensitivity of the capacitive measurement is limited by the quality factor Q of the differential transformer, whose losses produce thermal noise. To achieve the required sensitivity with the sensing noise density of 1 aF/√Hz (equivalent to 1.8 nm/√Hz), transformer must have large inductance and large Q (> 150) [4]. This requires low inter- and intra-winding stray capacitances. In addition, the challenging requirement of < 50 ppm is needed for inductance imbalance between primary windings to achieve sensing offset better than 600 aF (1 μm) in the ±200 μm High-Resolution (HR) sensing range.

For stability reasons, a planar winding design with balance tuning capability of the two primaries was selected. Each winding is made of 16-layer Printed Circuit Board (PCB). All three windings are glued together and this stack is then precisely positioned and glued against the gap of the Ferrite core. The transformer is designed by Swiss industry (HES-SO Sion). A similar transformer winding design, made by ETH, is shown on figure 2.

Figure 2. An example of the planar differential transformer design, i.e. a stack of two primary windings and one secondary winding separated by Teflon spacers. Each winding has 80 turns, which is constructed by stacking 16 layer coils, each with 5 turns. A gapped Ferrite core surrounds the stack to form the transformer.

Any instability in the sensing bias amplitude (100 kHz injection signal) mimics TM motion and causes low frequency noise proportional to the TM out-off-center position. The amplitude stability of the bias must therefore, be better than 50 ppm/√Hz not to compromise sensing noise performance in the measurement range of interest, i.e. at ±10 μm TM displacement (HR performance range). In LPF, the 50 ppm/√Hz stability requirement is fulfilled at 1 mHz.

Test of sensing performance requires a TM simulator, which is a 6 fF differential capacitance simulator (10 μm equivalent TM displacement) with nominal capacitance in each arm of ~1 pF (equivalent to 4 mm gap between TM and electrode). With the presence of such a low capacitance, the fluctuation of stray capacitances in the simulator can produce low frequency noise and thus mask the true FEE sensing performance. The ground measurements of sensing performance already showed an excess noise at 1 mHz indicating that an improved simulator design is needed for LISA band of 0.1 mHz.

The sensing performance check was performed in flight with TM grabbed by mechanical fingers in z-axis. Figure 3 shows that this provided better differential capacitance stability than the on-ground simulators (note 10 nm equivalent TM drift during 20 hours). Sensing noise performance with TM translated by 2 μm in z-axis was found to be 1.4 nm/√Hz and flat down to 1 mHz.

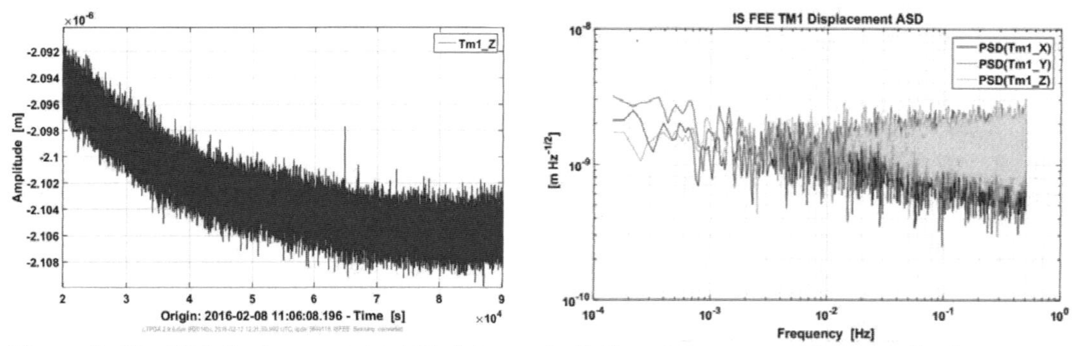

Figure 3. The TM displacement (stability) in z-axis (left) and a corresponding amplitude spectrum density in all three axes (right) measured during GRS FEE commissioning in space.

3. Design Upgrades for LISA

LISA performance in sensing and actuation must extend at least to 0.1 mHz for which some design changes are needed. For non-zero TM displacement the injection bias amplitude stability must improve since the amplitude noise density on LPF FEE is 150 ppm/√Hz at 0.1 mHz. The design improvement of the TM simulator is already mentioned in the text, as a prerequisite for accurate on-ground verification of sensing performance below 1 mHz. Furthermore, the 2 ppm/√Hz actuation amplitude stability has to be achieved at 0.1 mHz since the LPF FEE performance is 4-7 ppm/√Hz at 1 mHz and 12-60 ppm/√Hz at 0.1 mHz.

As the voltage reference circuit is the main performance driver for the amplitude stability of the sensing bias and actuation waveform, the voltage reference circuit has to be modified. Additional

IOP Conf. Series: Journal of Physics: Conf. Series **840** (2017) 012040 doi:10.1088/1742-6596/840/1/012040

modifications are also needed in the actuation electronics where zero-offset (auto-zero) amplifiers and other drift cancelling techniques shall be implemented to reduce the low frequency noise.

A prototype of the LISA GRS FEE has already been manufactured by ETH Zurich, RUAG Zurich and HES-SO Sion. HES-SO designed the sensing and actuation circuits, RUAG the control electronics, box and integrated the unit. ETH designed the Reference Unit with stable voltage reference and performed the performance testing of the integrated electronics. The sensing bias and actuation amplitude stability is already fulfilled for LISA prototype as shown on figure 4 and figure 5.

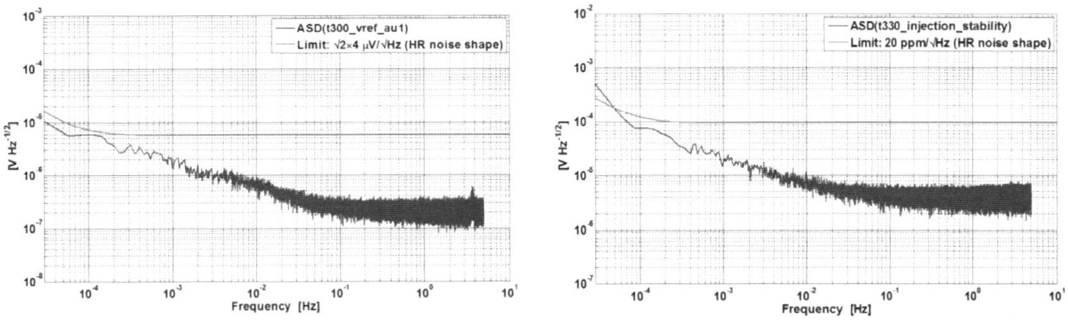

Figure 4. The LISA prototype voltage reference amplitude (left) and sensing bias, 100 kHz amplitude (right) stability spectrum densities. The voltage reference stability limit is equivalent to √2 x 1.6 ppm/√Hz for the 2.5 V reference. The sensing bias stability was measured at nominal 4.83 V level.

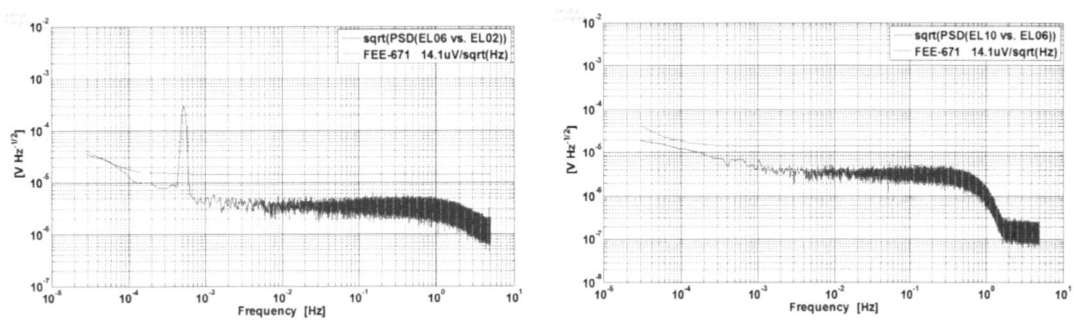

Figure 5. The LISA prototype actuation in-band DC noise (left) and 120 Hz actuation waveform amplitude stability (right) spectrum densities. The actuation DC noise limit is set to √2 x 10 μm/√Hz and measurement is performed at 5 V DC and 5 V AC simultaneous actuation. The actuation waveform amplitude stability limit is equivalent to √2 x 2 ppm/√Hz for the 5 V AC actuation.

For the voltage reference and the actuation noise / stability testing a differential measurement technique was used for which noise limits were multiplied by factor √2. The voltage reference and actuation DC noise test used uncorrelated voltage reference circuits. The phase of two actuation waveforms in differential measurement setup could not be perfectly equalized, which generated an aliasing peak in the DC noise plot. This was a measurement artifact, which does not affect the true performance. The differential actuation stability test (right trace on figure 5) required a locking amplifier with tuned amplitudes of two waveforms with small residual differential input. A single reference circuit was used, with the consequence that the correlated reference noise was removed from measurement, thus leaving only the uncorrelated noise sources. In real operation, the correlated noise will be cancelled as well since a common voltage reference will be used on all actuation channels.

4. Conclusion

GRS FEE technology was tested during LPF main mission phase and the required performance has been achieved in 1 mHz frequency bandwidth, as set for LPF. The requirements for a space-borne gravitational waves observatory are more stringent and extend at least to 0.1 mHz for which a LISA GRS FEE prototype has been developed and tested in Switzerland. The major modification is the design of a single voltage reference unit common for all DoF, compared to a LPF design with distributed reference for each quadruple of electrodes. The voltage reference, stable to 2 ppm/√Hz at 0.1 mHz is a prerequisite for the GRS performance. This level of performance has been achieved and it is essential for the performance of the sensing and actuation circuits that rely on voltage reference. The sensing and actuation noise performance has been verified down to 0.1 mHz.

References

[1] Danzmann K et al. *LISA Laser Interferometer Space Antenna - A proposal in response to the ESA call for L3 mission concepts* (2017)
[2] Armano M et al. *Sub-femto-g free fall for space-based gravitational wave observatories: LISA Pathfinder results,* Phys. Rev. Lett. **116**, 231101 (2016).
[3] Weber W J et al. *Position sensors for flight testing of LISA drag-free control,* SPIE Proc. **4856**, 31 (2002)
[4] Mance D *Development of electronic system for sensing and actuation of test mass of the inertial sensor LISA*, PhD Thesis University of Split (2012)

IOP Conf. Series: Journal of Physics: Conf. Series **840** (2017) 012041 doi:10.1088/1742-6596/840/1/012041

Gravitational Reference Sensor Front-End Electronics Simulator for LISA

Neda Meshksar[1], Luigi Ferraioli[1], Davor Mance[1], Jan ten Pierick[2,1], Peter Zweifel[1] and Domenico Giardini[1] for the LISA Pathfinder colaboration

[1] Department of Earth Sciences, ETH Zurich, Sonneggstrasse 5, 8092 Zurich, Switzerland
[2] Physik-Institut, University of Zurich,Winterthurerstr. 190, 8057 Zurich, Switzerland

E-mail: neda.meshksar@erdw.ethz.ch

Abstract. At the ETH Zurich we are developing a modular simulator that provides a realistic simulation of the Front End Electronics (FEE) for LISA Gravitational Reference Sensor (GRS). It is based on the GRS FEE-simulator already implemented for LISA Pathfinder. It considers, in particular, the non-linearity and the critical details of hardware, such as the non-linear multiplicative noise caused by voltage reference instability, test mass charging and detailed actuation and sensing algorithms. We present the simulation modules, considering the above-mentioned features. Based on the ETH GRS FEE-simulator for LISA Pathfinder we aim to develop a modular simulator that provides a realistic simulation of GRS FEE for LISA.

1. Introduction

The LTP Data Analysis (LTPDA) toolbox provides a closed loop simulation of the Laser Interferometer Space Antenna (LISA) Technology Package (LTP). It is composed of modular units, each of which is implemented as a linear state space model [1].

Gravitational Reference Sensor Front End Electronics (GRS FEE) is modelled linearly in order to provide fast simulations at 10 Hz and 1 Hz frequencies. However, the linear model prevents reproduction of some complicated specific hardware details. These are for instance the non-linear multiplicative noise caused by the reference voltage instability and the thermal noise in dispersive elements of the circuit.

At the ETH Zurich we are improving the simulator to obtain a more realistic model for LISA Pathfinder GRS FEE. We introduce the main features of our model in section 2 and give a detailed description in section 3.

2. GRS FEE simulator for LISA Pathfinder

GRS FEE is responsible for capacitive actuation and sensing of the position of the proof masses in the LTP. Each proof mass is surrounded by an electrode housing [2]. Figure 1 shows the arrangement of the electrodes around the proof mass and illustrates the simplified electronic circuits of the actuation and sensing units inside FEE.

In the advanced simulator each of the actuation and sensing units is split into blocks. Splitting into blocks provides more flexibility to feed the complicated hardware details into the simulator,

IOP Conf. Series: Journal of Physics: Conf. Series **840** (2017) 012041 doi:10.1088/1742-6596/840/1/012041

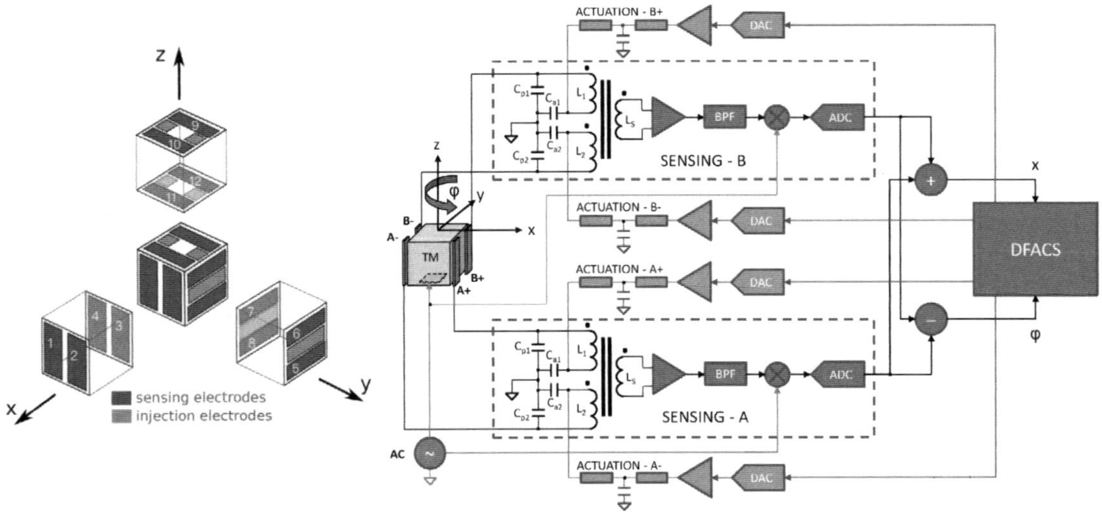

Figure 1. Arrangement of the sensing/actuation electrodes and the injection electrodes around the proof mass (left-hand side) and electronic circuit for sensing and actuation (right-hand side).

which also allows the non-linear simulations. The units and the corresponding division into blocks are elaborated as follows.

- Actuation unit
  The actuation unit is responsible for applying the forces commanded from the Drag-Free Attitude Control System (DFACS) to the proof mass. In modelling of the actuation unit the first step is to calculate the commanded voltages from the commanded forces ($F_{cmd} \rightarrow V_{cmd}$). Afterwards the commanded voltages are converted to the applied voltages ($V_{cmd} \rightarrow V_{appl}$) and finally the applied forces are computed from the applied voltages ($V_{appl} \rightarrow F_{appl}$). On board LISA Pathfinder the first step is the task of the onboard computer. In the second step the conversion from commanded to applied voltages is executed through the actuation electronics of the GRS FEE (see figure 1). The applied voltage to the electrodes causes a force and consequently the desired proof mass displacement.
- Sensing unit
  The sensing unit is mainly divided into the following blocks: First, the conversion from physical coordinates to the capacitance ($q_{ph} \rightarrow C$) and second the calculation of the readout coordinates from the capacitance ($C \rightarrow q_r$). On board LISA Pathfinder any displacement of the proof mass changes the distance to the corresponding electrode, which in turn results in the changes of capacitance. Thus, the readout coordinates can be calculated from the measured electrode capacitances.

Crosscorrelations of the blocks and also crosstalks between the units can be implemented in a separate block. This is specially important for noise identification. Figure 2 summarises the breakdown of the actuation and sensing units and depicts their correlation in a block diagram.

3. Detailed description of the blocks and development for LISA

To develop the GRS FEE simulator for LISA additional details should be considered and implemented. At the first stage of the development we propose modifications, which in particular

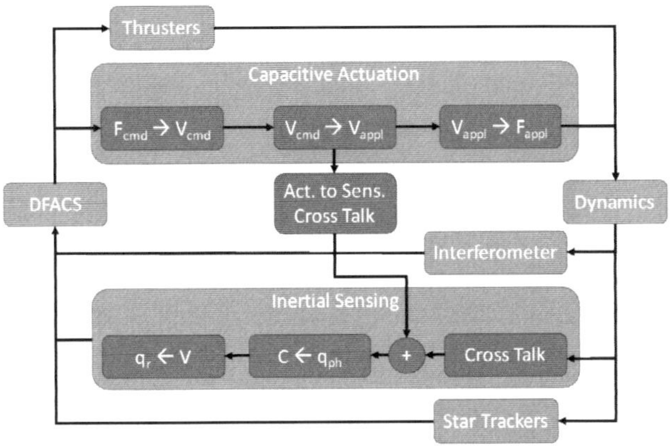

Figure 2. Sensing and actuation units of the GRS FEE.

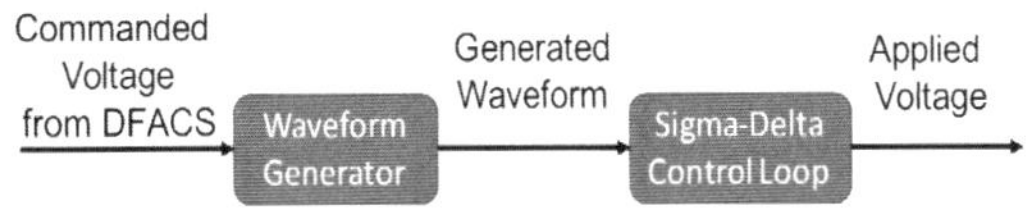

Figure 3. Detailed description of the conversion of commanded to applied voltage in the actuation unit

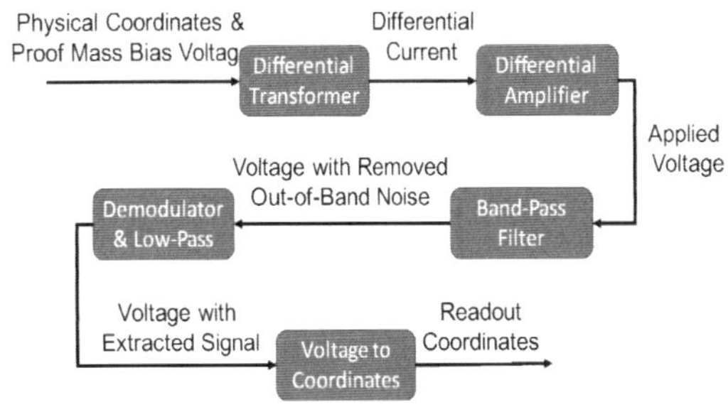

Figure 4. Detailed description of the sensing unit

refer to the actuation unit. At this stage the sensing unit is inherited from the GRS FEE simulator for LISA Pathfinder. The details on each block are described below.

3.1. *Actuation unit*

The most challenging part of the actuation model refers to the simulation of the block that converts the commanded voltages to the applied voltages. More details about this block are summarised in Figure 3 and described as follows.

The DFACS produce commanded forces (and consequently commanded voltages) in 10 Hz frequency and in double precision, however the electronics accept the inputs in 16 bit integers. Thus, the commanded voltages are truncated, such that the sample values accomplish the precision of the electronics.

The force applied to the proof mass is proportional to the square of the voltages injected to the electrodes surrounding the proof mass. A convenient way to apply a force with desired magnitude is to generate sinusoidal waves. This is done by a waveform generator. The commanded voltages are then amplitudes of the sinusoidal waves. In order to distinguish between actuation at the different degrees of freedom (DoF) a different frequency has been assigned to the generated wave at each DoF.

The generated waves enter a sigma-delta loop, which provides an analog signal at the electrodes. The sigma-delta loop includes a digital controller, a digital to analog converter, an amplifier and a feedback analog to digital converter, all operating at high frequency, in order to process the high resolution DFACS data more accurately [3].

3.2. *Sensing unit*

The sensing unit takes at the input the physical coordinates of the proof mass position and provides at the output the readout coordinates. Figure 4 depicts a detailed description of this unit.

IOP Conf. Series: Journal of Physics: Conf. Series **840** (2017) 012041 doi:10.1088/1742-6596/840/1/012041

The proof mass displacement changes the capacities at the surrounding electrodes. Two pairs of electrodes are responsible for sensing the translation and rotation. A differential transformer senses the differential current of the one electrode pair. The differential current is amplified by a trans-impedance amplifier. The amplified voltage is filtered to remove the out-of-band noise. Moreover, an AC voltage (bias) of 100 kHz is applied to the proof mass as a carrier signal. The proof mass displacement generates a low frequency signal. This is modulated to the 100 kHz carrier. Thus, the output of the bandpass filter has to be demodulated, in order to extract the voltages caused by the proof mass displacement. Finally, the readout coordinates are computed from the voltages.

4. Conclusion and outlook

Based on the ETH GRS FEE-simulator for LISA Pathfinder we are developing a modular simulator that provides a realistic simulation of the GRS FEE for LISA. The simulator considers, in particular, the non-linearity and the critical details of hardware. The simulation design and its features are presented. We aim to develop the non-linear FEE-simulator for LISA and finally to integrate it in LISA-Code [4].

References

[1] Hewitson M and et al 2009 *Classical and Quantum Gravity* **26** 094003 ISSN 0264-9381 1361-6382

[2] Weber W J and et al 2003 *Gravitational-Wave Detection* **4856** 31–42 ISSN 0277-786x URL `<Go to ISI>://WOS:000182225800004`

[3] Mance D 2012 *Development of Electronic System for Sensing and Actuation of Test Mass of the Inertial Sensor LISA* Thesis University of Split

[4] Petiteau A and et al 2008 *Physical Review D* **77** ISSN 1550-7998 URL `<Go to ISI>://WOS:000252864000008`

GRS vs. OMS Calibration in LISA Pathfinder Data Analysis

Neda Meshksar[1], Luigi Ferraioli[1], Davor Mance[1], Jan ten Pierick[2,1], Peter Zweifel[1] and Domenico Giardini[1] for the LISA Pathfinder Colaboration

[1] Department of Earth Sciences, ETH Zurich, Sonneggstrasse 5, 8092 Zurich, Switzerland
[2] Physik-Institut, University of Zurich,Winterthurerstr. 190, 8057 Zurich, Switzerland

E-mail: neda.meshksar@erdw.ethz.ch

Abstract. On board LISA Pathfinder spacecraft the test mass displacement along the main measurement axis is sensed in two different ways: optically and electrostatically. We have monitored the relative calibration between the two measurements during the mission science phase. The trend sensitivity of the relative calibration has been computed for different physical parameters, such as temperature, magnetic field, test mass bias voltage and current.

1. Introduction

LISA Pathfinder (LPF) spacecraft (SC) uses two test masses (TM) as proof bodies of the gravitational reference sensor (GRS). TM movement on board LISA Pathfinder SC is sensed according to the optical and electrostatic readout methods [1]. Based on optical readout method the optical metrology system (OMS) measures the displacement of the drag-free TM relative to the SC and also relative to the other TM, along the main measurement axis, labeled x. It also measures the rotation of both test masses around y and z. On the other hand, the GRS provides the electrostatic readout of the displacement and rotation along all degrees of freedom for both test masses.

We have monitored the data resulted from system identification experiments during the LPF science operation (March - June 2016) and we present the calibration of GRS measurements with respect to OMS for displacements along x-axis. The relative gain and relative offset have been obtained by a linear fit of GRS vs. OMS readout. We have excluded any straightforward correlation with temperature and non-linearity in the GRS electronics. We observed that the relative calibration gain does not obey a clear trend. The origin of such oscillation dependency is still under investigation.

2. Method overview

GRS provides an electrostatic measurement of the displacements of both test masses with respect to the SC. Along the main measurement axis, x, we label the TM1 and TM2 displacements x_1^{GRS} and x_2^{GRS}, respectively. The differential displacement along x is thus calculated as

$$x_{12}^{\mathrm{GRS}} = x_2^{\mathrm{GRS}} - x_1^{\mathrm{GRS}}. \tag{1}$$

IOP Conf. Series: Journal of Physics: Conf. Series **840** (2017) 012042 doi:10.1088/1742-6596/840/1/012042

On the other hand, the OMS does not directly sense the displacement of TM2. It measures the position of TM1 with respect to the SC and the differential displacement between the two test masses. The latter is labeled x_{12}^{OMS}. Thus, the displacement of TM2 along x with respect to the spacecraft is given by

$$x_2^{\mathrm{OMS}} = x_1^{\mathrm{OMS}} + x_{12}^{\mathrm{OMS}}. \tag{2}$$

Figure 1 illustrates GRS and OMS in the LISA Technology Package (LTP). A description of the coordinates is given in figure 2.

In order to compare the measured displacement by GRS and OMS we calibrate the data by means of a linear fit. A linear fit of the GRS data to the OMS data provides a gain and an offset, which are parameters for matching the GRS data to the OMS data:

$$x_i^{\mathrm{GRS}} = \mathrm{gain} \cdot x_i^{\mathrm{OMS}} + \mathrm{offset}, \quad i = 1, 2, 12 \tag{3}$$

Figure 1. Illustration of GRS and OMS in the LTP. In the nominal state the test masses are at the centre of the inertial sensors.

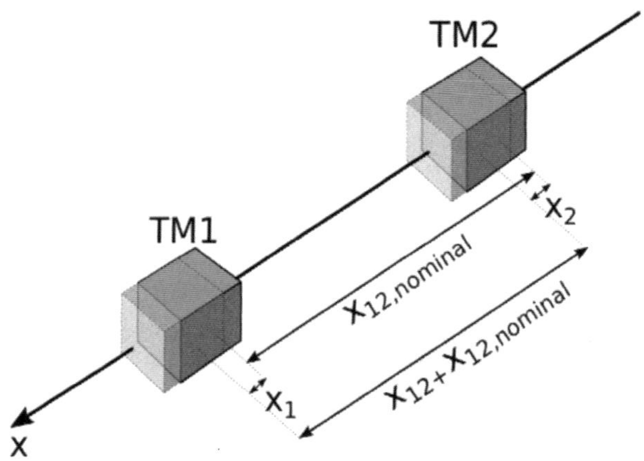

Figure 2. Illustration of test mass displacements and labelling the coordinates along the main measurement axis, x.

3. Observation

In order to understand any correlation of the gain with the other parameter, we have monitored the data resulted from the system identification experiments during the LPF science operation (March - June 2016). To calibrate the displacement for TM1 we used the data of drag-free injections [2]. For TM2 and also for the differential displacement, x_{12}, we used the data of suspension injections. The gain trend for x_1, x_2 and x_{12} obtained by the linear fit are reported in figure 3. The gain does not obey a clear trend as shown in the figure.

The gain dependency on the different parameters, such as the temperature, magnetic field, current and the TM bias voltage are investigated. Figure 4 shows the results for x_1. We cannot recognise any gain correlation with the temperature, magnetic field and the TM bias voltage. However, it seems that the gain has a correlation with the +5 V digital line consumption current in GRS. This correlation is still under investigation.

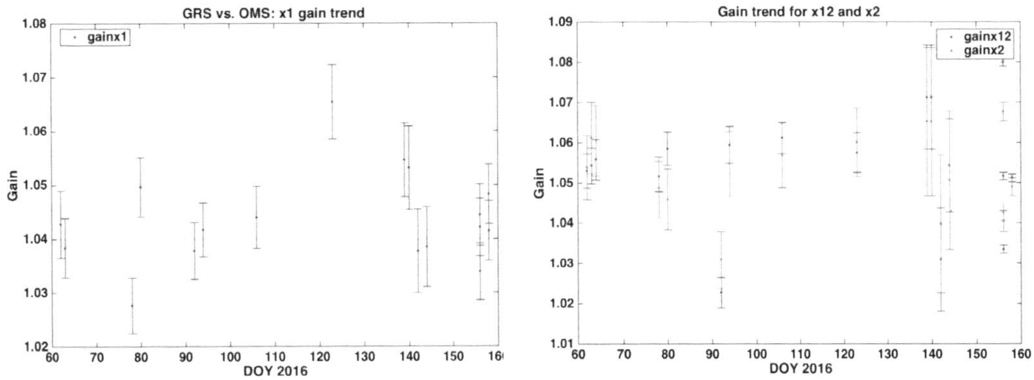

Figure 3. Gain trend resulted from the linear fit of GRS data to OMS data for TM1 (left-hand side), TM2 (right-hand side, red) and the differential displacement (right-hand side, blue).

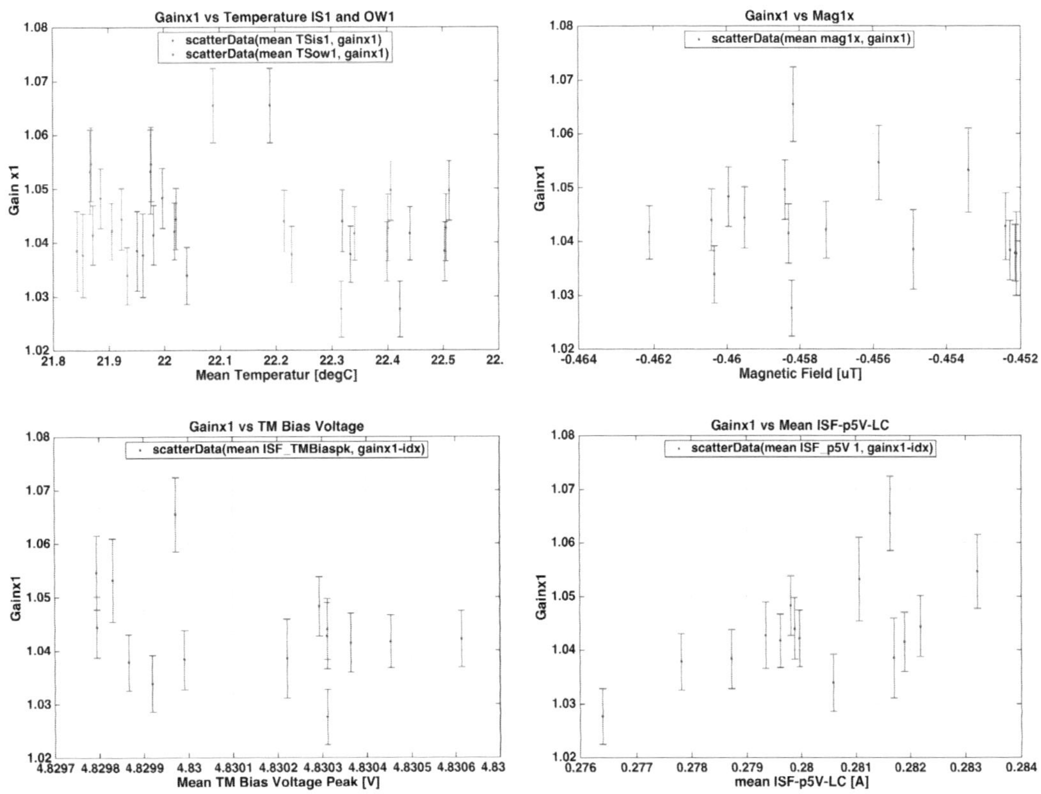

Figure 4. Gain trend for TM1 displacements with respect to the following parameters: temperature of the inertial sensors (left-hand side, top, blue), temperature of the optical window (left-hand side, top, red), TM bias voltage (left-hand side, bottom), magnetic field (right-hand side, top) and +5 V digital line consumption current in GRS (right-hand side, bottom).

4. Conclusion and outlook

We have investigated the time estimation of the relative calibration between GRS and OMS readout on board LISA Pathfinder spacecraft. The relative gain does not obey a clear trend. The observations do not show any straightforward correlation with the temperature, magnetic field and the test-mass bias voltage. The fluctuation of the gain trend and especially the effect of the GRS readout consumption current is still under investigation.

References

[1] Anza S and et al 2005 *Classical and Quantum Gravity* **22** S125–S138 ISSN 0264-9381 URL `<Go to ISI>://WOS:000230230100002` `http://iopscience.iop.org/article/10.1088/0264-9381/22/10/001/pdf`

[2] Armano M and et al 2015 *10th International Lisa Symposium* **610** ISSN 1742-6588 URL `<Go to ISI>://WOS:000358149000006` `http://iopscience.iop.org/article/10.1088/1742-6596/610/1/012006/pdf`

IOP Conf. Series: Journal of Physics: Conf. Series **840** (2017) 012043 doi:10.1088/1742-6596/840/1/012043

Preliminary results on the suppression of sensing cross-talk in LISA Pathfinder

Gudrun Wanner and Nikolaos Karnesis on behalf of the LISA Pathfinder collaboration

Max Planck Institute for Gravitational Physics (Albert Einstein Institute) and Institute for Gravitational Physics of the Leibniz Universitt Hannover

E-mail: gudrun.wanner@aei.mpg.de

Abstract. In the original paper describing the first measurements performed with LISA Pathfinder, a bulge in the acceleration noise was shown in the 200 mHz - 20 mHz frequency band. This bulge noise originated from cross-coupling of spacecraft motion into the longitudinal readout and it was shown that it is possible to subtract this cross-talk noise. We discuss here the model that was used for subtraction as well as an alternative approach to suppress the cross talk by realignment of the test masses. Such a realignment was performed after preliminary analysis of a dedicated cross-talk experiment, and we show the resulting noise suppression. Since then, further experiments have been performed to investigate the cross-coupling behaviour, however analysis of these experiments is still on-going.

1. Introduction

The well known LISA Pathfinder differential acceleration measurement published in [1] and displayed for convenience in Fig. 1(a), showed a bulge in the noise curve in the 200 mHz - 20 mHz frequency band. This bulge was caused by cross-coupling of spacecraft motion leaking into the main differential acceleration measurement and it was shown in [1] that it is possible to subtract the cross-coupling noise resulting in the blue curve in Fig. 1(a). This subtraction was performed by fitting the following model to the measured acceleration noise, i.e. to the red trace in Fig 1(a) yielding estimates of the coefficients C_i and δ_{ifo}:

$$\Delta g_{\text{x-talk}} = C_1\ddot{\bar{\phi}}[t] + C_2\ddot{\bar{\eta}}[t] + C_3\ddot{\bar{y}}[t] + C_4\ddot{\bar{z}}[t] + C_5\bar{y}[t] + C_6\bar{z}[t] + \delta_{\text{ifo}}\ddot{x}_1[t]\,. \tag{1}$$

Each term in this equation defines a cross-coupling of spacecraft motion to the differential acceleration measurement:

- All but the last summand are cross-coupling terms, because ϕ, η, y, z are orthogonal coordinates to the differential readout along x. See Fig. 2(a) for a definition of the coordinate system.
- The annotation $\bar{k}$ with $k \in \{\phi, \eta, y, z\}$, defines a mean motion of both test masses along k, which is a description of the spacecraft motion along $-k$.
- The last summand, describes a residual coupling of spacecraft motion along x entering the differential readout. As the term 'differential readout' indicates, Δg is defined as $\Delta g = \ddot{x}_2 - \ddot{x}_1$ (neglecting here corrections such as for applied or inertial forces). It should

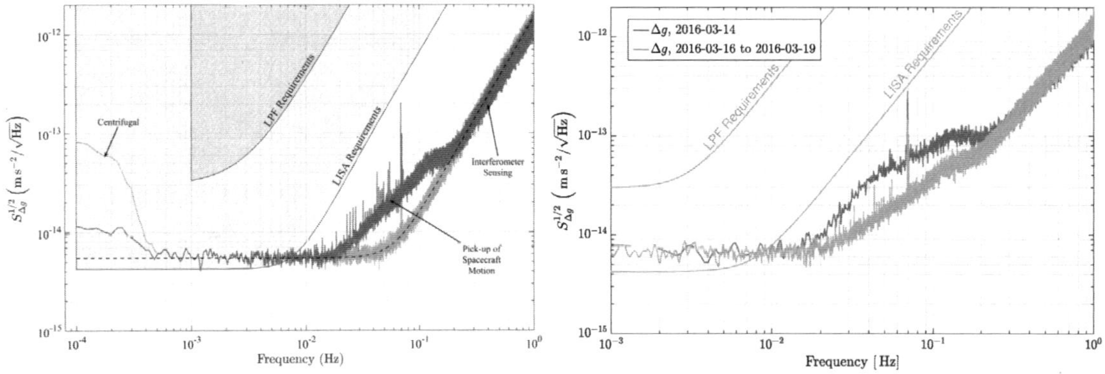

(a) Subtraction method (b) Suppression method

Figure 1. (a): Result taken from [1], showing that the bulge at 200-20 mHz can be subtracted. (b) In this figure, no subtraction was performed. Instead the reduction was achieved by a realignment of the test masses prior to the second noise run.[1]

therefore be insensitive to any common-mode motion, where $\ddot{x}_2 = \ddot{x}_1$. Since spacecraft motion effectively shows as common test mass motion, Δg should not sense spacecraft motion. However, there is a residual leakage, because the interferometer readout response is not identical to TM1 motion as to TM2 motion, such that $\Delta g = k_2\ddot{x}_2 - k_1\ddot{x}_1$ with $k_2 \approx k_1 \approx 1$. The small difference between k_1 and k_2 causes the residual cross-coupling.

Fitting eq. (1) to the measurement data and subtracting the resulting model works very well, as visible in Fig. 1(a). However, this technique has several disadvantages: First of all, the model provides no interpretation what the coefficients physically mean, and therefore no explanation how the cross-coupling could be suppressed. Secondly, several parameters show high correlation, such that the given model is not the only one allowing a fit and subtraction with low residuals. Instead, there is a set of models, performing comparably well. Finally, the coefficients vary over mission time, and this variation cannot be explained with the shown model.
For these reasons, there is an ongoing effort, to build a physical model, which can describe the cross-coupling over the entire mission duration with a constant set of parameters, and can be used without adaptation for any set of data to subtract the bulge. Additionally, such a model also allows a direct suppression of the cross-talk. In fact, Fig. 1(a) shows a noise run performed in April 2016 (2016-04-08 to 2016-04-14), which was after a first realignment of the test masses had already reduced the amount of cross-talk. This will be further discussed in the next section.

2. Cross-coupling suppression

Of all terms given in eq. (1), we will focus now on the first four summands. For these four terms exists a multitude of mechanisms contributing to the parameters C_1 to C_4. Some of these are geometric, others result from the interference of Gaussian beams and their phase detection (see e.g. [2, 3] for more information on some non-geometric cross-talk mechanisms). Unfortunately, the amount of cross-coupling originating from non-geometric effects needs to be regarded as unknown. The geometric effects however, are known very well and can be used to counteract any unknown effects.
In Fig. 2 we illustrate two related geometrical effects. If a test mass is tilted with respect to

[1] Please note: the slight difference in curvature around 3 mHz in the LPF requirements in the two figures results from different notations: while in Fig. 1(a) correlated noise sources were assumed, we chose in Fig. 1(b) the notation for uncorrelated noise sources.

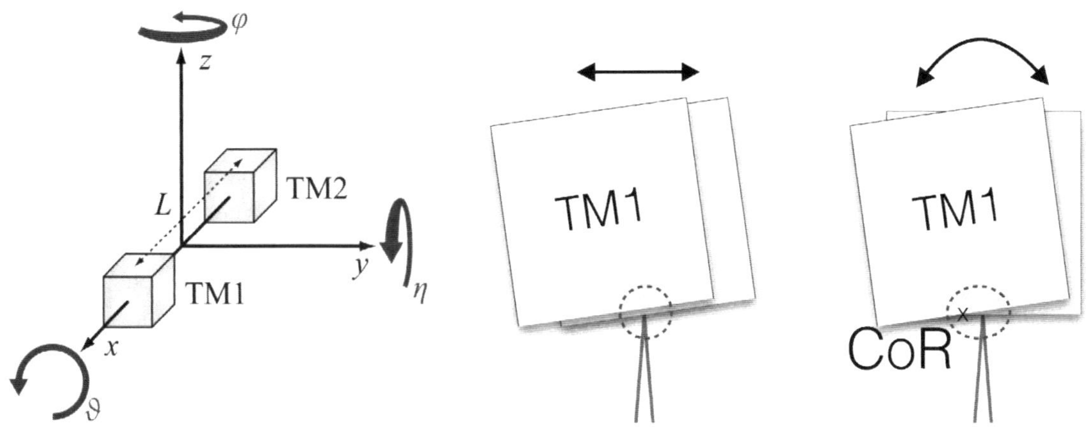

(a) Coordinate system (b) Lateral jitter coupling (c) Angular jitter coupling

Figure 2. Illustration of the displacement of the laser beam reflection point on test mass 1 (TM1). These are two related cases of geometrical cross-coupling.

the optical bench and performs a lateral motion, the reflection point on the test mass moves longitudinally, i.e. in x-direction. This causes a direct coupling of lateral motion into the longitudinal readout signal, since the optical pathlength of the laser beam changes.

A second geometric coupling effect occurs, if the test mass reflection point does not coincide with the center of rotation of the test mass - as shown in Fig. 2(c). In this case, a test mass tilt causes a longitudinal displacement of the reflection point and thereby a change in the optical pathlength. In both cases, there was no motion of the test mass along x present, but since the reflection point on the test mass was displaced in x-direction, the interferometer senses an apparent test mass x-motion.

In fact, these two coupling mechanisms are closely related since in both cases, the coupling occurs due to a combination of rotation and lateral displacement, and both can be described by the following simple equation:

$$x^{\mathrm{rp}} = y\varphi \, . \tag{2}$$

Here, x^{rp} annotates the displacement of the reflection point along x, and thereby the apparent longitudinal test mass displacement. The lateral shift between the center of rotation and the reflection point is labeled y, and φ is the test mass angle with respect to the optical bench. The same equation holds for the orthogonal plane:

$$x^{\mathrm{rp}} = z\eta \tag{3}$$

where z is a vertical displacement of a reflection point with respect to the center of rotation, and η a test mass tilt around the y-axis, measured relative to the optical bench (see Fig. 2(a) for an illustration of the coordinate system). Since eq. (2) and (3) are linear in each degree of freedom, they can be used to counter act all other cross-coupling mechanisms - at least to first order. For example, assume a cross-coupling of TM1 φ-jitter was measured to be $c_1 \cdot \varphi_1 + O(\varphi_1^2)$. In that case, a lateral displacement of TM1 by $y_1 = -c_1$ would suppress this measured cross talk to first order. This means, by a well chosen set of rotations and shifts of the test masses, the amount of cross-coupling leaking into Δg can be reduced. In order to find the required offsets, we injected sinusoidal motion in each degree of freedom and extracted the parameters via fitting a dedicated model to the measured data. Fig. 3 shows an excerpt of the injections performed

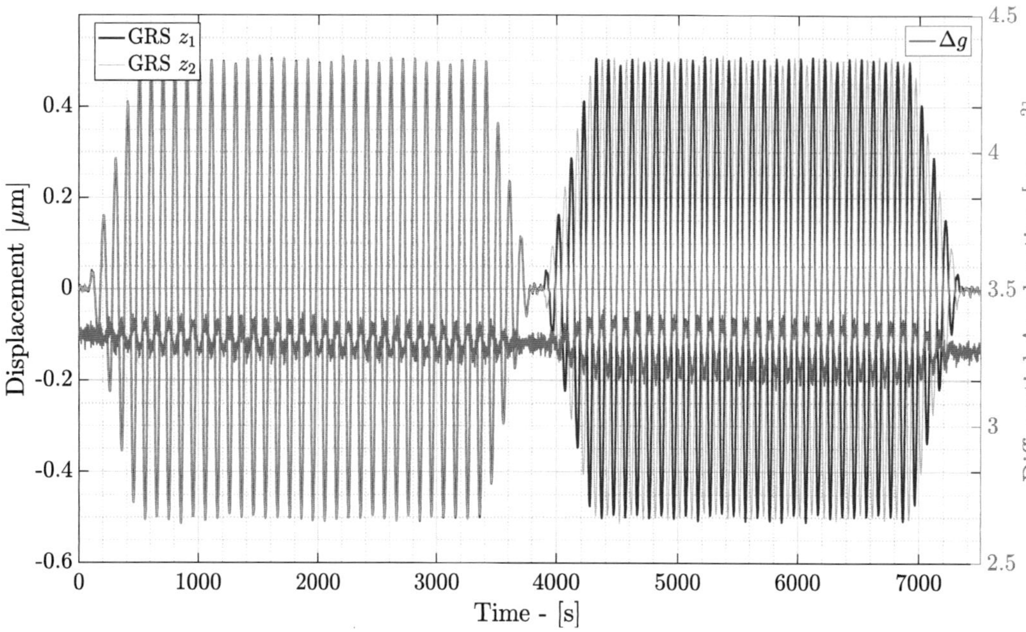

Figure 3. Example of injections performed during the 'engineering-day' cross-talk experiment performed on 2016-March-16. The labels 'GRS $z_{1,2}$' indicate, that the shown vertical test mass displacement (z_1, z_2) is taken from the 'Gravitational Reference Sensor' readout.

on 2016-March-16: the common-mode and counter-mode injections into the vertical test mass motion. To the right-hand axis, also the measured differential acceleration Δg is plotted, which shows a clear correlation to the injections and thereby the cross-coupling. A preliminary data analysis resulted in a first estimate of the parameters which were then used to realign the test masses. The resulting reduction of cross-coupling noise is given in Fig. 1(b). Please note, that eq. (2) and eq. (3) are not the physical model yet, which is needed to describe the cross-talk with constant parameters over the entire mission duration. Instead, these are toy-models describing two geometric cross-talk effects. Furthermore, the parameters in this model are written here not in terms of readout signals, which is necessary for applying the model to the data. The full model is still work in progress, the analysis of the cross-talk behaviour is therefore still ongoing and further information will follow in a separate paper.

3. Acknowledgments

This work has been made possible by the LISA Pathfinder mission, which is part of the space-science program of the European Space Agency. We gratefully acknowledge support by the European Space Agency (ESA) (22331/09/NL/HB, 16238/10/NL/HB), by the Deutsches Zentrum fur Luft- und Raumfahrt (DLR) with funding of the Bundesministerium fur Wirtschaft und Energie with a decision of the Deutschen Bundestag (DLR project reference numbers FKZ OQ 0501 and FKZ 50 OQ 1601), and thank the German Research Foundation for funding the Cluster of Excellence QUEST (Centre for Quantum Engineering and Space-Time Research).

References

[1] Armano, *et al*, *PRL* **116**, 231101 (2016), DOI: 10.1103/PhysRevLett.116.231101
[2] Schuster, *et al*, *Optics Express* **10466** (2016) DOI: 10.1364/OE.24.01046
[3] Schuster, *et al*, *Applied Optics* **54** (2015) DOI: 10.1364/AO.54.001010

LISA Pathfinder: Understanding DWS noise performance for the LISA mission

Lennart Wissel on behalf of the LPF collaboration

Max Planck Institute for Gravitational Physics (Albert Einstein Institute), Callinstraße 38, 30167 Hannover, Germany

E-mail: lennart.wissel@aei.mpg.de

Abstract. ESA's L3 Laser Interferometer Space Antenna (LISA) mission contains a mechanism to compensate for out-of-plane angles between the received and emitted beams of the three satellites. Depending on the configuration of this Point-Ahead Angle Mechanism (PAAM) it is expected to contribute readout noise through Differential Wavefront Sensing (DWS). This was investigated with LISA Pathfinder (LPF) through a dedicated investigation. One of the two free-falling test masses was rotated via the on-board electrostatic actuators while the resulting angular noise in the differential interferometer between the two test masses was measured. For angles between $-250\,\mu$rad to $250\,\mu$rad and corresponding contrast in the range of 59.4 % to 97.9 % an increased spectral density was found. The differential displacement noise remains almost unchanged for these misalignments.

1. Introduction

The performance of LPF's Optical Metrology System is about two orders of magnitude better than the requirement and is subject to ongoing investigations throughout the mission [1]. Differential Wavefront Sensing (see [2]) as a part of this subsystem is used to control four out of six angular degrees-of-freedom of the two test masses. Conceptually, the segments of Quadrant-Photodiodes (QPD) are demodulated to compare the relative phase difference between the wavefronts of the interfering measurement and reference beams for a group of two complementary quadrants. After processing all four channels using a Single-Bin Discrete Fourier Transformation (SBDFT) the resulting complex values, F, contain the encoded phase information which can be retrieved by subtracting left, right and up, down quadrants according to

$$\begin{aligned} \mathrm{DWS}^{\phi} &= \arg\left(F^{\mathrm{left}}\right) - \arg\left(F^{\mathrm{right}}\right), \\ \mathrm{DWS}^{\eta} &= \arg\left(F^{\mathrm{up}}\right) - \arg\left(F^{\mathrm{down}}\right), \end{aligned} \tag{1}$$

during which the equally distributed longitudinal phase is cancelling and the angular phase difference remains. These raw phase measurements can be expressed as equivalent test mass angles $\phi_i, \eta_i; i \in \{1, 2\}$ over a linear range of at least $-100\,\mu$rad to $100\,\mu$rad. To calibrate from phase differences to test mass angles, on-ground measurements were performed to determine the coupling coefficients and they are applied as calibration factors in the Data Management Unit (DMU) aboard LPF.

IOP Conf. Series: Journal of Physics: Conf. Series **840** (2017) 012044 doi:10.1088/1742-6596/840/1/012044

While the digitisation process has a sampling frequency of 50 kHz, further downsampling (to 10 Hz) is applied before transmitting the data to ground. This 10 Hz data is also used to drive the Drag-Free and Attitude Control System (DFACS), keeping the three body system as free from spurious forces as possible.

In the high frequency regime of the DWS data, a quasi-flat noise floor can be observed, as shown in figure 1, allowing a specific noise estimation. In LISA, non-vanishing DWS signals

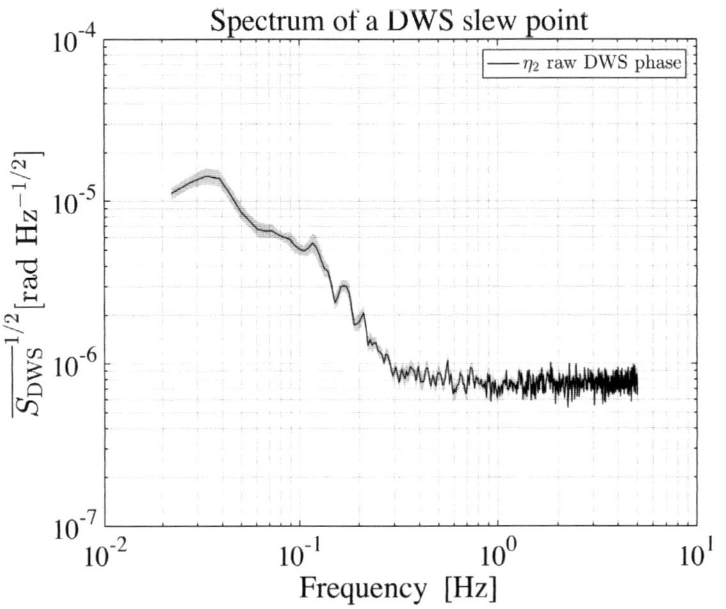

Figure 1. DWS spectral density for a specific offset, showing the high frequency flat noise floor. Below 300 mHz real test mass motion is dominating the spectrum.

(depending on the point ahead configuration) will originate from motion between the three spacecrafts in combination with their long arm lengths. Beam tilts are magnified by the telescope, causing relatively large DWS phase differences on the quadrants. This distorts the phase readout in a certain way that is investigated here.

2. The DWS step experiment

The DWS step experiment has been designed to measure the DWS noise behaviour and its positional dependence. For this, the electrostatic controllers are used to rotate test mass two around its y and z axes. Motivation for this investigation arises from the following observations: The test mass alignment along ϕ_1, η_1, ϕ_2 and η_2 does not only couple into the differential length measurement given by the x12 interferometer, it has also shown a different DWS noise floor for each of the angular set points that were commanded during the mission. The experiment is designed to quantify different noise levels and investigate their coupling. It assesses the phase noise in the DWS readout regarding its amplitude and phase (or positional) dependence, estimates possible noise sources and coupling mechanisms, allows to fit contrast models related to test mass slews and helps to draw conclusions for requirements on the LISA point-ahead setup.

2.1. Conception

The basic idea consists of commanding a set of angular test mass offsets via the electrostatic controller and calculating the resulting amplitude spectral noise density for each of the steps. LISA Pathfinder has proven to be quiet enough to reach a quasi-flat noise floor within the high frequency regime in the angular test mass readout, where no test mass motion is believed to be happening. For the slews test mass two was chosen because of its closer position to the photodiodes, thus minimizing any induced beam spot offset on the surface of the quadrants. Another benefit is that the differential interferometer is less susceptible to environmental disturbances. The somewhat randomly chosen test mass tilts are shown in figure 2. Larger tilts are not easily possible due to spacecraft safety regulations.

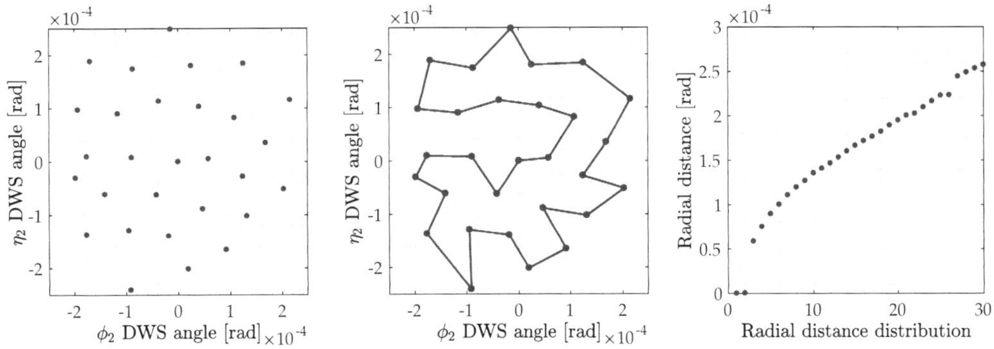

Figure 2. The commanded test mass two tilts are shown in the ϕ_2, η_2 parameter space. All points are randomly placed such that the maximum radial distance is smaller than 260 μrad. The middle plot shows the shortest connecting path of all points, calculated by a travelling salesman algorithm (courtesy of G. Heinzel). It reduces test mass trajectories between set points and the risk of larger overshoots.

2.2. In-orbit behaviour

The experiment was run on DOY 154 (day-of-year) during the mission. In figure 3 the resulting time series can be found. Shown is the angular readout (DWS) and the differential interferometer readout. Every step with a total duration of 45 min is reached in less than 30 min until the rotational degree of freedom stabilizes. In comparison tilt-to-length coupling in x12 motion, introduced by the commanded torques, experiences a slower impulse response of the suspension

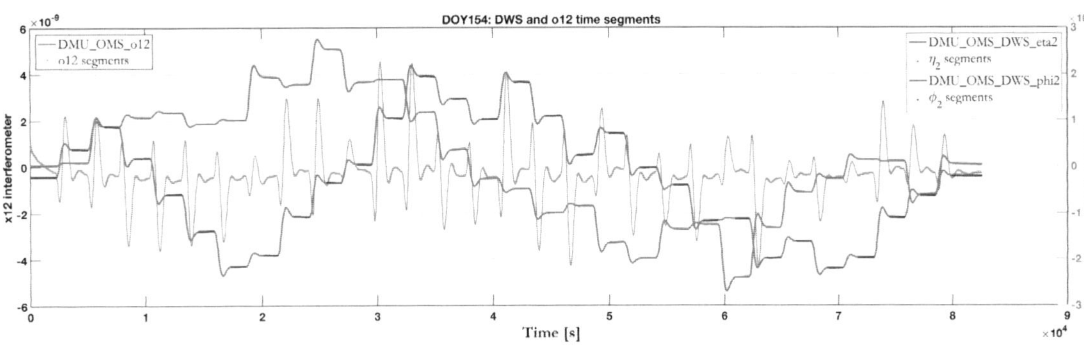

Figure 3. Time series of the measured slews in the DWS step experiment.

controller. Every set point shows an overshoot of various length, when the test mass rotates further than it should, due to the relatively high slew velocities that the controller can not quickly reduce. The maximal overshoot is just below 30 μrad in η_2. The analysis shows a different amplitude spectral density for every set-point. It increases for larger misalignments, i.e. larger phase differences on the measurement diodes.

3. Outlook

Models are used in the ongoing analysis to find the coupling parameters, which are believed to introduce Relative Intensity Noise (RIN) from the laser. A subsequent investigation was run during the mission extension, incorporating much larger slews while also compensating beamwalk effects with parallel commanded test mass one tilts. A publication with the results is in preparation.

4. Acknowledgments

The Albert-Einstein-Institut acknowledges the support of the German Space Agency, DLR. The work is supported by the Federal Ministry for Economic Affairs and Energy based on a resolution of the German Bundestag (FKZ 50OQ0501 and FKZ 50OQ1601).

References

[1] Armano M *et al.* 2016 *Phys. Rev. Lett.* **116**(23) 231101
[2] Morrison E, Meers B J, Robertson D I and Ward H 1994 *Appl. Opt.* **33** 5041–5049 URL `http://ao.osa.org/abstract.cfm?URI=ao-33-22-5041`

IOP Conf. Series: Journal of Physics: Conf. Series **840** (2017) 012045 doi:10.1088/1742-6596/840/1/012045

A proto-Data Processing Center for LISA

Cécile Cavet, Antoine Petiteau, Maude Le Jeune, Eric Plagnol, Etienne Marin-Martholaz, Jean-Baptiste Bayle

François Arago Center, APC, Univ. Paris Diderot, CNRS/IN2P3, CEA/Irfu, Obs. de Paris, Sorbonne Paris Cité, 13 rue Watt, 75013, Paris, France

E-mail: `ccavet@apc.in2p3.fr`

Abstract. The LISA project preparation requires to study and define a new data analysis framework, capable of dealing with highly heterogeneous CPU needs and of exploiting the emergent information technologies. In this context, a prototype of the mission's Data Processing Center (DPC) has been initiated. The DPC is designed to efficiently manage computing constraints and to offer a common infrastructure where the whole collaboration can contribute to development work. Several tools such as continuous integration (CI) have already been delivered to the collaboration and are presently used for simulations and performance studies. This article presents the progress made regarding this collaborative environment and discusses also the possible next steps towards an on-demand computing infrastructure. This activity is supported by CNES as part of the French contribution to LISA.

1. Data Processing Center

The LISA Data Processing Center (DPC), as described in [Amaro-Seoane et al.(2017)], is the entity that receives calibrated data (level 1 data) from the Science Operations Center (SOC) at ESA, processes them to identify gravitational wave sources and their parameters, and sends the results (level 2 and 3 data) back to the SOC. It is also tasked with identifying transient events to provide the outside community with alerts to search for electro-magnetic counterparts. The DPC will be delivered by the LISA Consortium under the responsibility of France [Amaro-Seoane et al.(2017), CNES Phase 0].

In order to begin the DPC activities and to provide the consortium with the tools to advance the LISA detector definition, the LISA proto-DPC [DPC] has been initiated in 2014. These tools will be used for the implementation of data analysis softwares, for the development of the LISA simulation package and to enhance the collaborative environment of these tasks.

1.1. Goal

The observations LISA will make will be the first of its kind (i.e. the simultaneous observation of a possibly large number of GW sources) and will explore a new way of observing the Universe. Therefore innovative data analysis (DA) techniques have to be implemented. During the mission operation, the DPC will have large fluctuations of CPU load due to the need of rapid processing of transient events and the regular reprocessing of the data with optimized DA techniques, new calibrations, consistency checks, etc. Moreover the source event rate is uncertain and can vary from a few events to a few tens thousand events per year. All these factors make the dimensioning of the pipeline and of the necessary resources a difficult challenge.

Some basic methods have been developed during past Mock LISA Data Challenges (MLDCs), demonstrating that the LISA DA challenge should be within reach. However details have to be studied using MLDCs of increasing complexity both for gravitational wave source and instrument modeling coming from the recent results of LISAPathFinder and from the implementation of future technological developments.

During the development of the DPC as well as during the operation phase, the pipelines will strongly evolve to integrate updated developments in a short loop cycle. For the operation phase, the processing of the data is expected to be done on a daily and weekly basis and should ensure non-regression on short timescales.

Because of all these complexities, new challenges have to be handle using innovative IT technologies such as virtualization and DevOps (see following sections). In addition, due to the very long term nature of these activities and the rapidly evolving IT solutions, the DPC has to be kept flexible and easily upgradeable until the end of the mission.

1.2. Why now?
There are a number of reasons pushing for an immediate start of the DPC:

- A framework to start collaborative work on DA and simulation is needed on the short term.
- An infrastructure to support the DA challenges has to be implemented soon.
- An infrastructure to support the end-to-end simulations that will be used to produce realistic data, evaluate mission performances and assess the industrial proposals is needed by 2018.
- A structure hosting the various softwares used during the development of LISA, in particular performance management tools, is also necessary in the near future.

The proto-DPC will be the framework that will support the LISA simulations and the next data analysis collaborative activities such as MLDCs.

2. Continuous integration
Continuous integration (CI) has become a standard development practice in the recent years, answering some of the needs that have emerged with Agile [Agile] and DevOps [DevOps] movements in the software development community.

Besides its trending methodological context, CI is now embodied into software systems which provide tools to help software development and collaborative work. In this section, we focus on the description of such CI environments, highlighting how they can help and impact the collaborative software development, in the context of the LISA DA pipeline construction.

2.1. Automatic tests
The main ingredient in CI is the development of a suite of non-regression tests, which can be automatically and systematically run by the CI system. Those tests can address multiple problems that are usually encountered in big software projects such as compatibility between versions, performances, validations, etc.

A schematic view of the CI cycle is given in Figure 1. The CI system is fed by code commits coming from a version control system (VCS) which hosts the source code and its version history. Commits automatically trigger the execution by the CI system of the suite of non-regression tests, starting with code compilation, building and installation on the CI environment. Test results are stored in the CI database, and exposed by the CI web interface. At any moment, developers can check those results, correct and improve the source code if needed, or retrieve a working version of the code (possibly associated with its working environment in a DevOps style project).

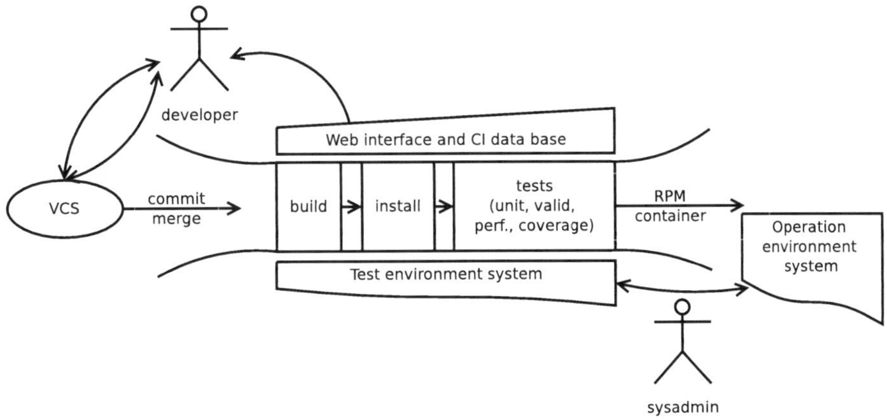

Figure 1. The CI system supports automatic code building and testing. By checking the test results using the web interface, developers can improve their codes. The integration in the production environment can be done either by cloning the CI environment or using software containers.

The automation of those tasks provides great benefits, especially for projects with a high number of interdependent sections of code that rapidly evolve, such as DA pipelines. Scientific developers are able to keep improving any part of the source code, even during operation phases, as the CI system guarantees the provision of a working version of the pipeline.

2.2. Towards good practices

The second main advantage of CI is more subtle. One has to realize that the CI system is by default without any intelligence: it is merely a robot that runs what has been implemented. CI relies on tests that the developer has to design and provide to the system. Those tests have to be sensitive to the particular (possibly evolving) objectives that the code must fulfill (precision, speed performance, lack of side effects, etc.): they have to be well-written and kept up-to-date. This is where the weakness and the force of the CI system resides: the system by itself strongly impacts the development methods adopted by the developer by suggesting (but not imposing) good practices such as a test-driven development (TDD).

Another kind of tests that can be use to feed the CI system is source code syntactic analysis. On this aspect, one can use the standard publicly available test rules (like Python PEP 8 standard coding rules) which enforce code readability.

As a first version of such a CI environment for the LISA DPC, some of the most popular software solutions at this time, namely Jenkins [Jenkins] for CI, and SonarQube [SonarQube] for code quality checking have been chosen.

3. Virtualization

Virtualization is a major new technique that has a big impact on the way hardware and configurations are managed.

3.1. Full virtualization and cloud computing

Full virtualization is a technology that allows to isolate an entire guest Operating System (OS) from the host OS using an hypervisor (Xen, KVM ..., see Fig. 2). The Infrastructure-as-a-Service (IaaS) cloud solutions, such as Amazon Web Services (AWS) and OpenStack, provide a stack of components including dynamical managers of hypervisors. Cloud computing is very interesting for data analysis due to the fact that instantiation of infrastructures provide on-demand virtual resources: virtual machines with virtual CPU, memory and network; virtual but persistent disk volumes; object storage. In other words, one can arbitrary request hardware configurations instead of buying CPU units. Started by the AWS company in 2006, public IaaS cloud infrastructures have also emerged in the academic community. Researchers have now access to the European e-Infrastructures [EGI] and from a national point of view, to the French federation [FG-cloud] to which the Computing Center of IN2P3 (CC-IN2P3) takes part.

The dynamic provision of cloud's virtual machines has resonated with the DevOps approach. Indeed, system administrators usually develop scripts to automatize complex workflows in order to facilitate service and software deployment. With cloud computing, this configuration automation step also opens a window on the usage of on-demand resources: with an automatic configuration, it is easier to deploy services and codes on virtual instances. Several configuration frameworks (SlipStream, Heat, Juju ...) are dedicated to the cloud while others (Ansible, Puppet, Chief ...) were developed for general platforms (bare metal or virtual machines).

These tools will be put in use for the LISA DPC in order to handle CPU fluctuation needs, and to pool hardware resources from all partners.

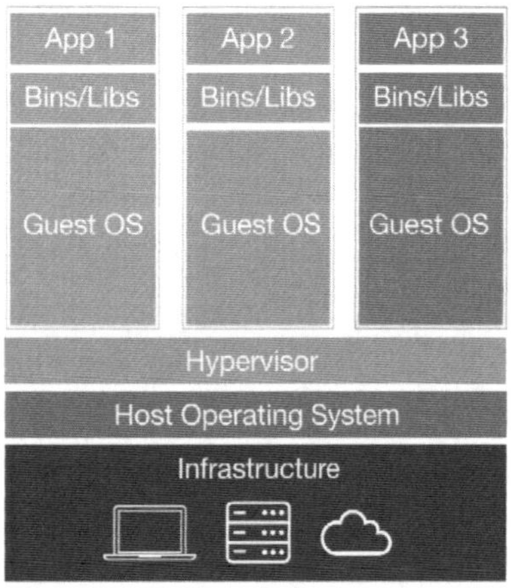

Figure 2. Virtual machine working.

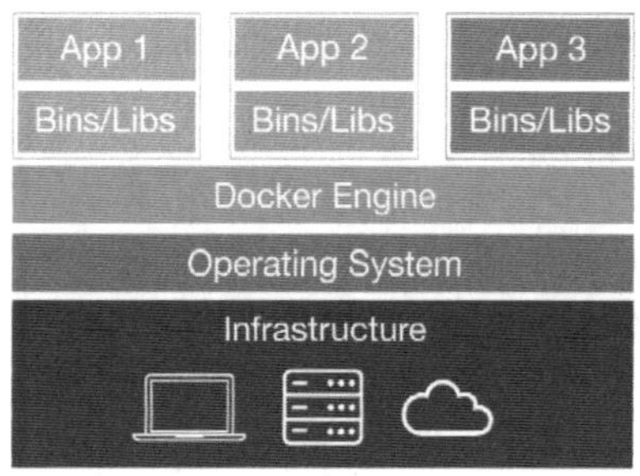

Figure 3. Docker working.

3.2. Containers

Recently, the re-discovery of Linux containers has given rise to an innovative way to encapsulate and share codes and services. Containers are based on a specific Linux process (namespaces) which allows to isolate one task execution from another (see Fig. 3). The emergence of several solutions (LXC, Docker, rkt, Singularity, etc.) has changed developer workflow on code prototyping and production phase. As for virtual machines in the cloud, an ecosystem of components is provided to handle specific properties of containers. Furthermore, hubs and catalogs of images have been created to offer a central place for sharing images between users. In a production environment, orchestration of containers is managed by a specific tool (Mesos/Chronos, Swarm, Kubernetes) which allows to combine cluster and cloud infrastructure in a hybrid way [Poncet et al.(2016)].

4. How to contribute?

Our philosophy is to provide the LISA community with a suited stack of tools, which should help the consortium simulation and analysis code development and execution. In order to do so, we aim at providing such tools in the near future, to support the mission preparation phase and initialize an improvement *virtuous circle* within the LISA community.

We are looking forward to getting feedback from the LISA data scientists willing to start using these tools. The current entry point is the DPC website [DPC] which gathers all the information. In particular, it provides technical details on how to use code hosting and continuous integration platform. It will be updated with new information through the DPC development.

Acknowledgments
The proto-DPC activity is supported by the François Arago Centre (Paris) and CNES. CC acknowledge the financial support from the UnivEarthS Labex program of Sorbonne Paris Cité (ANR-10-LABX-0023 and ANR-11-IDEX-0005-02). JBB was supported by CNES.

References

[Amaro-Seoane et al.(2017)] Amaro-Seoane, P., Audley, H., Babak, S., et al. 2017, arXiv:1702.00786
[Agile] The Agile manifesto:
`http://www.hristov.com/andrey/fht-stuttgart/The_Agile_Manifesto_SDMagazine.pdf`
[CNES Phase 0] CNES, "NGO Ground segment - Phase 0 report", CNES Report, DCT/ME/EU - 2014.0015466, 2016
[DevOps] DevOps: `http://radar.oreilly.com/2012/06/what-is-devops.html`
[DockerHub] DockerHub: `https://hub.docker.com/u/elisadpc`
[DPC] Proto-DPC platform: `http://elisadpc.in2p3.fr/home`
[EGI] EGI Advanced Computing Services for Research: `https://www.egi.eu/`
[FG-cloud] FG-cloud: `http://www.france-grilles.fr/services-catalogue/fg-cloud/`
[Jenkins] Jenkins: `https://jenkins.io/`
[Poncet et al.(2016)] Poncet, M., Faure, T., Cavet, C., Petiteau, A., et al. 2016, BiDSO16
`http://esaconferencebureau.com/custom/16M05/bids/ALL/D2_1210_Poncet.pdf`
[SonarQube] SonarQube: `https://www.sonarqube.org/`

From LPF to eLISA: new approach in payload software

Ll. Gesa, V. Martin, A. Conchillo, J.A. Ortega, I. Mateos, A. Torrents, J.P. Lopez-Zaragoza, F. Rivas, I. Lloro, M. Nofrarias, CF. Sopuerta

Space Sciences Institut (CSIC-IEEC) - Barcelona, Spain

E-mail: gesa@ieec.cat

Abstract. eLISA will be the first observatory in space to explore the Gravitational Universe. It will gather revolutionary information about the dark universe. This implies a robust and reliable embedded control software and hardware working together. With the lessons learnt with the LISA Pathfinder payload software as baseline, we will introduce in this short article the key concepts and new approaches that our group is working on in terms of software: multiprocessor, self-modifying-code strategies, 100% hardware and software monitoring, embedded scripting, Time and Space Partition among others.

1. Introduction

The Gravitational Wave Astronomy Research and Technology group of CSIC-IEEC has developed the LISA Pathfinder Data Management Unit (DMU). During several years, our group has implemented its embedded software following the ESA guidelines and standards to ensure each development stage along its quality. Now it's time to, from lessons learnt from more than 8700 hours of real operativity in flight, start to define the new software approach for the future space based gravitational wave observatory: **e**LISA.

2. Lisa Pathfinder Software conclusions

The LPF DMU Software has shown its complete robustness during the LPF operations. But from the daily operative and monitoring we have learnt that things could be improved:

- Hardware status could be better informed to ground and a best internal diagnostic would be useful.
- A more flexible upgrade/modification mechanism it is desired: Now science operation mode must be stopped to have time for upgrades for example. On ground observatories the non-productive time should be avoided or at least reduced, therefore is desired to have the same aim in a space obervatory.
- Historicals: An internal status history could help to analyze problems reducing the offline time needed for that.

Consequently, if the future software development adds improvements in those points it will maximize the support of the payload software for the scientific return. Hardware development also should take care of it, for example with more data room and CPU speed.

IOP Conf. Series: Journal of Physics: Conf. Series **840** (2017) 012046 doi:10.1088/1742-6596/840/1/012046

3. Our approach

Next figure deploys our approach; multiple applications running together in parallel sharing hardware resources and data. Each one with its own operating system and isolated security, and each one covering some specific functionality needed to control the payload system.

Our proposal is to group each needed behavior in separate applications to be executed isolated from the rest, using a technology called Time and Space Partitioning (TSP). Our solution is to have next applications:

- The **Basis partition**, running at highest security level, is in charge of overall system monitoring

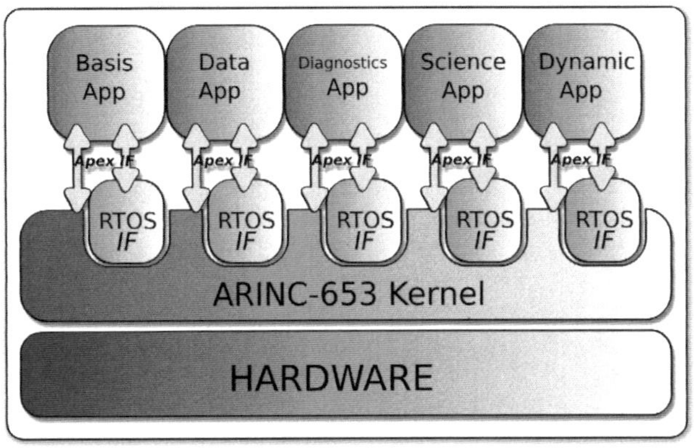

Figure 1. Architecture under study

- **Data application** is handling the internal data and its telemetry packing. The subsystem manager and housekeeping monitor is perfomed by the **Diagnostics** part.
- The **Science application** is the main core performing the specific science.
- And our innovative bet; the **Dinamic**, an application partition enabled to mirror any other partition, adding the capacity of system upgrades while system is in operational mode and also enabled to run scripting sequences, useful for debugging or in-place testing of new functionalities before to be official.

All of them running over a hardware abstraction layer in charge of managing the resources, for example the physical communication layers or controlling the memory data sharing mechanism between applications. These layers are known as bare metal hypervisors. Several solutions are already present, none of them in a maturity level to be used in space mission. Several among them are curently under study such as: AIR, an ESA sponsored project based on RTEMS, and XtratuM hypervisor system. Other commercial solutions could be VxWorks 653 or PikeOS.

In relation with the real-time operating systems on partitions, LISA Pathfinder Payload Software has been based on RTEMS, so our first approach is to still using it, but other options like Xluna, or the provided with XtratuM: LithOS are not discarded. Specially the use of standard GNU/Linux could be useful on the dynamic partition to enable the use of scripting execution for fast development, execution and diagnostic.

4. The key : ARINC-653

To acomplish our approach, the use of the standard ARINC-653 is under study. This Avionics Application Standard Software Interface is a specification for Time and Space Partitioning (TSP) in safety-critical avionics Real-Time Operating Systems (RTOS). It allows the hosting of multiple applications of different software levels on the the same hardware in the context of an Integrated Modular Avionics (IMA) architecture. This technology is recently under ESA studies to be applied in space [1].

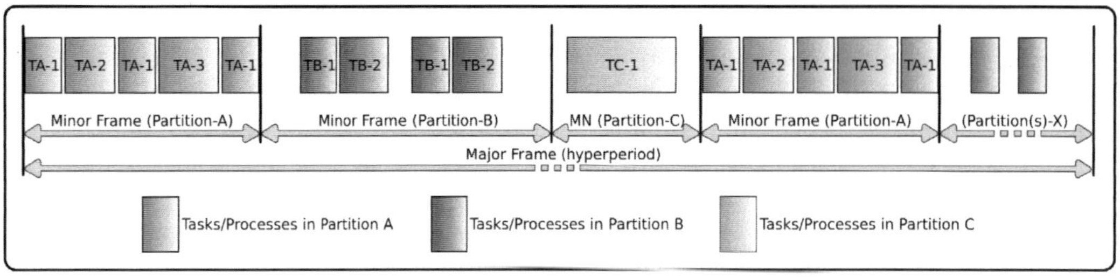

Figure 2. TSP scheduling diagram

Several software applications with its own characteristics and security levels can co-exist over same hardware exchanging (or not) information and the failure of one of them doesn't affect the operativity of the others.

5. Development & Qualification

Development should follow ESA quality guidelines, the ECSS standards defines the technological development for space missions. Specificaly ECSS-E-40 and Q-80 for software. In new researches in TSP area from a whole system architecture the guidelines and processes proposed by Space AVionics Open Interface aRchitecture (SAVOIR) should be followed. Since some ARINC-653 implementations are part of several ESA studies, projects and researches, full space qualification is not yet reached. Our work could be a partner in this process. Our approach of Dynamic APP partition is a new concept not yet studied, therefore a further work in terms of development and qualification will be required.

6. Conclusion

The work for eLISA is just starting. Many open points and much work ahead, but the challenges of all these new software technologies and technical proposals give us the energy to kick off as soon as possible to start to see results in short. Study phase is not yet finished, and this is important to enter in the developing phase ensuring the final goal. Validation and acceptance phase is the expected more complex part. These technologies are new in space and the qualification process will take its time.

7. Acknowledgments

We acknowledge support from Project AYA2010-15709-Espacio of the Spanish Ministry of Science and Innovation (MICINN)

References

[1] V. Bos, P. Mendham, P. Kauppinen, N. Holsti, A. Crespo, M. Masmano, J.A. de la Puente, J. Zamorano, "Time and Space Partitioning the EagleEye Reference Mission" in Data Systems In Aerospace (DASIA 2013), Porto, Portugal:, pp. 1-7, May 2013.

IOP Conf. Series: Journal of Physics: Conf. Series **840** (2017) 012047 doi:10.1088/1742-6596/840/1/012047

An optical read-out system for the LISA gravitational reference sensor: present status and perspectives.

A Grado[1,3], R De Rosa[1,2], L Di Fiore[1], F Garufi[1,2], L Milano[1,2], G Russano[4] and V Spagnuolo[1]

[1] INFN Sez. Napoli, Via Cintia, Napoli, Italy, I-80126.
[2] University of Napoli "Federico II" - Dept. of Physical Science, Napoli, Italy, I-80126.
[3] INAF Osservatorio Astronomico di Capodimonte, Salita Moiariello 16, Napoli, Italy, I-80131
[4] Universitá di Trento and INFN-TIFPA.

E-mail: agrado@na.astro.it

Abstract.
Since a few years, the LISA-PF group in Napoli has been working to the development of an optical read-out system, based on optical levers and position sensitive detectors, for the LISA Gravitational Reference Sensor (GRS). This is intended as a more sensitive extra sensing device, in addition to capacitive readout that is the reference solution already tested on flight by the LISA-Pathfinder mission. The reliability of the proposed ORO device and the fulfillment of the sensitivity goals have been already demonstrated in bench-top measurements and tested with torsion pendulum facilities. In this paper we report on the present status of this activity, presenting the results obtained so-far and the perspectives for the future LISA mission.

1. Introduction

We report on the status of the development of an Optical Read-Out (ORO) [1] system for the LISA [2] inertial sensor. The present design of the Gravitational Reference Sensor (GRS) [3] is based on capacitive sensor; our goal is to integrate the ORO in the capacitive one without changing the present set-up and electrode design. The motivation is twofold: a) risk mitigation and b) to improve the sensitivity. For the first aspect the ORO could be a back-up sensor in case the capacitive sensor fails after the launch. For the second aspect we should note that any reduction of the noise of the sensor used for Drag Free and Attitude Control System (DFACS) would relax specifications on cross couplings (CC). Present requirement on CC is 0.1 % that is a very strong specification. In the following sections we will describe the proposed ORO set-up, we report on sensitivity for tests on bench top and on a two stages torsion pendulum facility [4] and we describe possible layouts for the implementation in LISA. First results and next steps for space qualification are finally discussed.

2. Sensor Set-up

Figure 1 shows the schematic drawing of the ORO sensor. A light beam is sent trough a single mode optical fibre to the test mass and the position of the reflected beam is measured with a Quadrant Photo-Diode (QPD) or (Position Sensing Device (PSD). The sensitivity depends on input power, measurement range (beam size for QPD or detector size for PSD) and geometry.

IOP Conf. Series: Journal of Physics: Conf. Series **840** (2017) 012047 doi:10.1088/1742-6596/840/1/012047

With a suitable combination of beams and sensors we can recover the six degree of freedom (DOF) of the test mass.

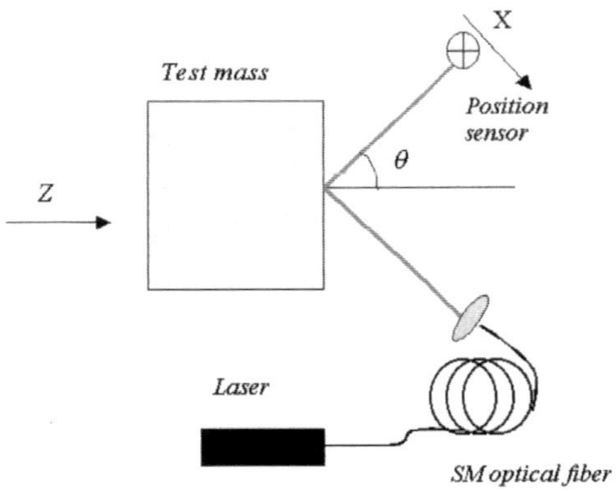

Figure 1. Schematic of the ORO sensor.

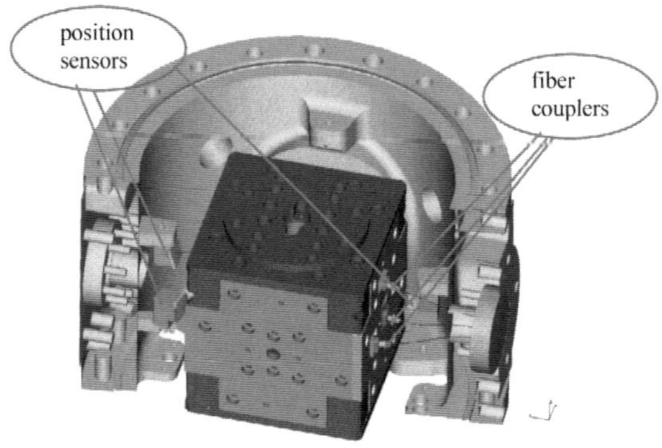

Figure 2. Possible implementation in the GRS.

3. Sensitivity

The figure 3 shows the ORO sensitivity for the two types of photodetectors PSD (upper curve) and QPD (lower curve) measured in a bench-top setup. The measurement ranges are 0.4 mm for QPD and 4.7 mm for PSD. In both cases the ORO sensitivity is limited by the input trans-impedance amplifier current noise. For comparison also the sensitivity of the capacitive sensor is reported.

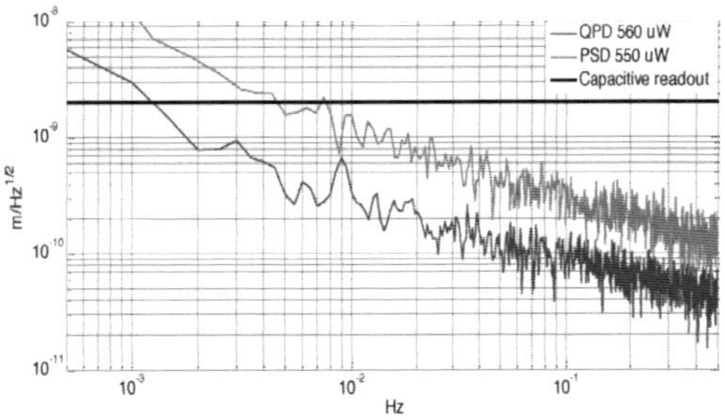

Figure 3. ORO sensitivity with different photodetectors. The black line shows the capacitive readout sensitivity

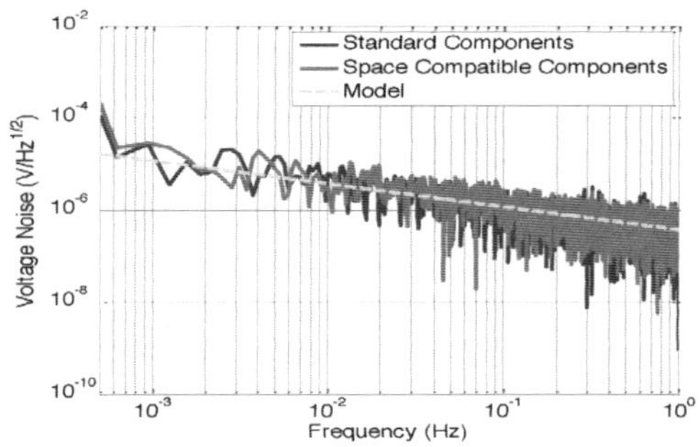

Figure 4. Comparison between standard and space qualified front end electronics.

4. Space qualified front-end electronics

We have verified the noise characteristics of Space Qualified (SQ) front-end electronics (see figure 4). The electronic used for processing QPD signals is based on OP27EP operational amplifier. There is a SQ equivalent component OP27AJ/QMLR. The noise performance is exactly the same as for the standard components.

5. Possible implementation in LISA

The idea is to use the electrodes as mirrors for directing the beams to the test mass surface and to the sensors. We tested the 3 beams configuration with a prototype (figures 5, 6). We used a two degrees of freedom angular piezoelectric (PZT) stage and a 3 axial linear PZT stage to check 5 DOFs out of 6 (Z, Y, α, η, θ). The analytical model was validated moving the dummy test mass with PZT actuators equipped with calibrated capacitive readout. The measured matrix elements coincide, within a few percent, with the analytical model (compatibly with the prototype machining and assembling tolerances). Figure 2 shows a possible implementation of the ORO in the GRS used for LISA-PF.

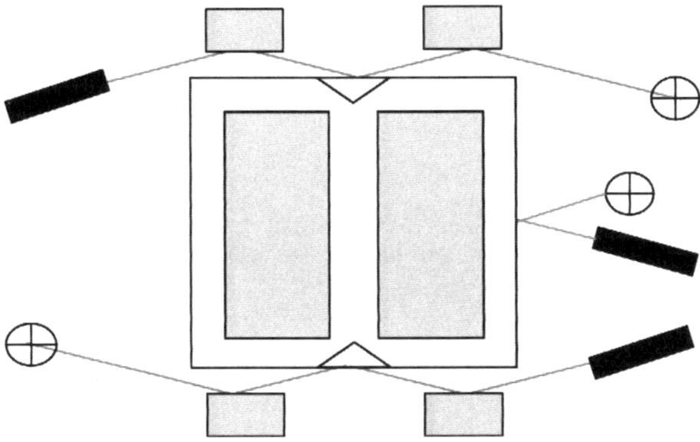

Figure 5. Schematics of 5 DOF (Z, Y, α, η, θ) readout system using 3 beams. Front view.

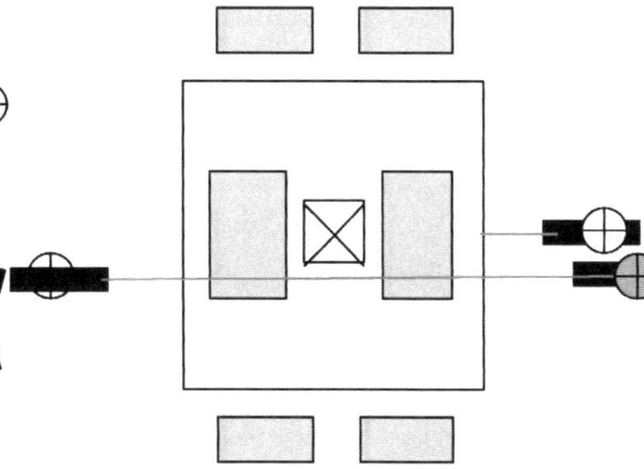

Figure 6. Schematics of 5 DOF (Z, Y, α, η, θ) readout system using 3 beams. Top view.

6. Conclusion

Both tests on bench top and on engineering model suspended to a torsion pendulum confirm that the ORO sensitivity can be better than the capacitive sensor, above 1 mHz. The noise level is well characterised and allows to make predictions and trade-off between sensitivity and measurement range. It has been identified a possible layouts for the integration, with minimal impact on the present design, into the GRS. Study of space compatible parts is just started: read-out electronics is not a problem. Next step would be the search for space qualified optical components. The ORO has been developed up to TRL 4 and is a good candidate as a back-up sensor for the LISA inertial sensors, with possible sensitivity improvement.

References

[1] De Rosa R, Di Fiore L, Garufi F, Grado A, La Rana A and Milano L , 2011 An optical readout system for the drag free control of the LISA spacecraft *Astroparticle Physics* **34** Issue 6 394-400,

[2] P. Bender et al, LISA laser interferometer space antenna: a cornerstone mission for the observation of gravitational waves ESA-SCI, **11** (2000).

[3] Cavalleri A, Dolesi R, Fontana G, Hueller M, Turneaure J, Vitale S and W Weber 2001 *Classical and Quantum Gravity* Progress in the development of a position sensor for lisa drag-free control **18** Issue 19 4133?4144.

[4] Bassan M, Cavalleri A, De Laurentis M, De Marchi F, De Rosa R, Di Fiore L, Dolesi R, Finetti N, Garufi F, Grado A et al. 2016 Approaching Free Fall on Two Degrees of Freedom: Simultaneous Measurement of Residual Force and Torque on a Double Torsion Pendulum *Physical Review Letters* **116** Issue 5 id.051104.

Optimizing orbits for (e)LISA

Hubert Halloin

APC, AstroParticule et Cosmologie, Université Paris Diderot, CNRS/IN2P3, CEA/Irfu , Observatoire de Paris, Sorbonne Paris Cit, 10, rue Alice Domon et Léonie Duquet, 75205 Paris Cedex 13, France

E-mail: hubert.halloin@apc.in2p3.fr

Abstract. Earth and other planets gravitational fields induce perturbations of the (e)LISA constellation geometry (mostly flexing and Doppler frequency shifts). We present here a numerical optmization method minimizing these distrubances, subject to constraints (maximum distance to Earth and launch mass).

1. Introduction

The three, free-falling, satellites of (e)LISA form an (quasi) equilateral triangle trailing the Earth by about 20°. While orbiting around the Sun, the distances between the S/C evolve, leading to breathing and flexing effects, which should be accommodated by the satellite design, e.g telescope steering (or laser in-field pointing capability) and wide photodetector bandwidth (due to the Doppler frequency shifts). These fluctuations can be efficiently minimized analytically for pure keplerian orbits [1]. However, the constellation is also perturbed by the gravitational influence of the Earth and, to a lesser extent, Jupiter and Mars. Numerical optimization is required to find the optimal initial orbital parameters and, possibly, velocity increments for orbit control during the mission lifetime.

The optimization process is also subject to two other constraints. The first constraint is due to the data link budget which requires a maximum distance to Earth of $\approx 70 \times 10^6$ km[2]. The capability of the launcher sets a second constraint on the totalwet mass (i.e. including propellant for orbital maneuvers) of the constellation.

2. Optimization Method

In the present work, the numerical optimization process combines an orbit propagator and the minimization of a cost function. The orbit propagator is based on a VEFRL (Velocity Extended Forest-Ruth Like) integrator [3], which is a constant step size, quasi 6th order symplectic integrator and optimized for a weak perturbation of a central force potential. The chosen minimization method is an Active Covariance Matrix Adaptation Evolution Strategy (aCMAES) [4] optimizer, developed for non-linear, non-convex, black-box optimization and especially used for continuous cost functions with sharp edges, local minima, outliers, etc.

The cost function is expressed as a linear combination of various penalties:

$$C(\mathcal{P}) = \alpha_A \cdot C_A(\mathcal{P}) + \alpha_X \cdot C_X(\mathcal{P}) + \alpha_F \cdot C_F(\mathcal{P}) + \alpha_M \cdot C_M(\mathcal{P}) + \alpha_D \cdot C_D(\mathcal{P})$$

$$C_A(\mathcal{P}) = \frac{1}{3}\sum_{i=1}^{3}\left\langle\left(\frac{L_i(t) - L_{ref}}{L_{ref}}\right)^2\right\rangle_t \quad ; \quad C_X(\mathcal{P}) = \max_{i\in[1:3]} \frac{\max_t \theta_i(t) - \min_t \theta_i(t)}{60^\circ}$$

$$C_F(\mathcal{P}) = \max_{i\in[1:3]} \frac{\max_t \Delta F_i(t) - \min_t \Delta F_i(t)}{1\ \mathrm{MHz}} \quad ; \quad C_M(\mathcal{P}) = P_{Th}\left(\frac{\sum_{i=1}^{3} M_i - M_{max}}{M_{spread}}\right)$$

$$C_D(\mathcal{P}) = P_{Th}\left(\frac{\max_{i,t} D_i(t) - D_{max}}{D_{spread}}\right) \quad ; \quad P_{Th}(x) = \frac{e^{4x}}{1+e^{4x}} + 0.025\log\left(1+e^{4x}\right)$$

, where :

- $\mathcal{P}$ is the set of initial orbital parameters and maneuver velocity increments to be optimized
- $C_A(\mathcal{P})$ is the armlength penalization, minimizing the mean quadratic difference between the lenth of arm i at time t ($L_i(t)$) and the desired arm lenth (L_{ref})
- $C_X(\mathcal{P})$ is the flexing penalization, minimizing the amplitude of the fluctuations of the inner angle at vertex i ($\theta_i(t)$)
- $C_F(\mathcal{P})$ is the Doppler shift penalization, minimizing the amplitude of frequency fluctuations on arm i ($\Delta F_i(t)$)
- $C_M(\mathcal{P})$ is the mass penalization, with M_{max} being the maximum possible launched mass, M_{spread} the allowed mass dispersion and M_i the computed wet mass of satellite i. $P_{Th}(x)$ is a non-linear (threshold) penalization function.
- $C_D(\mathcal{P})$ is the distance to Earth penalization, with D_{max} being the maximum possible distance to Earth, D_{spread} the allowed distance dispersion and $D_i(t)$ the distance to Earth of satellite i at time t.

$P_{Th}(x)$ is used to define a threshold function : $P_{Th}(x)$ exhibits a smooth transition from 0 to 1 with a slope of 1 at $x = 0$ and an additional small linear increase (slope of 0.1) for $x > 0$.

The hyper-parameters α_u are used to tune the relative importance of the different penalizations factors.

In the present work (based on the former NGO mission profile [2]), the estimation of the propellant mass is done using the following scenario :

- Launch on a Geostationary Transfer Orbit (200 km x 36 000 km)
- Escape maneuvre on the ecliptic plane computed for a cruise time of 15 months
- 6 to 12 months Lambert transfer maneuver at the longitude of the ascending or descending node of the targeted orbital plane towards the initial position (and time) of the science mission.
- 'Stopping' maneuver to reach the desired initial orbital parameters at the beginning of the science mission.

Additional periodic velocity increments can also be included in the optimization process.

3. Optimization results

As an example, the above optimization scheme has been applied with the following constellation configuration : 3×10^6 km armlength (L_ref) ; 10 years mission lifetime (2030 - 2040) ; 8 t maximum launch mass (compatible with Ariane 64 capability to geostationnary transfer orbit) ; 70×10^6 km maximum distance to Earth. The combined gravitational perturbing effects of the Earth, Moon, Mars and Jupiter have been considered.

3.1. Optimization with no additional orbital maneuvers

In a first optimization run, maneuvers are only used to position the S/C on their initial orbits, with no further orbit control maneuver during the duration of the mission.

Figure 1 represents the resulting time evolution of flexing angle, armlength, frequency shift and distance to Earth. In this case, the optimized initial orbital parameters lead to a maximum

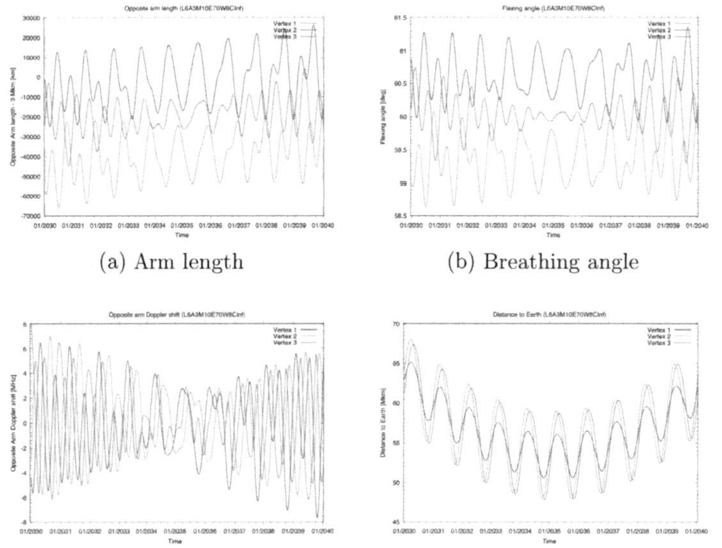

(a) Arm length (b) Breathing angle

(c) Frequency shift (d) Distance to Earth

Figure 1: Evolution of constellation geometry for optimized initial orbital parameters (without additional orbit control maneuvers)

vertex angle fluctuations amplitude of $\approx$ 1.5 deg peak-to-peak (for each vertex), a maximum Doppler frequency shift of $\approx$ 14 MHz peak-to-peak and a total wet mass of the constellation of about 7835 kg.

Noticeably, due to the relatively large wet mass allowed by an Arinae 6.4 launch, a full stopping maneuver can be performed on the 3 spaceraft, so that they are subsequently following a ballistic trajectory towards the Earth. Actually, the S/C are slightly faster than the Earth. Its pulling effect increases the semi-major axis of each S/C orbit, hence slowly decreasing the S/C orbital velocity, and finally inverting the relative movement between the Earth and the satellites (see figure 1d). This numerical solution is similar to the semi-analytic 'projectile' solution from [5].

Moreover, the geometrical center of the constellation is oscillating by $\pm 8 \times 10^6$ km above and below the ecliptic plane. This movement mitigates the pulling effect of the Earth on the S/C (and therefore reduces its perturbing influence).

3.2. Optimization with periodic orbital maneuvers

A second optimization run was performed including orbital maneuvers (i.e. velocity increments) every 18 months (i.e. 6 maneuvers for each S/C in the mission lifetime), see fig. 2.

Thanks to these additional maneuvers, the amplitude of the vertex angle fluctuations is reduced to $\approx$ 0.6 deg peak-to-peak (for each vertex) and the maximum Doppler frequency shift is of $\approx$ 4 MHz peak-to-peak. The total wet mass of the constellation is slightly reduced to about 7431 kg, taking into account the additional propellant required for the orbit control maenuvers

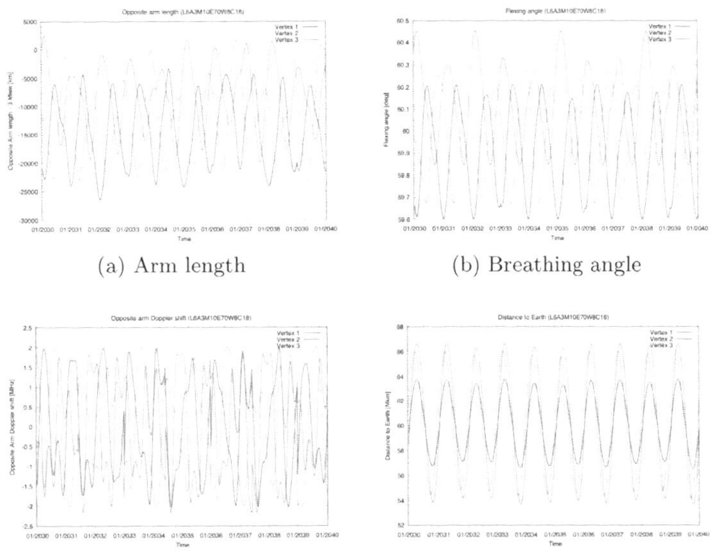

(a) Arm length (b) Breathing angle

(c) Frequency shift (d) Distance to Earth

Figure 2: Evolution of constellation geometry for optimized initial orbital parameters, with additional periodic orbit control maneuvers (every 18 months)

(using ΔV impulses ranging from 70 to 140 m/s). Moreover, these maneuvers result in a globally constant distance to Earth (see figure 2d), hence allowing a longer mission lifetime (regardless of propoellant consumption).

4. Conclusion

The optimization method described above allows to find the orbital parameters minimizing the disturbances due to the gravitational field of the Earth and other planets. This method can also be used to optimize and compute the impact of periodic additional orbital maneuvers during the mission lifetime. Clearly, a few control maneuvers can significantly reduce the beam pointing constraint and Doppler effect, with no mass penalty. These maneuvers could also be useful to compensate for previous maneuvers uncertainties, unmodeled perturbations and velocity dispersion from the initial positioning maneuvers.

As drawbacks, orbit control maneuvers will require to suspend the science acquisition, grab the test masses and, probably, re-acquire the laser links after the maneuvers. They might also induce significant changes in the mass distribution (hence the local gravitational field) of the S/C.

References

[1] Nayak K R, Koshti S, Dhurandhar S V and Vinet J-Y 2006 On the minimum flexing of LISA's arms *Classical and Quantum Gravity* **23** 1763

[2] ESA 2011 *NGO Assesment Study Report* ESA/SRE(2011)19

[3] Omelyan I P, Mryglod I M and Folk R 2002 Optimized ForestRuth-and Suzuki-like algorithms for integration of motion in many-body systems *Computer Physics Communications* **146** 188

[4] Jastrebski G A and Arnold D V 2006 Improving evolution strategies through active covariance matrix adaptation. *Proc. IEEE World Congress on Computational Intelligence* p 9719 and `https://github.com/beniz/libcmaes`

[5] Dhurandhar S V, Nayak K R and Vinet J-Y 2008 Optimising LISA orbits: The projectile solution. *arXiv preprint* arXiv:0809.1935

A simulation of weak-light phase-locking for space laser interferometer

Y Q Li[1], Y H Dong[1,2], H S Liu[1,2], Z R Luo[4,5] and G Jin[1,3]

[1] Institute of Mechanics, Chinese Academy of Sciences, Beijing 100190, China;
[2] University of Chinese Academy of Sciences, Beijing 100190, China;
[3] School of Engineering Science, University of Chinese Academy of Sciences, Beijing 100049, China;
[4] Leibniz Universität Hannover, Hannover 30167, Germany;
[5] Max-Planck-Institude for Gravitational Physics, Hannover 30167, Germany
* Corresponding author: gajin@imech.ac.cn

Abstract. A simulation was investigated to better understand the impacts and effects of the additional technical noises on weak-light phase-locking for space laser interferometer. The result showed that the locking precision was limited by the phase readout noise when the laser frequency noise and clock jitter noise were removed, and this result was then confirmed by a benchtop experimental test. The required space laser interferometer noise floor was recovered from the simulation which proved the validity of the simulation program.

1. Motivation

The space-borne gravitational wave (G.W.) antennas were designed to use space laser interferometers to detect the G.W. signal at the frequencies from 0.1 mHz to 1 Hz. The LISA (Laser Interferometer Space Antenna) mission, due to the long distance transmission and the finite aperture of the telescope, only very few fraction of the transmitted light (≈100 pW) can be received by the remote satellite [1]. A technique called'weak-light phase-locking' (WLPL) is adopted by the remote satellite and makes it a 'phase-transponder' with the local satellite [2]. In this paper, a simulation was performed to better understand the impacts and effects of the additional technical noises on WLPL for LISA.

2. Investigation of noises

To simulate WLPL for LISA in a systematic manner, the noise sources categorized have to be taken into account as the interferometric measurement system (IMS) noise and the acceleration noise. The IMS noise mainly includes shot noise, beam pointing noise, laser frequency noise, clock jitter noise and so on [3-7]. Since the simulation will be run in time series, the above LISA noises given by smooth function in ASD (Amplitude Spectral Density) should be converted into random noise in time domain. An algorism based on Fourier transform was invented to convert the

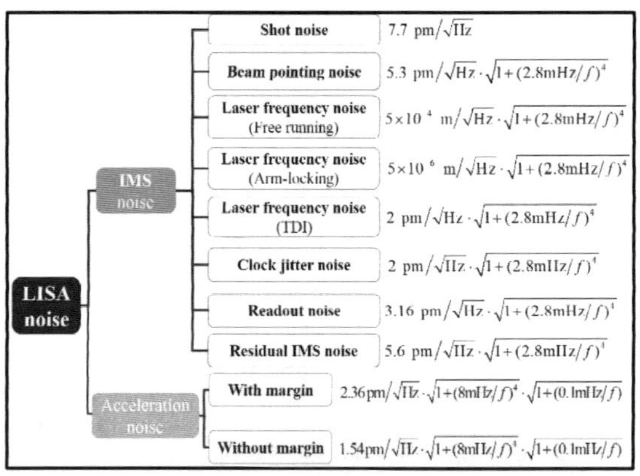

Fig.1 Diagram of the LISA noise

noise function into real time data with random characteristics.

3. Condition and simulation

A simulation program using Matlab Simulink is established and the schematic diagram of the program is shown in Fig. 2. The transmitting light received by the remote satellite is represented by 'add1'. A single frequency 'G.W. signal' (100 pm/√Hz at 30 mHz) is also added as an indicator to analyze the impacts of various noises. The slave laser in the remote satellite is given as 'add2'. The slave laser and the received light "interfere" at 'add3' and the phase difference is sent to and read out by a phasemeter which is represented by 'add4'. Then, the PID (Proportional-Integral-Derivative) controller is introduced to feedback control the slave laser to phase-locking the transmitting light.

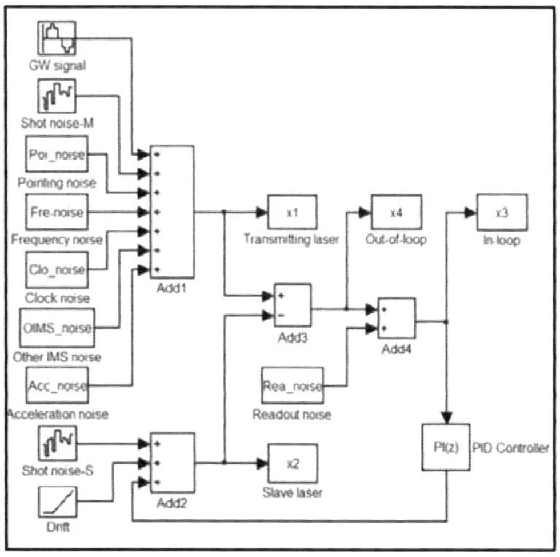

Fig.2 The Simulink model of WLPL

Considering all of the noises in Fig.1, supposing the frequency noise and the clock jitter noise have been well suppressed by arm-locking, TDI (Time Delay Interferometer) and other post process technics. It could be seen from the Fig.3 that the out-of-loop signal is mainly limited by the readout noise and the in-loop signal is well below 10^{-12} m/√Hz. Under this situation, the 'G.W. signal' emerges out.

An experimental demonstration is also performed in order to test the control scheme. The schematic diagram and the result of the experimental test are given in Fig. 4. Here, the beat frequency is 1 MHz, and the power of the slave laser and the master laser are about 1 mW. Limited by the readout noise and optical pathlength noise, the phase lock loop is working at nanometer level. It could be seen that the phase difference between the slave laser and the master laser is suppressed from 10^{-7} to 3×10^{-10} m/√Hz at 1 mHz after the phase locking loop is turned on. This directly demonstrates the functionality of our control scheme.

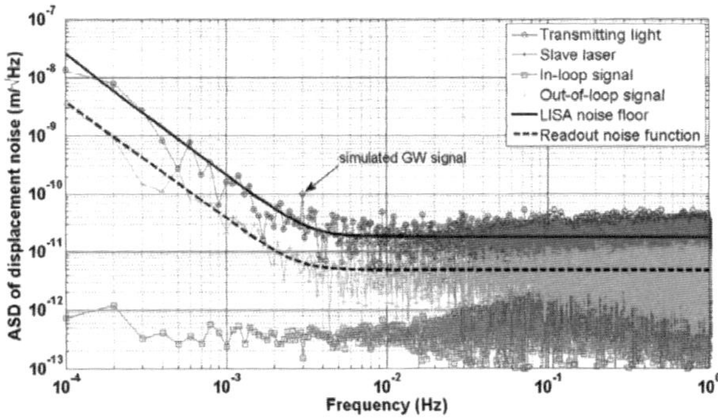

Fig.3 ASD noise performance of the simulation

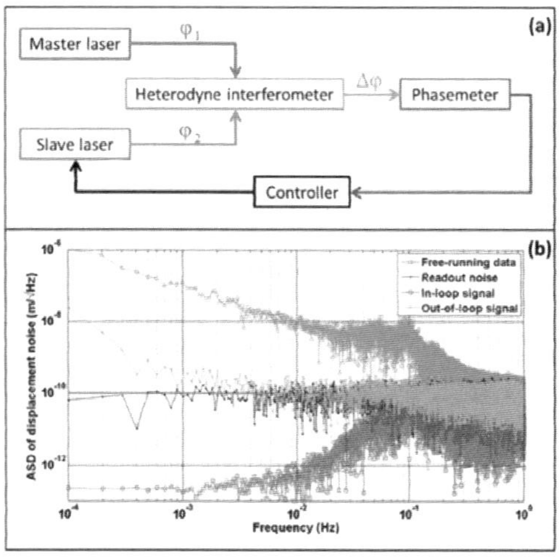

Fig.4 Schematic diagram (a) and result of the experimental test (b)

4. Results

The result showed that the locking precision was limited by the phase readout noise when the laser frequency noise and clock jitter noise were removed, and this result was also confirmed by a benchtop experimental test. The required LISA noise floor was recovered from the simulation which proved the validity of the simulation program.

Acknowledgements
This work was supported by the National Natural Science Foundation of China, Grant No. 61575209, and the Strategic Priority Research Program of the Chinese Academy of Sciences, Grant No. XDB23030200.

References

[1] Diekmann C *et al* 2009 *J. Phys. Conf. Series* **154** 012020
[2] Esteban J J *et al.* 2011 *Opt. Express* **19** 15937–15946
[3] Jennrich O 2009 *Class. Quantum Grav.* **26** 153001
[4] Yu Y, Mitryk S and Mueller G 2014 *Phys. Rev. D* **90** 062005
[5] Otto M, Heinzel G and Danzmann K 2012 *Class. Quantum Grav.* **29** 205003
[6] Danzmann K *et al* 2011 *Assessment Study Report ESA/SRE* 54–72
[7] Heinzel G *et al* 2011 *Class. Quantum Grav.* **28** 094008

Iodine frequency references for space

Thilo Schuldt[1], **Klaus Döringshoff**[2], **Markus Oswald**[3,4], **Ulrich Johann**[4], **Achim Peters**[2], **and Claus Braxmaier**[1,3]

[1] German Aerospace Center (DLR), Linzer Straße 1, 28359 Bremen, Germany
[2] Humboldt-University Berlin, Newtonstr. 15, 12489 Berlin, Germany
[3] University Bremen, Center of Applied Space Technology and Microgravity (ZARM), Am Fallturm 1, 28359 Bremen, Germany
[4] Airbus DS GmbH, Claude-Dornier-Straße, 88090 Immenstaad, Germany

E-mail: thilo.schuldt@dlr.de

Abstract. Optical frequency references are a key element for the realization of future space missions. They are needed for missions related to tests of fundamental physics, gravitational wave detection, Earth observation and navigation and ranging. In missions such as GRACE follow-on or LISA the optical frequency reference is used as light source for high-sensitivity inter-satellite distance metrology. While cavity-based systems are current baseline e.g. for LISA, frequency stabilization on a hyperfine transition in molecular iodine near 532 nm is a promising alternative. Due to its absolute frequency, iodine standards crucially simplify the initial spacecraft acquisition procedures. Current setups fulfill the GRACE-FO and LISA frequency stability requirements and are realized near Engineering Model level. We present the current status of our developments on Elegant Breadboard (EBB) and Engineering Model (EM) level taking into account specific design criteria for space compatibility such as compactness (size iodine spectroscopy EM: $38 \times 18 \times 10\,\mathrm{cm}^3$) and robustness. Both setups achieved similar frequency stabilities of $\sim 1 \cdot 10^{-14}$ at an integration time of 1 s and below $5 \cdot 10^{-15}$ at integration times between 10 s and 1000 s. Furthermore, we present an even more compact design currently developed for a sounding rocket mission with launch in 2017.

1. Introduction

Future space missions rely on the availability of high performance optical frequency references. They are either used as laser source for high-sensitivity inter-spacecraft optical metrology, as part of a payload enabling tests of fundamental physics or as a high accuracy timebase e.g. for global navigation satellite systems. While different methods for laser frequency stabilization are conceivable including stabilization to an optical resonator and an atomic or molecular transition, Iodine-based frequency references using modulation transfer spectroscopy (MTS) at a wavelength near 532 nm is a well-known technology used in laboratories for many years. Such systems offer high long-term stability and the provision of an absolute frequency reference. With respect to future applications in space, two setups on elegant breadboard and engineering model level, respectively, were realized during the last years.

2. Compact and Ruggedized Iodine Spectroscopy Setups for Space

A specific assembly-integration technology is applied where the optical components are bonded onto a glass baseplate using a space-qualified two-component epoxy. This results in a compact and ruggedized semi-monolithic optical assembly. Two spectroscopy setups have been realized using a Nd:YAG solid state laser with an output wavelength of 1064 nm as light source, frequency doubled to 532 nm. The 532 nm output laser beam is split into pump and probe for iodine spectroscopy and both beams are fiber coupled to the spectroscopy units. Phase/frequency modulation is either carried out by an acousto-optic or an electro-optic modulator.

The EBB spectroscopy unit uses a 550 mm x 250 mm x 50 mm low thermal expansion baseplate made of OHARA Clearceram-Z HS [2]. The optics are made of fused silica and integrated on the baseplate using adhesive bonding technology. A commercial 30 cm long iodine cell is used in triple-pass configuration. Mechanical mounts for fiber outcoupler, waveplates and polarizers are made of Invar. Four pairs of wedged glass plates enable an alignment of the beam overlap in the gas cell after integration of the optical setup. Noise cancelling detection is implemented for residual amplitude modulation (RAM) stabilization and MTS signal detection.

In a subsequent activity on EM level, the design of the EBB setup was further optimized with respect to compactness and mechanical and thermal stability, cf. Figure 1. For realizing a compact and ruggedized setup, a special designed 10 cm x 10 cm x 3 cm fused silica multi-pass gas cell was developed. The spectroscopy unit was subjected to thermal cycling from -20°C to +60°C and vibrational loads with sine vibration up to 30 g and random vibration up to 25.1 g_{rms}. The frequency stability was measured before and after the tests where no degradation was observed.

The frequency stabilities of the two setups were determined from a beat-note measurement with a ULE cavity stabilized laser. The Allan deviation shows a frequency stability of $1 \cdot 10^{-14}$ at an integration time of 1 s and below $5 \cdot 10^{-15}$ for integration times between 10 s and 1000 s where a linear drift was removed from the beat time record. The frequency stabilities of both setups fulfill the LISA and GRACE follow-on requirements.

Within the project JOKARUS (Jod-Kammresonator unter Schwerelosigkeit), a spectroscopy setup for use on a sounding rocket mission is currently implemented using micro-integrated external cavity diode lasers (ECDLs) as light source. An iodine-based reference is compared to a microwave reference via an optical frequency comb, resulting in a test of local position invariance (LPI).

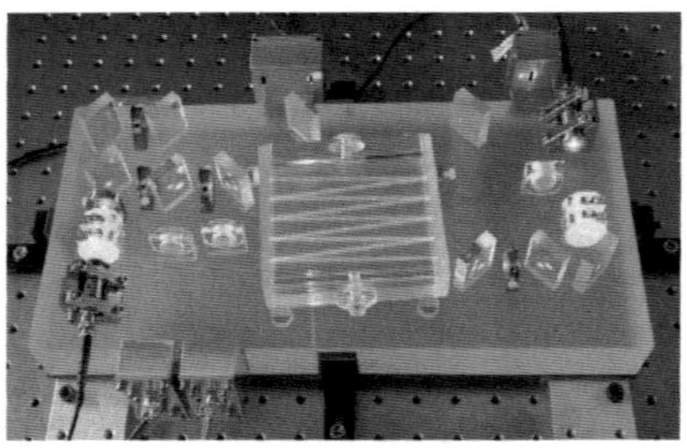

Figure 1. Iodine spectroscopy unit on engineering model (EM) level: the optical components are integrated on a 18 cm x 38 cm x 4 cm fused silica baseplate [1].

Acknowledgments
Financial support by the German Space Agency DLR with funds provided by the Federal Ministry of Economics and Technology (BMWi) under grant numbers 50 QT 1102, 50 QT 1201 and 50WM1646 is highly appreciated.

References
[1] Schuldt T, Döringshoff K, Milke A, Sanjuan J, Gohlke M, Kovalchuk E, Gürlebeck N, Peters A and Braxmaier C 2016 *Journal of Physics: Conference Series* **723** 012047
[2] Schuldt T, Döringshoff K, Kovalchuk E, Ketmann A, Pahl J, Peters A and Braxmaier C accepted for publication *Applied Optics*

IOP Conf. Series: Journal of Physics: Conf. Series **840** (2017) 012051 doi:10.1088/1742-6596/840/1/012051

Bayesian treatment of prospective LISA parameter estimation for massive black hole mergers

John G Baker[1] **and Sylvain Marsat**[2]

[1] NASA – Goddard Space Flight Center, 8800 Greenbelt Road, Greenbelt, MD 20771, USA
[2] Albert Einstein Institut, Am Muehlenberg 1, 14476 Potsdam-Golm, Germany

Abstract. A full understanding of LISA's science capability will require accurate models of incident waveform signals and the instrumental response. While Fisher matrix analysis is useful for some estimates, a full Bayesian treatment is needed for important cases at the limit of LISA's capability. We will apply fast analysis algorithms enabling accurate treatment with EOB waveforms and the full-featured LISA response to study the significance of higher spherical harmonics and mergers in LISA analysis.

1. Introduction

A LISA-like space-based gravitational-wave instrument is currently planned as part of the European Space Agency's Cosmic Vision Programme with NASA planning to participate as a minority partner. Programmatically labeled L3, the mission follows two other missions in the large mission category with launch expected in the mid 2030s

A key target of a LISA-like mission will be the direct observation of signals from merging MBHs. Previous studies have shown that despite cosmological distances, $2 < z < 10$ typically, LISA-like mission variants could precisely measure the component MBHs, measuring masses to typically better than 1% and spins within a few %. Distance and sky position measurements are more challenging and more sensitive to design choices and subtle features, thus motivating Bayesian analysis. Among the challenges of this analysis, although details of the waveform response encode important information, studies to date have all left out some features. Computationally the signals cover a large dynamic range with frequencies up to 1 Hz for signals lasting sometimes longer than a year. Bayesian analysis requires millions of likelihood samples demanding very fast codes.

2. Fast and accurate likelihoods

To bring computation times down to milliseconds per likelihood we need fast methods for waveform computation, for the LISA response and for the innerproducts to compare signals. Our implementation begins with a computationally fast Fourier domain reduced order treatment in amplitude/phase form for each of several EOB spherical harmonic modes.

The LISA response includes both modulations $F(t)$ representing the change of orientation of the constellation and delays $d(t)$ representing the motion of the detectors in the wavefront itself, for a signal of the form $s(t) = F(t)h(t + d(t))$. In the Fourier domain this becomes

$$\tilde{s}(f) = \int df' \tilde{h}(f - f')G(f, f').$$

To improve speed we take advantage of the binary MBH waveform property allowing it to be written as sum of a few components $\tilde{h}_{lm}(f) = A(f)\exp(-i\Psi(f))$, each with gently varying amplitude and phase. Then drawing on local properties of A and Ψ we can factor the signal as $\tilde{s}(f) = T(f)\tilde{h}(f)$ where $T(f)$ provides an excellent approxmation of the convolved result, but can be computed quickly. At leading order, this resembles a stationary phase approximation treatment but the approach is applicable more generally and additional correction terms are also available.

The last component of our fast-likelihood computation is to analytically integrate the signal overlaps as piecewise polynomials between relatively few evaluation points.

3. LISA performance

We apply these methods to study the capability of a 5-year classic LISA instrument to measure nonspinning mergers. Fig. 1 shows the SNR for these mergers. We apply the Multinest/BAMBI nested-sampling code or our own MCMC code to generate posterior distributions for simulated observations with no added noise; see example in Fig. 2. On a set of 10 broadly selected example cases we achieve typical speeds of about 15 − 20 ms per likelihood in practice in runs with order 10 million samples each.

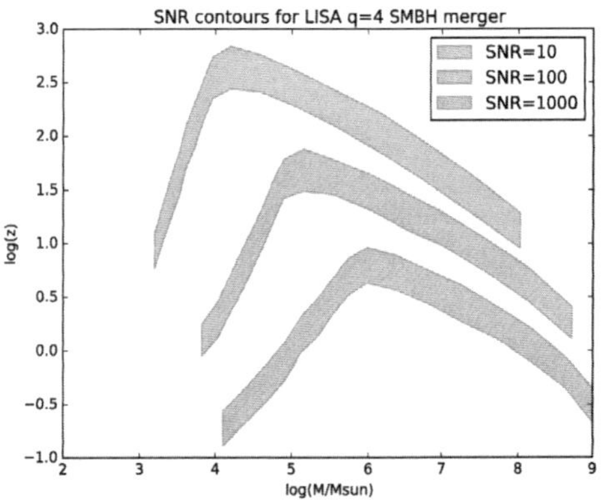

Figure 1. Typical redshifts (log scale) for non-spinning 4:1 mass ratio mergers at various LISA SNRs and total masses. Bands indicate 1-σ logarithmic range over 120 random sets of angle params.

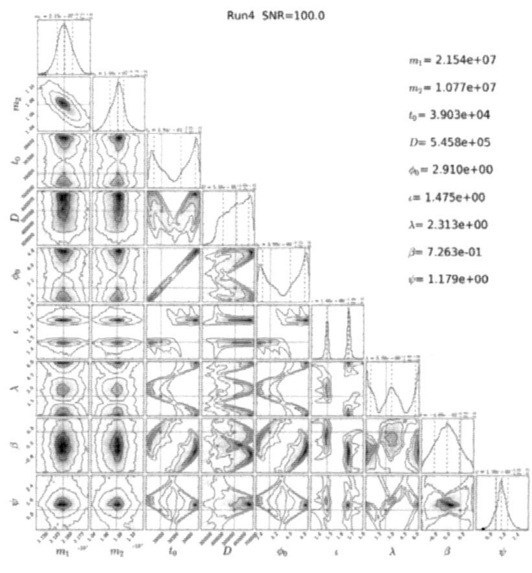

Figure 2. Example of posterior distribution sections over the 9 nonspinning (and non-eccentric) merger parameters.

Acknowledgments
This research was supported by NASA grant 13-ATP13-0077.

Understanding the importance of transient resonances in extreme mass ratio inspirals

C P L Berry[1,2], R H Cole[2], P Cañizares[3,2] and J R Gair[4,2]

[1] School of Physics and Astronomy, University of Birmingham, Birmingham, B15 2TT, UK
[2] Institute of Astronomy, Madingley Road, Cambridge, CB3 0HA, UK
[3] Institute of Mathematics, Astrophysics and Particle Physics, Radboud University, Heyendaalseweg 135, 6525 AJ Nijmegen, Netherlands
[4] School of Mathematics, University of Edinburgh, Peter Guthrie Tait Road, Edinburgh EH9 3FD, UK

E-mail: `cplb@star.sr.bham.ac.uk`

Abstract. Extreme mass ratio inspirals (EMRIs) occur when a compact object orbits a much larger one, like a solar-mass black hole around a supermassive black hole. The orbit has 3 frequencies which evolve through the inspiral. If the orbital radial frequency and polar frequency become commensurate, the system passes through a *transient resonance*. Evolving through resonance causes a jump in the evolution of the orbital parameters. We study these jumps and their impact on EMRI gravitational-wave detection. Jumps are smaller for lower eccentricity orbits; since most EMRIs have small eccentricities when passing through resonances, we expect that the impact on detection will be small. Neglecting the effects of transient resonances leads to a loss of $\sim 4\%$ of detectable signals for an astrophysically motivated population of EMRIs.

The first observations of gravitational waves came from comparable mass stellar-mass binary black holes [1]. The evolving Laser Interferometer Space Antenna (eLISA) provides the chance to observe gravitational waves from stellar-mass black holes orbiting supermassive black holes [2]. These extreme mass ratio inspirals (EMRIs) emit $\sim 10^4$–10^5 gravitational-wave cycles in eLISA's frequency band, allowing exquisite measurements to be made [3]. To detect and analyse EMRIs accurate waveform templates are required.

Flanagan and Hinderer [4] highlighted a previously overlooked phenomenon that complicates EMRI waveform modelling, that of transient resonances. During an inspiral, a resonance occurs when the radial frequency Ω_r and polar frequency Ω_θ are a rational ratio of each other. Then, the orbit does not cover the whole allowed r–θ plane, but cycles over a single loop [5]. The orbital phase can be expressed as

$$\varphi \sim (n_r\Omega_r - n_\theta\Omega_\theta)\,t + \left(n_r\dot{\Omega}_r - n_\theta\dot{\Omega}_\theta\right)t^2 + \ldots = \Omega t + 2\pi\frac{t^2}{\tau_{\text{res}}^2} + \ldots \qquad (1)$$

On resonance $\Omega = 0$ so the first term, which normally dominates, goes to zero; the second term then governs dynamics for a duration set by the resonance time τ_{res} [6, 7]. The evolution of the inspiral is determined by the gravitational self-force [8]; across a resonance, terms that usually average to zero can combine coherently, significantly impacting the orbital motion [4].

For previous EMRI studies, adiabatic waveforms have been used. These average the self-force over the r–θ plane and do not include resonances. Adiabatic waveforms quickly dephase from true EMRI waveforms if passing through resonance causes a jump in the orbital parameters. The mismatch between adiabatic and the true waveforms could prevent signals from being detected or lead to biased parameter estimates.

In [7], we investigated jumps in the orbital parameters from passing through a resonance. We used the same approximate self-force as Flanagan and Hinderer [4]. The size of jumps depends upon the magnitude of the self-force, the resonance time and the relative r–θ phase [7, 9, 10]. Jumps are most significant for low-order resonances, like the 2:3 resonance, because higher-order resonances come closer the covering the entire r–θ plane (like nonresonant orbits). The jump magnitude is sensitive to the orbital parameters on resonance, and decreases for lower eccentricity or lower inclination orbits [11]: there can be no resonances for circular orbits.

To assess the impact of resonances on detectability, we studied an astrophysical population of EMRIs that could be detectable with eLISA, assuming a two-year mission [12]. We considered $10M_{\odot}$ black holes inspiralling into 10^4–$10^7M_{\odot}$ supermassive black holes with a uniform distribution of spins. The initial eccentricity distribution followed Hopman and Alexander [13]. As gravitational-wave emission circularises orbits [14], we found that when important low-order resonances are encountered the eccentricity is small. Accordingly, the resonant jumps are also small. The match with adiabatic waveforms are good enough to still detect EMRIs in most cases [7]. Considering the distribution of signal-to-noise ratios ρ for the EMRIs before accounting for resonances and after including the reduction from imperfect waveforms, we find that $\sim 4\%$ of signals drop below the assumed detection threshold $\rho_{\text{thres}} = 15$.

In summary, transient resonances are a generic feature of EMRIs. Passing through resonance causes a jump in the evolution: the jump depends upon the phase on resonance, its size depends upon the orbit and the resonance. Ignoring resonances leads to waveform mismatch, but jumps should be small for most EMRIs because of their eccentricity. Therefore, detectability is not significantly reduced. The effect of resonances on parameter estimation is still to be investigated.

Acknowledgments
This work was supported in part by STFC and the Cambridge Philosophical Society. This is LIGO document reference LIGO-P1600349.

References

[1] Abbott B P *et al.* 2016 *Phys. Rev. X* **6** 041015 (*Preprint* 1606.04856)
[2] Amaro-Seoane P, Gair J R, Freitag M, Miller M C, Mandel I, Cutler C J and Babak S 2007 *Class. Quantum Grav.* **24** R113–R169 (*Preprint* 0703495)
[3] Barack L and Cutler C 2004 *Phys. Rev. D* **69** 082005 (*Preprint* 0310125)
[4] Flanagan É É and Hinderer T 2012 *Phys. Rev. Lett.* **109** 071102 (*Preprint* 1009.4923)
[5] Grossman R, Levin J and Perez-Giz G 2012 *Phys. Rev. D* **85** 023012 (*Preprint* 1105.5811)
[6] Ruangsri U and Hughes S A 2014 *Phys. Rev. D* **89** 084036 (*Preprint* 1307.6483)
[7] Berry C P L, Cole R H, Cañizares P and Gair J R 2016 *Phys. Rev. D* **94** 124042 (*Preprint* 1608.08951)
[8] Barack L 2009 *Class. Quantum Grav.* **26** 213001 (*Preprint* 0908.1664)
[9] Kevorkian J 1987 *SIAM Rev.* **29** 391–461
[10] van de Meent M 2014 *Phys. Rev. D* **89** 084033 (*Preprint* 1311.4457)
[11] Flanagan E E, Hughes S A and Ruangsri U 2014 *Phys. Rev. D* **89** 084028 (*Preprint* 1208.3906)
[12] Amaro-Seoane P *et al.* 2012 *Class. Quantum Grav.* **29** 124016 (*Preprint* 1202.0839)
[13] Hopman C and Alexander T 2005 *Astrophys. J.* **629** 362–372 (*Preprint* 0503672)
[14] Peters P C 1964 *Phys. Rev.* **136** B1224–B1232

IOP Conf. Series: Journal of Physics: Conf. Series **840** (2017) 012053 doi:10.1088/1742-6596/840/1/012053

Gravitational wave backgrounds

C Conneely

Department of Physics, Imperial College London, Prince Consort Rd, Kensington, London SW7 2BB

E-mail: c.conneely14@imperial.ac.uk

Abstract. This paper gives an introduction into how to apply techniques from the analysis of cosmic microwave background polarisation fields to a gravitational wave background.

The power spectra of the cosmic microwave background (CMB) polarisation and temperature anisotropies have provided significant cosmological constraints (see, e.g., [1]). It is hoped that similar techniques can be applied to a gravitational wave stochastic background to learn about the properties of the sources, whether they are cosmological or astrophysical in origin.

The gravitational wave tensor $h_{\mu\nu}(f,\hat{k})$ is a symmetric, trace-free, rank-2 tensor field on the sphere and so can be expanded in terms of tensor spherical harmonics $Y^G_{(\ell m)\mu\nu}(\hat{k})$ and $Y^C_{(\ell m)\mu\nu}(\hat{k})$ [2] as

$$\begin{aligned} h_{\mu\nu}(f,\hat{k}) &= h_+(f,\hat{k})h^+_{\mu\nu}(\hat{k}) + h_\times(f,\hat{k})h^\times_{\mu\nu}(\hat{k}) \\ &= \sum_{\ell=2}^{\infty}\sum_{m=-\ell}^{\ell} a^G_{(\ell m)}(f)Y^G_{(\ell m)\mu\nu}(\hat{k}) + \sum_{\ell=2}^{\infty}\sum_{m=-\ell}^{\ell} a^C_{(\ell m)}(f)Y^C_{(\ell m)\mu\nu}(\hat{k}), \end{aligned} \tag{1}$$

where $h_{+/\times}(f,\hat{k})$ are the amplitudes (as opposed to power) of the + and × polarisations in the $h^{+/\times}_{\mu\nu}(\hat{k})$ basis and are functions of frequency f and direction $\hat{k}$. The coefficients $a^{G/C}_{(\ell m)}(f)$ can be used to calculate power spectra

$$\hat{C}^{PP'}_\ell(f) = \frac{1}{2\ell+1}\sum_{m=-\ell}^{\ell} a^P_{(\ell m)}(f)a^{P'*}_{(\ell m)}(f), \qquad P,P' \in \{G,C\}, \tag{2}$$

where we expect the $\hat{C}^{GC}_\ell(f)$ and $\hat{C}^{CG}_\ell(f)$ spectra to be 0 by parity conservation.

In contrast to CMB measurements, which involve a similar calculation using the electromagnetic Stokes parameters Q^{EM} and U^{EM}, this method uses amplitudes of the signal rather than its power. An alternative approach uses gravitational Stokes parameters Q^{GW} and U^{GW} [3],

$$Q^{\mathrm{GW}} = \langle |h_+|\rangle^2 - \langle |h_\times|\rangle^2 \tag{3a}$$

$$U^{\mathrm{GW}} = -2\mathbb{R}(\langle h_+ h^*_\times\rangle), \tag{3b}$$

which can be expanded as

$$\frac{1}{2}\left(Q^{\mathrm{GW}} \pm iU^{\mathrm{GW}}\right)(f,\hat{k}) = \sum_{l=4}^{\infty}\sum_{m=-\ell}^{\ell} {}_{\pm 4}b_{\ell m}(f)\,{}_{\pm 4}Y_{\ell m}(\hat{k})\,. \tag{4}$$

Here ${}_{\pm 4}Y_{lm}(\hat{k})$ are the spin-4 spherical harmonics, the ${}_{\pm 4}b_{\ell m}(f)$ coefficients can be combined to give

$$b^{E}_{\ell m}(f) = -\frac{{}_{+4}b_{\ell m}(f) + {}_{-4}b_{\ell m}(f)}{2} \tag{5a}$$

$$b^{B}_{\ell m}(f) = i\frac{{}_{+4}b_{\ell m}(f) - {}_{-4}b_{\ell m}(f)}{2}\,, \tag{5b}$$

and EE, BB and EB power spectra can be computed from the $b^{E/B}_{\ell m}(f)$ coefficients analogously to Eq. (2).

The two decompositions have different advantages and disadvantages. The fact that the first method (G/C) is a decomposition of amplitude and the second method (E/B) one of power means that the power spectra are linear combinations of $\langle h^2\rangle$ and $\langle h^4\rangle$ respectively and so any measurement of the EE, BB or EB power spectra may be sensitive to non-Gaussianity. A consequence of the C/G decomposition is that the complex phases of the $h_{+/\times}$ amplitudes may be uncorrelated between directions meaning that the correlations of $h_{+/\times}$ between two different directions can be 0, even if the magnitudes $|h_{+/\times}|$ are correlated.

The E/B decomposition is well designed for studying the correlators of h_+ and $h_\times$ and of $h_{L/R} = (h_+ \pm ih_\times)/\sqrt{2}$ and can be used to probe interactions between gravitons. This method, however, does not include the overall power Stokes parameter,

$$I^{\mathrm{GW}} = \langle |h_+|\rangle^2 + \langle |h_\times|\rangle^2\,, \tag{6}$$

and, therefore, will not be able consider all types of anisotropy without introducing another harmonic expansion. This is an issue because any directly detected gravitational wave signal is likely to involve a term like I^{GW}. This is in contrast to the C/G decomposition which, in principle, is sensitive to any Gaussian contribution.

The frequency dependence of the gravitational waves will be significant. For example, it may be difficult to extract phase information (and so amplitude, as opposed to power) from high-frequency gravitational waves. Another issue is the fact that low-frequency gravitational wave detectors, such as pulsar timing arrays, are insensitive to C modes [2]. This should not be an issue for higher frequency detectors such as LISA. The frequency-dependent nature of gravitational wave detectors to E/B is a work in progress [4], though certain correlations of h_R and h_L can, in principle, affect CMB polarisation patterns [3].

The main conclusion is that the choice of decomposition is not always immediately obvious. For example, the E/B formalism is appropriate for studies of certain types of anisotropy — including those involving correlations of h_R and h_L — while C/G is well suited to others — such as those involving the overall power. Other considerations, including the observability of each power spectrum, are left for future work.

References

[1] Planck Collaboration Planck 2015 results. XIII. Cosmological parameters 2016 *A&A* **594** A13.
[2] Gair J, Romano J D, Taylor S and Mingarelli C M F 2014 Mapping gravitational-wave backgrounds using methods from CMB analysis: application to pulsar timing arrays *Phys. Rev.* D **90** 082001.
[3] Gubitosi G and Magueijo M 2016 Correlation between opposite-helicity gravitons: imprints on gravity-wave and microwave backgrounds *Preprint* gr-qc/1610.05702.
[4] Conneely C and Jaffe A H, *in preparation*.

IOP Conf. Series: Journal of Physics: Conf. Series **840** (2017) 012054 doi:10.1088/1742-6596/840/1/012054

Constraining inflationary physics with primordial gravitational waves at small scales

Maria Chiara Guzzetti

Dipartimento di Fisica e Astronomia "G. Galilei", Università degli Studi di Padova, Via Marzolo 8, I-35131, Padova, Italy

INFN, Sezione di Padova, via Marzolo 8, I-35131, Padova, Italy

E-mail: mariachiara.guzzetti@pd.infn.it

Abstract. Any inflationary model predicts the production of a stochastic gravitational-wave background. Experiments of direct gravitational-wave detection at small scales represent a promising way of extracting several exciting information encoded in such a primordial signal.

1. Theoretical picture: information carried by inflationary gravitational waves

Any inflationary model predicts the production of a stochastic gravitational wave (GW) background which turns out to represent an exciting window on the physics of the early universe; see [1] for a recent review and refs. therein.

Inflationary GWs from quantum fluctuations: because of quantum fluctuations of the gravitational field the production of a GW background takes place during inflation. In many cases, the corresponding GW power spectrum turns out to be well described by a power-law:

$$P_T(k) = A_T(k_*)\left(\frac{k}{k_*}\right)^{n_T}, \tag{1}$$

with $A_T(k_*)$ the amplitude at at chosen pivot scale k_*, and n_T the spectral index. Such a stochastic GW background represents a smoking gun for the cosmic inflationary model and carries unique information about such a stage of the universe. For example, according to single-field slow-roll inflation, the GW amplitude reflects the energy scale of the inflationary mechanism, while n_T measures the deviation with respect to a de-Sitter background.

Inflationary GWs from classical production: quantum fluctuations of the gravitational field are not the only mechanism by which GWs can be generated during inflation. The presence of further fields besides the gravitational one, gives rise to source terms in the GW equation of motion. Depending on the efficiency of the source, this mechanism can provide significant extra contributions to the inflationary GW background.

Quantum and classical GW contributions carry different information about the inflationary process: the first reflects the properties of the underling theory of gravity, while the latter the features of the involved fields. Therefore, inflationary GWs represent a promising way of discriminating among the variety of inflationary models; see tab.2 of [1].

IOP Conf. Series: Journal of Physics: Conf. Series **840** (2017) 012054 doi:10.1088/1742-6596/840/1/012054

Inflationary consistency relation: Single-field slow-roll inflation predicts a consistency relation between the tensor-to-scalar ratio r and the GW spectral index: $r = -8n_T$. However, several inflationary models are expected to violate such an equality, making it a powerful test for single-field slow-roll inflation, see tab.3 of [1]. Clearly, in order to test such a consistency relation, GW measurements on different scales are required.

2. Observational prospects and the role of experiments at small scales

Modeling the evolution of the universe, the inflationary GW spectral-energy density expected to fill the universe at the present time can be calculated [2]. The obtained GW background turns out to reflect the primordial amplitude and spectral index and to carry the imprints of the history of the universe it underwent. Notice that the resulting GW spectrum covers a wide range of frequencies: from $f \sim 10^{-16}$Hz, up to $f \sim 10^3$Hz or more, i.e. the range of the so-called small scales ($f = k/2\pi a$). Figure 1 provides an updated overview of the sensitivity curves for current and future experiments of direct GW detection at small scales.

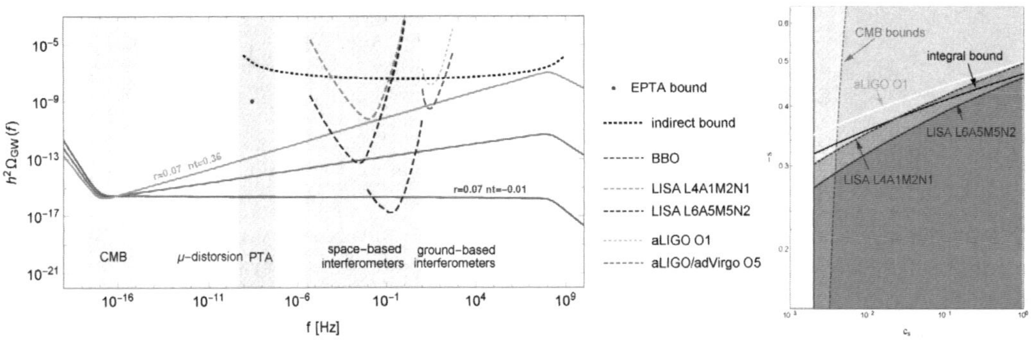

Figure 1. Left (taken from [1] *with kind permission of Società Italiana di Fisica* ©): current bounds and sensitivity curves of direct GW detection experiments. The three lines represent possible inflationary signals with $r = 0.07$ at $k = 0.05\text{Mpc}^{-1}$ and different n_t values. Right (taken from [3]): constraints on the parameter-space $c_s - s$ of a specific inflationary model (see main text).

Such experiments are expected to provide new constraints on the GW parameters. In particular, LISA [4] should significantly improve current bounds obtained by limits on the integral amount of GW [5] and by aLIGO O1 [6] (see also [7]).

More precisely, the combination of current CMB measurements and the data from experiments at small scales will provide new constraints on the GW spectral index and then new significant hints in the direction of testing the inflationary consistency relation. Moreover, the new bounds on GW features provided by upcoming detectors at small scales could be translated into limits on the parameter-space of specific inflationary models (see [3], w.r.t. LISA). For example, if scalar *spectator* fields are present during inflation [3] (and refs. therein), they can give rise to a significant, classical GW production. The amplitude and the spectral index of such an extra contribution turns out to depend on the propagation speed of the scalar perturbation of the spectator field c_s, and to its variation with respect to time s, respectively. Expected limits on the parameter space $c_s - s$ are shown in fig.1.

In conclusion, experiments of GW detection represents a promising way of probing inflationary physics. In particular, experiments related to GW at small scales are expected to provide new information with respect to CMB experiments and to give new hints with respect to the test of the inflationary consistency relation.

Acknowledgments
The author would like to acknowledge the LISA Cosmology working group for exciting discussions and Nicola Bartolo and Sabino Matarrese for supervising this work. The author thank financial support from a Cariparo foundation grant.

References
[1] Guzzetti M C, Bartolo N, Liguori M and Matarrese S 2016 *Riv. Nuovo Cim.* **39** 399–495 (*Preprint* 1605.01615)
[2] Watanabe Y and Komatsu E 2006 *Phys. Rev.* **D73** 123515 (*Preprint* astro-ph/0604176)
[3] Bartolo N *et al.* 2016 *JCAP* **1612** 026 (*Preprint* 1610.06481)
[4] Seoane P A *et al.* (eLISA) 2013 (*Preprint* 1305.5720)
[5] Pagano L, Salvati L and Melchiorri A 2016 *Phys. Lett.* **B760** 823–825 (*Preprint* 1508.02393)
[6] Abbott B P *et al.* (Virgo, LIGO Scientific) 2016 *Phys. Rev. Lett.* **116** 131102 (*Preprint* 1602.03847)
[7] Abbott B P *et al.* (Virgo, LIGO Scientific) 2016 (*Preprint* 1612.02029)

IOP Conf. Series: Journal of Physics: Conf. Series **840** (2017) 012055 doi:10.1088/1742-6596/840/1/012055

Testing Fundamental Properties of Space with the Fermilab Holometer

Brittany Kamai

California Institute of Technology, 1200 E. California Ave., MC 100-36, Pasadena, CA, 91125
Fermi National Accelerator Laboratory, MS 127, P.O. Box 500, Batavia, IL, 60510

E-mail: bkamai@ligo.caltech.edu

Abstract. Precision length measurements provide valuable insights about the fundamental properties of space-time. The Holometer is a research program to both experimentally probe signatures of the Planck scale and to extend the accessible frequency range from kHz up to MHz for gravitational wave searches. The instrument consists of separate yet identical 39-meter Michelson interferometers operated at Fermi National Accelerator Laboratory, which can reach length sensitivities better than $10^{-20}\mathrm{m}/\sqrt{\mathrm{Hz}}$ within the 1-10 MHz frequency range. The Holometer is fully operational with 130 of hours of science quality data obtained during the first observational campaign.

1. Introduction

The Holometer experiment was built and operated at Fermi National Accelerator Laboratory. The instrument consists of two 39 meter power-recycled Michelson Interferometers. Each interferometer has separate yet identical vacuum systems, lasers, electronics, signal detectors and digitizers [1]. By 2014, the construction and commission was completed. Each interferometer operates with cavity powers greater than 2 kW, which produces an instantaneous length sensitivity of $10^{-20}\mathrm{m}/\sqrt{\mathrm{Hz}}$ at frequencies above 1 MHz. The first observing campaign obtained 130 hours of science data during the summer of 2015. This dataset was used to perform two different scientific searches - Holographic Noise and MHz gravitational waves.

2. Holographic Noise

A new noise has been proposed as a consequence of fluctuations at the Planck-scale, which would introduce an imprecision in the location of an object. The predicted additional RMS length noise would be $2{\times}10^{-17}$ m, for a 39 m Michelson Interferometer due to the scaling relationship of the Planck length (10^{-35} m) and the length of the instrument[2]. The first model of the spectral shape and scaling amplitude of how this new noise will show up in a 39 m Michelson Interferometer was tested. This model was ruled out at 5σ significance by using 130 hours of science data [3].

3. Gravitational Waves

The unique frequency range of the Holometer enables an extension of the gravitational wave spectrum from kHz up to a broad range of MHz frequencies. Early universe remnants such as primordial black holes and cosmic super-strings are capable of radiating at these high of frequencies. Constraints on the stochastic gravitational wave background and a first search for primordial black holes binaries at MHz frequencies has been performed using the 130 hour dataset. [4]

3.1. Acknowledgments

The author wishes to acknowledge the entire Holometer collaboration for their hard-work on this experiment *(Aaron S. Chou, Richard Gustafson, Craig Hogan, Ohkyung Kwon, Robert Lanza, Shane L. Larson, Lee McCuller, Stephan S. Meyer, Jonathan Richardson, Chris Stoughton, Raymond Tomlin, Rainer Weiss)*. The author wishes to acknowledge the support from the National Science Foundation (DGE-0909667), Ford Foundation, Universities Research Association Fermilab Visiting Scholars Program, and the John Templeton Foundation.

4. References

[1] Chou A S *et al.* (Holometer Collaboration) 2016 *Class. Quantum Grav.*, submitted (arxiv:1611.08265)
[2] Hogan C J 2012 *Phys Rev. D* **85**, 064007
[3] Chou A S *et al.* (Holometer Collaboration) 2016 *Phys. Rev. Lett.* **117**, 111102
[4] Chou A S *et al.* (Holometer Collaboration) 2016 *Phys. Rev. D*, submitted (arxiv:1611.05560)

IOP Conf. Series: Journal of Physics: Conf. Series **840** (2017) 012056 doi:10.1088/1742-6596/840/1/012056

A frequency-domain implementation of the particle-without-particle approach to EMRIs

Marius Oltean[1,2,3,4,5], **Carlos F Sopuerta**[1] **and Alessandro D A M Spallicci**[3,4,5]

[1] Institut de Ciències de l'Espai (CSIC-IEEC), Campus Universitat Autònoma de Barcelona Carrer de Can Magrans s/n, 08193 Cerdanyola del Vallès, Spain

[2] Departament de Física, Facultat de Ciències, Universitat Autònoma de Barcelona Edifici C, 08193 Cerdanyola del Vallès, Spain

[3] Observatoire des Sciences de l'Univers en région Centre (OSUC), Université d'Orléans 1A rue de la Férollerie, 45071 Orléans, France

[4] Pôle de Physique, Collegium Sciences et Techniques (CoST), Université d'Orléans Rue de Chartres, 45100 Orléans, France

[5] Laboratoire de Physique et Chimie de l'Environnement et de l'Espace (LPC2E) Centre National de la Recherche Scientifique (CNRS) 3A Avenue de la Recherche Scientifique, 45071 Orléans, France

E-mail: `oltean@ice.cat`

Abstract. The gravitational waves emitted by binary systems with extreme mass ratios carry unique astrophysical information expected to be probed by the next generation of gravitational wave detectors such as LISA. The detection of these binaries rely on an accurate modeling of the gravitational self-force that drives their orbital evolution. Although the theoretical formalism to compute the self-force has been largely established, the mathematical tools needed to implement it are still under development, and the self-force computation remains an open problem. We present here a frequency-domain implementation of the particle-without-particle (PwP) technique previously developed for the computation of the scalar self-force – a helpful testbed for the gravitational self-force.

1. Introduction and Motivation

Extreme-Mass-Ratio Inspirals (EMRIs) are one of the main sources of gravitational waves (GWs) for space-based detectors like the LISA mission. EMRIs are binary system which consist of a stellar compact object (SCO; with a mass range $m_* \sim 1 - 50 M_\odot$) orbiting a massive black hole (MBH; with a mass range $M_\bullet \sim 10^{5-7} M_\odot$). In the regime where the dynamics is driven by GW emission, the SCO inspirals into the MBH sweeping through the LISA frequency band, and mapping the MBH spacetime onto the structure of the GWs in great detail. However, due to the complexity of EMRI GW signals, we need precise theoretical waveform templates to extract the physical parameters of the system from the detector data stream.

The challenge in modeling EMRIs is to compute the perturbations generated by the SCO in the (background) gravitational field of the MBH, and how these perturbations affect the motion of the SCO itself. The most extended approach [1] consists in modelling the SCO by using a point-like description, and then, to describe the backreaction effects on the dynamics as the action of a local *self-force* that is responsible for the deviations from geodesic motion.

While tremendous progress has been made on this problem in recent years, significant obstacles remain – stemming, in large part, from the singular nature of the gauge where the metric perturbations are usually computed, which renders the self-force calculation computationally challenging. To the effect of developing and testing an effective method for circumventing these issues in the full gravitational case, we consider here the simpler problem of the self-force due to a scalar field in the frequency domain, via a previously developed technique called the particle-without-particle (PwP) approach [2, 3]; for other similar approaches using jump conditions, see [4].

2. Setup

In our simplified EMRI model, the SCO is a charged scalar particle, with charge q associated to a scalar field Φ, orbiting a non-rotating MBH with a fixed geometry, the Schwarzschild-Droste metric. Because of spherical symmetry in this background, the field harmonic modes $\Phi^{\ell m}(t,r)$ decouple, and the scalar field equation reduces to wave-like PDEs for $\psi^{\ell m} = r\Phi^{\ell m}$,

$$(\Box - V_\ell(r))\,\psi^{\ell m} = S^{\ell m}\delta(r - r_p(t))\,, \tag{1}$$

where $\Box = -\partial_t^2 + \partial_{r_*}^2$, $r_* = r + 2M_\bullet \ln(r/2M_\bullet - 1)$, $V_\ell(r)$ is the so-called Regge-Wheeler potential, and $S^{\ell m}$ is a source term proportional to q and dependent on the particle location $r_p(t)$. Once the field is solved for, its singular part must be subtracted (via "mode-sum regularisation"), leaving us with its "regular" part from which the value of the self-force can be calculated [1].

The full solution has to be found numerically, but the presence of singularities makes the task difficult. To circumvent this, the idea of the PwP method is to split the computational domain into two disjoint regions whereby any non-singular quantity $\mathcal{Q}(t,r)$ is decomposed as $\mathcal{Q} = \mathcal{Q}_-\Theta_p^- + \mathcal{Q}_+\Theta_p^+$, where $\Theta_p^\pm = \Theta(\pm(r - r_p))$ is the Heaviside step function. Quantities that are not continuous will have jumps across the SCO trajectory: $[\mathcal{Q}]_p = \lim_{r\to r_p(t)} \mathcal{Q}_+(t,r) - \mathcal{Q}_-(t,r)$. In this setup then, the problem (1) with a singular source is effectively replaced with homogeneous equations to the left and right of the SCO, subject to certain jump conditions across it.

3. Frequency-domain numerical implementation and results

Previously, the PwP method has been used to solve this problem (as a PDE) directly in the time domain. But this approach generically leads to slow computational times, and may not be adaptable to more realistic spacetimes such as Kerr. To ameliorate these issues, we can tackle the problem in the frequency domain: we expand the fields in discrete Fourier series, transforming the scalar field PDEs into Schrödinger equation-like ODEs – supplemented by appropriate (non-reflecting) boundary conditions as well as the jump conditions.

We use a pseudospectral collocation method to find numerical solutions for the Fourier modes, which are then matched by imposing the jump conditions. Thus far, using this approach, the known value of the self-force has been recovered for circular orbits (in agreement with the PwP in the time domain and other methods in the literature); we are working on extending this to generic (eccentric) orbits, and we aim to also adapt this method to the Kerr spacetime.

Acknowledgements Due by MO to the Natural Sciences and Engineering Research Council of Canada, by MO and ADAMS to LISA France-CNES, and by CFS to the Ministry of Economy and Competitivity of Spain, MINECO, contracts ESP2013-47637-P and ESP2015-67234-P.

References

[1] Poisson E, Pound A and Vega I 2011 *Living Rev. Relativity* **14** 7
[2] Canizares P and Sopuerta C F 2009 *Phys. Rev.* D **79** 084020
[3] Canizares P and Sopuerta C F 2011 *Class. Quant. Grav.* **28** 134011
[4] Aoudia S and Spallicci A D A M 2011 *Phys. Rev.* D **83** 064029

IOP Conf. Series: Journal of Physics: Conf. Series **840** (2017) 012057 doi:10.1088/1742-6596/840/1/012057

Detecting additional polarization modes with LISA

L Philippoz and P Jetzer

Physik-Institut, Universität Zürich, Winterthurerstrasse 190, 8057 Zürich, Switzerland

E-mail: plionel@physik.uzh.ch

Abstract. Within the frame of Einstein's General Relativity, gravitational waves are expected to possess two tensorial polarizations, namely the well-known h_+ and $h_\times$ modes, whereas more general metric theories of gravity however allow the existence of additional modes (up to two vector and two scalar modes). The (non-)observation of those additional polarizations could put constraints on alternative theories and consequently provide a further test for General Relativity. We address the question of whether a given LISA configuration can provide a sufficient sensitivity to detect additional polarization modes and then allow the extraction of the latter in order to determine the GW spectrum for each mode.

1. Introduction

The recent detections of gravitational waves (GW) by the LIGO Collaboration [1, 2] represent a milestone in GW research and open new perspectives in the study of general relativity and astrophysics. Many projects are still under way, one of them being the next ESA L3 mission, namely the Laser Interferometer Space Antenna (LISA)[3]. The scope of LISA is to detect and study low-frequency gravitational radiation in the range from 0.1 mHz to 1 Hz, offering a complementary window of observation to the earth-based experiments. Phenomena such as the merger of supermassive black holes at cosmological distances, or binary systems composed of close white dwarves would produce a signal within the reach of LISA.

Alternative gravitation theories can influence the dynamics of those mergers, and LISA is thus expected to be able to measure the inprints of some alternative theories since this future space-borne detector will offer access to an unprecedented signal sensitivity. The proposed design [4] will allow the detection of possible additional polarization modes of GW, which is an invaluable tool to explore GW in alternative theories of gravity, e.g. $f(R)$ and scalar-tensor theories. Depending if additional polarizations are found or not in a detected signal, our knowledge of gravitation could have to be extended beyond GR, but we could in any case exclude some theoretical models according to which modes are actually detected. Moreover, the detection of a stochastic GW background (SGWB) could tell us more about the early stages of the universe. The polarization content of such a signal would also be of use to discriminate between existing gravitation theories.

2. Network of detectors

Within the frame of GR, gravitational waves possess two tensorial polarizations, the so-called h_+ and $h_\times$ modes, whereas more general metric theories of gravity predict the existence of additional modes: up to 2 vector and 2 scalar modes. The perturbed metric corresponding to a propagating gravitational wave can be expressed as:

$$h_{ij}(\omega t - \mathbf{k}\cdot\boldsymbol{x}) = \sum_A h_A(\omega t - \mathbf{k}\cdot\boldsymbol{x})e^A_{ij} \tag{1}$$

with $A = +, \times, x, y, b, l$ the six possible polarization modes, h_A the GW amplitude of the mode A, and the following polarization tensors (tensor $(+, \times)$, vector (x, y) and scalar (b, l) modes):

$$e^+_{ij} = \begin{pmatrix} 1 & 0 & 0 \\ 0 & -1 & 0 \\ 0 & 0 & 0 \end{pmatrix} \quad e^x_{ij} = \begin{pmatrix} 0 & 0 & 1 \\ 0 & 0 & 0 \\ 1 & 0 & 0 \end{pmatrix} \quad e^b_{ij} = \begin{pmatrix} 1 & 0 & 0 \\ 0 & 1 & 0 \\ 0 & 0 & 0 \end{pmatrix}$$

$$e^\times_{ij} = \begin{pmatrix} 0 & 1 & 0 \\ 1 & 0 & 0 \\ 0 & 0 & 0 \end{pmatrix} \quad e^y_{ij} = \begin{pmatrix} 0 & 0 & 0 \\ 0 & 0 & 1 \\ 0 & 1 & 0 \end{pmatrix} \quad e^l_{ij} = \begin{pmatrix} 0 & 0 & 0 \\ 0 & 0 & 0 \\ 0 & 0 & 1 \end{pmatrix}$$

Without assuming a particular theory of gravitation, a GW signal is expected to contain a mixture of up to 6 polarizations, and it is consequently important to determine the detection threshold for a SGWB as well as find a way of extracting its polarization content.

The determination of the sensitivity to additional polarization modes was discussed for an earlier version of the LISA project [5, 6, 7, 8, 9, 10] but it is necessary to investigate this question for several time-delay interferometric (TDI) combinations (i.e. different combinations of the signals depending on the number of laser links between the satellites) and establish the appropriate sensitivity curves for the chosen future design. We already found that the noise spectrum of the future detector presents large similarities with the older project, especially in the frequency range $f \in [10^{-2}, 1]$ Hz, and expect the sensitivity spectrum to be quite similar in that frequency domain where, in particular, the sensitivity to the vector and longitudinal-scalar modes is higher.

We then need to look for a way of extracting the polarization content of a given signal produced by a SGWB

$$h(t, \boldsymbol{x}) = \sum_A \int_{S^2} d\hat{\boldsymbol{\Omega}} \int_{-\infty}^{\infty} df\ \tilde{h}_A(f, \hat{\boldsymbol{\Omega}}) e^{2\pi i f(t - \hat{\boldsymbol{\Omega}}\boldsymbol{x}/c)} F_A(\hat{\boldsymbol{\Omega}}), \tag{2}$$

with F_A the antenna pattern function of a single detector (which describes its geometry), and h_A the GW amplitude of the mode A. In its current proposed design (3 satellites with 3 arms, i.e. 3 times 2 laser links between the satellites), LISA is actually equivalent to two single detectors; we will describe this arrangement of three satellites as a cluster. In a single detector, the vectors $\hat{u}$ and $\hat{v}$ give the direction of each arm and one can define the detector tensor D as

$$D_{ij} = \frac{1}{2}(\hat{u}_i\hat{u}_j - \hat{v}_i\hat{v}_j), \tag{3}$$

which gives the response of that detector to a signal.

If one now considers a network of several identical clusters [11, 12, 13], the first step consists in determining the so-called overlap reduction functions (ORFs), defined for a pair of detectors I and J separated by $\Delta\boldsymbol{x}$ as

$$\begin{aligned} \gamma^M_{IJ}(f) &= \frac{1}{\sin^2(\chi)} \Big(\rho^M_1(\alpha) D^{ij}_I D^J_{ij} + \rho^M_2(\alpha) D^i_{I,k} D^{kj}_J \hat{d}_i \hat{d}_j \\ &\quad + \rho^M_3(\alpha) D^{ij}_I D^{kl}_J \hat{d}_i \hat{d}_j \hat{d}_k \hat{d}_l \Big), \end{aligned} \tag{4}$$

with M denoting the tensor (T), vector (V) or scalar (S) polarization modes, $\sin^2(\chi) = 1-(\hat{u}\cdot\hat{v})^2$ a geometry factor (which is simply $\frac{3}{4}$ for an equilateral-triangle-shaped cluster such as LISA), $\rho_i^M = f(j_0(\alpha), j_2(\alpha), j_4(\alpha))$ a linear combination of spherical Bessel functions, D_I^{ij} the detector tensor of the interferometer I, $\hat{d}_i = \dfrac{\Delta \boldsymbol{x}}{|\Delta \boldsymbol{x}|}$, $\alpha = \dfrac{2\pi f\,|\Delta \boldsymbol{x}|}{c}$. An ORF tells how much degree of correlation is preserved when one correlates the output of two detectors, according to their relative orientation.

Note that, schematically, the signal $h(t) + n(t)$ measured by a detector is composed of the GW signal $h(t)$ as well as the noise $n(t)$. Next, we can thus consider the one-sided power spectral density S_h^A:

$$\langle \tilde{h}_A^*(f, \hat{\boldsymbol{\Omega}})\tilde{h}_{A'}(f', \hat{\boldsymbol{\Omega}}')\rangle = \delta(f - f')\frac{1}{4\pi}\delta^2(\hat{\boldsymbol{\Omega}}, \hat{\boldsymbol{\Omega}}')\delta_{AA'} \cdot \frac{1}{2}S_h^A(|f|), \tag{5}$$

as well as the noise spectrum $P_I(f)$:

$$\langle \tilde{n}_I(f)\tilde{n}_J(f')\rangle = \frac{1}{2}\delta(f - f')\delta_{IJ} \cdot P_I(|f|), \tag{6}$$

and the GW background energy density

$$\Omega_{\rm gw}^M(f) \propto f^3 S_h^A(f), \tag{7}$$

with e.g. $\Omega_{\rm gw}^T = \Omega_{\rm gw}^+ + \Omega_{\rm gw}^\times$ (and similarly for $M = V, S$), it is possible to find the optimal signal-to-noise ratio (SNR) necessary to separately detect the modes:

$$SNR^M \propto \int_0^\infty df \left[\frac{(\Omega_{\rm gw}^M(f))^2 \det \mathbf{F}(f)}{f^6 \mathcal{F}_M(f)}\right]^{(1/2)}, \tag{8}$$

where $\mathbf{F}$ is a (3×3)-matrix whose elements are given by

$$F_{MM'} = \sum_{\substack{\text{detector}\\ \text{pairs } (I,J)}} \int_0^{T_{\rm obs}} \mathrm{d}t \frac{\gamma_{IJ}^M(t,f)\gamma_{IJ}^{M'}(t,f)}{P_I(f)P_J(f)},$$

($T_{\rm obs}$ is the mission duration) and $\mathcal{F}_M(f)$ is the determinant of the matrix obtained by removing all the M-elements from the matrix $\mathbf{F}$. The separation of the modes can then be achieved by inverting the so-called correlation matrix; the details of the procedure can be found in [11].

This method thus gives the detection threshold for each mode and is valid for a network of independant detectors in space, i.e. several clusters (for instance 2 independant LISA-like clusters, or 4 clusters such as in the DECIGO project [14]), in the low frequency limit and for a full polarized GW background.

3. Single detector

A previous analysis of the sensitivity to additional polarization modes has been performed for an earlier version of LISA (cluster with 3 arms) and the sensitivity curves for each mode and various TDI simply need to be updated regarding the new proposed design [9]. However, it is nevertheless worth investigating a minimal version consisting of a cluster with only 2 arms, which could be useful to consider in case of hypothetical technical difficulties reducing the number of usable arms. Such an alternative solution focusing only on a single 2-arm detector in space makes use of the so-called autocorrelation method [15]. If, as previously, we write the output data of the detector as $h(t) + n(t)$, with $n(t)$ the noise and $h(t)$ the GW signal, the autocorrelation of the signal reads

$$\langle \tilde{h}(f)\tilde{h}^*(f')\rangle = \frac{1}{2}\delta(f - f')S_h(|f|), \tag{9}$$

and similarly for the noise density P_n:

$$\langle \tilde{n}(f)\tilde{n}^*(f')\rangle = \frac{1}{2}\delta(f - f')P_n(|f|), \tag{10}$$

and in that case, it is possible to find the optimal SNR to detect a signal as

$$SNR = \left(\frac{T_{\rm obs}}{2}\int_{-\infty}^{\infty} df \; \frac{S_h(|f|)^2}{[S_h(|f|) + P_n(|f|)]^2}\right)^{1/2}. \tag{11}$$

This method applies to a single-detector and is valid in the high-frequency limit. Since this analysis did not assume a particular polarization content, it only gives a detection threshold for a GW signal, but it will be necessary to generalize it in order to take into account all the possible modes, similarly to the network analysis.

4. Conclusion

As we have seen, the current analysis method for a network of space-borne detectors requires the use of multiple LISA-like clusters and presents the advantage of considering a general polarization of the signal. A second method involves only the exploitation of a single cluster. A study was already performed for a 3-arm detector and simply needs to be adapated to the most recent design; one can also focuse on a single 2-arm detector, but this method does not yet address the polarization of the signal and needs to be generalized in order to fully treat the polarization content.

With the proposed LISA design as well as future possible project of space-borne inteferometers, it is therefore necessary to investigate both methods in order to set limits on the detectability of each polarization mode potentially present in a SGWB.

References

[1] Abbott B P *et al.* 2016 *Phys. Rev. Lett.* **116** 061102
[2] Abbott B P *et al.* 2016 *Phys. Rev. Lett.* **116** 241103
[3] Amaro-Seoane P *et al.* 2013 *GW Notes* **6** 4–110
[4] Amaro-Seoane P *et al.* 2017 (*Preprint* `1702:00786`)
[5] Armstrong J W, Estabrook F B and Tinto M 1999 *ApJ* **527** 814
[6] Estabrook F B, Tinto M and Armstrong J W 2000 *Phys. Rev.* D **62** 042002
[7] Tinto M, Armstrong J W and Estabrook F B 2000 *Phys. Rev.* D **63** 021101
[8] Tinto M, Estabrook F B and Armstrong J W 2002 *Phys. Rev.* D **65** 082003
[9] Tinto M and Alves M E 2010 *Phys. Rev.* D **82** 122003
[10] Tinto M and Dhurandhar S V 2014 *Living Rev Relativity* **17** 6
[11] Nishizawa A, Taruya A, Hayama K, Kawamura S and Masa-aki S 2009 *Phys. Rev.* D **79** 082002
[12] Nishizawa A and Hayama K 2013 *Phys. Rev.* D **88** 064005
[13] Nishizawa A, Taruya A and Kawamura S 2010 *Phys. Rev.* D **81** 104043
[14] Kawamura S *et al.* 2006 *Classical Quantum Gravity* **23** S125–S131
[15] Tinto M and Armstrong J W 2012 (*Preprint* `gr-qc/1205.4620v1`)

IOP Conf. Series: Journal of Physics: Conf. Series **840** (2017) 012058 doi:10.1088/1742-6596/840/1/012058

Continuous gravitational wave searches with pulsar timing arrays: Maximization versus marginalization over pulsar phase parameters

Yan Wang[1], **Soumya D. Mohanty**[2] **and Yi-Qian Qian**[1]
[1] School of Physics, Huazhong University of Science and Technology, 1037 Luoyu Road, Wuhan, Hubei Province 430074, China
[2] Department of Physics, The University of Texas Rio Grande Valley, One West University Blvd, Brownsville, TX 78520, USA

E-mail: ywang12@hust.edu.cn

Abstract. Resolvable Supermassive Black Hole Binaries are promising sources for Pulsar Timing Array based gravitational wave searches. Search algorithms for such targets must contend with the large number of so-called pulsar phase parameters in the joint log-likelihood function of the data. We compare the localization accuracy for two approaches: Maximization over the pulsar phase parameters (MaxPhase) against marginalization over them (AvPhase). Using simulated data from a pulsar timing array with 17 pulsars, we find that for weak and moderately strong signals, AvPhase outperforms MaxPhase significantly, while they perform comparably for strong signals.

1. Introduction

Pulsar timing arrays (PTAs) aim to detect gravitational waves (GWs) in the $10^{-9} - 10^{-6}$ Hz range. Among the most promising sources in this frequency band are Supermassive Black Hole Binaries (SMBHBs). (See [1] for a review of the current status of PTA based SMBHB searches.)

The pulsar timing residual induced by GWs from a non-evolving SMBHB crossing an Earth-pulsar line of sight depends on the so-called pulsar phase parameter. This parameter arises from the time-of-flight of a radio pulse crossing the perturbed space-time between a pulsar and Earth. For a PTA consisting of N pulsars, this leads to at least N unknown parameters in the joint log-likelihood function of all the timing residuals.

Following the Maximum Likelihood (ML) prescription for parameter estimation, the joint log-likelihood function must be maximized over all signal parameters, including the N pulsar phases. Direct numerical maximization over pulsar phases [2] is not scalable when N becomes large. A scalable algorithm (MaxPhase), where the pulsar phases are maximized semi-analytically, was proposed in [3]. (The code is available from `https://github.com/yanwang2012/RAAPTR`). This algorithm was applied to a prospective PTA in the Square Kilometer Array (SKA) era in [4]. While the above algorithms follow the Frequentist canon of maximum likelihood, an alternative approach, which is natural in Bayesian statistics, is to treat pulsar phases as nuisance parameters and marginalize over them. Maximizing the marginalized likelihood over the remaining parameters provides an *ad hoc* Frequentist estimator, which we call AvPhase. In this paper, we present a brief comparison of MaxPhase and AvPhase.

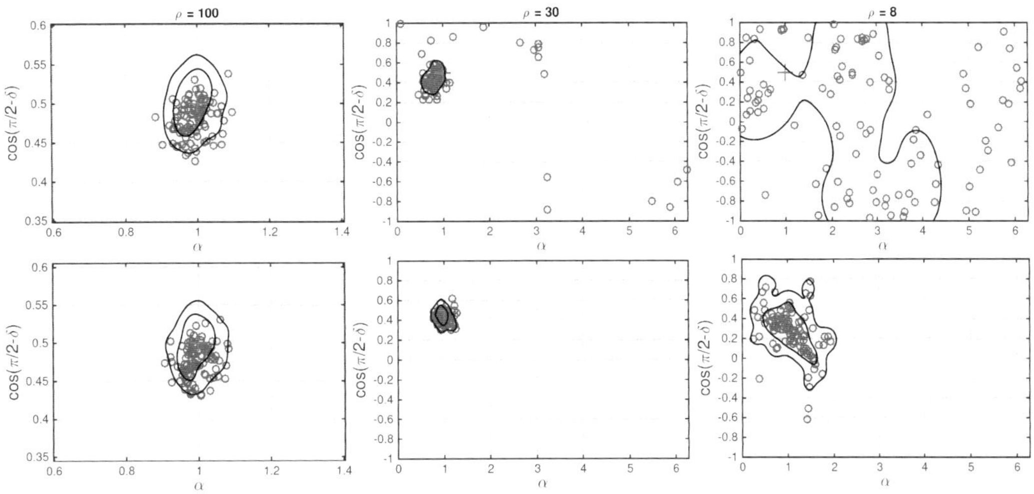

Figure 1. Estimated sky locations (circles) obtained from MaxPhase (top row) and avPhase (bottom row). A red plus sign marks the true location of the source, with 100 independent data realizations used for each location. The Right Ascension and Declination of the source are denoted by α and δ respectively. The 68% and 95% confidence contours are overlaid.

2. Results and discussion

We simulated a PTA data set with 17 pulsars which are observed for 5 years with biweekly cadence. The timing precision for each pulsar is set to be 100 ns. The strength of the GW signal is characterized by the network signal-noise-ratio ρ, which is chosen to be 100, 30 and 8 corresponding to strong, moderate and weak signal scenarios.

Fig. 1 shows the estimated sky locations of the GW source for the three scenarios. In order of decreasing ρ, the 68% (95%) confidence contours for MaxPhase cover 18 deg^2 (49 deg^2), 354 deg^2 (all sky), and 14785 deg^2 (all sky) respectively; while the corresponding ones for avPhase cover 18 deg^2 (46 deg^2), 111 deg^2 (369 deg^2), and 1162 deg^2 (3535 deg^2) respectively.

Our results show that the localization of avPhase performs significantly better than MaxPhase in the weak and moderate signal scenario, and its performance is comparable to MaxPhase in the strong signal scenario.

Acknowledgments

Y.W. is supported by the National Natural Science Foundation of China under grants 11503007, 91636111 and 11690021. The contribution of S.D.M. to this paper is supported by NSF awards PHY-1505861 and HRD-0734800.

References

[1] Lommen A N 2015 *Reports on Progress in Physics* **78** 124901

[2] Wang Y, Mohanty S D and Jenet F A 2014 *Astrophysical Journal* **795** 96 (*Preprint* 1406.5496)

[3] Wang Y, Mohanty S D and Jenet F A 2015 *Astrophysical Journal* **815** 125 (*Preprint* 1506.01526)

[4] Wang Y and Mohanty S D 2016 *ArXiv e-prints* (*Preprint* 1611.09440)

IOP Conf. Series: Journal of Physics: Conf. Series **840** (2017) 012059 doi:10.1088/1742-6596/840/1/012059

After-dinner speech: the path of LISA to become 'L3'

M C E Huber

Former Head of ESA's Space Science Department
Glärnischstrasse 20, 8700 Küsnacht, Switzerland

mceh@bluewin.ch

Abstract. We describe the state of fundamental-physics experiments in space in the mid 1980s on both sides of the Atlantic, and then follow the development of this field within ESA's Scientific Programme. The special case of LISA — a fundamental physics experiment in space also supported by astronomers, and now, following the ground-based detection of gravitational waves, the means to extend the observable spectrum of such waves towards lower frequencies and long-lasting signals — is then followed in its rise to become the 'Large Project L3' of ESA.

LISA is a well-deserved piece of luck! It has its roots in the quest for fundamental physics in space, has met the interest of astronomers and is now the preferred model for ESA's so-called L3 mission. Its current, high profile is based on two recent, impressive experimental achievements:

- the direct detection of gravitational waves by the advanced LIGO interferometers and
- the LISA-Pathfinder mission, whose performance moreover exceeded the specifications.

In his address yesterday, Alvaro Gimenez emphasised that ESA-missions, such as PLANCK, GAIA and EUCLID, make contributions to fundamental physics and pointed out that ESA supports space missions, which specifically address fundamental physics — among those MICROSCOPE and ACES (the Atomic Clock Ensemble in Space), both tests of the equivalence principle, under French leadership.

When I joined ESA in the mid-eighties, a mission's contribution to testing the foundations of physics was not yet considered an asset worth mentioning. At that time ESA's Science Directorate and its advisory structure were strictly focused on addressing two areas of space science, namely 'Solar System Science' and 'Astronomy'.

The astrometry mission HIPPARCOS, for example — a mission close to launch in the mid-eighties —, had to account for the deflection of light by the Sun's gravitation, and thus was promising to deliver another, competitive test of General Relativity. Pointing this out was deemed expedient only after the idea of 'Fundamental Physics in Space' had been introduced into ESA's Science Programme.

Things had been different on the other side of the Atlantic; there fundamental physics was actively pursued by experiments on rocket flights and orbiting satellites.

A test of the gravitational red shift provided by NASA's Gravity Probe A had been published in 1980 already. Bob Vessot and his colleagues at the Smithsonian Astrophysical Observatory in Cambridge, Mass., had confirmed the gravitational redshift predicted by General Relativity to 60 ppm by launching a rocket with a hydrogen maser on board to an altitude of 10'000 km.

IOP Conf. Series: Journal of Physics: Conf. Series **840** (2017) 012059 doi:10.1088/1742-6596/840/1/012059

A second, more complex Gravity Probe, GP-B, had in the mid-eighties been under development for 20 years. The aim of GP-B was to establish the existence of the Lense-Thirring effect — also referred to as 'gravito-magnetic' effect, or simply called 'frame-dragging'. GP-B was going to measure this 'drag' by observing the deflection of the axes of rotation of cryogenic gyroscopes orbiting in a polar orbit at 642 km altitude, as a consequence of the Earth's rotation underneath. The deflection predicted by General Relativity, is a minute 39.2 milli-arc-second per year!

I had the opportunity to visit the W. W. Hansen Experimental Physics Laboratory at Stanford, where Francis Everitt and his team worked on the many technical challenges that they had to overcome in order to achieve GP-B's goal. There I encountered a wonder-world of high technology dealing with cryogenics, drag-free operation, magnetics, SQUIDs and gyroscopes. The latter, for example, were the most round objects on Earth at the time — five times rounder than the best industrial spheres. GP-B, launched in 2004, did indeed confirm the existence of the Lense-Thirring effect and directly measured its minuscule size, with an accuracy of 19 %.

Alvaro's mentioning yesterday that ESA's Science Programme Committee had nearly cancelled LISA in 2011, reminded me of a nice story about Francis Everitt. While ESA generally shies away from cancelling missions, rescissions are more frequent in NASA. Indeed, over the course of its development, Gravity Probe B has been cancelled several times. But Francis regularly managed to reinstate it. While overcoming these recurring hurdles, he seems to have developed a particular sense of humour. So, when he had been invited to lecture on the development of Gravity Probe B in front of a prominent audience, he took the opportunity to welcome a person in the audience, whom he praised as a real pioneer — this was the person, who had cancelled GP-B for the first time.

Now back to ESA in the mid-80s: in contrast to NASA, we were nowhere with 'Fundamental Physics in Space'. Notably though, Fundamental Physics had been included as a discipline in 'Space Science – Horizon 2000' (H2000), the long-term programme championed by the then new ESA-Director of Science, Roger Bonnet. (I should perhaps underline here the importance of 'Horizon 2000' for Science in ESA: in the ten years following the approval of H2000 in 1984, the funding of ESA's Science Directorate grew by about a factor of two.)

Studying and, later, implementing fundamental-physics experiments, it turned out, led to a paradigm shift in mission-design within ESA's Science Directorate. The reason was that most such experiments require a pico-gravity environment, and thus drag-free technology.

Traditionally designing, building and testing a spacecraft and its scientific experiments could be neatly separated. The spacecraft, which provided the pointing, was built by Industry under the surveillance of the Project Manager and his engineers in the Project Department. The Project Scientist from the Science Department, on the other hand, closely monitored the development of the observing instruments, which were being built in Research Institutes and Universities across Europe, to assure that they accomplished the mission goals.

The instruments were eventually inserted into the spacecraft according to the interface documentation. In other words: the development of the satellite and its instruments could proceed in parallel, with conflict resolution, where needed, between Project Manager and Project Scientist, i.e. within ESA internally.

With missions in 'Fundamental Physics in Space' there was a game-change. As performance in the pico-gravity range required drag-free conditions, it was now an experiment rather than the spacecraft that contained the sensor that delivered the error signal to point and steer the spacecraft. Some people now suggested replacing the word spacecraft by 'sciencecraft'. In any case, the staff of the Project and Science Departments had to collaborate much more closely, sometimes involving the Principal Investigators and parts of their Science Team in the discussions with the Project Team.

One of the best things I did in my life — even before this paradigm shift became obvious — was to ask Rüdeger Reinhard to promote the topic of 'Fundamental Physics in Space' within ESA's Science Directorate. Rüdeger was an experienced Project Scientist. He had been Project Scientist of Giotto, the mission that had a close encounter (596 km) with comet Halley in 1986. This mission attracted worldwide attention to ESA. In a way, Giotto put ESA on the map.

IOP Conf. Series: Journal of Physics: Conf. Series **840** (2017) 012059 doi:10.1088/1742-6596/840/1/012059

In no time Rüdeger had assembled a Fundamental Physics Advisory Committee (FPAC) consisting of the best people in the field — many of them here with us tonight. There was a sequence of excellent chairs: Jean-Pierre Blaser of ETHZ, Maurice Jacob of CERN, Stefano Vitale of Trento and Bernard Schutz of the Albert Einstein Institute in Potsdam. And within a short time we could hire an additional staff member, Yusuf Jafry, to give support to Rüdeger. Yusuf came from Stanford, his thesis adviser, Dan de Bra, had flown the first drag-free spacecraft in 1973.

Rüdeger then encouraged people to submit mission proposals with topics in Fundamental Physics that were to be evaluated by FPAC. As a consequence, the response to a call for mission proposals consisted from now on of about a third of the proposals dealing with Fundamental Physics, while the remaining proposals were distributed more or less evenly between 'Solar System Science' and 'Astronomy'. As a result of such a call, the fundamental-physics proposal 'LISA' was selected for further study. While it dealt with the direct detection of low-frequency gravitational waves — an outstanding problem in Fundamental Physics — it eventually met the interest of astronomers, too.

Rüdeger knew that building a community was absolutely crucial in the European context. ESA is built on a grass-roots system — we have no 'National Academy of Sciences' (yet) in Europe, and there is no country in Europe, where the Academy has a say beyond the borders. The first grass roots efforts were followed by the backing through important institutions, such as Max Planck Institutes and Universities in France, Great Britain and Italy, and so we are on track today! As mentioned by Karsten Danzmann, it was Rüdeger who insisted on organising LISA Symposia to maintain the community's interest in the mission. In this 11th LISA Symposium, we celebrate this week an important milestone in LISA's way to become the 3rd Large Mission of ESA's long-term programme!

Rüdeger died in January 2015. It is a pity that he cannot see how his ideas came to fruition.

To conclude let me make a plea regarding the LISA launch date, which is currently planned for the year 2034. When I was working at ESA in the Netherlands, my neighbour in Leiden happened to be the LISA Project Manager. So, I asked him to invite me to the launch. Although I doubt that this rendezvous remains valid, I still want to attend the launch of LISA. The problem is that I will be 98 years old in 2034. Although I seem to be in good shape today, I reckon that another 18 years is stretching it a bit. So, please work hard to bring the launch forward by half a decade for example!